Molecular Pathology of Type 1 Diabetes mellitus

Current Directions in Autoimmunity

Vol. 4

Series Editor   *A.N. Theofilopoulos*, La Jolla, Calif.

Basel · Freiburg · Paris · London · New York · New Delhi · Bangkok · Singapore · Tokyo · Sydney

Molecular Pathology of Type 1 Diabetes mellitus

Volume Editor   *Matthias G. von Herrath*, La Jolla, Calif.

24 figures and 10 tables, 2001

Basel · Freiburg · Paris · London · New York ·
New Delhi · Bangkok · Singapore · Tokyo · Sydney

· ·
Matthias G. von Herrath, MD

Associate Professor
Department of Neuropharmacology and Immunology
Division of Virology
The Scripps Research Institute
La Jolla, Calif.

Library of Congress Cataloging-in-Publication Data

Molecular pathology of type 1 diabetes mellitus / volume editor, Matthias G. von Herrath.
    p.; cm. – (Current directions in autoimmunity; vol. 4)
    Includes bibliographical references and indexes.
    ISBN 3805572409
    1. Diabetes–Molecular aspects. I. Title: Molecular pathology of type one diabetes mellitus. II.
Von Herrath, Matthias G. III. Series.
    [DNLM: 1. Diabetes Mellitus, Insulin-Dependent–genetics. 2. Autoimmune
Diseases–immunology. 3. Diabetes Mellitus, Insulin-Dependent–immunology. WK 810
    M71894 2001]
    RC660.M568 2001
    616.4'6207–dc21
                                                                            2001038131

© Copyright 2001 by S. Karger AG, P.O. Box, CH–4009 Basel (Switzerland)
www.karger.com
Printed in Switzerland on acid-free paper by Reinhardt Druck, Basel
ISSN 1422–2132
ISBN 3–8055–7240–9

Contents

Preface

I am very pleased to present the 4th volume in the series *Current Directions in Autoimmunity* with a focus on insulin-dependent diabetes. Type 1 diabetes (T1D) as well as multiple sclerosis (MS) are thought to be T-cell-mediated autoimmune diseases, where autoaggressive lymphocytes lead to destruction of β cells or oligodendrocytes, respectively. Despite more than two decades of research, there are still many issues in autoimmune diabetes that are unclear and we do not understand. For example, although genetic factors clearly predispose to T1D, there have to be other environmental factors that tie into disease pathogenesis to explain the limited concordance of diabetes in monozygotic twins (30–40%). Further, it is not well understood how β cells die in vivo. Inflammatory cytokines, perforin-mediated lysis by cytotoxic T lymphocytes and FAS-dependent apoptosis have all been implemented through various experimental systems. Most importantly, the crucial autoantigens targeted by the autoaggressive response are still not well defined. This makes induction of antigen-specific tolerance as a strategy to prevent autoimmune diabetes difficult. However, other interventions relying on the induction of regulatory circuits have been used in animal models with good success. In the following I will give a brief overview over the chapters in this book and how they relate to these pertinent issues.

Genetic factors involving MHC susceptibility alleles are known to predispose to T1D from studies in humans (Bach, Garchon and van Endert) as well as the nonobese diabetic (NOD) mouse model (Serreze and Leiter). Initial enthusiasm has been somewhat dampened by the realization that the genetic etiology of autoimmune diabetes is complex and multifactorial involving the interaction of many susceptibility and protective loci. However, it will likely be possible to gradually determine the function of each gene in relation to disease in conjunction with the other susceptibility or protective alleles. It is an important realization that a significant proportion of the genes that correlate with disease incidence and severity have no clearly defined immunological function at this point. Additionally, some of them might be directly protective for β cells or enhance their regenerative capacity. The links between genes and autoreactive T-cell responses are explored by Ridgeway and Fathman, who present some intriguing novel findings in their chapter. With modern technology we will be able to further unravel this complex network and establish the links between genetically determined functions, the environment and the immune system.

Environmental factors such as viral infections, nutrition and the gut as the major interface between our surroundings and the immune system are thought to have an important influence on disease development from studies in animal models (Solly, Honeyman and Harrison) and based on the fact that concordance of diabetes in monozygotic twins at risk is around 30–40%. Indeed, interesting novel results indicate that certain viral infections of the gut, for example rotaviruses, can have a statistically significant association with diabetes development in young children. The underlying mechanisms are not clear at this point. It is possible that similarity between viral and self-determinants ('molecular mimicry') leads to enhancement of autoaggressive responses. However, there is no direct in vivo proof for mimicry in T1D to date. Additionally, viral infections are excellent inducers of inflammation and antigen-presenting cell activation. This could lead to enhanced attraction of lymphocytes to an immune-privileged site such as the islets and increased presentation of self-antigens. Although these considerations make viruses excellent candidates for inducing or enhancing diabetes development, direct proof has remained scarce. One explanation for this failure to establish a direct link to date is that a viral infection can largely differ in terms of its effect on T1D depending on dose, strain and timing of infection (von Herrath, Oldstone, Homann and Christen). These findings underline the need for precise and in-depth immunological analyses that will have to accompany ongoing clinical trials or prospective studies. It is well possible that different infection patterns prevailing in regions of the world can explain the geographic divergence that has been noted for T1D. Another issue worth mentioning is that incidence of T1D is on the increase in industrialized countries. Environmental factors such as nutrition, use of antibiotics and hygiene standards might be partly responsible, but better understanding of T1D immune pathogenesis will be required to evaluate this hypothesis.

Autoreactive lymphocytes with the ability to attack β cells can escape thymic selection and are found in lymphoid organs and the blood. How is tolerance to autoantigens maintained? Important findings relevant to this theme are presented by Kreuwel and Sherman, who have investigated this issue in an antigen-specific model for T1D. An issue linked to the breaking of self-tolerance is, which effector pathways are utilized by autoaggressive lymphocytes once they have been activated, to destroy β cells. Studies from Thomas and Kay have provided us with more insight into the importance of perforin produced by activated cytotoxic T cells and interferon-γ as well as TNF-α in killing the β cell. The current knowledge supports the notion that multiple detrimental immunological influences can act on the β cell simultaneously and that it will be a challenging task to build a death-resistant islet cell as an interventive approach. A crucial step could be the identification of essential 'bottlenecks' in death pathways and their targeted inhibition.

Not all components of an ongoing autoreactive process are necessarily damaging to the targeted organ. Indeed, autoaggressive and autoreactive regulatory responses have been described in many experimental models for autoimmune diseases. These coexist in a relatively fragile equilibrium, which is usually shifted in favor of the aggressive response before clinically manifested autoimmunity (in diabetes associated with destruction of more than 90% of all β cells) develops. Several differential 'regulatory circuits probably exist (Quinn, Kumar, Jensen and Sercarz) and not all of them are antigen-specific. Enhancing the regulatory component by external immunization with the 'regulatory' autoantigens via the oral, nasal or intramuscular (DNA vaccines) routes has been successful in preventing diabetes in several animal models (Arreaza, Sharif, Cameron, Chen and Delovitch/von Herrath et al.). In these experiments the curative effect extended over the life of the animal without requiring continuous immunizations and interleukin-4 was an important mediator of protection. Induction of autoreactive regulatory cells is therefore an attractive strategy to prevent T1D, because it allows for antigen-specific regulation. However, in order to bring this intervention closer to a potential application in humans, we still have to overcome several obstacles. For example, it is not precisely known how to optimize efficacy, which effector mechanisms are used by autoreactive regulatory cells and how to safely choose the appropriate autoantigen suited for immune intervention.

Cytokines are a crucial regulator of autoimmunity at several levels. Timing of their induction, expression levels and precise localization appear to be very important in addition to the class of cytokine (Green and Flavell). Cytokines can act directly on the β cell, influence systemic activation and death of autoaggressive T cells and can help to induce or maintain regulatory T cells. Cytokine networks contain a certain redundancy and are therefore difficult to dissect using transgenic and knockout approaches. Novel regulatory promoters such as those employed successfully in Green's models are of help to overcome some of this complexity.

It is important to constantly compare the features of animal diabetes models with the facts we really know about the human disease in order to establish their validity. Autoantibodies to islet cell antigens are commonly found in prediabetic individuals and can serve as a marker to assess the risk to develop

T1D in humans (Pietropaolo and Eisenbarth). Autoreactive T cells are more difficult to assess and Sønderstrup and Durinovitch-Bello offer us insight into this theme. It is known that antibodies do not contribute to diabetes pathogenesis, a finding that holds true in both humans and various animal models. Further, autoreactive T cells are found in humans as well as mouse models. Evidence suggests that the autoreactive process spreads from targeting certain initial self-antigens to a more diverse antigenic repertoire that involves aggressive as well as regulatory components. To date, it is not clear which antigen (insulin, GAD, heat-shock protein or IA-2) is the crucial primary diabetes antigen.

Therapeutically, since the primary antigens are still unknown and individuals at risk are frequently identified late during pathogenesis, when the autoreactive process has already spread to secondary islet antigens, induction of tolerance specific to one or a few autoantigens is not easily feasible. Therefore, current promising interventions to inhibit recurrent autoimmunity or rejection of islet grafts involve either more generalized inhibition of lymphocyte activation for example by the use of costimulatory blockers or induction of regulatory cells. The chapters of Gaglia and Harlan, Chatenoud and Delovitch/von Herrath discuss the most important aspects of such interventions. For example, blockade of CD40/CD40L interactions has been giving preliminary successful results in transplantation tolerance. Induction of CD25-positive regulatory T cells is a current important area of investigation and finally, autoantigen-specific (i.e. insulin) regulatory T cells are promising possibilities that are discussed in this book.

I would like to express my sincere thanks to all of the contributors for their excellent efforts, and to the Series Editor, A.N. Theofilopoulos, for commissioning me to edit this fourth volume. Due to their input, the book presents a state-of-the-art overview on the pathogenesis and current interventions of T1D and will be of value for investigators interested in understanding autoimmunity.

Matthias G. von Herrath, MD

von Herrath MG (ed): Molecular Pathology of Type 1 Diabetes mellitus.
Curr Dir Autoimmun. Basel, Karger, 2001, vol 4, pp 1–30

Genetics of Human Type 1 Diabetes mellitus

J.F. Bach, H.J. Garchon, P. van Endert

INSERM U25, Hôpital Necker, Paris, France

Type 1 diabetes mellitus (T1D), former designation insulin-dependent diabetes mellitus (IDDM), is an autoimmune disease due to the aggression of the β cells of the islets of Langerhans by islet-reactive T cells [1]. The existence of disease genetic control has long been known, since the disease involves a strong hereditary component. The problem is complicated, however, by the multiplicity of predisposing genes, by the existence of protector genes and by the relatively low penetrance of predisposition. Additionally, the disease is heterogeneous with variable rapidity of progression, exemplified by the difference in age of onset. There may even exist particular subsets of patients in whom the pathophysiology (and consequently the genetics) are clearly distinct from the bulk of other patients. This is the case of patients with autoimmune polyendocrinopathy-candidiasis-ectodermal dystrophy (APECED) [2] where T1D is part of a multifaceted autoimmune disease and where a single gene mutation is incriminated. It might also be the case of some non-DR3 non-DR4 patients without signs of islet-specific autoimmunity in whom one may question the autoimmune nature of T1D [3].

The search for predisposition genes first progressed through the testing of candidate genes. It is by using this approach that the major candidate set of genes (HLA genes) were discovered [4], in fact much before the function of HLA molecules on antigen presentation to T cells was unraveled. Two other candidate genes have been pinpointed, the insulin [5] and the CTLA4 [6] genes but it is fair to recognize that the association of these genes with T1D remains weak and that more work is needed before incriminating the genes in question in the predisposition rather than a gene located in the close vicinity, particularly for the CTLA4 gene.

The other approach is that of the systematic screening of the genome using polymorphic markers (successively VNTR, microsatellites, and biallelic probes (SNP)). A number of chromosomal areas have been associated with disease predisposition but the association is usually weak and sometimes the matter of controversial results. A similar approach has been successfully taken in the non-obese diabetic (NOD) mouse [7] with the important opening in the mouse to generate congenic resistant mice allowing to reduce the chromosomal interval and to study the gene significance. The results derived from NOD mouse genome screening provide an interesting source of candidate genes to be tested in human T1D.

The importance of the detailed knowledge of the genetics of human T1D is clearly apparent both at the pathophysiology level and at that of disease prediction, a necessary step for designing new immunointervention procedures. The progress in genetic technology and in the knowledge of the human genome will rapidly generate reliable new data. It will then be even more crucial to evaluate the relative importance of the various genes and their biological significance which might prove to be a difficult task.

Familial Transmission of the Disease

T1D has long been known as a hereditary disease on the basis of the relatively high rate of familial transmission: the risk of becoming diabetic is approximately 7% for a sibling and 6% for a child of a diabetic [8]. The disease concordance rate is approximately 35–40% in identical twins [8, 9] but penetrance of genetic factors evaluated from the identical twin concordance rate is probably less than 40% for the following reasons: (1) twins share more environmental factors than unrelated individuals; (2) there is a tendency for disease-concordant identical twins to respond more to population calls than non-concordant twins; and (3) a significant percentage of twins carrying the whole set of predisposing genes are both resistant to the disease.

An intriguing observation has been made concerning the relative risk of developing diabetes when the mother or the father has the disease, especially if they developed T1D before the age of 11. It appears that the risk is 2- to 4-fold higher when the diabetic parent is the father than when it is the mother [10–12]. No clear explanation has been provided to this difference which could relate to the changes in the immune system of the fetus under the influence of the mother's islet-reactive immune response.

Penetrance

The penetrance of a genetically controlled disease relates to the influence of environmental factors on the expression of disease predisposition genes. This influence is the resultant of triggering and protective events.

Level of Penetrance

The best data for the determination of the penetrance derives from the study of concordance rates in identical (monozygotic) twins mentioned above, which indicates that 60% of pairs of monozygotic twins are discordant for the disease (a figure submitted to the reservations discussed above). It should be mentioned, however, that the penetrance is not equivalent to concordance rates in twins since pairs of twins where the two twins do not show the disease in spite of the presence of genetic predisposition are missed. A simple calculation indicates that for intermediate value (20–80%) concordance rate in twins and penetrance are close. It should also be mentioned that the penetrance varies according to the strength of genetic predisposition. Thus, the concordance rate is higher in DR3/4 twins than in DR3/X or DR4/X twins [13].

Role of Environmental Factors

Several lines of evidence point to a major role of environmental factors in the pathogenesis of T1D. First, as just discussed, more than 60% of identical twins are discordant for the disease, and it is quite unlikely that this is due to differential somatic rearrangement of T-cell receptors. Second, disease frequency varies enormously from country to country [14], and these differences cannot simply be explained by ethnic genetic differences since migrants from countries with a low T1D frequency to countries with a high frequency are more susceptible than their compatriots [15]. Intriguingly, northern countries are more exposed to the disease than southern countries [14]; it will be critical to discover the factor(s) responsible for this North/South gradient. Third, a number of apparently nonimmunological interventions can increase or decrease the disease rate in animal models [1].

Lastly, disease incidence is on the increase in most countries (a 2-fold rise has occurred in Finland [16] and more generally in Europe [17] over the last 15 years), strongly pointing to an environmental influence; this holds true even in areas with a distinct genetic background such as Sardinia, where the incidence has recently increased dramatically to values much higher than those in surrounding regions [18].

Triggering Environmental Factors

Animal Models. Diabetes onset may be accelerated in the NOD mouse and the BB rat by protein-rich or casein-enriched diets [19]. It is also accelerated in the BB rat by infection with Kilham's virus [20]. Lastly, low-dose streptozotocin accelerates diabetes onset in NOD mice at doses that do not induce diabetes in diabetes-prone mice [21]. It is also accelerated by the nonspecific pancreatic inflammation caused by Coxsackie B virus infection [22].

Human T1D. Much interest was devoted for several decades to the possible triggering influence of viral infections. Very provocative data had been reported for Coxsackie B virus [23] but one must admit that the responsibility of Coxsackie B virus in the etiology of human T1D has never been established on a firm basis, notably at the epidemiological level.

The role of cow milk protein has been the matter of controversy. Some data has been generated that might indeed suggest that T1D is less common in children who were initially fed with breast milk [24] but the data has not yet been confirmed. The data showing the cross-reactivity and amino acid sequence homology between β-lactalbumin and p69, a β-cell autoantigen, are intriguing but indirect [25].

Protective Environment

Animal Models. The incidence and rapidity of appearance of diabetes in NOD mice and BB rats tightly depends on the sanitary environment; the cleaner the animals, the higher the disease incidence [26, 27]. Many pathogens probably contribute to the protection since it has been obtained with a large number of pathogens including bacteria, viruses and parasites [1]. The model which has been the most thoroughly studied is that of protection by mycobacteria [28]. In this model, the protective role of regulatory CD4 T cells has been well documented. One should also mention the protection afforded by low-protein diets in NOD mice [29].

Human T1D. There is no direct data demonstrating that some environmental factors may protect from the development of human T1D. However, when one considers together the NOD mouse protection afforded by infectious agents as well as the higher incidence of the human disease in northern countries [30] and more precisely in regions with high socioeconomical level [31], one may hypothesize that the improvement of hygiene and medical care in highly developed countries has led to the loss of the protection brought by infections.

HLA Genes

HLA Associations with T1D

Although a large number of genes may modulate the risk of contracting T1D [32, 33], about half of the genetic risk can be attributed to products of genes in the HLA complex found on the short arm of chromosome 6 (6p21). This estimate is based on studies of the disease risk among siblings according to HLA identity with an affected family member: this risk is 1% for an HLA nonidentical sibling, but 5% if one haplotype is shared, and 16% for full HLA identity [34, 35]. Considering that genetic factors account for about 50% of

T1D risk, the overall contribution of HLA polymorphism to this risk can be estimated at 25%. This prominent role of HLA as a risk factor renders HLA typing useful as one parameter for identifying individuals with elevated disease risk, for example in surveys of selected populations such as schoolchildren in high-risk countries, or siblings of T1D patients.

In initial studies using HLA serotyping, HLA class I alleles such as A1 or B8 were proposed to confer increased T1D risk. Subsequently, HLA class II genes in linkage disequilibrium with the mentioned class I alleles were identified as the true culprits in T1D pathogenesis [36]. Here again, increased risk was first thought to be associated with products of HLA-DR genes, before more refined methods revealed that HLA-DQ genes in linkage disequilibrium with the former have a stronger effect on T1D risk [37, 38]. Finally, the current consensus holds that although HLA-DQ polymorphism generally has a dominant effect on risk, HLA-DR has a modulating or sometimes even dominant influence. These conclusions have been confirmed in a large number of ethnic groups [39].

In Caucasians, two haplotypes predispose strongly to T1D: DRB1*04-DQA1*0301/B1*0302 and DRB1*03-DQA1*0501/0201. More than 80% of patients carry one or both of these haplotypes. Their effect on risk is synergistic, with a relative risk of 15–25 for 'DR3/4' individuals. Some other haplotypes (e.g. the DR1 haplotype) confer less pronounced elevated risks [40]. Importantly, HLA polymorphism can also strongly protect from T1D. Individuals carrying the haplotype DRB1*15-DQA1*0102/B1*0602 have a 6-fold lower risk of contracting T1D than the population average. This protection is dominant, i.e. its effect unchanged even in the presence of the DR3 or DR4 haplotypes on the second chromosome [41]. Like predisposition, protection is linked mainly to HLA-DQ polymorphism. Thus, among the rare DR1501$^+$ patients (less than 1%), a significant number may carry recombined chromosomes with DQ molecules other than DQ602, while patients with inverse recombinations (DR15$^+$, DQ602$^-$) have not been described [34, 42, 43].

HLA-DR molecules can also confer predisposition for as well as protection from T1D. DR effects have been best studied in the case of the subtypes of HLA-DRB1*04 which modulate the risk conferred by DQA1*0301/B1*0302. While the subtypes 02, 05 and (to a lesser extent) 01 predispose, the subtypes 03, 04 and 06 confer dominant protection [44–46].

The molecular features of HLA class II molecules that distinguish high- and low-risk molecules have been subject to much study. A landmark event was the discovery by McDevitt and co-workers [47] that predisposing HLA-DQ β chains carry small neutral residues (alanine, serine, valine) at position 57 while an aspartic acid (Asp) is found in resistant alleles. For example, while the DQB1*03 subtypes 01 (Asp 57) and 02 (alanine 57) are evenly distributed among DR4$^+$ individuals in the general population, DR4$^+$ T1D patients carry almost exclusively the subtype 02 [48, 49]. These observations were subsequently confirmed in numerous ethnic groups, although some exceptions have to be noted [50–52]. The significance of this finding is enhanced by the fact that, among inbred laboratory mice, the NOD strain, which develops spontaneous autoimmune diabetes resembling human T1D, is almost the only one that expresses an MHC class II molecule (I-A^{g7}) without an Asp in position 57 [53]. The critical role of position 57 for disease risk was confirmed by the demonstration that transgenic expression of an I-A^{g7} molecule with an Asp 57 protects NOD mice from autoimmune diabetes [54, 55].

The location of residue 57 in HLA-DQ and DR β chains suggested that T1D predisposing and resistant alleles differ with respect to peptide binding and presentation. HLA class II molecules bind peptides in a groove formed by two α helices sitting on a platform of β sheets [56]. Peptides are bound by hydrogen bonds with their backbone and by interactions of side chains in certain positions with pockets in the groove; binding specificity (the 'motif') is based on the latter type of interaction. Crystallographic studies have shown that most MHC class II molecules possess 4–5 pockets of variable depth, generally in the relative positions 1, 4, 6, 7 and 9 [57]. Pocket size, hydrophobicity and charge determine the nature of accepted 'anchor' residues. While hydrophobic pockets are very frequent, especially in aminoterminal anchor positions, preferences for charged amino acids can also be found and appear to be crucial in HLA association with T1D.

Residue 57 of class II β chains is part of the pocket accepting anchor residue 9 of bound peptides. In diabetes-resistant alleles, the Asp 57 forms a salt bridge with a conserved arginine residue in position 79 of the α chain. This salt bridge has been proposed to stabilize class II molecules [58], although two recent crystallographic studies of I-A^{g7} show that other interactions between α and β chains stabilize class II molecules in the absence of Asp 57 [59, 60]. While class II stabilization by the salt bridge is therefore controversial, Asp 57 clearly has an effect on the nature of the amino acids accepted in position 9, as demonstrated in several peptide-binding studies [61–63]. Both I-A^{g7} and DQB1*0302 prefer peptides with acidic residues (aspartic or glutamic acid) in position 9 while such peptides cannot bind to the acidic pocket formed in the presence of Asp 57 [61]. The crystal structure of I-A^{g7} also reveals an unusually open pocket 9, whose ready accessibility to T-cell receptors may also play a role in T1D pathogenesis [59]. Strongly predisposing HLA-DR alleles such as DRB1*0405 also lack Asp 57, suggesting that the capacity of binding peptides with acidic carboxyterminal residues may be a general hallmark of T1D-associated class II molecules. It should be noted that charged anchor residues also seem to play a role in susceptibility to another autoimmune disease, rheumatoid arthritis [64]. However, in this case, the nature of the charges accepted in position 4 is critical, and acceptance of basic instead acidic residues seems to be linked to susceptibility [65]. In conclusion, T1D-associated class II molecules clearly differ from resistant ones with respect to accepted charges in position 9 of bound peptides, and perhaps also to lower stability. However, the mechanism by which these particularities contribute to T1D pathogenesis remains to be established (see also below).

While discrete polymorphic features in HLA-DQ and DR molecules clearly modify T1D risk, it is possible that other genes in the large HLA complex have an independent effect on T1D susceptibility. These include HLA-DP [44], products of the central class III region such as TNF [66–68], and also HLA class I genes [69, 70]. Support for a role of additional susceptibility genes in the HLA class III region, close to the TNF genes, has recently been provided by an analysis of the Belgian diabetes registry [71]. Moreover, a recent family study using microsatellite markers suggested a protective effect of an additional gene located within the telomeric HLA class I region, between the genes coding for HLA-F and HFE [72, 73]. Although polymorphism in antigen processing genes encoded in the class II region, such as TAP, LMP and HLA-DM, has been proposed by some authors to be linked to IDM susceptibility [74, 75], it has subsequently become clear that such associations are overwhelmingly or entirely due to linkage disequilibrium between these genes and neighboring HLA class II genes [76, 77].

Apart from determining T1D risk, HLA genes can also modulate clinical features of the disease, such as age of disease onset or outcome of active cellular autoimmunity. Thus, the combination of the DR3 and DR4 haplotypes not only predisposes strongly to T1D but also accelerates disease onset, i.e. is found much more frequently among patients with T1D onset during childhood [78]. Conversely, the rare individuals contracting T1D despite the presence of the protective DQB1*0602 allele generally show a very late onset of disease [41]. In most cases, this allele appears to prevent overt disease even in the presence of active autoimmunity. DQB1*0602$^+$ first-degree relatives of T1D patients very rarely develop T1D, even after many years of β-cell-targeted autoimmune responses indicated by circulating antibodies to islet cells (ICA). Finally, HLA class I polymorphism may influence the rate of progression to overt disease exclusively among individuals having developed autoimmune insulitis (T1D is thought to evolve in three phases entailing priming of an autoreactive repertoire, nondestructive mononuclear islet infiltration or peri-insulitis, and finally destructive intrainsulits). Among ICA$^+$ relatives of T1D patients, HLA-A24 predisposes to overt disease [79–81]. The same class I allele is associated with early age of disease onset, a low rate of 'honeymoon' remissions, and diabetic retinopathy.

As outlined above, molecular and structural analysis of MHC class II molecules suggests that peptide ligand selection and presentation by class II is directly involved in its effect on T1D risk. This conclusion is strongly supported by the finding that, in mice prone to autoimmunity due to a number of non-MC genes, MHC class II molecules with distinct ligand preferences can target autoimmunity to distinct target organs. Thus, congenic NOD mice expressing I-A^k instead of I-A^{g7} molecule develop thyroiditis instead of diabetes, while I-A^b NOD mice are prone to destructive sialitis [82]. This demonstrates that peptide binding preferences of class II molecules direct autoimmune diseases to specific target organs. It is likely that only a small number of autoantigenic peptides meet the (so far undefined) specific binding and presentation requirements for a single MHC class II molecule that ultimately lead to destructive autoimmunity. This finding underlines the interest of attempts to define 'crucial' autoantigens and their epitopes. At the same time, it suggests that association of individual class II molecules with specific autoimmune diseases is not based on global functional defects but rather on class II interaction with a very limited number of peptide antigens.

Unfortunately, the site and nature of such crucial MHC class II/peptide interactions remain largely to be defined. Several mutually nonexclusive hypotheses on the mechanism of MHC class II implication have been proposed. Some of these focus on thymic T-cell education while others consider autoantigen presentation in affected target organs. Some theories propose that general structural features of T1D-associated molecules are implicated, while others envision a critical role for presentation of specific epitopes by a given MHC class II molecule.

One theory postulates that low stability of T1D-associated MHC class II molecules, presumably due to the lack of the salt bridge between Asp β57 and Arg α79 or to poor peptide binding capacity, is at the root of their link to autoimmunity [58, 83, 84]. These authors contend that rapid dissociation of peptides from such molecules reduces the average avidity of TCR/peptide/MHC interaction during thymic negative selection, so that many autoreactive T cells escape into the periphery. This phenomenon is not necessarily dangerous for the organism, since activation of such T cells in the periphery should also be hampered by low stability of peptide/MHC complexes, and since a larger number of peptide/MHC complexes is required for peripheral T-cell activation than for negative selection in the thymus. However, in certain circumstances, such as abundant availability of autoantigen in an inflammatory environment, a large autoreactive repertoire may indeed pose a threat for the organism.

Evidence for low stability of human and murine T1D predisposing class II molecules has been reported. Thus, in a comparison to several other I-A molecules, cell surface I-A^{g7} dissociates more rapidly, and immunoprecipitated I-A^{g7} is less stable in the presence of the strong detergent SDS [58]. Similarly, the T1D-associated DQB1*0302 is less resistant to SDS than the closely related Asp 57$^+$ T1D-resistant molecule DQB1*0301 [85]. A role of Asp β57 for stability has more directly been suggested by a mutagenesis analysis of HLA-DQB1*0602 whose stability in SDS is reduced by introduction of a valine in position 57 [86]. Moreover, two groups have reported that NOD mice have significantly more autoreactive T cells than other mouse strains [87]. However, the hypothesis of a role of low class II stability in T1D susceptibility remains controversial. Firstly, multiple class II molecules are associated with human T1D, but there is no evidence that all of these are unstable. Secondly, the relationship between SDS stability and efficacy of antigen presentation during thymic selection is not clear, and very little information on stability of various class II molecules on the cell surface of dendritic cells (mediating thymic deletion) is available. Furthermore, two recent crystallographic studies did not provide evidence for low stability or functional efficacy of I-A^{g7}, since other unusual interactions between α and β chains may compensate for the absence of the salt bridge Aspβ57/Argα79 [59, 60]. Binding studies by Teyton and co-workers [63] suggested very recently that I-A^{g7} is a normally stable molecule that differs from other molecules by binding an unusually large promiscuous peptide repertoire. The conundrum of how unstable class II molecules can active low avidity autoreactive T cells in the periphery remains to be resolved. Finally, the observation that NOD mice develop autoimmune syndromes other than T1D when expressing presumably stable I-A alleles other than g7 also argues against an essential role of class II stability [82].

Another simple hypothesis postulates that T1D predisposing class II molecules can present crucial autoantigenic epitopes in the target organ while protective alleles cannot. Since MIIC-associated protection from T1D is dominant, this mechanism alone cannot explain MHC association with T1D. Nevertheless, as long as such 'crucial' autoantigens and epitopes remain undefined, it is difficult to prove or disprove a role for this mechanism. Based on substantial evidence assigning a significant though not necessarily crucial role in T1D to the autoantigens GAD65 and IA-2, we have recently screened comprehensive sets of GAD and IA-2 peptides for binding to HLA-DR and DQ molecules predisposing for, or protecting from T1D [88]. We found that both types of molecules bind a similar number of peptides with similar efficiency. This suggests that binding affinity for autoantigenic peptides may not be related to disease implication. However, besides uncertainty with regard to crucial autoantigens, two considerations limit the bearing of binding studies. Firstly, due to constraints in the T-cell receptor repertoire and in antigen processing, high binding affinity does not always translate into antigenicity of an epitope. Secondly, 'efficient' presentation of an autoantigen, i.e. presentation ultimately leading to destruction of self tissues, may be different from efficient presentation of a foreign antigen derived from a pathogen that needs to be cleared from the organism. While efficient responses of the latter type usually require TCR/peptide/MHC inter-actions of high avidity, low avidity interactions may sometimes be more important in initiating or sustaining autoimmune responses. This is exemplified in the case of experimental allergic encephalitis (EAE), a model of human multiple sclero-sis induced by injection of susceptible mouse strains with peptide 1–9 of myelin basic protein. Low binding affinity of this peptide for I-A^u leads to inefficient tolerization of autoreactive T cells in the thymus and is a prerequisite for sus-ceptibility to EAE, since injection of analogs with higher binding affinity prevents the disease [89, 90]. At least in the case of autoantigens available for presentation in the thymus, the search for candidate key epitopes should therefore include peptides with low binding affinity.

Variable capacity of deleting autoreactive T cells has also been proposed to underlie MHC-associated variation in T1D risk. This hypothesis postulates that T1D-associated class II molecules fail to delete potentially autoreactive T cells during thymic education while protective molecules are particularly effi-cient at this. Inefficient deletion of autoreactive T cells may concern either a large part of the total autoreactive repertoire, for example as a result of intrin-sically unstable MHC class II molecules (see above), or individual T-cell speci-ficities. As a result, a large repertoire of autoreactive T cells, or alternately a small number of T cells with a perilous specificity would escape from the thymus. Limited evidence for a role of a failure to delete individual T-cell speci-ficities has been reported in mouse models. Thus, reduced diabetes incidence in mice expressing a diabetogenic transgenic T-cell receptor together with a 'protective' MHC class II haplotype was associated with partial thymic deletion of diabetogenic T cells [91]. While this example shows that deletion can play a role, most reports argue against an important role of this mechanism on bulk T-cell repertoires. For example, in an analysis of NOD mice protected from T1D by transgenic introduction of the 'resistant' I-A^k molecule, Slattery et al. [92] found no evidence for thymic deletion or anergy of autoreactive T cells. On the contrary, autoreactive cells were present and functional and could transfer dia-betes to NOD recipients lacking a protective MHC class II molecule. More recent studies have provided additional evidence that – even in animals express-ing diabetogenic transgenic T-cell receptors – MHC-associated protection is likely to be based on selection of regulatory T cells rather than on deletion of autoreactive T cells (see below).

Competition for binding of autoantigenic peptides between susceptible and resistant molecules has also been proposed to explain modification of T1D risk by MHC class II. This hypothesis, also referred to as determinant capture and originally proposed by Nepom [93] envisages that protective molecules with high affinity for crucial autoantigenic peptides 'capture' these (i.e. bind all available material), thereby preventing their presentation by predisposing alleles [94]. This scenario may take place in the thymus, where selection of low avidity T cells restricted by predisposing MHC would be prevented, or in the target organ or its regional lymph nodes, where activation of autoreactive T cells would be avoided. The theory implies that autoantigen presentation by the protective molecule is not dangerous for the organism. This may be due either to efficient thymic deletion of reactive T cells, or to peripheral mechanisms preventing harm, such as induction of regulatory T cells or ignorance. Determinant capture can occur if predisposing and protective molecules bind identical or overlapping peptides, but may also apply to neighboring epitopes contained in processing intermediates produced in class II processing compartments. 'Promiscuous' epitopes (binding to numerous MHC class II molecules) as well as epitope clusters in autoantigens are most likely to be involved in phenomena of determinant capture. In our recent analysis of the important autoantigen GAD65, we found peptides binding to almost all of the 14 tested (high- and low-risk) class II molecules, and protein regions containing numerous peptides binding to multiple class II molecules [88]. Although the biological potential of determinant capture has been demonstrated in a study of a model response to hen egg lysozyme, a role of 'epitope stealing' in a context of autoimmunity remains to be demonstrated [95].

Regulatory T cells or their absence are highly likely to be involved in MHC effects on T1D susceptibility. This hypothesis assumes that protective class II molecules favor production of T cells limiting autoimmunity and preventing organ destruction, while predisposing molecules fail to do so. Thus, it is postulated that autoreactive responses are generated in the presence of predisposing and protective molecules but differ with respect to their outcome. This hypothesis readily explains dominance of protective MHC class II molecules. Regulatory cells may result from T-cell education in the thymus, or from recognition of autoantigens presented by protective class II molecules in the target organ or regional lymph nodes. Evidence for the latter hypothesis stems from recent studies of autoimmunity-prone rats. In this model, generation of regulatory T cells preventing autoimmune thyroiditis depended strictly on the presence of the affected target organ beyond the postnatal period [96]. NOD mice protected from diabetes by MHC have also provided evidence for regulatory T cells. While the demonstration that mice protected by I-A^k harbor diabetogenic T cells already suggested the presence of regulatory cells [92], adoptive transfer experiments with lymphocytes from I-A^b transgenic NOD mice provided direct evidence for this hypothesis [97]. In another study, NOD mice protected by expression of I-E showed similar proliferative responses to autoantigens as nontransgenic littermates [98]; however, autoreactive T cells from protected mice produced IL-4, a cytokine generally considered 'protective' in the NOD mouse, instead of IFNγ. Finally, even in mice expressing a highly diabetogenic T-cell receptor in the vast majority of CD4+ T cells, selection on a protective MHC class II molecule of a limited population of T cells with a second TCR resulted in a strong reduction in diabetes incidence [99]. Thus, regulatory T cells clearly can be highly efficient in limiting destructive auto-immunity. The antigens recognized by such cells, the site of antigen recognition and the molecular features of antigen recognition that determine regular vs. destructive responses remain to be determined. Effector mechanisms involved in regulation include bystander suppression by cytokines such as IL-4 and/or IL-10, but also direct inhibitory cellular interaction between T lymphocytes and/or antigen-presenting cells. Regulatory mechanisms are discussed in other sections of this book.

Non-Class II HLA Genes

Long before a detailed knowledge of class II gene organization and nucleotide sequence was available, several lines of evidence led to question the idea that class II genes by themselves could fully explain the contribution of the HLA region to the genetic control of T1D. As mentioned above, analysis of extended haplotypes including markers from the class II gene region (and also the GLO gene, centromeric to the HLA region on the short arm of chromosome 6) to the class I region and including class III genes, notably those encoding complement factors (C4A, C4B, C2, factor B) (fig. 1), suggested that haplo-types with identical class II gene alleles but with different polymorphisms of their complement genes were associated with varying risks of developing T1D [100–102]. Likewise, other candidate genes, including genes of the tumor necrosis factors [103–105], which are also located in the class III region, between complement and class I genes, and the genes encoding the various components of the antigen-processing machinery [76], which map centromeri-cally to class II genes, were assessed. No firm conclusion, however, could be drawn because of the strong linkage disequilibrium between HLA genes. In a slightly different vein, recent studies, taking advantage of a better definition of the polymorphism of class I genes, have shown that HLA-A24, although not being a proper risk factor for T1D, was associated with an earlier age at onset of disease and with a more aggressive course [80, 81].

Linkage disequilibrium is the primary but not the only explanation that accounts for the often contradictory findings in HLA studies. Most of the above studies were based on a case-control design, i.e. on the comparison of

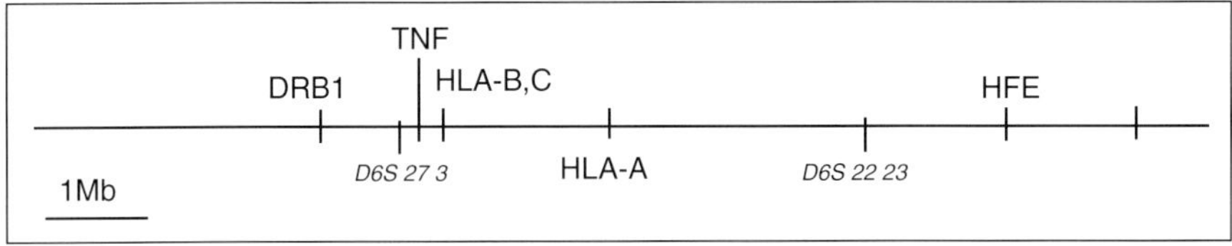

Fig. 1. Schematic organization of the HLA complex.

unrelated patients and controls. Such a design is sensitive to the heterogeneity of the population samples. The lack of biological correlation (gene expression and function) with the genetic polymorphisms also results in limited power when interpreting purely statistical data. The recent years have seen the emergence of new tools that should help overcome these difficulties. First is the availability of the complete genomic organization and nucleotide sequence of the MHC in man, allowing the precise positioning of many genes and polymorphisms whose location was hitherto unknown [106]. This knowledge also provides the basis for exhaustive identification of genomic polymorphisms, including microsatellite markers and biallelic single nucleotide polymorphisms (SNP). A fairly complete list of multiallelic microsatellite markers mapping to the MHC is now available [107] and keeps being updated. Their use is of great interest for the initial scanning of the entire MHC and to narrow down the regions of potential interest. However, their average spacing of 1/30 kb and the lack of known biological function associated with these short tandem repeats is a limitation for the eventual identification of the disease loci. In contrast, SNPs are abundant, at least one in every kilobase, and may be located in genes, either in the coding exons where they may result in alteration of their protcin prod ucts, or in the regulatory regions, where they may influence their expression level. These properties make them not only useful for gene mapping, but also candidates for disease loci. Another advantage of SNPs is the ease with which they can be typed. Constantly improving typing techniques, that do not require gel electrophoresis, guarantee both reliability and high throughput [108].

The second tool is a set of statistical methods of analysis of genotype data using linkage disequilibrium [109]. These methods are looking for a bias of the transmission of parental alleles to their affected offsprings at a test marker. When present, such a bias indicates both linkage and allelic association of the marker with the disease locus. The main advantage of these methods is 2-fold: they require only single-case or so-called simplex families, and they are not sensitive to the structure of the population. In addition, they are often more sensitive than classical, association-independent methods of linkage analysis (see below the case of IDDM2). The development of these analytical tools is currently a topic of intense research in statistical genetics as their use appears nondispensable for the fine mapping of disease loci of complex genetic diseases. A popular and early implementation is known as the transmission-disequilibrium test (TDT) [110] but more powerful variants have been proposed since.

Using these tools, at least two groups were able to provide evidence for additional susceptibility HLA-linked loci in DR3-DR4 or DR3 homozygous patients in the past 2 years. One of these loci would be associated with the D6S273 microsatellite marker, which is located at the boundary between the class II and the class III region [111]. A second locus was associated with the D6S2223 microsatellite [73] and thus would map outside the proper MHC, telomerically to HLA-A and not far from the HFE gene, which is mutated in patients with hemochromatosis. These data are currently based on the use of microsatellites. They certainly await reproduction in different patient samples and extension using SNPs. They are nevertheless promising because they echo similar findings in the NOD mouse model that were obtained using congenic lines and that formally demonstrated the existence of non-class II susceptibility genes closely linked to the murine MHC [112, 113].

MHC-Unlinked Genes

As discussed above, MHC-linked genes account for only half of the overall genetic predisposition to T1D. Thus, collectively, non-MHC genes account for the other half [114]. Their search was the first instance of a systematic genome scan in a polygenic disease. These studies were based on the analysis of affected sib pairs (ASP). In principle, if a locus is not involved in the genetic predisposition to T1D, parental alleles at this locus are inherited by affected siblings following the Mendelian laws: 25% of sib pairs share two parental alleles, 50% share one allele, and 25% do not share any allele at this locus. An excess of allele sharing by ASP at a given marker will then reflect the influence of a disease closely linked to that marker. The method is independent of any model of inheritance, i.e. it does not require to specify the mode of transmission and the penetrance of the disease gene. It is therefore well suited for searching genes contributing to polygenic diseases in small-size families. However, it extracts only a small amount of information from each family, which must be compensated by increasing the number of families studied. Given that polymorphic markers that cover the entire genome are available, it is theoretically possible to explore the human genome for disease loci exhaustively.

Table 1. Predisposition loci in human T1D

Locus[a]	Chromosomal region	Dependency of HLA[b]	Candidate genes	NOD equivalent[c]	Ref.
IDDM1	6p21		Class II genes	Idd1	4
IDDM2	11p15		INS		5, 126
IDDM3	15q26		IGF1R	Idd13	120
IDDM4	11q13		ICAM, CD3, FGF3	Idd2	32, 119
IDDM5	6q25		ESR		124
IDDM6	18q21	DR4/X		Idd5	127
IDDM7	2q31		HOXD8	Idd5	121
IDDM8	6q27				124
IDDM9	3q21–q25	DR3/4			117
IDDM10	10cen				159
IDDM11	14q24–q31				125
IDDM12	2q33		CTLA4	Idd5	151
IDDM13	2q35		IGFBP2, IGFBP5	Idd5	123
IDDM15	6q21				124
D14S70-D14S276	14q12–q21				117
D16S515-D16S520	16q22–q24				117
D19S247-D19S226	19p13				117
D19S225	19q13			Idd16	117
D1S1617	1q				116
DXS1068	Xp	DR3/X			118

[a] The IDDM14 locus was assigned but not published.

[b] Only strict dependency on HLA for significance is reported. DR4/X: X is not DR4. DR3/X: X is not DR3.

[c] Type 1 diabetes predisposition loci in the NOD mouse strain in equivalent chromosomal regions.

The main outcome of these studies altogether is that the MHC, also designated IDDM1, is the major T1D locus and that no other chromosomal region matches it. Multipoint maximum lod scores (MLS), a measurement of the strength of the linkage that is considered significant when >3.6 [115], have reached 32 for IDDM1 [116, 117]. This level is 10 times more significant than that of any non-MHC locus. Beyond this consensus statement, there is a debate about the number and the chromosomal location of the non-MHC loci which could be identified by the various groups that have undertaken this endeavor (table 1). Thus the group of John Todd [117], who pioneered this kind of study in human T1D, was able to identify nine regions in addition to the MHC [32]. Additional loci were detected only after partitioning the families according to HLA and the insulin (INS) gene polymorphism. It was reasoned that the dependency of a locus on HLA or INS should reflect different pathogenetic pathways. In this regard, most striking was the finding of an X-linked locus in DR3 patients, corresponding to an excess of males among patients with this DR haplotype [118].

Other groups detected different regions, some of which, but not all, were characterized in Todd's studies [119–125]. Most puzzling was the outcome of a genome scan of comparable importance to that of Todd's, carried out by Concannon et al. [116]. These authors detected none of the loci found by Todd's group, except IDDM1, but instead disclosed only one locus, on distal chromosome 1, that was not significant at all in Todd's latest genome survey. Altogether, it is difficult to compare the various studies between them because they are often based on partially overlapping sets of family samples.

Almost all these loci (except IDDM10), also, have been detected at low levels of statistical significance, which are considered to be suggestive only. They may therefore represent false-positives given the huge number of statistical tests that were performed. Alternatively, they could correspond to true but weak loci. Their large number, however, is inconsistent with the fairly high risk of recurrence of T1D in first-degree relatives of patients (6%). The variable findings could also reflect the heterogeneity of the disease, with different loci being involved in different families or in different ethnic groups and corresponding to different pathogenetic mechanisms.

In some instances, the low level of statistical significance may result from a lack of power of ASP methods which use only information relevant to linkage. Methods based on linkage disequilibrium, such as the TDT mentioned above, may be powerful. The most striking example concerns the insulin gene, which historically was the second locus to be implicated in T1D, hence its designation as IDDM2. This was initially based on case-control studies [5]. And although this result could be easily reproduced in many populations, linkage could be proven by ASP methods only later [126]. Analysis of parental transmission of insulin polymorphisms, however, allowed to confirm association and also to prove linkage [110]. This example illustrates the greater sensitivity of linkage disequilibrium analysis compared to that of linkage analysis and justifies to develop linkage disequilibrium analysis of loci linked at a low level of significance. Another example of a weakly linked locus with successful followup using association methods was IDDM6 on chromosome 18q12–q21 [127].

The IDDM2 Locus and the INS Gene

The insulin gene was a logical candidate for investigating the genetic basis of diabetes. Early on, a polymorphic short tandem repeat, also called minisatellite or variable number tandem repeat (VNTR), located in the upstream region of the gene was associated with T1D [128, 129]. This repeat consists of a variable number of short motifs of 14–15 nucleotides and falls into two main discrete classes: class I (26–63 repeats) and class III (141–209 repeats), whereas class II alleles are very rare. Homozygosity for class I alleles is associated with an increased risk for T1D whereas class III alleles are protective [5]. The high frequency of class I alleles in the normal population (51% homozygous subjects vs. 73% in T1D patients) accounts for the difficulty to demonstrate linkage, independently of association. It has also been argued that association of this chromosomal region was conditioned by the HLA region, namely the DR4 haplotype [130]. However, this could not be confirmed in subsequent studies [131].

Extensive analysis of the nucleotide sequence spanning the entire insulin gene has disclosed a limited number of polymorphisms that all mapped to the noncoding region [132]. These polymorphisms were associated in two major haplotypes, themselves in tight linkage disequilibrium with the class I and class III VNTR. Association of these biallelic polymorphisms appeared to be secondary to that of the VNTR. It remains that the increase in the relative risk associated with class I alleles is only moderate. Indeed, there is microheterogeneity within class I VNTR alleles, resulting from variation of both the number of repeats and their nucleotide sequence and it was disputed whether only certain allele numbers might be associated with the disease [133].

The VNTR sequence might influence T1D predisposition by modulating the transcription of downstream sequences [134]. Decrease in steady-state INS mRNA levels were found in certain, but not all, individuals with class I alleles [135, 136], contrasting with an increased expression in the thymus [137, 138]. It is understood that increased expression in the thymus would facilitate the deletion of high-affinity autoreactive T cells. It is also conceivable that decreased expression of the autoantigen in the pancreas could lead to less efficient activation of the autoreactive T-cell clones and therefore be protective against development of diabetes. It remains to be explained how the same haplotype may result in opposite variation of gene expression in different tissues. At the clinical level, it has been proposed that the class I alleles were associated with lower insulin secretory response [139, 140], a feature that is predictive of diabetes in at-risk subjects. However, this finding has not been reproduced [141].

Regulation of INS gene expression should also be viewed in the more general context of the 11p15 chromosomal region (fig. 2). This region is subject to genomic imprinting, a mechanism of gene regulation by which the parental origin of the chromosomal region dictates expression of the genes it contains [142]. This mechanism is based on methylation of DNA which usually results in repression of transcription. Indeed, the first example of an imprinted gene was

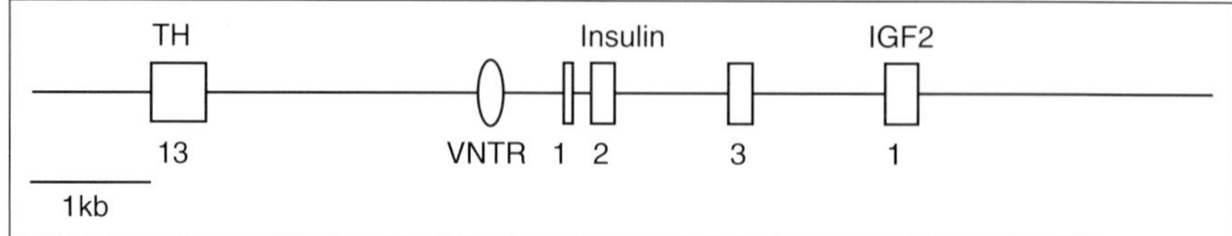

Fig. 2. The insulin (INS) gene on chromosome 11p15 is flanked by the tyrosine hydroxylase (TH) and IGF2 genes. The VNTR (variable number tandem repeat) is located in the upstream region of the INS gene. Boxes denote exons. Numbers indicate exon numbers. Only the last exon of the TH gene and the first exon of IGF2 are represented.

the IGF2 locus which is located immediately downstream to the INS gene. Imprinting of the equivalent chromosomal region in the mouse (on chr. 7) has been taken as a study model. It was reported in several populations that the INS VNTR class I allele conferred predisposition to T1D only when it was inherited from the father, whereas the maternally inherited allele was neutral [143–146]. Imprinting, however, does not seem to occur in the pancreas. Thus it could occur in the thymus, which remains to be demonstrated, but would make this organ the target of the susceptibility caused by the IDDM locus.

Two additional lines of evidence argue against restricting the IDDM2 locus to the sole INS gene. The tyrosine hydroxylase gene, which is also immediately adjacent to INS but on its 5′ side, seems to be associated to T1D in a manner at least partly independent of INS [147]. The INS class III VNTR is associated with disorders affecting cell growth and development as often are imprinted genes, including type 2 diabetes, polycystic ovary syndrome and increased size at birth [148–150].

Other Candidate Genes

Other candidate genes that were investigated in human T1D include IL-1α, IL-1β and IL1-RA (IDDM7), CTLA4 (IDDM12), IFNγ, glucokinase and glutamic acid decarboxylases (GAD) [151–156]. Polymorphisms of these genes, except GAD, could be associated directly with T1D. None of these polymorphisms, however, explains how the gene could predispose to the disease for the time being. The CTLA4 gene is of particular interest, given the demonstration of its major role in the regulation of lymphocyte activation in mice and its possible implication in murine diabetes [157]. Detailed analysis of the polymorphisms in the surrounding region points to the CTLA4 gene itself as being the disease locus and notably rules out a role of its neighbor, CD28 [158]. Three polymorphisms are currently known in CTLA4, including a A/G SNP in exon I resulting in a Ala/Thr exchange in the leader peptide, a C/T SNP in the first intron and a microsatellite repeat in the 3′ untranslated region. However, an alteration of CTLA4 gene expression in T1D patients has not been demonstrated yet. No phenotype that could result from an altered expression of CTLA4 was described either. Similar comments can be made for the IL-1 gene cluster and the IFNγ gene. As to the GAD65 gene, it maps to the T1D10 locus on chromosome 10p11–q11, a region which was detected at a significant level in British T1D patients [159], but not in other populations [156].

Significance

The multiplicity of genes predisposing to or protecting from T1D explains the heterogeneity of the disease, which is increasingly recognized. At the basic level, one would like to understand the cellular and molecular mechanisms that relate to the expression of these various genes, making a clear distinction between genes that are used by the pathologic process as part of the current cellular machinery and those whose expression is either abnormal in T1D patients or uniquely adapted to interaction with environmental factors. At the clinical level, one would like to correlate genetic polymorphism with clinical heterogeneity and possibly use the most disease-specific genes for prediction of diabetes onset in families at risk.

How Many Genes?

The number of genes predisposing to T1D is very high both in the NOD mouse and in human T1D (>15 and probably significantly more). Such a large number does not fit well with the relatively high disease concordance rate observed in siblings of diabetic patients (7%) particularly if one notes (1) that the concordance rate is even higher (~16%) if one considers HLA identical siblings and (2) that the penetrance is of the order of 35% (concordance rate in monozygotic twins). Two main explanations can be provided to this paradox:

(1) Different patients' families use different genes (except MHC genes which are used by most patients), although not necessarily all since 10–20% of patients with adulthood onset are neither DR3 nor DR4 [3]. This hypothesis might explain the difficulty met to identify T1D predisposing chromosomal areas through the analysis of a large number of families, which provides a selective advantage to widely used genes. This difficulty is exemplified by the low odds ratios observed for all regions identified by genome scanning so far. The inconsistent usage of the various genes in different families might also explain

the differences in results obtained in some studies. One must realize that in extreme cases, several diabetic syndromes may get the generic name of T1D but are fundamentally distinct entities except for the common denominator of the autoimmune origin (and the HLA association).

(2) There is a hierarchy of predisposing genes with a limited number of 'major genes' and a multitude of 'minor genes'. This situation is well illustrated in autoimmune diseases by Wakeland's studies [160] in systemic lupus erythematosus showing the disease control by four major genes. This would fit with the disease concordance discussed above. A crucial step will be to identify these major genes (in addition of MHC genes). One must admit that there is not yet any indication for major genes in the presently available data. A predisposition gene may only have a minor influence on the genetic risk for two very different orders of reasons. The influence may appear as very limited just due to the fact that the gene is expressed in a large fraction of the general population. In this case, the gene may be important for disease predisposition, eventually mandatory, but its presence in a large number of healthy subjects results in a low odds ratio. In fact, in the latter situation, the absence of the gene confers protection and we have seen examples of such protective genes both in man (HLA DR2 DQB1*0602) and in the NOD mouse (Fcgr II). Alternatively, the predisposing gene is not commonly expressed in the general population but its contribution to disease onset is not very important except in certain settings where its interaction with other environmental factors or other predisposing genes is unfavorable.

Mechanisms of Gene Expression

HLA genes are tightly associated with diabetes predisposition (and protection). The mechanisms underlying this association have been extensively discussed in the section 'HLA Genes'. It would be interesting to know if some of the non-MHC T1D predisposing genes directly interact with MHC genes as has been suggested in another autoimmune disease, myasthenia gravis, where HLA genes interact with the acetylcholine receptor α-chain gene [161]. Such interaction could involve molecules implicated in antigen recognition, such as the β-cell autoantigens. In fact the insulin gene, or more precisely its promoter, has been shown to be associated with T1D (see 'Non-Class II HLA Genes'). It has even been suggested that the T1D-associated insulin gene was associated with a decreased expression of proinsulin in the thymus, perhaps leading to decreased intrathymic negative selection [137, 138]. One may note however that no association has been reported for the GAD gene [156, 162] that encodes for another major β-cell autoantigen.

The other predisposing gene discussed above is the CTLA4 gene. It remains to be determined if it is indeed the CTLA4 gene that is directly incriminated. If this were the case, one would like to know the level of CTLA4 expression in lymphocytes from diabetic patients. It may be relevant to mention here that CTLA4+ regulatory T cells have been reported to play a major role in the control of diabetes onset in the NOD mouse. $B7^{-/-}$ or $CD28^{-/-}$ NOD mice show fulminant disease that can be mitigated by administration of CD25+ CTLA4+ T cells [163] suggesting the possible role of CTLA4 in the regulation, a role also advocated in the polyautoimmune syndrome induced by neonatal thymectomy [164] and in the colitis induced by the transfusion of $CD45RB^{lo-}$ T cells in scid mice [165].

No other information is available whatsoever on the nature (and function) of other T1D predisposition genes. If one extrapolates data obtained in the NOD mouse to human T1D, one may expect that some of the non-MHC genes might control expression of some cytokines (like IL-2 (Idd3) or IL-4 (Idd9) in the NOD mouse).

Clinical Heterogeneity

Diabetes is presently classified in two main clinical groups, type 1 and type 2, on the basis of insulin dependency. It is clearly apparent that this separation does not account for a number of clinical presentations that would be more readily grouped according to physiopathology and consequently on genetic basis.

One may thus identify seven subsets of patients with type 1 or type 2 related disease: (1) conventional T1D including childhood and adulthood onset; (2) fulminant diabetes of the infancy, that may just represent the most severe form of the previous subset; (3) delayed and slowly developing T1D, preceded by a long phase of non-insulin dependency (LADA for latent autoimmune diabetes of the adult) [166, 167]; (4) familial diabetes as reported in a limited number of families [168]; (5) non-DR3 non-DR4-associated diabetes often not associated with islet cell autoantibodies and in which consequently one may question, at least for some patients, the autoimmune origin [3]; (6) the exceptional cases of diabetes association with anti-insulin receptor antibodies [169]; and (7) APECED [2].

One would like to know more about the genes controlling these different clinical presentations. Some data are available for HLA genes. DR3/4 heterozygotes are prone to develop early and severe disease onset [78]. LADA is only associated with HLA DR3 or DR4 in approximately 50% of cases.

Conclusions

T1D is the autoimmune disease in which the most documented genetic data is available both in terms of gene mapping and gene identification. The existence of two highly relevant animal models consolidates the human data and opens interesting perspectives for future research. It remains though that with the exception of HLA genes, little is known yet on the nature of the predisposition genes and even less on their significance. One may hope that the recent knowledge of the entire human genome will speed up the understanding of this highly complex genetic predisposition and break it down into well-defined genetic systems amenable to reverse genetic approach as done in monogenic diseases.

References

1 Bach JF: Insulin-dependent diabetes mellitus as an autoimmune disease. Endocr Rev 1994;15: 516–542.
2 Bjorses P, Aaltonen J, Horelli-Kuitunen N, Yaspo ML, Peltonen L: Gene defect behind APECED: A new clue to autoimmunity. Hum Mol Genet 1998;7:1547–1553.
3 Dubois-Laforgue D, Timsit J, Djilali-Saiah I, Boitard C, Caillat-Zucman S: Insulin-dependent diabetes mellitus in non-DR3/non-DR4 subjects. Hum Immunol 1997;57:104–109.
4 Nerup J, Platz P, Andersen OO, Christy M, Lyngsoe J, Poulsen JE, Ryder LP, Nielsen LS, Thomsen M, Svejgaard A: HL-A antigens and diabetes mellitus. Lancet 1974;ii:864–866.
5 Bell GI, Horita S, Karam JH: A polymorphic locus near the human insulin gene is associated with insulin-dependent diabetes mellitus. Diabetes 1984;33:176–183.
6 Marron MP, Raffel LJ, Garchon HJ, Jacob CO, Serrano-Rios M, Larrad MTM, Teng WP, Park Y, Zhang ZX, Goldstein DR, Tao YW, Beaurain G, Bach JF, Huang HS, Luo DF, Zeidler A, Rotter JI, Yang MCK, Modilevsky T, MacLaren NK, She JX: Insulin-dependent diabetes mellitus is associated with CTLA4 polymorphisms in multiple ethnic groups. Hum Mol Genet 1997; 6:1275–1282.
7 Wicker LS, Todd JA, Peterson LB: Genetic control of autoimmune diabetes in the NOD mouse. Annu Rev Immunol 1995;13:179–200.
8 Rotter JI, Vadheim CM, Rimoin DL: Genetics of diabetes mellitus; in Rifkin H, Porte D (eds): Diabetes mellitus. Theory and Practice, ed 4. Amsterdam, Elsevier, 1990, pp 378–413.
9 Lo SS, Tun RY, Hawa M, Leslie RD: Studies of diabetic twins. Diabetes Metab Rev 1991;7: 223–238.
10 Rjasanowski I, Heinke P, Michaelis D, Kurajewa TL: The higher frequency of type I (insulin-dependent) diabetes in fathers than in mothers of type I-diabetic children. Exp Clin Endocrinol 1990;95:91–96.
11 Warram JH, Krolewski AS, Gottlieb MS, Kahn CR: Differences in risk of insulin-dependent diabetes in offspring of diabetic mothers and diabetic fathers. N Engl J Med 1984;311:149–152.
12 Dosch HM, Martin JM, Robinson BH, Akerblom HK, Karjalainen J: An immunological basis for disproportionate diabetes risks in children with a type I diabetic mother or father. Diabetes Care 1993;16:949–951.
13 Johnston C, Pyke DA, Cudworth AG, Wolf E: HLA-DR typing in identical twins with insulin-dependent diabetes: Difference between concordant and discordant pairs. Br Med J 1983;286: 253–255.
14 Diabetes Epidemiology Research International Group: Geographic patterns of childhood insulin-dependent diabetes mellitus. Diabetes 1988;37:1113–1119.
15 Patrick SL, Moy CS, Laporte RE: The world of insulin-dependent diabetes mellitus: What international epidemiologic studies reveal about the etiology and natural history of IDDM. Diabetes Metab Rev 1989;5:571–578.
16 Tuomilehto J, Karvonen M, Pitkaniemi J, Virtala E, Kohtamaki K, Toivanen L, Tuomilehto-Wolf E: Record-high incidence of type I (insulin-dependent) diabetes mellitus in Finnish children. The Finnish Childhood Type I Diabetes Registry Group. Diabetologia 1999;42:655–660.

17 EURODIAB ACE Study Group: Variation and trends in incidence of childhood diabetes in Europe. Lancet 2000;355:873–876.

18 Songini M, Muntoni S: High incidence of type-I diabetes in Sardinia. Lancet 1991;337:1047.

19 Elliott RB, Martin JM: Dietary protein: A trigger of insulin-dependent diabetes in the BB rat? Diabetologia 1984;26:297–299.

20 Guberski DL, Thomas VA, Shek WR, Like AA, Handler ES, Rossini AA, Wallace JE, Welsh RM: Induction of type I diabetes by Kilham's rat virus in diabetes-resistant BB/Wor rats. Science 1991;254:1010–1013.

21 Orlow S, Yasunami R, Boitard C, Bach JF: Induction précoce du diabète chez la souris NOD par la streptozotocine. CR Acad Sci III 1987;304:77–78.

22 Horwitz MS, Bradley LM, Harbertson J, Krahl T, Lee J, Sarvetnick N: Diabetes induced by Coxsackie virus: Initiation by bystander damage and not molecular mimicry. Nat Med 1998;4:781–785.

23 Yoon JW: Role of viruses in the pathogenesis of IDDM. Ann Med 1991;23:437–445.

24 Virtanen SM, Laara E, Hypponen E, Reijonen H, Rasanen L, Aro A, Knip M, Ilonen J, Akerblom HK: Cow's milk consumption, HLA-DQB1 genotype, and type 1 diabetes: A nested case-control study of siblings of children with diabetes. Childhood Diabetes in Finland Study Group. Diabetes 2000; 49:912–917.

25 Miyazaki I, Gaedigk R, Hui MF, Cheung RK, Morkowski J, Rajotte RV, Dosch HM: Cloning of human and rat p69 cDNA, a candidate autoimmune target in type 1 diabetes. Biochim Biophys Acta 1994;1227:101–104.

26 Ohsugi T, Kurosawa T: Increased incidence of diabetes mellitus in specific pathogen-eliminated offspring produced by embryo transfer in NOD mice with low incidence of the disease. Lab Anim Sci 1994;44:386–388.

27 Hansen AK, Josefsen K, Pedersen C, Buschard K: Neonatal stimulation of beta-cells reduces the incidence and delays the onset of diabetes in a barrier-protected breeding colony of BB rats. Exp Clin Endocrinol 1993;101:189–193.

28 Qin HY, Sadelain MW, Hitchon C, Lauzon J, Singh B: Complete Freund's adjuvant-induced T cells prevent the development and adoptive transfer of diabetes in nonobese diabetic mice. J Immunol 1993;150:2072–2080.

29 Elliott RB, Reddy SN, Bibby NJ, Kida K: Dietary prevention of diabetes in the non-obese diabetic mouse. Diabetologia 1988;31:62–64.

30 Bach JF: Predictive medicine in autoimmune diseases: from the identification of genetic predisposition and environmental influence to precocious immunotherapy. Clin Immunol Immunopathol 1994;72:156–161.

31 Patterson CC, Carson DJ, Hadden DR: Epidemiology of childhood IDDM in Northern Ireland 1989–1994: Low incidence in areas with highest population density and most household crowding. Diabetologia 1996;39:1063–1069.

32 Davies JL, Kawaguchi Y, Bennett ST, Copeman JB, Cordell HJ, Pritchard LE, Reed PW, Gough SC, Jenkins SC, Palmer SM, et al: A genome-wide search for human type 1 diabetes susceptibility genes. Nature 1994;371:130–136.

33 Dorman JS, Laporte RE, Trucco M: Genetics of diabetes. Genes and environment. Baillières Clin Endocrinol Metab 1991;5:229–245.

34 Noble JA, Valdes AM, Cook M, Klitz W, Thomson G, Erlich HA: The role of HLA class II genes in insulin-dependent diabetes mellitus: Molecular analysis of 180 Caucasian, multiplex families. Am J Hum Genet 1996;59:1134–1148.

35 Zamani M, Cassiman JJ: Reevaluation of the importance of polymorphic HLA class II alleles and amino acids in the susceptibility of individuals of different populations to type I diabetes. Am J Med Genet 1998;76:183–194.

36 Nepom GT: HLA and type I diabetes. Immunol Today 1990;11:314–315.

37 Baisch JM, Weeks T, Giles R, Hoover M, Stastny P, Capra JD: Analysis of HLA-DQ genotypes and susceptibility in insulin-dependent diabetes mellitus. N Engl J Med 1990;322: 1836–1841.

38 Erlich HA, Bugawan TL, Scharf S, Nepom GT, Tait B, Griffith RL: HLA-DQ beta sequence polymorphism and genetic susceptibility to IDDM. Diabetes 1990;39:96–103.

39 She JX: Susceptibility to type I diabetes: HLA-DQ and DR revisited. Immunol Today 1996;17: 323–329.

40 Thomson G: HLA DR antigens and susceptibility to insulin-dependent diabetes mellitus. Am J Hum Genet 1984;36:1309–1317.

41 Pugliese A, Gianani R, Moromisato R, Awdeh ZL, Alper CA, Erlich HA, Jackson RA, Eisenbarth GS: HLA-DQB1*0602 is associated with dominant protection from diabetes even among islet cell antibody-positive first-degree relatives of patients with IDDM. Diabetes 1995; 44: 608–613.

42 Erlich HA, Griffith RL, Bugawan TL, Ziegler R, Alper C, Eisenbarth G: Implication of specific DQB1 alleles in genetic susceptibility and resistance by identification of IDDM siblings with novel HLA-DQB1 allele and unusual DR2 and DR1 haplotypes. Diabetes 1991;40: 478–481.

43 Sterkers G, Zeliszewski D, Chaussee AM, Deschamps I, Font MP, Freidel C, Hors J, Betuel H, Dausset J, Levy JP: HLA-DQ rather than HLA-DR region might be involved in dominant nonsusceptibility to diabetes. Proc Natl Acad Sci USA 1988;85:6473–6477.

44 Erlich HA, Zeidler A, Chang J, Shaw S, Raffel LJ, Klitz W, Beshkov Y, Costin G, Pressman S, Bugawan T, et al: HLA class II alleles and susceptibility and resistance to insulin-dependent diabetes mellitus in Mexican-American families. Nat Genet 1993;3:358–364.

45 Harfouch-Hammoud E, Timsit J, Boitard C, Bach JF, Caillat-Zucman S: Contribution of DRB1*04 variants to predisposition to or protection from insulin-dependent diabetes mellitus is independent of DQ. J Autoimmun 1996;9:411–414.

46 Reijonen H, Ilonen J, Akerblom HK, Knip M, Dosch HM: Multi-locus analysis of HLA class II genes in DR2-positive IDDM haplotypes in Finland. The 'Childhood Diabetes in Finland' (DiMe) Study Group. Tissue Antigens 1994;43:1–6.

47 Todd JA, Bell JI, McDevitt HO: HLA-DQ beta gene contributes to susceptibility and resistance to insulin-dependent diabetes mellitus. Nature 1987;329:599–604.

48 Owerbach D, Gunn S, Gabbay KH: Primary association of HLA-DQw8 with type I diabetes in DR4 patients. Diabetes 1989;38:942–945.

49 Robinson DM, Holbeck S, Palmer J, Nepom GT: HLA DQ beta 3.2 identifies subtypes of DR4+ haplotypes permissive for IDDM. Genet Epidemiol 1989;6:149–154.

50 Awata T, Kuzuya T, Matsuda A, Iwamoto Y, Kanazawa Y: Genetic analysis of HLA class II alleles and susceptibility to type 1 (insulin-dependent) diabetes mellitus in Japanese subjects. Diabetologia 1992;35:419–424.

51 Chuang LM, Jou TS, Hu CY, Wu HP, Tsai WY, Lee JS, Hsieh RP, Chen KH, Tai TY, Lin BJ: HLA-DQB1 codon 57 and IDDM in Chinese living in Taiwan. Diabetes Care 1994;17:863–868.

52 Yamagata K, Nakajima H, Hanafusa T, Noguchi T, Miyazaki A, Miyagawa J, Sada M, Amemiya H, Tanaka T, Kono N: Aspartic acid at position 57 of DQ beta chain does not protect against type 1 (insulin-dependent) diabetes mellitus in Japanese subjects. Diabetologia 1989;32:762–764.

53 Todd JA, Bell JI, McDevitt HO: A molecular basis for genetic susceptibility to insulin-dependent diabetes mellitus. Trends Genet 1988;4:129–134.

54 Lund T, O'Reilly L, Hutchings P, Kanagawa O, Simpson E, Gravely R, Chandler P, Dyson J, Picard JK, Edwards A, Kioussis D, Cooke A: Prevention of insulin-dependent diabetes mellitus in non-obese diabetic mice by transgenes encoding modified I-A beta-chain or normal I-E alpha-chain. Nature 1990;345:727–729.

55 Quartey-Papafio R, Lund T, Chandler P, Picard J, Ozegbe P, Day S, Hutchings PR, O'Reilly L, Kioussis D, Simpson E, Cooke A: Aspartate at position 57 of nonobese diabetic I-Ag7 beta-chain diminishes the spontaneous incidence of insulin-dependent diabetes mellitus. J Immunol 1995; 154:5567–5575.

56 Stern LJ, Wiley DC: Antigenic peptide binding by class I and class II histocompatibility proteins. Structure 1994;2:245–251.

57 Rammensee HG, Friede T, Stevanoviic S: MHC ligands and peptide motifs: First listing. Immunogenetics 1995;41:178–228.

58 Carrasco-Marin E, Shimizu J, Kanagawa O, Unanue ER: The class II MHC I-Ag7 molecules from non-obese diabetic mice are poor peptide binders. J Immunol 1996;156:450–458.

59 Corper AL, Stratmann T, Apostolopoulos V, Scott CA, Garcia KC, Kang AS, Wilson IA, Teyton L: A structural framework for deciphering the link between I-Ag7 and autoimmune diabetes. Science 2000;288:505–511.

60 Latek RR, Suri A, Petzold SJ, Nelson CA, Kanagawa O, Unanue ER, Fremont DH: Structural basis of peptide binding and presentation by the type I diabetes-associated MHC class II molecule of NOD mice. Immunity 2000;12:699–710.

61 Kwok WW, Domeier ME, Johnson ML, Nepom GT, Koelle DM: HLA-DQB1 codon 57 is critical for peptide binding and recognition. J Exp Med 1996;183:1253–1258.

62 Kwok WW, Domeier ML, Raymond FC, Byers P, Nepom GT: Allele-specific motifs characterize HLA-DQ interactions with a diabetes-associated peptide derived from glutamic acid decarboxylase. J Immunol 1996;156:2171–2177.

63 Stratmann T, Apostolopoulos V, Mallet Designe V, Corper AL, Scott CA, Wilson IA, Kang AS, Teyton L: The I-Ag7 MHC class II molecule linked to murine diabetes is a promiscuous peptide binder. J Immunol 2000;165:3214–3225.

64 Wucherpfennig KW, Strominger JL: Selective binding of self peptides to disease-associated major histocompatibility complex (MHC) molecules: A mechanism for MHC-linked susceptibility to human autoimmune diseases. J Exp Med 1995;181:1597–1601.

65 Hammer J, Gallazzi F, Bono E, Karr RW, Guenot J, Valsasnini P, Nagy ZA, Sinigaglia F: Peptide binding specificity of HLA-DR4 molecules: Correlation with rheumatoid arthritis association. J Exp Med 1995;181:1847–1855.

66 Caplen NJ, Patel A, Millward A, Campbell RD, Ratanachaiyavong S, Wong FS, Demaine AG: Complement C4 and heat shock protein 70 (HSP70) genotypes and type I diabetes mellitus. Immunogenetics 1990;32:427–430.

67 Ilonen J, Merivuori H, Reijonen H, Knip M, Akerblom HK, Pociot F, Nerup J: Tumour necrosis factor-beta gene RFLP alleles in Finnish IDDM haplotypes. The Childhood Diabetes in Finland (DiMe) Study Group. Scand J Immunol 1992;36:779–783.

68 Pociot F, Ronningen KS, Nerup J: Polymorphic analysis of the human MHC-linked heat shock protein 70 (HSP70-2) and HSP70-Hom genes in insulin-dependent diabetes mellitus (IDDM). Scand J Immunol 1993;38:491–495.

69 Bonifacio E: HLA-A associations with IDDM – A case of numbers? Diabetologia 1995;38:751–752.

70 Fujisawa T, Ikegami H, Kawaguchi Y, Yamato E, Takekawa K, Nakagawa Y, Hamada Y, Ueda H, Shima K, Ogihara T: Class I HLA is associated with age-at-onset of IDDM, while class II HLA confers susceptibility to IDDM. Diabetologia 1995;38:1493–1495.

71 Moghaddam PH, de Knijf P, Roep BO, Van Der Auwera B, Naipal A, Gorus F, Schuit F, Giphart MJ: Genetic structure of IDDM1: Two separate regions in the major histocompatibility complex contribute to susceptibility or protection. Diabetes 1998;47:263–269.

72 Lie BA, Sollid LM, Ascher H, Ek J, Akselsen HE, Ronningen KS, Thorsby E, Undlien DE: A gene telomeric of the HLA class I region is involved in predisposition to both type 1 diabetes and coeliac disease. Tissue Antigens 1999;54:162–168.

73 Lie BA, Todd JA, Pociot F, Nerup J, Akselsen HE, Joner G, Dahl-Jorgensen K, Ronningen KS, Thorsby E, Undlien DE: The predisposition to type 1 diabetes linked to the human leukocyte antigen complex includes at least one non-class II gene. Am J Hum Genet 1999;64:793–800.

74 Caillat-Zucman S, Bertin E, Timsit J, Boitard C, Assan R, Bach JF: Protection from insulin-dependent diabetes mellitus is linked to a peptide transporter gene. Eur J Immunol 1993;23:1784–1788.

75 Deng GY, Muir A, MacLaren NK, She JX: Association of LMP2 and LMP7 genes within the major histocompatibility complex with insulin-dependent diabetes mellitus: Population and family studies. Am J Hum Genet 1995;56:528–534.

76 Caillat-Zucman S, Daniel S, Djilali-Saiah I, Timsit J, Garchon HJ, Boitard C, Bach JF: Family study of linkage disequilibrium between TAP2 transporter and HLA class II genes – Absence of TAP2 contribution to association with insulin-dependent diabetes mellitus. Hum Immunol 1995;44:80–87.

77 Van Endert PM, Liblau RS, Patel SD, Fugger L, Lopez T, Pociot F, Nerup J, McDevitt HO: Major histocompatibility complex-encoded antigen processing gene polymorphism in IDDM. Diabetes 1994;43:110–117.

78 Caillat-Zucman S, Garchon HJ, Timsit J, Assan R, Boitard C, Djilali-Saiah I, Bougneres P, Bach JF: Age-dependent HLA genetic heterogeneity of type 1 insulin-dependent diabetes mellitus. J Clin Invest 1992;90:2242–2250.

79 Honeyman MC, Harrison LC, Drummond B, Colman PG, Tait BD: Analysis of families at risk for insulin-dependent diabetes mellitus reveals that HLA antigens influence progression to clinical disease. Mol Med 1995;1:576–582.

80 Mizota M, Uchigata Y, Moriyama S, Tokunaga K, Matsuura N, Miura J, Juji T, Omori Y: Age-dependent association of HLA-A24 in Japanese IDDM patients. Diabetologia 1996;39:371–373.

81 Nakanishi K, Kobayashi T, Murase T, Nakatsuji T, Inoko H, Tsuji K, Kosaka K: Association of HLA-A24 with complete beta-cell destruction in IDDM. Diabetes 1993;42:1086–1093.

82 Wicker LS: Major histocompatibility complex-linked control of autoimmunity. J Exp Med 1997; 186:973–975.

83 Nepom GT, Kwok WW: Molecular basis for HLA-DQ associations with IDDM. Diabetes 1998; 47:1177–1184.

84 Ridgway WM, Fasso M, Fathman CG: A new look at MHC and autoimmune disease. Science 1999;284:749–751.

85 Buckner J, Kwok WW, Nepom B, Nepom GT: Modulation of HLA-DQ binding properties by differences in class II dimer stability and pH-dependent peptide interactions. J Immunol 1996;157:4940–4945.

86 Ettinger RA, Liu AW, Nepom GT, Kwok WW: Beta57-Asp plays an essential role in the unique SDS stability of HLA-DQA1*0102/DQB1*0602 alphabeta protein dimer, the class II MHC allele associated with protection from insulin-dependent diabetes mellitus. J Immunol 2000;165:3232–3238.

87 Ridgway WM, Fasso M, Lanctot A, Garvey C, Fathman CG: Breaking self-tolerance in nonobese diabetic mice. J Exp Med 1996;183:1657–1662.

88 Harfouch-Hammoud E, Walk T, Otto H, Jung G, Bach JF, Van Endert PM, Caillat-Zucman S: Identification of peptides from autoantigens GAD65 and IA-2 that bind to HLA class II molecules predisposing to or protecting from type 1 diabetes. Diabetes 1999;48:1937–1947.

89 Anderton S, Burkhart C, Metzler B, Wraith D: Mechanisms of central and peripheral T-cell tolerance: Lessons from experimental models of multiple sclerosis. Immunol Rev 1999;169:123–137.

90 Liu GY, Fairchild PJ, Smith RM, Prowle JR, Kioussis D, Wraith DC: Low avidity recognition of self-antigen by T cells permits escape from central tolerance. Immunity 1995;3:407–415.

91 Schmidt D, Verdaguer J, Averill N, Santamaria P: A mechanism for the major histocompatibility complex-linked resistance to autoimmunity. J Exp Med 1997;186:1059–1075.

92 Slattery RM, Miller JF, Heath WR, Charlton B: Failure of a protective major histocompatibility complex class II molecule to delete autoreactive T cells in autoimmune diabetes. Proc Natl Acad Sci USA 1993;90:10808–10810.

93 Nepom GT: A unified hypothesis for the complex genetics of HLA associations with IDDM. Diabetes 1990;39:1153–1157.

94 Sheehy MJ: HLA and insulin-dependent diabetes. A protective perspective. Diabetes 1992;41: 123–129.

95 Deng H, Apple R, Clare-Salzler M, Trembleau S, Mathis D, Adorini L, Sercarz E: Determinant capture as a possible mechanism of protection afforded by major histocompatibility complex class II molecules in autoimmune disease. J Exp Med 1993;178:1675–1680.

96 Seddon B, Mason D: Peripheral autoantigen induces regulatory T cells that prevent autoimmunity. J Exp Med 1999;189:877–882.

97 Singer SM, Tisch R, Yang XD, McDevitt HO: An Abd transgene prevents diabetes in nonobese diabetic mice by inducing regulatory T cells. Proc Natl Acad Sci USA 1993;90:9566–9570.

98 Hanson MS, Cetkovic-Cvrlje M, Ramiya VK, Atkinson MA, MacLaren NK, Singh B, Elliott JF, Serreze DV, Leiter EH: Quantitative thresholds of MHC class II I-E expressed on hemopoietically derived antigen-presenting cells in transgenic NOD/Lt mice determine level of diabetes resistance and indicate mechanism of protection. J Immunol 1996;157:1279–1287.

99 Luhder F, Katz J, Benoist C, Mathis D: Major histocompatibility complex class II molecules can protect from diabetes by positively selecting T cells with additional specificities. J Exp Med 1998; 187:379–387.

100 Raum D, Awdeh Z, Yunis EJ, Alper CA, Gabbay KH: Extended major histocompatibility complex haplotypes in type I diabetes mellitus. J Clin Invest 1984;74:449–454.

101 Thomsen M, Molvig J, Zerbib A, de Preval C, Abbal M, Dugoujon JM, Ohayon E, Svejgaard A, Cambon Thomsen A, Nerup J: The susceptibility to insulin-dependent diabetes mellitus is associated with C4 allotypes independently of the association with HLA-DQ alleles in HLA-DR3,4 heterozygotes. Immunogenetics 1988;28:320–327.

102 Degli-Esposti MA, Abraham LJ, McCann V, Spies T, Christiansen FT, Dawkins RL: Ancestral haplotypes reveal the role of the central MHC in the immunogenetics of IDDM. Immunogenetics 1992;36:345–356.

103 Badenhoop K, Schwarz G, Trowsdale J, Lewis V, Usadel KH, Gale EA, Bottazzo GF: TNF-alpha gene polymorphisms in type 1 (insulin-dependent) diabetes mellitus. Diabetologia 1989;32:445–448.

104 Pociot F, Wilson AG, Nerup J, Duff GW: No independent association between a tumor necrosis factor-alpha promotor region polymorphism and insulin-dependent diabetes mellitus. Eur J Immunol 1993;23:3050–3053.

105 Deng GY, MacLaren NK, Huang HS, Zhang LP, She JX: No primary association between the 308 polymorphism in the tumor necrosis factor alpha promoter region and insulin-dependent diabetes mellitus. Hum Immunol 1996;45:137–142.

106 The Mhc Sequencing CONSORTIUM: Complete sequence and gene map of a human major histocompatibility complex. Nature 1999;401:921–923.

107 Foissac A, Cambon-Thomsen A: Microsatellites in the HLA region: 1998 update. Tissue Antigens 1998;52:318–352.

108 Syvanen AC: From gels to chips: 'Minisequencing' primer extension for analysis of point mutations and single nucleotide polymorphisms. Hum Mutat 1999;13:1–10.

109 Spielman RS, Ewens WJ: The TDT and other family-based tests for linkage disequilibrium and association. Am J Hum Genet 1996;59:983–989.

110 Spielman RS, McGinnis RE, Ewens WJ: Transmission test for linkage disequilibrium: the insulin gene region and insulin-dependent diabetes mellitus. Am J Hum Genet 1993;52:506–516.

111 Hanifi Moghaddam P, de Knijf P, Roep BO, Van Der Auwera B, Naipal A, Gorus F, Schuit F, Giphart MJ: Genetic structure of IDDM1: Two separate regions in the major histocompatibility complex contribute to susceptibility or protection. Belgian Diabetes Registry. Diabetes 1998;47:263–269.

112 Ikegami H, Makino S, Yamato E, Kawaguchi Y, Ueda H, Sakamoto T, Takekawa K, Ogihara T: Identification of a new susceptibility locus for insulin-dependent diabetes mellitus by ancestral haplotype congenic mapping. J Clin Invest 1995;96:1936–1942.

113 Hattori M, Yamato E, Itoh N, Senpuku H, Fujisawa T, Yoshino M, Fukuda M, Matsumoto E, Toyonaga T, Nakagawa I, Petruzzelli M, McMurray A, Weiner H, Sagai T, Moriwaki K, Shiroishi T, Maron R, Lund T: Cutting edge: Homologous recombination of the MHC class IK region defines new MHC-linked diabetogenic susceptibility gene(s) in nonobese diabetic mice. J Immunol 1999;163:1721–1724.

114 Risch N: Assessing the role of HLA-linked and unlinked determinants of disease. Am J Hum Genet 1987;40:1–14.

115 Lander E, Kruglyak L: Genetic dissection of complex traits: Guidelines for interpreting and reporting linkage results. Nat Genet 1995;11:241–247.

116 Concannon P, Gogolin-Ewens KJ, Hinds DA, Wapelhorst B, Morrison VA, Stirling B, Mitra M, Farmer J, Williams SR, Cox NJ, Bell GI, Risch N, Spielman RS: A second-generation screen of the human genome for susceptibility to insulin-dependent diabetes mellitus. Nat Genet 1998;19:292–296.

117 Mein CA, Esposito L, Dunn MG, Johnson GCL, Timms AE, Goy JV, Smith AN, Sebag-Montefiore L, Merriman ME, Wilson AJ, Pritchard LE, Cucca F, Barnett AH, Bain SC, Todd JA: A search for type 1 diabetes susceptibility genes in families from the United Kingdom. Nat Genet 1998;19:297–300.

118 Cucca F, Goy JV, Kawaguchi Y, Esposito L, Merriman ME, Wilson AT, Cordell HJ, Bain SC, Todd JA: A male-female bias in type 1 diabetes and linkage to chromosome Xp in MHC HLA-DR3-positive patients. Nat Genet 1998;19:301–302.

119 Hashimoto L, Habita C, Beressi JP, Delepine M, Besse C, Cambon-Thomsen A, Deschamps I, Rotter JI, Djoulah S, James MR, et al: Genetic mapping of a susceptibility locus for insulin-dependent diabetes mellitus on chromosome 11q. Nature 1994;371:161–164.

120 Field LL, Tobias R, Magnus T: A locus on chromosome 15q26 (IDDM3) produces susceptibility to insulin-dependent diabetes mellitus. Nat Genet 1994;8:189–194.

121 Owerbach D, Gabbay KH: The HOXD8 locus (2q31) is linked to type I diabetes. Interaction with chromosome 6 and 11 disease susceptibility genes. Diabetes 1995;44:132–136.

122 Luo DF, Bui MM, Muir A, MacLaren NK, Thomson G, She JX: Affected-sib-pair mapping of a novel susceptibility gene to insulin-dependent diabetes mellitus (IDDM8) on chromosome 6q25–q27. Am J Hum Genet 1995;57:911–919.

123 Morahan G, Huang DX, Tait BD, Colman PG, Harrison LC: Markers on distal chromosome 2q linked to insulin-dependent diabetes mellitus. Science 1996;272:1811–1813.

124 Delepine M, Pociot F, Habita C, Hashimoto L, Froguel P, Rotter J, Cambon-Thomsen A, Deschamps I, Djoulah S, Weissenbach J, Nerup J, Lathrop M, Julier C: Evidence of a non-MHC susceptibility locus in type I diabetes linked to HLA on chromosome 6. Am J Hum Genet 1997;60: 174–187.

125 Field LL, Tobias R, Thomson G, Plon S: Susceptibility to insulin-dependent diabetes mellitus maps to a locus (IDDM11) on human chromosome 14q24.3–q31. Genomics 1996;33:1–8.

126 Owerbach D, Gabbay KH: Linkage of the VNTR/insulin-gene and type I diabetes mellitus: Increased gene sharing in affected sibling pairs. Am J Hum Genet 1994;54:909–912.

127 Merriman T, Twells R, Merriman M, Eaves I, Cox R, Cucca F, McKinney P, Shield J, Baum D, Bosi E, Pozzilli P, Nistico L, Buzzetti R, Joner G, Ronningen K, Thorsby E, Undlien D, Pociot F, Nerup J, Bain S, Barnett A, Todd J: Evidence by allelic association-dependent methods for a type 1 diabetes polygene (IDDM6) on chromosome 18q21. Hum Mol Genet 1997;6:1003–1010.

128 Bell GI, Karam JH, Rutter WJ: Polymorphic DNA region adjacent to the 5′ end of the human insulin gene. Proc Natl Acad Sci USA 1981;78:5759–5763.

129 Bell GI, Selby MJ, Rutter WJ: The highly polymorphic region near the human insulin gene is composed of simple tandemly repeating sequences. Nature 1982;295:31–35.

130 Julier C, Hyer RN, Davies J, Merlin F, Soularue P, Briant L, Cathelineau G, Deschamps I, Rotter JI, Froguel P, Boitard C, Bell JI, Lathrop GM: Insulin-IGF2 region on chromosome 11p encodes a gene implicated in HLA-DR4-dependent diabetes susceptibility. Nature 1991;354:155–159.

131 Bain SC, Prins JB, Hearne CM, Rodrigues NR, Rowe BR, Pritchard LE, Ritchie RJ, Hall JR, Undlien DE, Ronningen KS, Dunger DB, Barnett AH, Todd JA: Insulin gene region-encoded susceptibility to type 1 diabetes is not restricted to HLA-DR4-positive individuals. Nat Genet 1992;2: 212–215.

132 Lucassen AM, Julier C, Beressi JP, Boitard C, Froguel P, Lathrop M, Bell JI: Susceptibility to insulin-dependent diabetes mellitus maps to a 4.1 kb segment of DNA spanning the insulin gene and associated VNTR. Nat Genet 1993;4:305–310.

133 Bennett ST, Lucassen AM, Gough SC, Powell EE, Undlien DE, Pritchard LE, Merriman ME, Kawaguchi Y, Dronsfield MJ, Pociot F: Susceptibility to human type 1 diabetes at IDDM2 is determined by tandem repeat variation at the insulin gene minisatellite locus. Nat Genet 1995;9: 284–292.

134 Hammond-Kosack MC, Kilpatrick MW, Docherty K: The human insulin gene-linked polymorphic region adopts a G-quartet structure in chromatin assembled in vitro. J Mol Endocrinol 1993;10:121–126.

135 Lucassen AM, Screaton GR, Julier C, Elliott TJ, Lathrop M, Bell JI: Regulation of insulin gene expression by the IDDM associated, insulin locus haplotype. Hum Mol Genet 1995;4:501–506.

136 Kennedy GC, German MS, Rutter WJ: The minisatellite in the diabetes susceptibility locus IDDM2 regulates insulin transcription. Nat Genet 1995;9:293–298.

137 Vafiadis P, Bennett ST, Todd JA, Nadeau J, Grabs R, Goodyer CG, Wickramasinghe S, Colle E, Polychronakos C: Insulin expression in human thymus is modulated by INS VNTR alleles at the IDDM2 locus. Nat Genet 1997;15:289–292.

138 Pugliese A, Zeller M, Fernandez A JR, Zalcberg LJ, Bartlett RJ, Ricordi C, Pietropaolo M, Eisenbarth GS, Bennett ST, Patel DD: The insulin gene is transcribed in the human thymus and

transcription levels correlated with allelic variation at the INS VNTR-IDDM2 susceptibility locus for type 1 diabetes. Nat Genet 1997;15:293–297.

139 Owerbach D, Poulsen S, Billesbolle P, Nerup J: DNA insertion sequences near the insulin gene affect glucose regulation. Lancet 1982;i:880–883.

140 Halminen M, Veijola R, Reijonen H, Ilonen J, Akerblom HK, Knip M: Effect of polymorphism in the insulin gene region on IDDM susceptibility and insulin secretion. Eur J Clin Invest 1996;26: 847–852.

141 Ahmed S, Bennett ST, Huxtable SJ, Todd JA, Matthews DR, Gough SC: INS VNTR allelic variation and dynamic insulin secretion in healthy adult non-diabetic Caucasian subjects. Diabet Med 1999;16:910–917.

142 Pfeifer K: Mechanisms of genomic imprinting. Am J Hum Genet 2000;67:777–787.

143 Pugliese A, Awdeh ZL, Alper CA, Jackson RA, Eisenbarth GS: The paternally inherited insulin gene B allele (1,428 FokI site) confers protection from insulin-dependent diabetes in families. J Autoimmun 1994;7:687–694.

144 Polychronakos C, Kukuvitis A, Giannoukakis N, Colle E: Parental imprinting effect at the INS-IGF2 diabetes susceptibility locus. Diabetologia 1995;38:715–719.

145 Bui MM, Luo DF, She JY, MacLaren NK, Muir A, Thomson G, She JX: Paternally transmitted iddm2 influences diabetes susceptibility despite biallelic expression of the insulin gene in human pancreas. J Autoimmun 1996;9:97–103.

146 Vafiadis P, Bennett ST, Colle E, Grabs R, Goodyer CG, Polychronakos C: Imprinted and genotype-specific expression of genes at the IDDM2 locus in pancreas and leucocytes. J Autoimmun 1996;9:397–403.

147 Doria A, Lee J, Warram JH, Krolewski AS: Diabetes susceptibility at IDDM2 cannot be positively mapped to the VNTR locus of the insulin gene. Diabetologia 1996;39:594–599.

148 Weaver JU, Kopelman PG, Hitman GA: Central obesity and hyperinsulinaemia in women are associated with polymorphism in the 5′ flanking region of the human insulin gene. Eur J Clin Invest 1992;22:265–270.

149 Waterworth DM, Bennett ST, Gharani N, McCarthy MI, Hague S, Batty S, Conway GS, White D, Todd JA, Franks S, Williamson R: Linkage and association of insulin gene VNTR regulatory polymorphism with polycystic ovary syndrome. Lancet 1997;349:986–990.

150 Ong KK, Phillips DI, Fall C, Poulton J, Bennett ST, Golding J, Todd JA, Dunger DB: The insulin gene VNTR, type 2 diabetes and birth weight. Nat Genet 1999;21:262–263.

151 Copeman JB, Cucca F, Hearne CM, Cornall RJ, Reed PW, Ronningen KS, Undlien DE, Nistico L, Buzzetti R, Tosi R: Linkage disequilibrium mapping of a type 1 diabetes susceptibility gene (IDDM7) to chromosome 2q31–q33. Nat Genet 1995;9:80–85.

152 Nistico L, Buzzetti R, Pritchard LE, Van Der Auwera B, Giovannini C, Bosi E, Larrad MT, Rios MS, Chow CC, Cockram CS, Jacobs K, Mijovic C, Bain SC, Barnett AH, Vandewalle CL, Schuit F, Gorus FK, Tosi R, Pozzilli P, Todd JA: The CTLA-4 gene region of chromosome 2q33 is linked to, and associated with, type 1 diabetes. Hum Mol Genet 1996;5:1075–1080.

153 Donner H, Rau H, Walfish PG, Braun J, Siegmund T, Finke R, Herwig J, Usadel KH, Badenhoop K: CTLA4 alanine-17 confers genetic susceptibility to Graves' disease and to type 1 diabetes mellitus. J Clin Endocrinol Metab 1997;82:143–146.

154 Awata T, Matsumoto C, Urakami T, Hagura R, Amemiya S, Kanazawa Y: Association of polymorphism in the interferon-gamma gene with IDDM. Diabetologia 1994;37:1159–1162.

155 Rowe RE, Wapelhorst B, Bell GI, Risch N, Spielman RS, Concannon P: Linkage and association between insulin-dependent diabetes mellitus susceptibility and markers near the glucokinase gene on chromosome 7. Nat Genet 1995;10:240–242.

156 Rambrand T, Pociot F, Ronningen K, Nerup J, Michelsen BK: Genetic markers for glutamic acid decarboxylase do not predict insulin-dependent diabetes mellitus in pairs of affected siblings. The Danish Study Group of Diabetes in Childhood. Hum Genet 1997;99:177–185.

157 Oosterwegel MA, Greenwald RJ, Mandelbrot DA, Lorsbach RB, Sharpe AH: CTLA-4 and T cell activation. Curr Opin Immunol 1999;11:294–300.

158 Marron MP, Zeidler A, Raffel LJ, Eckenrode SE, Yang JJ, Hopkins DI, Garchon HJ, Jacob CO, Serrano-Rios M, Larrad MTM, Park Y, Bach JF, Rotter JI, Yang MCK, She JX: Genetic and

physical mapping of a type 1 diabetes susceptibility gene (IDDM12) to a 100-kb phagemid artificial chromosome clone containing D2S72-CTLA4-D2S105 on chromosome 2q33. Diabetes 2000;49: 492–499.

159 Reed P, Cucca F, Jenkins S, Merriman M, Wilson A, McKinney P, Bosi E, Joner G, Ronningen K, Thorsby E, Undlien D, Merriman T, Barnett A, Bain S, Todd J: Evidence for a type 1 diabetes susceptibility locus (IDDM10) on human chromosome 10p11–q11. Hum Mol Genet 1997;6: 1011–1016.

160 Morel L, Croker BP, Blenman KR, Mohan C, Huang G, Gilkeson G, Wakeland EK: Genetic reconstitution of systemic lupus erythematosus immunopathology with polycongenic murine strains. Proc Natl Acad Sci USA 2000;97:6670–6675.

161 Djabiri F, Caillat-Zucman S, Gajdos P, Jais JP, Gomez L, Khalil I, Charron D, Bach JF, Garchon HJ: Association of the AChR alpha-subunit gene (CHRNA), DQA1*0101, and the DR3 haplotype in myasthenia gravis. Evidence for a three-gene disease model in a subgroup of patients. J Autoimmun 1997;10:407–413.

162 Wapelhorst B, Bell GI, Risch N, Spielman RS, Concannon P: Linkage and association studies in insulin-dependent diabetes with a new dinucleotide repeat polymorphism at the GAD65 locus. Autoimmunity 1995;21:127–130.

163 Salomon B, Lenschow DJ, Rhee L, Ashourian N, Singh B, Sharpe A, Bluestone JA: B7/CD28 costimulation is essential for the homeostasis of the CD4+ CD25+ immunoregulatory T cells that control autoimmune diabetes. Immunity 2000;12:431–440.

164 Takahashi T, Tagami T, Yamazaki S, Uede T, Shimizu J, Sakaguchi N, Mak TW, Sakaguchi S: Immunologic self-tolerance maintained by CD25(+)CD4(+) regulatory T cells constitutively expressing cytotoxic T lymphocyte-associated antigen 4. J Exp Med 2000;192:303–310.

165 Read S, Malmstrom V, Powrie F: Cytotoxic T lymphocyte-associated antigen 4 plays an essential role in the function of CD25(+)CD4(+) regulatory cells that control intestinal inflammation. J Exp Med 2000;192:295–302.

166 Zimmet PZ, Tuomi T, MacKay IR, Rowley MJ, Knowles W, Cohen M, Lang DA: Latent autoimmune diabetes mellitus in adults (LADA): The role of antibodies to glutamic acid decarboxylase in diagnosis and prediction of insulin dependency. Diabet Med 1994;11:299–303.

167 Tuomi T, Groop LC, Zimmet PZ, Rowley MJ, Knowles W, MacKay IR: Antibodies to glutamic acid decarboxylase reveal latent autoimmune diabetes mellitus in adults with a non-insulin-dependent onset of disease. Diabetes 1993;42:359–362.

168 Verge CF, Vardi P, Babu S, Bao F, Erlich HA, Bugawan T, Tiosano D, Yu L, Eisenbarth GS, Fain PR: Evidence for oligogenic inheritance of type 1 diabetes in a large Bedouin Arab family. J Clin Invest 1998;102:1569–1575.

169 Flier JS, Kahn CR, Roth J: Receptors, antireceptor antibodies and mechanisms of insulin resistance. N Engl J Med 1979;300:413–419.

Prof. Jean-François Bach, INSERM U25, Hôpital Necker,
161, rue de Sèvres, F–75743 Paris Cedex 15 (France)
Tel. +331 44 49 53 71, Fax +331 43 06 23 88, E-Mail bach@necker.fr

von Herrath MG (ed): Molecular Pathology of Type 1 Diabetes mellitus.
Curr Dir Autoimmun. Basel, Karger, 2001, vol 4, pp 31–67

Genes and Cellular Requirements for Autoimmune Diabetes Susceptibility in Nonobese Diabetic Mice

David V. Serreze, Edward H. Leiter

The Jackson Laboratory, Bar Harbor, Me., USA

The Genetics of Diabetes in Nonobese Diabetic Mice

Overview: What Kinds of Genes Make Nonobese Diabetic Mice Diabetic?

There are well over 300 inbred mouse strains extant world-wide. Of these, only one, the nonobese diabetic (NOD) mouse, initially expected to be a normoglycemic strain [1] spontaneously develops an autoimmune, T-cell-mediated type 1 diabetes (T1D; former designation insulin-dependent diabetes mellitus, IDDM) syndrome. Considerable excitement was generated by initial genetic analysis of this strain. Sequencing of its MHC class II Aβ chain allele revealed what initially was thought to be a unique (for mouse) sequence encoding a neutral serine instead of an aspartic acid (asp) residue at position 57 [2]. This finding of the 'non-asp' substitution generated considerable excitement at the time because similar 'non-asp' substitutions at residue 57 were reported in the 'diabetogenic' human homologs (*HLA-DQβ* alleles such as DQβ0302) [2]. As will be noted below, it is not only the MHC class II region that contributes susceptibility in NOD mice, but rather, multiple loci in the extended haplotype. Since the MHC complex contains genes essential for antigen processing and presentation as well as for immune cell function, it was both satisfying and logical that the NOD MHC would be a primary determinant of T1D susceptibility. Initial genetic segregation analyses in which NOD was outcrossed with other inbred strains confirmed major diabetogenic contributions of the NOD MHC haplotype (*H2^{g7}*). These early studies further suggested that only one or several non-MHC loci might also be necessary to mediate diabetogenesis in segregants that were $H2^{g7}$ homozygous [3]. The observation that stocks of NOD mice congenic for MHC haplotypes from T1D-resistant strains were strongly T1D-resistant confirmed the major diabetogenic contribution made by the $H2^{g7}$ haplotype [4, 5]. However, congenic transfer of the $H2^{g7}$ haplotype onto T1D-resistant strains failed to elicit T1D, or else did so at very low frequencies, demonstrating that the diabetogenic $H2^{g7}$ alone was generally insufficient to elicit T1D and that contributions from undefined numbers of non-MHC genes were also necessary for full T1D susceptibility [5, 6].

It was initially hoped that, if diabetogenesis in NOD mice were an oligogenic disease, with MHC contributing at least a third of the genetic variance, then identification of one or two major non-MHC linkages might reveal, through synteny, the homologs for human T1D [7]. Unfortunately, current studies indicate a much more complex multigenic control of T1D in the NOD model [5, 6]. The human 'T1D-permissive' HLA alleles are relatively common in Caucasians. Yet T1D develops sporadically in human populations. One attractive explanation, the so-called 'Nerup hypothesis', is that T1D development reflects the stochastic assortment of high-risk HLA haplotypes into 'unfavorable combinations' with common variants of non-HLA genes [8]. The HLA variants extant today probably reflect the historical contributions of these haplotypes to human survival from epidemics. Recent retrospective analysis of the gene pool of Dutch colonists who survived the multiple tropical diseases in Surinam and later returned to The Netherlands indicates that they inherited T1D-permissive or other autoimmune disease-permissive HLA alleles [9, 10]. Epidemiology of T1D development in colonies of NOD mice reinforces the notion that, in mice at least, there may be a relationship between genetic resistance to pathogens and increased risk for T1D development [11].

The production of any inbred strain from outbred progenitors inevitably leads to the fixation of gene mutations in homozygous state. Certain of these are recessive, loss of function mutations while others represent polymorphic variation in the coding or regulatory sequences of genes that may or may not affect function. The NOD strain is not different from other inbred strains in this regard; known mutations include a defective hemolytic complement (*Hc*) gene [12], a null mutation in the *H2-Ea* MHC class II gene [13], and defects in the Fc receptor γ1 (*Fcrg1*) and 2 (*Fcrg2*) genes [14–16]. Arguably, what sets this strain uniquely apart from all other extant strains is not a specific loss-of-function mutation such as the null *H2-Eab* gene, which is common to a variety of unrelated and nonautoimmune-prone inbred strains such as C57BL/6J (B6). Instead, we believe the NOD genome very well supports the 'Nerup hypothesis' by representing an 'unfavorable combination' of relatively common gene variations that, in themselves, are not diabetogenic, but collectively maintain tolerogenic immune functions at the borderline of adequacy [17]. Genetic analysis of T1D causation in this model (and probably in humans as well) is complicated by the realization that the genetic 'cross-talk' among this imperfectly-meshing collection of genes is not so severely impaired that tolerogenic mechanisms cannot be achieved if the mice are sufficiently challenged by extrinsic immunostimulation.

The NOD immune system confers a remarkable resistance to a variety of murine pathogenic agents [18]. When prediabetic NOD mice are exposed to any of a number of different murine viruses or microbial agents, they survive and the incidence of spontaneous T1D is drastically suppressed, sometimes to zero. Certain other NOD strain-specific characteristics (other than spontaneous T1D development) provide valuable genomic insights into this phenomenon of environmental suppression of genetic susceptibility. NOD mice breed early; their litter sizes are unusually large for an inbred strain, and most of the pups survive to weaning. Adults, as noted above, are resistant to many infectious agents. Hence, it appears as if the strain is balancing the future development of a life-shortening disease by early production of large numbers of pathogen-resistant offspring. What has just been referred to as an 'unfavorable collection of imperfectly-interacting genes' is only 'unfavorable' when the physical environment is made specific pathogen-free (SPF). When the immune system of the NOD mouse is deprived of normal immunoregulatory cues that the mouse would receive in a natural (wild) setting, its repertoire of MHC and non-MHC genes fail to autoregulate normally, in large part due to defects in stimulatory signals provided to T cells by antigen-presenting cells (APC) [17]. In SPF environments in which NOD mice exhibit high T1D incidence, the most effective method for maintaining NOD breeding pairs in a diabetes-free condition as long as possible is to immunostimulate them with a single injection of complete Freund's adjuvant before they are mated. In addition to microbial antigens and dictary factors [19], a plethora of other treatments will modulate T1D frequency in NOD mice [20]. Because of certain genus-specific differences distinguishing humans from mice, coupled with the ease by which the NOD genetic program for T1D can be deviated by so many different environmental manipulations, the relevance of NOD mice to human T1D has been questioned. If comparably strong environmental effects modulate the penetrance of T1D-predisposing genes in humans, the study of human T1D genetics becomes infinitely more complicated. What the NOD mouse provides is a 'living laboratory' for understanding the complex interactions between an 'unfavorable collection of imperfectly-interacting genes' and environment in the causation of clinical T1D. The first half of this chapter will detail the complexities of T1D genetics in the NOD model, and the second half will provide potential mechanisms whereby the complex genetics predisposes to a dysregulated immune system.

The 'Idd' Genes: What Are They and How Many Are Required for Diabetogenesis?

In order to describe murine chromosomal regions carrying genes capable of modulating T1D susceptibility when NOD/Lt mice were outcrossed to the related NON/Lt strain, we originally proposed the provisional designation of '*Idd*' followed by a number to describe any chromosomal segment showing significant evidence of linkage to the trait of clinical T1D (either chronic hyperglycemia or glycosuria) [21]. This nomenclature has been widely adopted to describe linkages obtained when NOD is outcrossed to various mouse strains; the combined results from outcross to a limited number of inbred strains indicate linkage of 19 loci on 10 chromosomes [reviewed in 6], with certain of the susceptibility linkages deriving from the genomes of T1D-free outcross partner strains. In almost all cases, the segment identified initially as containing a single '*Idd*' gene yields multiple loci when subcongenic analysis is conducted. Indeed, because the '*Idd*' nomenclature is limited to descriptions of loci identified by classical genetic segregation analysis, it greatly underestimates the actual numbers of genes required for T1D development. Current transgenic and gene targeting technologies being applied to analyze the genetic control of T1D development in NOD mice are providing an ever-expanding panoply of contributory '*Idd*' loci, a few of which will be discussed. Although most genomic manipulation by transgenic or gene 'knockout' technology has focused on genes expressed in the immune system, it is now clear that genes expressed systemically, including within the pancreatic β cell, can also be important modifiers of susceptibility [22]. We will begin with a detailed analysis of the MHC-linked diabetogenic contributions ('*Idd1*') as an illustration of the complexity of '*Idd*' loci and their interactions.

Dissection of the 'Idd1' Locus

H2-A^{g7} denotes the NOD class II molecule expressed on APC and encoded by a common *Aa^d* allele and a relatively rare *Ab^{g7}* allele in which a nested set of 5 nucleotide substitutions between position 248–252 converts a conserved proline residue at amino acid position 56 to histidine and the aforementioned aspartic acid 57 to serine [23]. Following the original report linking T1D susceptibility to the MHC region of Chr. 17 [13], '*Idd1*' was assumed to be these rare polymorphisms in the *H2-Ab^{g7}* allele rather than the common *H2-Ea^b* (null) mutation. Production of transgenic stocks of NOD mice expressing *Ab^{g7}* alleles modified by site-specific mutagenesis to convert either the histidine 56 residue to proline, or, in separate transgenics, the serine 57 residue to aspartic acid, indeed confirmed the diabetogenic relevance of both amino acids. However, T1D development was also drastically suppressed or completely prevented in stocks of NOD mice expressing *H2-Ea* transgenes that restored H2-E expression on APC [24–27]. This demonstrated that '*Idd1*' was comprised of at least two linked, but separable susceptibility components. Because of this complexity, it is difficult to discuss whether the haplotype contributes susceptibility in a dominant or recessive fashion. Segregants heterozygous for $H2^{g7}$ may develop subclinical levels of insulitis (suggesting dominant $H2^{g7}$ contributions to this important diabetes subphenotype), and may even develop overt T1D following cyclophosphamide treatment [5, 28]. However, it is quite rare for spontaneous T1D to present in such heterozygotes [5], especially when an *Ea* product is expressed by the other H2 haplotype. In this regard, the NOD model diverges from humans, where certain HLA heterozygous haplotypes increase rather than suppress T1D susceptibility [29].

Extended haplotype analysis of human genes in linkage disequilibrium with T1D-susceptibility conferring *HLA-DR* and *-DQ* alleles now provides evidence for additional susceptibility components toward the class I region [30]. A similar result appears to apply for $H2^{g7}$. The CTS/Shi strain is NOD-related, but T-lymphocytopenic and T1D-free [31, 32]. This strain shares MHC class II gene identity with NOD, and extending proximally, identity through a common $H2\text{-}K^d$ class I allele. Proximally from the class II region, the CTS alleles are identical with those of NOD through the Heat-shock protein 70 (*Hsp70*) complex. The CTS MHC haplotype is differentiated from $H2^{g7}$ by DNA sequence divergence distal to *Hsp70* and extending through the *H2-D* class I allele (where a different allele, designated $H2\text{-}D^{dx}$ has been found) [33]. This haplotype, originally designated $H2^{ct}$ and now designated $H2^{gx}$ since it is shared with the related ALR/Lt strain [33, 34], was transferred from CTS/Shi onto the NOD/Shi inbred background. T1D frequency was reduced and T1D onset delayed in this NOD.CTS-$H2^{gx}$ congenic stock; female frequency was 44% as compared to a control incidence of 90% in NOD/Shi females [31]. Since MHC class II alleles were not different from the NOD alleles, the MHC-linked resistance associated with $H2^{gx}$ was designated '*Idd16*'. Although it might be assumed that '*Idd16*' resides distal to the class II loci, a proximal location is equally plausible based upon analysis of NOD stocks congenic for recombinant haplotypes generated by cross-over between $H2^{g7}$ and $H2^{209}$ [35]. These congenic stocks provide evidence of additional '*Idd*' contributors centromeric to an *Lmp2* recombinatorial hotspot that both includes and extends proximal to the *H2K* gene. Even though the MHC class I alleles of NOD mice ($H2K^d$, $H2D^b$) are common in nonautoimmune-prone strains, they acquire a diabetogenic function in NOD mice – the selection and targeting of CD8+ T cells essential for initiation of the diabetogenic process [36]. As will be detailed below, the susceptibility conferred by class I alleles requires coexpression of the common β_2-microglobulin 'a' allelic variant ($B2m^a$, Chr. 2) expressed in NOD mice. A comparison of the MHC haplotypes of NOD and certain of the related strains is shown in table 1. To summarize, what is referred to as '*Idd1*' is a

Table 1. MHC alleles of NOD and ICR-derived related strains

Strain	MHC gene				Haplotype designation
	H2-K	*H2-A*	*H2-E*	*H2-D*	
NOD and ILI	d	g7	null	b	*H2^{g7}*
CTS and ALR	d	g7	null	dx	*H2^{gx}*
NON and ALS	b	nb1	K-like	b	*H2^{nb1}*

collective descriptor for multiple disease-predisposing alleles within or linked to the MHC complex.

Overview of the Role of Non-MHC Genes

While the *H2^{g7}* haplotype is necessary, it is not sufficient to elicit T1D, and is also not unique to the NOD strain. NOD mice originated from inbreeding of outbred Swiss Jcl:ICR mice. ILI, another Swiss-derived inbred strain, also expresses this haplotype [37]. Further, as noted above, two other Swiss-derived strains, CTS and ALR, carry the *H2^{g7}*-related *H2^{gx}* haplotype [31, 33]. Indeed, the *Aβ^{g7}* allele has been found in another strain [38] and the *Ea^{b}* null mutation carried by NOD is quite common in nonautoimmune-prone inbred strains. What seems to set the NOD mouse apart from its MHC-matched or 'almost-matched' related strains is its collection of non-MHC genes, no single one of which is absolutely required for diabetogenesis, but which combinatorially create a threshold liability for developing T1D [39]. Although it is universally assumed that these non-MHC diabetogenic susceptibility modifiers are genes controlling immune functions, this need not be true for all loci. The progenitors of NOD, outbred ICR/CD-1 mice, are segregating for quantitative trait loci (QTL) that may predispose to impaired glucose tolerance and obesity. Susceptibility for pre-type 2 diabetes is exhibited by males of the NOD-kindred strains, NON [40], NSY [41] and ALS [Mathews and Leiter, unpubl. data], while CD-1 females may become obese with aging [42]. That the NOD strain has fixed an undefined subset of these QTL can be observed by outcrossing with SPRET/Ei, an inbred strain recently derived from wild *Mus spretus*. Backcross of T1D-resistant (NOD × SPRET)F$_1$ hybrids to NOD produced a complex comingling of genes predisposing to both T1D and type 2 diabetes associated with islet hyperplasia, hyperinsulinemia, and insulin resistance without insulitis [43].

The example of the development of type 1 and type 2 diabetes in segregating progeny from a NOD outcross to a completely unrelated mouse strain emphasizes that identification of non-MHC linkages is 'contextual'; that is, it is wholly dependent upon the genome of the outcross partner strain. Hence, the numbers and genetic map positions of the non-MHC '*Idd*' genes segregating in a given cross are not constant. In certain outcrosses, but not in others, the partner strain may share susceptibility alleles with NOD and thus no segregation at those loci can be discerned. In other cases, the outcross partner strain may contribute strain-unique sets of resistance or susceptibility alleles not previously identified. Since certain of these genes may be epistatic in their action (requiring NOD genome at MHC and at certain other loci), the collection of genes eliciting T1D among genetically-diverse individuals generated in an F_2 or first backcross are not necessarily even the same from one diabetic proband to another. Many of the non-MHC loci summarized in table 2 below were identified by first transferring the diabetogenic *H2^{g7}* haplotype onto the outcross partner strain inbred background to fix susceptibility at '*Idd1*'. Certain of the linkages to be described encompass chromosomal areas implicated in susceptibility to a variety of other autoimmune diseases in mice [44]. This has led to speculation that the same '*Idd*' gene predisposing NOD mice to T1D might, in another strain's genome, predispose to a different, usually induced, form of autoimmunity such as experimental allergic encephalomyelitis (EAE) [45–47], experimental autoimmune gastritis [48], or experimental orchitis [49]. However, most of the '*Idd*' regions contain multiple contributory genes so that linkage overlap of multiple diseases to a common chromosomal region need not imply a common gene is contributing to all of the different disease entities. Indeed, the reader should keep in mind that very significant differences distinguish spontaneous from experimentally-induced diabetes in mice [50], so that caution must be used in extrapolating genetic data from induced versus spontaneous disease unless the comparison is made directly in NOD and its congenic stocks. Space limitations prevent us from enumerating the evidence supporting the existence of the ever-growing list of '*Idd*' loci provided in table 2. The interested reader is referred to the specific references to observe the partner strain used in outcross to establish the linkage.

Genetic 'Cross-Talk' between 'Idd' *Loci*

A major T1D susceptibility complex on Chr. 2, initially designated '*Idd13*', was identified in an outcross of NOD/Lt to the closely-related NOR/Lt strain described above [64]. Subcongenic analysis suggests a minimum of two genes contributing to NOR resistance in the long congenic segment (~32 cM) identified in table 2 [58]. The structural gene for β_2-microglobulin (*B2m*) lay at the proximal end of one of these protective subcongenic segments. The common *B2m^a* allotype expressed by NOD, and replaced by the B6-derived *B2m^b* allotype in NOR/Lt, might seem to be an unlikely candidate gene within this protective segment, since the two allotypes differ only by a single amino acid. However, dimerization of NOD MHC class I K^d and D^b chains with the alternative *B2m* isoforms alters their structural conformation without changing total expression levels [58]. If the common $B2m^a$ gene product interacted epistatically with NOD's common MHC class I gene products to affect selection of diabetogenic CTL by APC and/or affect CTL ability to recognize cognate antigen on beta β-cell targets, then this would provide excellent support for the 'Nerup hypothesis' that common alleles can acquire diabetogenic functions in specific pairing with other genes [8].

The ability to genetically disrupt specific loci in the mouse genome provides the investigator with the means to test the candidacy of *B2m* or any other candidate gene that has been cloned. As will be discussed in detail below, NOD mice congenic for a disrupted *B2m* allele fail to select diabetogenic CD8 T cells and do not develop T1D. Thus, according to the original broad definition of what constituted an '*Idd*' (a locus affecting susceptibility to T1D), the *B2m* gene would seemingly qualify. This NOD.$B2m^{null}$ stock provided the means for a rigorous test of *B2m* as an '*Idd*' gene. The stock was recently employed as the recipient of separately-introduced transgenes encoding various *B2m* isoforms. The CD8 T-cell repertoire was positively selected in both lines, but only the line with the reconstituted expression of the $B2m^a$ isoform developed T1D [65]. This represents the first empirical demonstration of the molecular nature of a non-MHC '*Idd*' candidate. Previous analysis of the markedly different course of insulitis progression in NOR vs. NOD indicated that NOR APC were less able to attract and activate peri-insular T cells [66]. In the study described above where genetic mapping of intra-islet T-cell content of (NOD × NOR)F$_2$ mice indicated linkage of the NOD susceptibility allele in the '*Idd5.2*' region, a second NOD susceptibility locus was linked to the proximal '*Idd13*' support interval (containing the *B2m* candidate) [67]. This second Chr. 2 locus appeared to interact epistatically with the Chr. 1 allele to increase intensity of intra-islet insulitis [67].

What Does This Genetic Complexity in Mice Portend for the Search for Human 'Idd' *Genes?*

Unraveling the genetic basis for T1D development in the NOD mouse is complicated by interactions between the diabetogenic $H2^{g7}$ and the welter of non-MHC '*Idd*' loci, none of which alone is sufficient to elicit T1D. This can be demonstrated by developing B6 stocks congenic for long '*Idd*' susceptibility intervals from NOD [68]. Although some level of peri-insulitis can be seen, severe insulitis does not develop [68], even when bicongenic stocks are analyzed that carry both the non-MHC interval and the NOD '*Idd1*' genes on Chr. 17. The situation is made even more complex by the knowledge that strong environmental influences as well as intergenic interactions affect gene penetrances and ultimate presentation of clinical disease. The advantages of the mouse system are that the environment can be controlled, and controlled outcrosses can be conducted in which the genotypes of the partner strains are defined (and homogeneous). Because of the level of genetic and environmental control, new mathematical computational and other analytic approaches can be developed to dissect the intergenic interactions that define the diabetogenic process [69, 70]. Such computational methodology for looking at multiplex interactions are now being applied to the inheritance of type 2 diabetes in humans [71]. Similar types of analytic methods will be required for the understanding of T1D inheritance. The finding that outcross of NOD mice with related versus unrelated partner strains uncover different '*Idd*' linkages emphasize that the genes segregating are contingent upon the inbred strain paired for outcross with NOD. Thus, an '*Idd*' exerting a strong effect in one cross may not be detectable at all in another outcross combination. Genetic studies with NOD outcrosses further suggest the likelihood of interaction between genes predisposing to T1D and type 2 diabetes. This is particularly evident when NOD is outcrossed either to the NON/Lt strain, a model for impaired glucose tolerance, or to a wild-derived inbred strain from *M. spretus* [39, 43].

Table 2 provides possible homologous T1D susceptibility regions in the NOD and human genomes. There are a number of reasons why the list of human *IDDM* loci in table 2 provides so few homologs for the mouse '*Idd*' loci. Despite striking similarities in organization of the mouse and human genomes, important genus-specific differences nevertheless distinguish mice from humans. For example, proinsulin/insulin is a major autoantigen in both humans and NOD mice [72, 73], and the human *IDDM2* locus has been defined by polymorphisms in a VNTR upstream of the human insulin *(INS)* gene [74]. Embryonic variation in the timing of thymic expression of the human insulin gene is hypothesized to be a diabetes susceptibility determinant [75]. The fact that the mouse homolog to the human *INS* gene (*Ins2* on distal Chr. 7) has not been detected in most segregation analyses might be expected because mice express two genes for insulin (the additional gene is *Ins1* on Chr. 19). However, in mice, it is reported that only the *Ins2* gene is expressed in both thymus and β cells [76]. In this regard, it is noteworthy that T1D protection provided by C57L-derived genome on distal Chr. 7 in the NOD.DR-4 strain (which also has a minor C57L segment 13 cM above the *Ins1* allele on Chr. 19) is associated with maintenance of high insulin content in the insulitic islets [53]. If any non-MHC could be considered a major autoimmunity regulator in the NOD mouse, it would be the '*Idd3*' (*Il2*) linkage. Thus far, there is no evidence that the homologous 4q26-q27 region in humans is a major T1D susceptibility locus. The '*Idd5.1*' region containing the *Cd152* (*Ctla4*) locus may be reflected by the *IDDM12* linkage to the homologous region on human 2q33. While the genes contributing to T1D development in humans and NOD mice do not completely overlap, they may contribute to common pathogenic dysfunctions.

To summarize, genetic analysis of the NOD mouse indicates that many of the diabetes susceptibility alleles present in this mouse's genome are not

Locus	Chr.	Interval markers	Candidate genes at locus	Potential human homolog	Ref.
Idd1	17	Entire MHC complex including H2-K and H2-D markers	Both $H2^{g7}$ class I and II genes confirmed	*IDDM1* (*HLA*, 6p21)	this chapter
Idd2	9	*D9Mit2-D9MitMit9* (17–48 cM)	*Cd3, Il18, Cyp19, Rapop*	*IDDM3* (15q26)	39
Idd3	3	*Il2* (*D3Nds2*, other sequence polymorphisms)	*Il2*		51
Idd4	11	*Il4-Acrb* (28–40 cM)	*Il3, Il4, Il5, Irf1, Csf2*		52
NOD.DR-3	11	(13–72 cM)	*Il12b, Nos2, Mip1a, Mip1b*	*IDDM18* (5q33-q34)	53
Idd5.1	1		*Casp8, Ctla4, Cflar, Cd28,*	*IDDM12* (2q31-33)	54
Idd5.2	1		*Nramp1, Cmkar2*	*IDDM13* (2q34)	54
Idd6	6	*D6Mit26-D6Mit14* (63–71 cM)	*Iapp, Tnfr1*		39, 52
Idd7	7	*Ckmm* (4.5 cM)	?		52
NOD.DR-4	7	*D7Mit283-D7Mit15* (57–71 cM)	*?Ins2?*	*IDDM2* (INS, 11p15.5)	53
Idd8	14	*Plau* (2.5 cM)	?		52
Idd9.1	4	*D4Mit28-D4Mit251* (42.5–66 cM)	*Jak1, Lck*		55
Idd9.2	4	*D4Nds23-D4Mit63* (66–69 cM)	*Tnrf2, Cd30*		55
Idd9.3	4	*D4Mit127-D4Mit42* (78–81 cM)	*Cd137*		55
Idd10	3	*D3Mit213-D3Mit189* (~49–50 cM)	*?*		56
Idd11	4	*D4Mit119-D4Mit69* (48–63 cM)	? possibly same as *Idd9.1*	*IDDM16* (1p36-p35)	57
Idd12	14	*Plau* (2.5 cM)	? possibly same as *Idd8*		52
Idd13	2	*D2Mit490-D2Mit144* (65–97 cM)	*β2m, Il1a, Il1b, Pcna*		58
Idd14	13	*D13Mit61* (22 cM)	?	*IDDM15* (6q21)	39
Idd15	5	*Xmv65* (1 cM)	?		39
Idd16	17	*D17Mit30-D17Mit192* (13.6–18.1 cM) or *D17Mit13-D17Mit104* (18.95–21.65 cM)	$H2^{g7}$-linked		59
Idd17	3	*D3Mit26-D3Mit40* (38–40 cM)	?		60
Idd18	3	*D3Nds86-D3Mit124* (52.5–55 cM)	*Csf1, CD53*		56

Locus	Chr.	Interval markers	Candidate genes at locus	Potential human homolog	Ref.
Idd19	6	*D6Mit11-D6Mit25* (49.4–65 cM)	?		28
Idd?	10	*Ifnrga-D10Mit87* (15–16 cM)	?		61
Idd?	15	Transgene insertional mutagenesis mapped by FISH	*Ptgerep4, Il7r*		62
Idd?	18	(~40–50 cM) meta-analysis for autoimmune responsiveness in different models	*Mad2, Mad4, Dcc*	*IDDM6* (18q12-q21)	63

'*Idd?*' descriptors without numbers indicate no assignments yet made. Readers are advised to consult the original references to see what partner strains were used in outcross to establish the linkages since many of the linkages are contingent on partner strain specification (e.g., C57BL/6, NON/Lt, NOR/Lt, CTS/Shi, PWK, 129/Sv, etc.). In almost all cases where subcongenic analysis of the longer congenic intervals shown above were performed, multiple loci were indicated. These subcongenic regions are not always given new '*Idd*' numbers (e.g., as for *Idd13*). Given the uncertainty surrounding the true position of *IDDM* loci in humans, the listing of potential syntenic regions is pure speculation by the authors. Descriptions of gene symbols may be accessed at the Mouse Genome Informatics database website (http://www.informatics.jax.org).

inherently diabetogenic, but can make diabetogenic contributions when present in certain combinations with a rather large collection of other genes. Further adding to the complexity of genetic analyses is the finding that not all of the susceptibility alleles can be expected to derive from the NOD genome. Even with the major component of susceptibility fixed within the MHC region, the numbers of non-MHC genes capable of facilitating or suppressing the diabetogenic process are sufficiently large such that predictions as to whether a segregant in an F_2 or first backcross population will develop diabetes is complicated. For these reasons, it is important that major phenotypes diagnostic of expression of sets of these diabetes susceptibility modifiers be defined.

Functional Genomics. How Do Susceptibility Loci Predispose to T1D?

Genetic Regulation of T-Cell Selection

The previous section has established that variable combinations of non-MHC '*Idd*' genes can interact with an inherently diabetogenic MHC to achieve a pathogenic threshold in mice. However, while different genetic subsets can interact to elicit T1D, they may do so by producing perturbations in common developmental or metabolic processes. Among the diabetogenic dysfunctions that could be elicited through variable combinations of '*Idd*' genes in both humans and NOD mice are disruptions in mechanisms that would normally block the development or functional activation of autoreactive T cells. An overview of some T-cell developmental pathways that could be disrupted by various combinations of '*Idd*' genes is provided below.

T lymphocytes are derived from precursor cells of bone marrow origin that subsequently differentiate within the thymus. Mature T lymphocytes that emigrate from the thymus and seed the blood and peripheral lymphoid organs can be divided into two major subsets based on cell surface expression of the CD4 or CD8 markers [reviewed in 77]. The ability of an individual T cell within either subset to recognize a specific antigen is imparted by clonally distributed T-cell receptor (TCR) molecules that are expressed as α/β chain heterodimers on the cell surface [reviewed in 78–80]. While the total mouse genome contains on the order of 10^5 genes, at least 10^{13} different TCR molecules can be generated. This broad diversity is generated as T cells differentiate in the thymus by splicing and rejoining of germline DNA sequences that encode various components of the TCR α and β chains followed by pairing of the resulting gene products.

The TCR α/β chain heterodimers generated by the genetic recombinatorial process described above recognize antigens which consist of peptide fragments bound to molecules encoded within the MHC. There are two primary types of MHC gene products. MHC class I gene products (*H2K* and *H2D* in mice) are expressed on virtually all cell types and present peptides of intracellular origin, such as those derived from replicating viruses, to T cells that express TCR α/β chain heterodimers complexed with CD8 molecules [reviewed in 81]. Such MHC class I-restricted CD8+ T cells usually exert a cytotoxic function. In contrast, MHC class II gene products (H2-A and H2-E in mice) which exist as α/β chain heterodimers, are generally only expressed on thymic epithelial cells and bone marrow-derived APC such as B lymphocytes, macrophages and dendritic cells. APC take up extracellular proteins and degrade them into peptide fragments which are then bound to MHC class II molecules for presentation to T cells that express TCR α/β chain heterodimers complexed with CD4 molecules [reviewed in 82]. Following antigenic activation, MHC class II-restricted CD4+ T cells usually provide helper functions that amplify other components of the immune response including cytotoxic CD8+ T cells and antibody production by B lymphocytes.

Neither MHC class I or class II molecules have an inherent capacity to discriminate between peptides derived from foreign pathogens or normal endogenous proteins [83]. Thus, to prevent the development of deleterious autoimmune responses, it is necessary to destroy or inactivate any T cells that express a rearranged TCR recognizing endogenous peptides bound to self MHC molecules. This normally occurs through several different mechanisms which select from the theoretical total pool of $\sim10^{13}$ TCR clonotypes, a subset of effectors that can respond to foreign, but not endogenous peptides bound to self MHC molecules. One such tolerogenic mechanism occurs in the thymus [reviewed in 84, 85]. The TCR molecules of T-cell precursors differentiating within the thymus interact with peptides presented by MHC gene products expressed on both thymic epithelial cells and hematopoietically derived APC. These interactions result in the positive selection of T cells capable of recognizing foreign antigens presented by self MHC gene products. Immature T cells whose TCR engages endogenous peptides bound to self MHC molecules are normally negatively selected in the thymus through an apoptotically-mediated activation-induced cell death (AICD) process. Several studies have demonstrated that the threshold of T-cell activation required to induce negative selection in the thymus is quantitatively greater than that required for positive selection [86, 87]. However, not all autoreactive T cells are negatively selected in the thymus. Such autoreactive T cells can undergo apoptotic cell death in the periphery when stimulated to a sufficiently high activation threshold by APC presenting large quantities of the appropriate antigen [88–92]. The activities of other autoreactive T cells escaping intrathymic deletion are suppressed in the periphery by immunoregulatory T cells which also must be stimulated by a highly activated APC in order to become functional [93].

Another mechanism by which APC have been proposed to down-regulate autoimmune responses is by inducing a change in the cytokine profile produced by CD4+ T cells reacting against self peptides [reviewed in 94–98]. It has been hypothesized that autoimmune tissue destruction is promoted when self peptide-reactive CD4+ T cells produce a Th1 pattern of cytokines including γ-interferon (IFNγ). These Th1 cytokines amplify cytotoxic CD8+ T-cell functions, as well as supporting macrophage activation and delayed-type hypersensitivity (DTH) responses. In contrast, it has been proposed that autoimmune tissue destruction is blocked when self peptide-reactive CD4+ T cells produce a Th2 pattern of cytokines including IL-4, IL-5, IL-6, IL-10 and IL-13 which provide help for the activation of B-lymphocyte-mediated humoral immunity. Of these cytokines, IL-4 appears to be most important in switching CD4+ T cells from a Th1 to Th2 response profile. Generally, CD4+ T cells switch from a Th1 to a Th2 profile as a function of increasing antigen dose presented by APC [99]. The cytokine profile produced by CD4+ T cells can also vary dependent upon the type of APC providing antigenic stimulation, with B lymphocytes tending to promote the activation of Th2 responses [100]. These response patterns can also be influenced by APC-produced cytokines, with macrophage production of IL-12 or IL-1 promoting Th1 and Th2 activation respectively [101–103].

A constant for all of the immunoregulatory mechanisms described above is that high levels of T-cell stimulation tend to promote tolerance, while lower levels tend to promote immunological effector responses. Thus, any genetic defects that compromise the stimulatory capacity of APC and/or impairs T-cell responsiveness could preferentially diminish any or all of these tolerogenic mechanisms without fully abrogating immunological effector responses. Indeed, as described later in this review, such defects are present in both APC and T cells from NOD mice and may be central to the development of auto-immune T1D.

The Diabetogenic Role of CD4+ versus CD8+ T Cells in NOD Mice

In NOD/Lt mice maintained at The Jackson Laboratory, T-cell infiltration of pancreatic islets (insulitis) initiates at approximately 5 weeks of age, with the first occurrence of overt disease observed several weeks later. However, it is becoming increasingly clear that the spectrum of autoreactive T-cell responses present at the onset of overt disease, which are also capable of mediating the rejection of subsequently implanted islet grafts, may be broader than that of the T effectors that initiate β-cell destruction in the earliest insulitic lesions. These differences are reflected by the extent to which CD4+ and CD8+ T cells must cooperate to mediate autoimmune β-cell destruction. It should be noted that some investigators employ the term 'benign' or 'nondestructive' to describe the early insulitis prevalent in young NOD mice in which most of the intra-islet infiltrates are localized just within the islet perimeters. As will be noted below, highly diabetogenic T cells have been isolated from such early islet lesions. Hence, the fact that overt T1D may be months away from diagnosis does not imply that the earliest infiltrating T cells are nondestructive, but rather that destruction of >90% of the β-cell mass required for hyperglycemia entails a time-dependent activation of an increasingly diverse spectrum of infiltrating CD4+ and CD8+ T cells.

Given they recognize antigens presented by the unusual H2-A^{g7} MHC class II molecule which is known to be a major contributor to T1D development in NOD mice, most early studies focused on the pathogenic role of CD4+ T cells. Indeed, a popular early hypothesis was that APC from NOD mice process soluble antigens common to all β cells, and bind these to H2-A^{g7} MHC class II molecules for presentation to autoreactive CD4+ T cells which then mediate β-cell destruction in a manner analogous to a DTH response. This hypothesis was partly based on the finding that diabetic NOD mice rapidly reject leukocyte-depleted islet, but not pituitary grafts, from allogeneic BALB/c donors [104]. The same group subsequently reported that diabetic NOD mice retain leukocyte-depleted BALB/c islet grafts if the recipients are first depleted of CD4+ T cells by monoclonal antibody treatment [105, 106]. Furthermore, cloned lines of CD4+ T cells isolated from spleens of overtly diabetic NOD mice can also passively transfer T1D and mediate islet graft rejection [107–110]. It is important to note that these early adoptive transfer studies used young prediabetic, sublethally irradiated NOD mice as recipients. We find that NOD/Lt mice are extremely radiation-resistant, requiring doses of γ-irradiation of 1,000 R or higher to completely ablate endogenous T cells. Hence, it was possible that the transferred CD4 T cells did not independently mediate T1D development, but rather recruited other effector populations. Indeed, evidence for such recruitment was provided when nonirradiated preweanling NOD mice were used as recipients of diabetogenic CD4 clones originally isolated from spleens of diabetic donors [111]. Hence, interpretation of earlier adoptive transfer studies was complicated by the possibility that the repertoire of diabetogenic T-cell clonotypes isolated from spleens (or pancreas) of overtly diabetic NOD mice may not necessarily reflect the initiating population.

Major insights into the effector mechanisms responsible for the initiation of autoimmune β-cell destruction have been provided by studies in which various T-cell populations have been passively transferred into a stock of NOD mice made T- and B-lymphocyte-deficient by congenic transfer of the severe combined immunodeficiency (*Prkdc^scid*) mutation [112–114]. Since these NOD-*scid* mice lack functional T lymphocytes, they remain diabetes-free. An obvious advantage of using T-cell-deficient NOD-*scid* mice as recipients for passive transfer studies is that the transferred T-cell population cannot activate T- or B-lymphocyte effectors endogenous to the host. Furthermore, processed β-cell autoantigens are present on the H2-A^{g7} MHC class II molecules of intra-islet APC from NOD-*scid* mice [115]. This indicates that NOD islets contain APC with H2-A^{g7} MHC class II molecules that are preloaded with β-cell autoantigens prior to the infiltration of diabetogenic CD4+ T cells. When CD4+ T cells isolated from the spleens of overtly diabetic NOD donors are transferred into NOD-*scid* recipients, both insulitis and overt T1D develop within 3–4 weeks [113]. Similarly, some cloned lines of islet-reactive CD4+ T cells isolated from the spleens of overtly diabetic NOD mice can also transfer T1D to NOD-*scid* recipients [116]. In addition, T1D does not develop in a congenic stock of NOD mice carrying a *Cd4* structural gene that has been functionally inactivated by homologous recombination [117]. Collectively, these results would seem to support the hypothesis that β-cell autoantigens bound to H2-A^{g7} MHC class II molecules on intra-islet APC of NOD mice elicit a DTH-like response by CD4+ T cells that is sufficient to induce T1D. However, passive transfer studies using NOD-*scid* recipients have demonstrated that while populations of CD4+ T cells present in *diabetic* NOD mice are sufficient to transfer T1D, CD4+ T cells from young prediabetic donors are unable to do so [113]. A mixture of CD4+ and CD8+ T cells from prediabetic donors could transfer T1D to NOD-*scid* recipients [113]. This suggested the possibility that in addition to CD4+ T-cell responses, the initiation of T1D development in NOD mice also requires contributions from CD8+ T cells that recognize antigens presented on the surface of pancreatic β-cells by the relatively common class I gene products (e.g., K^d and D^b) of the $H2^{g7}$ haplotype. Initial support for this model was provided by the finding that only an admixture of CD4+ and CD8+ T cells from *prediabetic donors* could adoptively transfer disease to NOD-*scid* recipients [113].

Several laboratories have unequivocally demonstrated an essential role for MHC class I-restricted CD8+ T cells in initiating β-cell destruction. MHC class I molecules must dimerize with B2m in order to be expressed in a stable fashion at the cell surface [reviewed in 81]. Thus, NOD.$B2m^{null}$ mice fail to express cell surface MHC class I molecules, and hence do not positively select CD8+ T cells [118–120]. These NOD.$B2m^{null}$ mice fail to develop either insulitis or T1D. This conclusively demonstrates that the relatively common MHC class I gene products encoded by the $H2^{g7}$ haplotype of NOD mice, in concert with the $B2m^a$ gene product, exert an autoimmune function that is essential for the initiation of autoimmune β-cell destruction. This function most likely entails the selection of autoreactive CD8+ T cells and their targeting to pancreatic β cells.

The initiation of β-cell destruction in NOD mice by a process requiring MHC class I-restricted CD8+ T cells, apparently results in the release of previously sequestered antigens which subsequently activate and amplify many additional effector T-cell populations (fig. 1). Thus, by the time overt diabetes has developed, NOD mice have accumulated a greatly expanded repertoire of β-cell-autoreactive T cells. While this 'autoimmune cascade' eventually results in the accumulation of CD4+ T cells in overtly diabetic NOD mice that can mediate autoimmune β-cell destruction in an independent fashion, their development is dependent upon prior responses requiring contributions from CD8+ T cells. Other studies examined the point in this 'autoimmune cascade' at which β-cell destruction becomes completely independent of MHC class I-restricted CD8+ T-cell responses [36, 121]. Splenic T cells from NOD donors of various ages were transferred into NOD-*scid* stocks that also carried either intact or functionally disrupted *B2m* genes. In the case of NOD-*scid* $B2m^{null}$ recipients (whose islets were MHC class I bare), only splenic T cells from overtly diabetic donors transferred T1D. Hence, these studies confirmed that the earliest initiation, and all but the final phases of autoimmune pancreatic β-cell destruction, required MHC class I-dependent T cells that were able to recognize cognate antigen on MHC class I-expressing β-cell targets.

β-Cell-autoreactive CD8+ T cells have been isolated from islets of young NOD donors that can rapidly transfer T1D to NOD-*scid* recipients in the

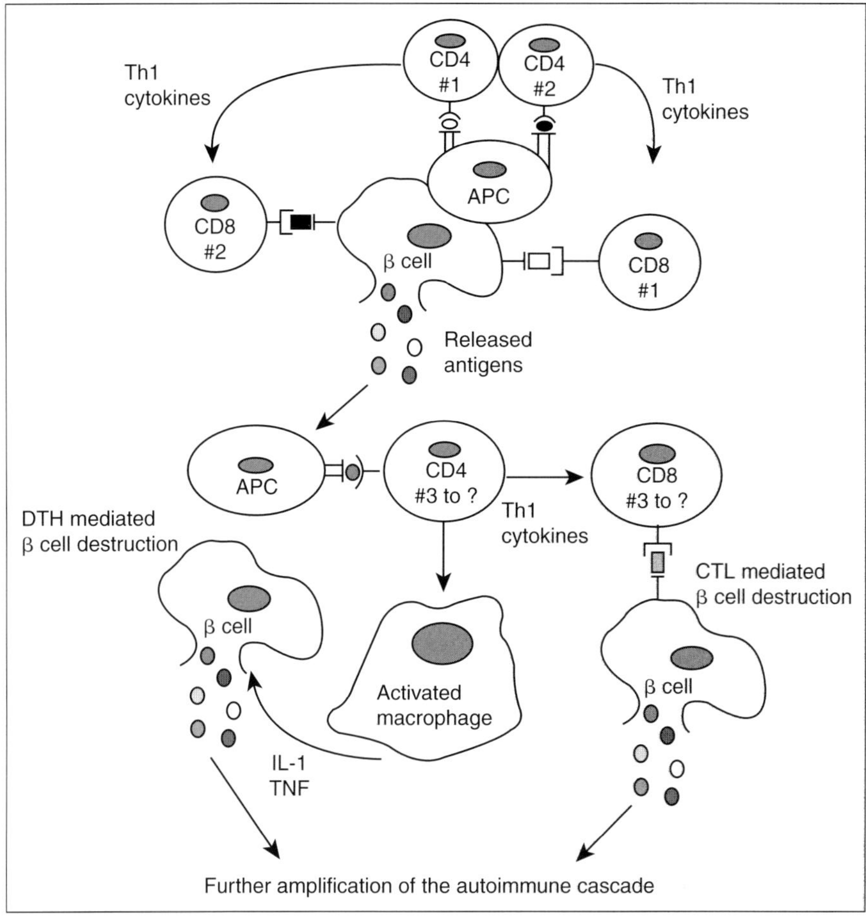

Fig. 1. Initiation and amplification of T-cell-mediated autoimmune β-cell destruction in NOD mice.

absence of CD4+ T cells. One such β-cell-autoreactive CD8+ T-cell clone (G9C8), now known to recognize the insulin B-chain amino acid 15–23 peptide, was derived from standard NOD islets, and amplified by activation in vitro using islets from a transgenic NOD stock which express B7-1 costimulatory molecules driven from an insulin promoter [122]. Two other groups produced stocks of NOD mice expressing rearranged TCR α and β chain transgenes derived from K^d-restricted β-cell-autoreactive CD8 clones present at the very early (AI4 clone) or mid-prodromal stages (NY8.3 clone) of T1D development [117, 123]. Due to the process of allelic exclusion, the vast majority of T cells (>90%) in these stocks express the transgenic TCR. Transgenic expression of the MHC class I-restricted AI4 or NY8.3 TCR on most T cells results in a greatly accelerated rate of T1D development. It was also determined whether T1D continued to develop at an accelerated rate in NOD AI4 or NY8.3 TCR transgenic mice when any residual nontransgenic CD4+ helper T cells were eliminated by introduction of the *scid* mutation or a functionally inactivated *Rag-2* gene. T1D development was attenuated, but not eliminated, in *Rag-2* deficient NOD NY8.3 TCR transgenic mice. Introduction of the *scid* mutation to NOD AI4 TCR transgenic mice did not impair their greatly accelerated rate of T1D development. Hence, while there is variation in the extent to which their activities are normally aided by CD4+ helper T cells, populations of MHC class I-restricted β-cell-autoreactive CD8+ T cells exist in NOD mice that, when present in sufficient numbers, can mediate T1D development without CD4+ T-cell involvement. However, as was noted for CD4 TCR transgenic NOD mice, T1D development in CD8 TCR transgenic NOD mice also does not faithfully duplicate the diversity of T-cell contributions that underlie the natural diabetogenic process.

TCR Gene Utilization and Antigenic Specificity of NOD
β-Cell-Autoreactive T Cells
As described above, the repertoire of T cells which initiate autoimmune β-cell destruction in NOD mice is less diverse than that present at the onset of overt hyperglycemia. Many investigators have analyzed whether the T cells proposed to initiate autoimmune β-cell destruction in NOD mice express a fixed and finite set of rearranged TCR α and β chain genes. Many of these efforts have been pursued with the hope that autoimmune T1D is initiated by a T cell that expresses a single or very limited range of TCR molecules specific for only a few pancreatic β-cell peptides. It was hoped such a finding would make it possible to design simple prevention therapies that either blocked the limited types of TCR molecules expressed by diabetogenic T cells, or alternatively induced tolerance to the few β-cell autoantigens recognized by these effectors. Unfortunately, although less diverse than the repertoire present at the onset of overt hyperglycemia, the repertoire of T clonotypes contributing to the initiation of autoimmune β-cell destruction in NOD mice appears to be quite broad.

The mosaic of TCR clonotypes that initiate and then amplify autoimmune β-cell destruction in NOD mice was initially assessed by reverse-transcription polymerase chain reaction (RT-PCR) analysis of islet RNA. Such analyses indicated that even at 3 weeks of age, islet infiltrating T cells of NOD mice utilize TCR genes in a polyclonal fashion [124, 125]. However, these early analyses could not distinguish between the TCR gene utilization patterns of diabetogenic

CD4+ versus CD8+ T cells. Hence, as an alternative approach, other investigators have isolated β-cell-autoreactive T-cell clones from NOD mice and then correlated their patterns of TCR gene utilization with their ability to passively transfer T1D.

A series of CD4+ T-cell clones have been isolated from NOD mice that proliferate in response to whole islets (or various subfractional components) or to recombinant proteins proposed to be autoantigenic targets in β cells. NOD CD4+ T-cell clones originally propagated on pancreatic islets have been found to utilize a very diverse range of TCR gene rearrangements [107, 123, 126, 127]. One interpretation of these findings is that multiple CD4+ T-cell clonotypes recognizing a quite diverse array of β-cell autoantigens contributes to T1D development in NOD mice. However, a single peptide-MHC antigenic complex can be recognized by T cells expressing multiple TCR gene rearrangements [128–131]. Thus, while the CD4+ T cells that mediate T1D development in NOD utilize a broad spectrum of TCR gene rearrangements, it does not necessarily mean that these effectors recognize an equally diverse set of β-cell autoantigens.

The breadth of β-cell antigens targeted by autoreactive T cells in NOD mice has been the subject of many investigations. Proposed to be among the initial antigens targeted are the 65- and 67-kD isoforms of glutamic acid decarboxylase (GAD). Their potential role as primary β-cell autoantigens is supported by two reports that CD4+ T cells in spleens of very young (3- to 4-week-old) prediabetic NOD mice demonstrate a spontaneous response against GAD65 [132, 133]. In older NOD mice with more pronounced insulitic lesions, splenic CD4+ T-cell reactivity was found to have spread to additional β-cell autoantigens including insulin, heat-shock proteins, peripherin, and carboxypeptidase-H. Other unidentified β-cell secretory granule proteins are also targeted by NOD CD4+ T-cell clonotypes activated in the latter stages of the autoimmune cascade [134, 135]. One of these unknown β-cell granule antigens is encoded by a gene on the region of NOD Chr. 6 that contains the T1D susceptibility locus '*Idd6*' [135]. The failure of NOD mice to establish T-cell tolerance to the early β-cell autoantigen GAD, can be overridden by injecting large quantities of this protein intrathymically [133] or intravenously [132] into 3- to 4-week-old recipients. These treatments not only blocked the development of primary NOD T-cell responses to GAD, but also the spread of T-cell reactivity to secondary β-cell autoantigens, and most importantly inhibited the development of T1D. Thus, the appearance of CD4+ T-cell reactivity to GAD may mark a key turning point in the development of autoimmune T1D in NOD mice. Indeed, there has been a report of a GAD-reactive CD4+ T-cell line with an ability to transfer T1D to NOD-*scid* recipients [136]. Interestingly, B lymphocytes have been shown to serve as a subpopulation of

APC in NOD mice that have a preferential ability to process and present certain β-cell antigens, including GAD, to autoreactive CD4+ T cells [137–139]. This most likely represents the mechanistic basis for the finding that NOD mice made genetically deficient in B lymphocytes are T1D-resistant [140–142]. Similarly, it has also been reported that the function of some non-MHC '*Idd*' gene(s) in NOD mice is to allow B lymphocytes to enter pancreatic islets prior to autoreactive T cells [66].

While the appearance of CD4+ T-cell reactivity to GAD may represent a key milestone in the pathogenesis of T1D, several factors indicate it is unlikely to represent the earliest autoantigen that is targeted in NOD pancreatic β cells. One of these is that while undetectable in spleen until ~12 weeks of age, insulin autoreactivity is reported to characterize a large proportion (~50%) of CD4+ T cells present in *pancreatic islets* of NOD mice as young as 4 weeks of age [143]. Another complicating factor is that both GAD isoforms are intracellular proteins [144], and thus should be immunologically presented by MHC class I molecules that trigger CD8+ rather than CD4+ T-cell responses. Therefore, it is likely that a CD4+ T-cell response to GAD is only elicited after β-cell lysis is initiated by T cells responding to other antigens.

As described earlier, the initiation of autoimmune β-cell destruction in NOD mice clearly requires contributions from MHC class I-restricted CD8+ T cells in addition to MHC class II-restricted CD4+ T cells. With the exception of one clone that was generated by an unusual peptide priming approach [145], all CD8+ T cells found to contribute to the earliest and later stages of pancreatic β-cell destruction in NOD mice have been restricted to the K^d rather than the D^b MHC class I gene product of the *H2^{g7}* haplotype [121, 122, 126, 146, 147]. These K^d-restricted β-cell-autoreactive CD8+ T cells are not characterized by a monoclonal TCR gene utilization pattern. However, the TCR gene utilization patterns of β-cell-autoreactive CD8+ T cells in NOD mice appear to be much less diverse than that of diabetogenic CD4+ T cells. Prevalent usage of a particular gene rearrangement is apparently restricted to the TCR α, rather than β chain of diabetogenic CD8+ T cells. Two separate groups have found that up to 30–35% of NOD K^d-restricted β-cell-autoreactive CD8 T cells utilize an identical Vα17 to Jα42 TCR gene rearrangement [121, 147]. In contrast, both groups found that these diabetogenic CD8+ T cells utilized a quite diverse array of TCR β chain gene rearrangements. These results indicate that the structure of the TCR α chain may be the primary determinant of what K^d-restricted β-cell autoantigenic complexes are recognized by NOD CD8+ T cells. Another important implication of these findings is that the array of β-cell antigens recognized by diabetogenic CD8+ T cells from NOD mice may be much more limited in scope than those targeted by autoreactive CD4+ clonotypes.

While the actual antigenic target remains unknown, a mimeotope peptide has been identified (NRP-A7) that is recognized in a K^d-restricted fashion by two separate NOD β-cell-autoreactive CD8+ T-cell clones expressing TCR molecules encoded by the same Vα17 to Jα42, but very diverse β-chain gene rearrangements [148, 149]. This indicates that diabetogenic CD8+ T cells from NOD mice characterized by the commonly observed Vα17 to Jα42 TCR gene rearrangement event are likely to recognize the same antigenic peptide even when their TCR β chains are completely unrelated. However, while it may represent a prevalent target, the peptide mimicked by NRP-A7 does not represent the sole antigen recognized by diabetogenic CD8+ T cells in NOD mice. Another group has found that a highly pathogenic β-cell-autoreactive CD8+ T-cell clone (G9C8) which does not utilize a Vα17 to Jα42 TCR gene rearrangement recognizes a K^d-restricted antigenic peptide consisting of amino acid residues 15–23 of the insulin-B chain [73]. Other studies have revealed that neither the peptide mimicked by NRP-A7 or the insulin-B chain 15–23 sequence represents the K^d-restricted β-cell autoantigen recognized by the highly pathogenic AI4 CD8+ T-cell clone discussed earlier which also does not utilize a Vα17 to Jα42 TCR gene rearrangement [117]. Thus, the repertoire of β-cell-autoreactive CD8+ T cells contributing to T1D development in NOD mice is not monoclonal in nature. However, given their indispensable pathogenic role and the fact that they appear to be less diverse than β-cell-autoreactive effectors in the CD4+ T-cell compartment, it can be argued that diabetogenic CD8+ T cells should be targets of future T1D inhibitory protocols as such therapies should be the easiest and most effective to develop.

MHC Gene Control of β-Cell-Autoreactive T-Cell Development in NOD Mice

The broad repertoire of T cells, especially in the CD4+ compartment, contributing to autoimmune β-cell destruction in NOD mice suggests this strain is characterized by broad-based rather than antigen-specific immunotolerogenic defects. This conclusion is further supported by the finding that NOD mice are characterized by a generalized defect in their responses to transplantation tolerance induction [150]. As described earlier, autoreactive T cells are normally physically deleted either in the thymus or periphery, functionally suppressed by immunoregulatory T cells, or rendered nonpathogenic by a shift in their cytokine production pattern from a Th1 to Th2 profile. For each of these immunoregulatory mechanisms, the threshold of T-cell activation required to induce tolerance is higher than that required to trigger an effector response. Thus, any genetically controlled defects that compromise the stimulatory capacity of APC and/or impair T-cell responsiveness could preferentially diminish tolerogenic mechanisms without fully abrogating immunological effector functions. Such dysfuctions are present in both APC and T cells from NOD mice, and appear to be major contributors to the development of autoimmune T1D.

The features of the unusual *H2^{g7}* MHC haplotype of NOD mice discussed above contribute to several APC dysfunctions that may lead to the development of β-cell-autoreactive T cells. T1D rarely develops (<3% incidence) in congenic stocks of NOD mice that heterozygously express MHC haplotypes from other strains [4, 151, 152]. Thus, the immunotolerogenic defects underlying the development of β-cell-autoreactive T cells in NOD mice occur most readily when *H2^{g7}* is homozygously expressed. Within the MHC, the diabetogenic contributions in part entail homozygous expression of the unusual H2-A^{g7} MHC class II gene product on APC in the absence H2-E MHC class II molecules. One mechanistic explanation for these findings is that when APC only express H2-A^{g7} MHC class II molecules, β-cell antigens may be presented in an inefficient fashion to autoreactive CD4+ T-cell clonotypes, and this preferentially results in the induction of effector rather than immunotolerogenic functions. Support for this model is provided by a report that the α/β chain complexes that comprise H2-A^{g7} MHC class II molecules in NOD mice do not dimerize in a stable fashion, potentially decreasing the efficiency by which they bind and present antigen [153]. In contrast, a recent study that established the crystal structure of H2-A^{g7} showed that this molecule could promiscuously bind a large number of GAD peptides in a stable fashion [154]. However, the ability to promiscuously bind a large number of peptides could prevent H2-A^{g7} from presenting any individual antigen in a fashion quantitatively sufficient to induce tolerance. Among the immunotolerogenic defects that result from homozygous expression of H2-A^{g7} and a lack of H2-E MHC class II molecules on NOD APC is a failure to induce the intrathymic negative selection of some β-cell-autoreactive CD4+ T cells. This was demonstrated by the finding that heterozygous expression of certain non-H2^{g7} MHC class II genes on NOD APC restored their ability to induce the intrathymic negative selection of a highly diabetogenic β-cell-autoreactive CD4+ T-cell clonotype [155, 156].

When particular transgene-encoded class II variants represent the sole diabetes-resistant MHC molecules expressed on NOD APC, some β-cell-autoreactive T cells continue to escape negative selection, but are held in a functionally-inactive state. This was demonstrated by the finding that NOD mice made T1D-resistant through expression of certain MHC class II transgenes, continue to harbor T cells that can induce disease when removed from the influence of the protective APC [27, 157, 158]. One of these studies presented evidence that as long as APC expressing a particular MHC class II transgene remain present, a population of immunoregulatory T cells are generated which can inhibit the functional activation of β-cell-autoreactive effectors [158]. This is consistent with previous findings that APC only expressing H2-A^{g7} MHC class II molecules have a greatly reduced ability to activate immunoregulatory T cells in a syngeneic mixed leukocyte reaction (SMLR) [159, 160]. Several lines of evidence indicate that the failure to activate immunoregulatory T cells in a SMLR contributes to the pathogenesis of autoimmune T1D in NOD mice. First is that the SMLR defect is overridden and T1D inhibited in NOD mice treated with recombinant IL-2 in vivo [161]. Secondly, T1D is inhibited in NOD mice injected at young age with cloned lines of immunoregulatory T cells propagated from a SMLR supplemented with IL-2 in vitro [162–164].

As described above, while able to prevent progression to overt T1D by other immunoregulatory mechanisms, APC from transgenic NOD stocks that solely express a disease-protective MHC class II variant are unable to mediate the global deletion of β-cell-autoreactive T cells. However, in a competitive bone marrow reconstitution system, it was found that the β-cell-autoreactive T cells which normally develop from NOD bone marrow are completely deleted or permanently inactivated when forced to mature in the presence of APC from a stock of NOD mice that congenically express an entire MHC haplotype associated with T1D resistance [165]. This may result from the protective APC population expressing class I as well as class II gene products from a diabetes-resistant MHC haplotype. Indeed, there is recent evidence that the $H2^{gx}$ MHC haplotype that shares all class II variants with $H2^{g7}$, but a different $H2\text{-}D$ class I allele, confers resistance to T1D [33, 34]. Collectively, these results indicate that as APC express increasing numbers of genes from a diabetes-resistant MHC haplotype, they acquire the ability to activate a wider array of immuno-tolerogenic mechanisms that limit both the development and function of β-cell-autoreactive T cells.

Regulation of β-Cell-Autoreactive T-Cell Development by Non-MHC Genes

Some non-MHC controlled defects also contribute to the reduced ability of NOD APC to activate immunoregulatory functions. A portion of these non-MHC controlled defects appear to entail a reduced ability of NOD APC to provide T-cell costimulatory signals at levels sufficient to induce tolerogenic rather than effector functions. One feature of NOD APC controlled by non-MHC genes which could contribute to tolerance induction defects is that compared to those from control strains, they express lower levels of the T-cell costimulatory molecule CD86 (aka B7.2) [166]. Another factor that contributes to the reduced T-cell costimulatory capacity of APC from NOD mice is that their macrophages produce significantly less IL-1 than those from T1D-resistant strains [159, 167]. The reduced ability of NOD macrophages to produce IL-1 results from their failure to fully differentiate from precursor cells in bone marrow [168–170]. This defect may be of pathogenic significance since T1D is blocked in NOD mice treated in vivo with either IL-1 or vitamin D which partially corrects their macrophage differentiation defect [167, 171]. Another consequence of NOD macrophages failing to fully differentiate is an impaired ability to up-regulate glutathione levels which is associated with a reduction in their capacity to process and present MHC class II-restricted antigens to CD4+ T cells [172]. Diminished levels of antigen presentation, CD86 expression, and IL-1 production may allow NOD APC to stimulate autoreactive T cells to an activation threshold sufficient to trigger effector, but not tolerance induction functions. It is possible that any or all of these factors contribute to the finding that T1D resistance in the NOD-related, and $H2^{g7}$-matched NOR/Lt strain may at least be partially explained by the presence of a non-MHC gene(s) that can induce the peripheral deletion of diabetogenic T cells [173].

High levels of prostaglandin E_2 (PGE_2) production by NOD macrophages represent an additional factor potentially contributing to the reduced T-cell costimulatory capacity of NOD APC [174]. When purified free of macrophages, NOD dendritic cells represent a highly efficient APC population for activating immunoregulatory T cells in a SMLR. However, in the presence of NOD macrophages, this function is suppressed by the high levels of macrophage-secreted PGE_2 [174]. PGE_2 can also down-regulate macrophage function in an autocrine fashion by elevating intracellular cAMP levels which in turn antagonize the protein kinase C (PKC)-mediated second messenger pathways required for IL-1 secretion [175–178]. As described above, the resulting decrease in IL-1-mediated T-cell costimulatory activity could then block the induction of various immunoregulatory functions. Similarly, a PGE_2-induced elevation of intracellular cAMP could block the induction of immunoregulatory mechanisms directly at the T-cell level by antagonizing TCR-coupled PKC second messenger activities. Indeed, the presence of PGE_2 has been shown to inhibit the ability of T cells to be stimulated to a level sufficient to induce AICD following TCR cross-linking [179].

If diminution of APC functions brought about by synergistic interactions between MHC and non-MHC genes is truly central to the pathogenesis of autoimmune T1D, this could provide an avenue for the development of prophylactic therapies. One possible approach could be the use of nonspecific immunostimulatory agents to up-regulate production of T-cell costimulatory factors by APC. Indeed, this may provide an explanation for the seemingly paradoxical finding that T1D is inhibited in NOD mice treated with a wide range of nonspecific immunostimulatory agents [reviewed in 20, 94, 97]. In addition, the reduced ability of $H2^{g7}$ MHC molecules to present β-cell autoantigens in a manner efficient enough to induce tolerogenic rather than effector T-cell responses, can be overridden by exposing NOD APC to large quantities of these antigens. This was demonstrated by the finding that the development of β-cell-autoreactive T cells and T1D are both inhibited in NOD mice intrathymically injected at a young age with whole syngeneic islet cells [180]. As described earlier, similar results were observed in NOD mice intrathymically injected with the β-cell autoantigen GAD [133]. Overriding inefficient presentation of β-cell autoantigens by $H2^{g7}$ MHC molecules would also account for the finding that T1D development is inhibited in NOD mice injected with dendritic cells purified from pancreatic lymph nodes (presumably presenting high levels of β-cell antigens), but not from cervical or axillary nodes [181]. Indirect evidence that NOD APC pulsed with high doses of β-cell autoantigens are capable of mediating T-cell tolerance is the finding that NOD T cells isolated from pancreatic lymph nodes are less efficient than those obtained from other anatomical sites in passively transferring T1D [182].

There is also evidence that some immunotolerogenic defects in NOD mice are intrinsic to T cells. Specifically, NOD thymocytes proliferate poorly in response to TCR cross-linking agents [183]. It was subsequently demonstrated that this defect results from a reduced ability of NOD thymocytes to activate TCR-coupled PKC second messenger pathways [184]. This could contribute to T1D development by inhibiting the ability of immature T cells with potential β-cell autoreactivity to be driven to an activation state sufficient to induce their deletion via apoptosis. Support for this possibility is provided by the finding that the NOD thymocyte response defect is at least partially controlled by a gene(s) that maps to the region of Chr. 11 previously shown to contain the diabetes susceptibility locus '*Idd4*' [185].

Role of Th1/Th2 Cytokines in Controlling T1D Development in NOD Mice: 'Buddy, Do You Have Change for a Paradigm?'

A currently popular hypothesis is that after avoiding deletional mechanisms, the pathogenic potential of β-cell-autoreactive CD4+ T cells in NOD mice is respectively enhanced or inhibited depending upon whether they produce cytokines of the Th1 (primarily IFNγ) or Th2 (primarily IL-4 and IL-10) type. This model is primarily based on reports that Th1 to Th2 cytokine shifts are often observed among β-cell-infiltrating T cells of NOD mice protected from overt T1D by many antigen-specific or nonspecific immunostimulation protocols [reviewed in 94, 97, 98]. An explanation for such putatively protective shifts may be provided by findings that at very high stimulation levels, CD4+ T cells can deviate from a Th1 to Th2 profile [99, 186].

T1D protection elicited through one GAD-based immune stimulation protocol clearly works through Th2 cytokine induction since it is ineffective in a stock of NOD mice made genetically-deficient in IL-4 [187]. However, while the induction of T1D resistance by many other protocols has also been associated with Th1 to Th2 cytokine shifts, several factors call into question whether the pathogenicity of β-cell-autoreactive CD4+ T cells in NOD mice can be strictly compartmentalized on the basis of these currently defined cytokine production profiles. These include reports of Th1 and Th2 clonotypic T cells isolated from NOD mice that contrary to expectations had a respective ability to inhibit or promote T1D development [164, 188, 189]. Furthermore, it was recently reported that prior skewing in vitro to either a Th1 or Th2 cytokine production profile did not alter the ability of the NOD-derived β-cell-autoreactive CD4+ T-cell clone BDC2.5 to passively transfer T1D [190]. There is also increasing evidence that the primary Th1 cytokine IFNγ does not play an obligatory role in T1D development in NOD mice. This conclusion was originally supported by a report that T1D developed normally in a stock of NOD mice congenic for a disrupted IFNγ gene [191]. Furthermore, as discussed earlier, signaling through the IFNγ receptor has also been found to be dispensable for T1D development [192]. Indeed, a genetically altered stock of NOD mice unable to respond to the key Th1 cytokine IFNγ, and characterized by greatly enhanced T-cell secretion of the Th2 cytokine IL-4, developed a high rate of T1D [192]. Collectively, these results indicate two things: first, that relative levels of Th1 and Th2 cytokines produced by splenic T cells do not necessarily reflect the pathogenic potential of islet-infiltrating T cells, and second, that the immune system is sufficiently well buffered such that simple conclusions cannot always be drawn from genetic disruption of a single cytokine gene.

A reduction in number and functionality of NKT cells is another NOD strain characteristic that has been cited as evidence that an impaired ability to produce Th2 cytokines might be critical to this strain's unique susceptibility to T1D [193, 194]. In nonautoimmune-prone strains, NKT cells represent a small population (<3%) of lymphoid cells that recognize glycolipid antigens presented by the nonclassical MHC class I-like CD-1 molecule [reviewed in 195]. Initial studies found that upon stimulation, NKT cells can quickly produce within several hours significant amounts of IL-4, and thus were proposed to be key initiators of Th2 responses. T1D development can be inhibited in NOD mice by several protocols that increase their NKT-cell levels [196, 197]. There is also evidence that increasing their NKT-cell levels may inhibit T1D development in NOD mice, at least in part, through Th2 cytokine production [197]. However, other studies have found that after stimulation, NKT cells actually produce much more IFNγ than IL-4 [198, 199]. Furthermore, upon stimulation, NOD NKT cells produce significantly less IFNγ than those from control strains [200]. Thus, it could be argued that a *diminished* ability to produce the Th1 cytokine IFNγ may represent the most important mechanism by which a deficiency in NKT cells contributes to T1D development in NOD mice. Supporting this possibility was the finding that a nonspecific immunostimulation protocol (BCG vaccine administration) which could possibly activate NKT cells, and had previously been proposed to inhibit T1D development by up-regulating Th2 cytokine production, protected NOD stocks congenic for disrupted IL-4 or IL-10 genes, but not the IFNγ-deficient stock [201].

The question remains why Th1 to Th2 cytokine shifts frequently characterize β-cell-autoreactive T cells remaining in NOD mice rendered T1D-resistant by various immunostimulatory protocols. One possibility is that an apparent Th1 to Th2 cytokine shift among β-cell-autoreactive T cells is a secondary outcome, rather than the cause of T1D resistance elicited by various immunostimulatory protocols. A frequently overlooked factor that could lead to the secondary appearance of such shifts is that higher rates of AICD occurs among CD4+ T cells producing Th1 than Th2 cytokines [202, 203]. Hence, some protocols may actually protect NOD mice from T1D by partially deleting β-cell-autoreactive T cells in a way that preferentially spares those producing Th2 cytokines. Such an 'unmasking' process could give a secondary appearance of being a Th1 to Th2 cytokine shift.

Conclusions

The genetic analysis of NOD mice has demonstrated that an essential set of MHC genes can interact with variable combinations of non-MHC genes to elicit T1D. However, while a rather broad spectrum of genes contribute to T1D pathogenesis, many may contribute to dysregulation of different biochemical steps in a common developmental or metabolic pathway. For example, sequential expression of hundreds, if not thousands, of genes would be expected in the development and functional maturation of a macrophage or dendritic cell from stem cell precursors. This process does not occur in a vacuum, but rather is contingent upon cues provided by the physical environment. In the case of APC development, the microfloral and dietary environments are crucial. When these environmental cues are minimized by maintaining NOD mice at a high SPF status, the NOD strain-specific peculiarities manifest as reduced ability of APC to provide tolerogenic signals to T cells, and weakened responsiveness of T cells to such signals. Such tolerance induction defects lead to a very broad array of CD4+, and a more restricted set of β-cell-autoreactive CD8+ T cells that are both essential to T1D development. If such a broad array of β-cell-autoreactive effectors also accumulate in humans, it will not be possible to develop antigen or T-cell clonotype-specific therapies for preventing T1D. However, studies in the NOD mouse indicate that it may be possible to prevent T1D with therapies that correct defects which underlie the genesis of β-cell-autoreactive T cells. Paradoxically, it appears that these therapies would have to be immunostimulatory in nature, and thus enable T-cell/APC interactions to occur at levels vigorous enough to preferentially activate tolerogenic rather than effector functions.

Acknowledgments

The authors gratefully acknowledge the receipt of prepublication manuscripts from Drs. John Todd and Linda Wicker. We thank Drs. Clayton Mathews and Pablo Silveira for their critical review. The authors were supported by the Juvenile Diabetes Foundation International, the American Diabetes Association, and National Institutes of Health grants DK27722, DK36175, DK46266, DK51090 and AI41469. Institutional shared services (T.J.L.) were supported by National Cancer Institute Center Support Grant CA-34196.

References

1 Makino S, Kunimoto K, Muraoka Y, Mizushima Y, Katagiri K, Tochino Y: Breeding of a non-obese, diabetic strain of mice. Exp Anim 1980;29:1–8.
2 Todd JA, Acha-Orbea H, Bell JI, Chao N, Fronek Z, Jacob CO, et al: A molecular basis for MHC class II-associated autoimmunity. Science 1988;240:1003–1009.
3 Ikegami H, Kawaguchi Y, Ueda H, Fukada M, Takakawa K, Fujioka Y, et al: MHC-linked diabetogenic gene of the NOD mouse: Molecular mapping of the 3′ boundary of the diabetogenic region. Biochem Biophys Res Commun 1993;192:677–682.
4 Prochazka M, Serreze DV, Worthen SM, Leiter EH: Genetic control of diabetogenesis in NOD/Lt mice: Development and analysis of congenic stocks. Diabetes 1989;38:1446–1455.
5 Wicker LS, Todd JA, Peterson LB: Genetic control of autoimmune diabetes in the NOD mouse. Annu Rev Immunol 1995;13:179–200.
6 Leiter EH: Genetics and immunogenetics of NOD mice and related strains; in Leiter EH, Atkinson MA (eds): NOD Mice and Related Strains: Research Applications in Diabetes, AIDS, Cancer and Other Diseases. Austin, Landes, 1998, pp 37–69.
7 Eisenbarth GS: Genes, generator of diversity, glycoconjugates, and autoimmune β-cell insufficiency in type I diabetes. Diabetes 1986;36:355–364.
8 Nerup J, Mandrup-Poulsen T, Helqvist S, Andersen HU, Pociot F, Reimers JI, et al: On the pathogenesis of IDDM. Diabetologia 1994;37(suppl 2):82–89.
9 De Vries RR, Meera Khan P, Bernini LF, van Loghem E, van Rood JJ: Genetic control of survival in epidemics. J Immunogenet 1979;6:271–287.
10 De Vries RR, van Rood JJ: HLA and infectious diseases. Arch Dermatol Res 1979;264:89–95.
11 Leiter EH: The NOD mouse: A model for analyzing the interplay between heredity and environment in the development of autoimmune disease. ILAR News 1993;35:4–14.
12 Baxter AG, Cooke A: Complement lytic activity has no role in the pathogenesis of autoimmune diabetes in NOD mice. Diabetes 1993;42:1574–1578.
13 Hattori M, Buse JB, Jackson RA, Glimcher L, Dorf ME, Minami M, et al: The NOD mouse: Recessive diabetogenic gene in the major histocompatibility complex. Science 1986;231:733–735.
14 Prins JB, Todd JA, Rodrigues NR, Ghosh S, Hogarth PM, Wicker LS, et al: Linkage on chromosome 3 of autoimmune diabetes and defective Fc receptor for IgG in NOD mice. Science 1993;260:695–698.
15 Gavin A, Hamilton J, Hogarth P: Extracellular mutations of non-obese diabetic mouse FcγRI modify surface expression and ligand binding. J Biol Chem 1996;271:17091–17098.

16 Luan J, Monteiro R, Sautes C, Fluteau G, Eloy L, Fridman W, et al: Defective Fc gamma RII gene expression in macrophages of NOD mice – Genetic linkage with up-regulation of IgG1 and IgG2b in serum. J Immunol 1996;157:4707–4716.

17 Serreze DV, Leiter EH: Insulin-dependent diabetes mellitus in NOD mice and BB rats: Origins in hematopoietic stem cell defects and implications for therapy; in Shafrir E (ed): Lessons from Animal Diabetes. V. London, Smith-Gordon, 1996, pp 59–73.

18 Leiter EH: The role of environmental factors in modulating insulin-dependent diabetes; in Vries RD, Cohen I, van Rood JJ (eds): Current Topics in Immunology and Microbiology. The Role of Microorganisms in Non-infectious Disease. Berlin, Springer, 1990, pp 39–55.

19 Coleman DL, Kuzava JE, Leiter EH: Effect of diet on the incidence of diabetes in non-obese diabetic mice. Diabetes 1990;39:432–436.

20 Atkinson MA, Leiter EH: The NOD mouse model of type 1 diabetes: As good as it gets? Nat Med 1999;5:601–604.

21 Prochazka M, Leiter EH, Serreze DV, Coleman DL: Three recessive loci required for insulin-dependent diabetes in NOD mice. Science 1987;237:286–289.

22 Mathews CE, Graser R, Savinov A, Serreze DV, Leiter EH: The NOD/Lt-related ALR/Lt strain: Unusual resistance of beta cells to autoimmune killing uncovers a role for beta-cell expressed resistance determinants. Proc Natl Acad Sci USA 2000;98:235–240.

23 Acha-Orbea H, McDevitt HO: The first external domain of the nonobese diabetic mouse class II I-Aβ chain is unique. Proc Natl Acad Sci USA 1987;84:2435–2439.

24 Miyazaki T, Uno M, Uehira M, Kikutani H, Kishimoto T, Kimoto M, et al: Direct evidence for the contribution of the unique I-Anod to the development of insulitis in non-obese diabetic mice. Nature 1990;345:722–724.

25 Lund T, O'Reilly L, Hutchings P, Kanagawa O, Simpson E, Gravely R, et al: Prevention of insulin-dependent diabetes mellitus in non-obese diabetic mice by transgenes encoding modified I-A β-chain or normal I-E α-chain. Nature 1990;345:727–729.

26 Slattery RM, Kjer-Nielsen L, Allison J, Charlton B, Mandel T, Miller JFAP: Prevention of diabetes in non-obese diabetic I-Ak transgenic mice. Nature 1990;345:724–726.

27 Hanson MS, Cetkovic-Cvrlje M, Ramiya VK, Atkinson MA, MacLaren NK, Singh B, et al: Quantitative thresholds of MHC class II I-E expressed on hematopoietically derived APC in transgenic NOD/Lt mice determine level of diabetes resistance and indicate mechanism of protection. J Immunol 1996;157:1279–1287.

28 Melanitou E, Joly F, Lathrop M, Boitard C, Avner P: Evidence for the presence of insulin-dependent diabetes-associated alleles on the distal part of mouse chromosome 6. Genome Res 1998;8:608–620.

29 Friday RP, Trucco M, Pietropaolo M: Genetics of type 1 diabetes mellitus. Diabetes Nutr Metab 1999;12:3–26.

30 Lie BA, Sollid LM, Ascher H, Ek J, Akselsen HE, Ronningen KS, et al: A gene telomeric of the HLA class I region is involved in predisposition to both type 1 diabetes and coeliac disease. Tissue Antigens 1999;54:162–168.

31 Ikegami H, Makino S, Yamato E, Kawaguchi Y, Ueda H, Sakamoto T, et al: Identification of a new susceptibility locus for insulin-dependent diabetes mellitus by ancestral haplotype congenic mapping. J Clin Invest 1995;96:1936–1942.

32 Yagi H, Matsumoto M, Nakamura M, Makino S, Suzuki R, Harada M, et al: Defect of thymocyte emigration in a T-cell deficiency strain of the mouse. J Immunol 1996;157:3412–3419.

33 Mathews CE, Graser RT, Serreze DV, Leiter EH: Reevaluation of the major histocompatibility complex genes of the NOD-progenitor CTS/Shi strain. Diabetes 2000;49:131–134.

34 Graser RT, Mathews CE, Leiter EH, Serreze DV: MHC characterization of ALR and ALS mice: Respective similarities to the NOD and NON strains. Immunogenetics 1999;49:722–726.

35 Hattori M, Yamato E, Itoh N, Senpuku H, Fujisawa T, Yoshino M, et al: Homologous recombination of the MHC class I K region defines new MHC-linked diabetogenic susceptibility gene(s) in nonobese diabetic mice. J Immunol 1999;163:1721–1724.

36 Serreze DV, Chapman HD, Varnum DS, Gerling I, Leiter EH, Shultz LD: Initiation of autoimmune diabetes in NOD/Lt mice is MHC class I-dependent. J Immunol 1997;158:3978–3986.

37 Hattori M, Fukuda M, Ichikawa T, Baumgartl HJ, Katoh H, Makino S: A single recessive non-MHC diabetogenic gene determines the development of insulitis in the presence of an MHC-linked diabetogenic gene in NOD mice. J Autoimmun 1990;3:1–10.

38 Liu G, Baker D, Fairchild S, Figueroa F, Quartey-Papafio R, Tone M, et al: Complete characterization of the expressed immune response genes in Biozzi AB/H mice: Structural and functional identity between AB/H and NOD A region molecules. Immunogenetics 1993;37:296–300.

39 McAleer MA, Reifsnyder PC, Palmer SM, Prochazka M, Love JM, Copeman JB, et al: Crosses of NOD mice with the related NON strain: A polygenic threshold model for type I diabetes. Diabetes 1995;44:1186–1195.

40 Leiter EH: The genetics of diabetes susceptibility in mice. FASEB J 1989;3:2231–2241.

41 Ueda H, Ikegami H, Kawaguchi Y, Fujisawa T, Yamato E, Shibata M, et al: Genetic analysis of late-onset type 2 diabetes in a mouse model of human complex trait. Diabetes 1999;48:1168–1174.

42 Pelleymounter MA, Cullen MJ, Healy D, Hecht R, Winters D, McCaleb M: Efficacy of exogenous recombinant murine leptin in lean and obese 10- to 12-month-old female CD-1 mice. Am J Physiol 1998;275:R290–R299.

43 Hattori M, Yamato E, Matsumoto E, Itoh N, Toyonaga T, Petruzzelli M, et al: Occurrence of pre-type I diabetes (pre-IDDM) and type II diabetes (NIDDM) in BC1 [(NOD × *Mus spretus*) F1 × NOD] mice; in Shafrir E (ed): Lessons from Animal Diabetes. VI. Boston, Birkhäuser, 1996, pp 83–95.

44 Vyse TJ, Todd JA: Genetic analysis of autoimmune disease. Cell 1996;85:311–318.

45 Encinas JA, Weiner HL, Kuchroo VK: Inheritance of susceptibility to experimental autoimmune encephalomyelitis. J Neurosci Res 1996;45:655–669.

46 Encinas JA, Wicker LS, Peterson LB, Mukasa A, Teuscher C, Sobel R, et al: QTL influencing autoimmune diabetes and encephalomyelitis map to a 0.15-cM region containing Il2. Nat Genet 1999;21:158–160.

47 Maron R, Hancock WW, Slavin A, Hattori M, Kuchroo V, Weiner HL: Genetic susceptibility or resistance to autoimmune encephalomyelitis in MHC congenic mice is associated with differential production of pro- and anti-inflammatory cytokines. Int Immunol 1999;9:1573–1580.

48 Silveira PA, Baxter AG, Cain WE, van Driel IR: A major linkage region on distal chromosome 4 confers susceptibility to mouse autoimmune gastritis. J Immunol 1999;162:5106–5111.

49 Meeker N, Hickey W, Korngold R, Hansen W, Sudweeks J, Wardell B, et al: Multiple loci govern the bone marrow-derived immunoregulatory mechanism controlling dominant resistance to autoimmune orchitis. Proc Natl Acad Sci USA 1995;92:5684–5688.

50 Leiter EH, Gerling IC, Flynn JC: Spontaneous insulin-dependent diabetes mellitus (IDDM) in nonobese diabetic mice: Comparisons with experimentally-induced IDDM; in McNeill JH (ed): Experimental Models of Diabetes. Boca Raton, CRC Press, 1999, pp 257–295.

51 Lyons PA, Armitage N, Argentina F, Denny P, Hill NJ, Lord CJ, et al: Congenic mapping of the type 1 diabetes locus, Idd3, to a 780-kb region of mouse chromosome 3: Identification of a candidate segment of ancestral DNA by haplotype mapping. Genome Res 2000;10:446–453.

52 Ghosh S, Palmer SM, Rodrigues NR, Cordell HJ, Hearne CM, Cornall RJ, et al: Polygenic control of autoimmune diabetes in nonobese diabetic mice. Nat Genet 1993;4:404–409.

53 McDuffie M: Derivation of diabetes-resistant congenic lines from the nonobese diabetic mouse. Clin Immunol 2000;96:119–130.

54 Hill NJ, Lyons PA, Armitage N, Todd JA, Wicker LS, Peterson LB: The NOD Idd5 locus controls insulitis and diabetes and overlaps the orthologous CTLA4/IDDM12 and NRAMP1 loci in humans. Diabetes 2000;49:1744–1747.

55 Lyons PA, Hancock WW, Denny P, Lord CJ, Hill NJ, Armitage N, et al: The NOD Idd9 genetic interval influences the pathogenicity of insulitis and contains molecular variants of Cd30, Tnfr2, and Cd137. Immunity 2000;13:107–115.

56 Podolin PL, Denny P, Armitage N, Lord CJ, Hill NJ, Levy ER, et al: Localization of two insulin-dependent diabetes (Idd) genes to the Idd10 region on mouse chromosome 3. Mamm Genome 1998;9:283–286.

57 Brodnicki TC, McClive P, Couper S, Morahan G: Localization of Idd11 using NOD congenic mouse strains: Elimination of Slc9a1 as a candidate gene. Immunogenetics 2000;51:37–41.

58 Serreze DV, Bridgett M, Chapman HD, Chen E, Richard SD, Leiter EH: Subcongenic analysis of the Idd13 locus in NOD/Lt mice: Evidence for several susceptibility genes including a possible diabetogenic role for β_2-microglobulin. J Immunol 1998;160:1472–1478.

59 Babaya N, Ikegami H, Kawaguchi Y, Fujisawa T, Ueda H, Fukuda M, et al: Congenic mapping and functional analysis of a second component of the MHC-linked diabetogenic gene, Idd16. Int J Diabetes 2000;8:1–7.

60 Podolin PL, Denny P, Lord CJ, Hill NJ, Todd JA, Peterson LB, et al: Congenic mapping of the insulin-dependent diabetes (Idd) gene, Idd10, localizes two genes mediating the Idd10 effect, and eliminates the candidate Fcgr1. J Immunol 1997;159:1835–1843.

61 Kanagawa O, Xu G, Tevaarwerk A, Vaupel BA: Protection of nonobese diabetic mice from diabetes by gene(s) closely linked to IFN-γ receptor loci. J Immunol 2000;164:3919–3923.

62 Bridgett MM, Cetkovic-Cvrlje M, Narayanswami S, Lambert J, O'Rourke R, Shi Y, et al: Differential protection in two transgenic lines of NOD/Lt mice hyperexpressing the autoantigen GAD65 in pancreatic beta cells. Diabetes 1998;47:1848–1856.

63 Merriman TR, Cordell HJ, Eaves IA, Danoy PA, Coraddu F, Barber R, et al: Suggestive evidence for association of human chromosome 18q12-q21 and its orthologue on rat and mouse chromosome 18 with several autoimmune diseases. Diabetes 2001;50:184–194.

64 Serreze DV, Prochazka M, Reifsnyder PC, Bridgett M, Leiter EH: Use of recombinant congenic and congenic strains of NOD mice to identify a new insulin-dependent diabetes resistance gene. J Exp Med 1994;180:1553–1558.

65 Hamilton-Williams EE, Serreze DV, Charlton B, Johnson EA, Marron MP, Mullbacher A, et al: A single amino acid polymorphism in β_2-microglobulin confers susceptibility to diabetes in NOD mice. Proc Natl Acad Sci USA 2001; in press.

66 Fox CJ, Danska JS: Independent genetic regulation of T-cell and antigen-presenting cell participation in autoimmune islet inflammation. Diabetes 1998;47:331–338.

67 Fox CJ, Paterson AD, Mortin-Toth SM, Danska JS: Two genetic loci regulate T-cell-dependent islet inflammation and drive autoimmune diabetes pathogenesis. Am J Hum Genet 2000; 67:67–81.

68 Yui MA, Muralidharan K, Moreno-Altamirano B, Perrin G, Chestnut K, Wakeland EK: Production of congenic mouse strains carrying NOD-derived diabetogenic genetic intervals: An approach for the genetic dissection of complex traits. Mamm Genome 1996;7:331–334.

69 Risch N, Ghosh S, Todd J: Statistical evaluation of multiple locus linkage data in experimental species and relevance to human studies: Application to murine and human IDDM. Am J Hum Genet 1993;53:702–714.

70 Reifsnyder P, Churchill G, Leiter E: Maternal environment and genotype interact to establish diabesity in mice. Genome Res 2000;10:1568–1578.

71 Cox NJ, Frigge M, Nicolae DL, Concannon P, Hanis CL, Bell GI, et al: Loci on chromosomes 2 (NIDDM1) and 15 interact to increase susceptibility to diabetes in Mexican Americans. Nat Genet 1999;21:213–215.

72 French MB, Allison J, Cram DS, Thomas HE, Dempsey-Collier M, Silva A, et al: Transgenic expression of mouse proinsulin II prevents diabetes in non-obese diabetic mice. Diabetes 1997;46:34–39.

73 Wong FS, Karttunen J, Dumont C, Wen L, Visintin I, Pilip IM, et al: Identification of an MHC class I-restricted autoantigen in type 1 diabetes by screening an organ-specific cDNA library. Nat Med 1999;5:1026–1031.

74 Vafiadis P, Bennett S, Todd J, Nadeau J, Grabs R, Goodyer C, et al: Insulin expression in human thymus is modulated by *INS* VNTR alleles at the *IDDM2* locus. Nat Genet 1997;15:289–292.

75 Pugliese A: Insulin expression in the thymus, tolerance and type 1 diabetes. Diabetes Metab Rev 1998;14:325–327.

76 Throsby M, Homo-Delarche F, Chevenne D, Goya R, Dardenne M, Pleau J: Pancreatic hormone expression in the murine thymus: Localization in dendritic cells and macrophages. Endocrinology 1998;139:2399–2406.

77 Von Boehmer H, Kisielow P: Lymphocyte lineage commitment: Instruction versus selection. Cell 1993;73:207–208.

78 Gilfillan S, Benoist C, Mathis D: Emergence of the T-cell receptor repertoire; in Bell JI, Owen MJ, Simpson E (eds). T-Cell Receptors. Oxford, Oxford University Press, 1995, pp 92–110.

79 Gascoigne NRJ: Genomic organization of T-cell receptor genes in the mouse; in Bell JI, Owen MJ, Simpson E (eds): T-Cell Receptors. Oxford, Oxford University Press, 1995, pp 288–302.

80 Hilyard KL, Strominger JL: The immune recognition unit: The TCR-peptide-MHC complex; in Bell JI, Owen MJ, Simpson E (eds): T-Cell Receptors. Oxford, Oxford University Press, 1995, pp 403–424.

81 Hansen TH, Lee DR: Mechanism of class I assembly with β_2-microglobulin and loading with peptide. Adv Immunol 1997;64:105–137.

82 Germain RN, Castellino F, Han RE, Sousa CR, Romagnoli P, Sadegh-Nasseri S, et al: Processing and presentation of endocytically acquired protein antigens by MHC class I and class II molecules. Immunol Rev 1996;151:5–30.

83 Von Boehmer H, Kisielow P: Self-nonself discrimination by T cells. Science 1990;248:1369–1373.

84 Jameson SC, Bevan MJ: T-cell selection. Curr Opin Immunol 1998;10:214–219.

85 Sebzda E, Mariathasan S, Ohteki T, Jones R, Bachmann MF, Ohashi PS: Selection of the T-cell repertoire. Annu Rev Immunol 1999;17:829–874.

86 Ashton-Rickardt PG, Bandeira A, Delany JR, Kaer LV, Pircher HB, Zinkernagel RM, et al: Evidence for a differential avidity model of T-cell selection in the thymus. Cell 1994;76:651–663.

87 Sebzda E, Wallace VA, Mayer J, Yeung RSM, Mak T, Ohashi PS: Positive and negative thymocyte selection induced by different concentrations of a single peptide. Science 1994;263:1615–1618.

88 Rocha B, Von Boehmer H: Peripheral selection of the T-cell repertoire. Science 1991;251:1225–1231.

89 Zhang L, Martin DR, Fung-Leung WP, Teh HS, Miller RG: Peripheral deletion of mature CD8+ antigen-specific T cells after in vivo exposure to male antigen. J Immunol 1992;148:3740–3745.

90 Ucker DS, Meyers J, Obermiller PS: Activation-driven T cell death. II. Quantitative differences alone distinguish stimuli triggering nontransformed T cell proliferation or death. J Immunol 1992;149:1583–1592.

91 Critchfield JM, Racke MK, Zuniga-Pflucker JC, Cannella B, Raine CS, Goverman J, et al: T cell deletion in high antigen dose therapy of autoimmune encephalomyelitis. Science 1994;263:1139–1143.

92 Pelfry CM, Tranquill LR, Boehme SA, McFarland HF, Lenardo MJ: Two mechanisms of antigen-specific apoptosis of myelin basic protein-specific T lymphocytes derived from multiple sclerosis patients and normal individuals. J Immunol 1995;154:6191–6202.

93 Ishikura H, Jayaraman S, Kuchroo V, Diamond B, Saito S, Dorf ME: Functional analysis of cloned macrophage hybridomas. VII. Modulation of suppressor T-cell-inducing activity. J Immunol 1989;143:414–419.

94 Rabinovitch A: Immunoregulatory and cytokine imbalances in the pathogenesis of IDDM: Therapeutic intervention by immunostimulation? Diabetes 1994;43:613–621.

95 Paul WE, Seder RS Lymphocyte responses and cytokines. Cell 1994;76:241–251.

96 Liblau RS, Singer SM, McDevitt HO: Th1 and Th2 CD4+ T cells in the pathogenesis of organ-specific autoimmune diseases. Immunol Today 1995;16:34–38.

97 Delovitch TL, Singh B: The nonobese diabetic mouse as a model of autoimmune diabetes: Immune dysregulation gets the NOD. Immunity 1997;7:291–297.

98 Tisch R, McDevitt H: Insulin-dependent diabetes mellitus. Cell 1996;85:291–297.

99 Hosken NA, Shibuya K, Heath AW, Murphy KM, O'Garra A: The effect of antigen dose on CD4+ T-helper cell phenotype development in a T-cell receptor-$\alpha\beta$-transgenic model. J Exp Med 1995;182:1579–1584.

100 Macaulay AE, DeKruyff RH, Goodnow CC, Umetsu DT: Antigen-specific B cells preferentially induce CD4+ T cells to produce IL-4. J Immunol 1997;158:4171–4179.

101 McAuthor JG, Raulet DH: CD28-induced costimulation of T-helper type 2 cells mediated by induction of responsiveness to interleukin 4. J Exp Med 1993;178:1645–1653.

102   Germann T, Gately MK, Schoehaut DS, Lohoff M, Mattner F, Fischer S, et al: Interleukin-12/T cell stimulating factor, a cytokine with multiple effects on T helper 1 (Th1) but not on Th2 cells. Eur J Immunol 1993;23:1762–1770.

103   Taylor-Robinson AW, Phillips RS: Expression of IL-1 receptor discriminates Th2 from Th1 cloned CD4+ T cells specific for *Plasmodium chabaudi*. Immunology 1994;81:216–221.

104   Nomikos IN, Prowse SJ, Carotenuto P, Lafferty KJ: Combined treatment with nicotinamide and desferrioxanine prevents islet allograft destruction in NOD mice. Diabetes 1987;35:1302–1304.

105   Wang Y, Hao L, Gill GR, Lafferty KJ: Autoimmune diabetes in NOD mouse is L3T4 T-lymphocyte-dependent. Diabetes 1987;36:535–538.

106   Wang Y, Pontesilli O, Gill RG, La Rosa FG, Lafferty KJ: The role of CD4+ and CD8+ T cells in the destruction of islet grafts by spontaneously diabetic mice. Proc Natl Acad Sci USA 1991;88:527–531.

107   Haskins K, Portas M, Bergman B, Lafferty K, Bradley B: Pancreatic islet-specific T-cell clones from nonobese diabetic mice. Proc Natl Acad Sci USA 1989;86:8000–8004.

108   Haskins K, McDuffie M: Acceleration of diabetes in young NOD mice with a CD4+ islet-specific T cell clone. Science 1990;249:1433–1436.

109   Bradley BJ, Wang Y, Lafferty KJ, Haskins K: In vivo activity of an islet-reactive T-cell clone. J Autoimmun 1990;3:449–456.

110   Bradley BJ, Haskins K, La Rosa FG, Lafferty KJ: CD8 T cells are not required for islet destruction induced by a CD4+ islet-specific T-cell clone. Diabetes 1992;41:1603–1608.

111   Peterson JD, Berg R, Piganelli JD, Poulin M, Haskins K: Analysis of leukocytes recruited to the pancreas by diabetogenic T cell clones. Cell Immunol 1998;189:92–98.

112   Prochazka M, Gaskins HR, Shultz LD, Leiter EH: The nonobese diabetic scid mouse: Model for spontaneous thymomagenesis associated with immunodeficiency. Proc Natl Acad Sci USA 1992;89:3290–3294.

113   Christianson SW, Shultz LD, Leiter EH: Adoptive transfer of diabetes into immunodeficient NOD-scid/scid mice: Relative contributions of CD4+ and CD8+ T lymphocytes from diabetic versus prediabetic NOD.NON-Thy 1a donors. Diabetes 1993;42:44–55.

114   Rohane PW, Shimada A, Kim DT, Edwards CT, Charlton B, Shultz LD, et al: Islet infiltrating lymphocytes from prediabetic NOD mice rapidly transfer diabetes to NOD-scid/scid mice. Diabetes 1995;44:550–554.

115   Shimizu J, Carrasco-Marin E, Kanagawa O, Unanue ER: Relationship between β cell injury and antigen presentation in NOD mice. J Immunol 1995;155:4095–4099.

116   Peterson JD, Haskins K: Transfer of diabetes in the NOD-scid mouse by CD4 T-cell clones: Differential requirement for CD8 T cells. Diabetes 1996;45:328–336.

117   Graser RT, DiLorenzo TP, Wang F, Christianson GJ, Chapman HD, Roopenian DC, et al: Identification of a CD8 T cell that can independently mediate autoimmune diabetes development in the complete absence of CD4 T-cell helper functions. J Immunol 2000;164:3913–3918.

118   Serreze DV, Leiter EH, Christianson GJ, Greiner D, Roopenian DC: MHC class I deficient NOD-B2m^{null} mice are diabetes and insulitis resistant. Diabetes 1994;43:505–509.

119   Wicker LS, Leiter EH, Todd JA, Renjilian RJ, Peterson E, Fischer PA, et al: β$_2$-Microglobulin-deficient NOD mice do not develop insulitis or diabetes. Diabetes 1994;43:500–504.

120   Sumida T, Furukawa M, Sakamoto A, Namekawa T, Maeda T, Zijlstra M, et al: Prevention of insulitis and diabetes in beta(2)-microglobulin-deficient non-obese diabetic mice. Int Immunol 1994;6:1445–1449.

121   DiLorenzo TP, Graser RT, Ono T, Christianson GJ, Chapman HD, Roopenian DC, et al: MHC class I-restricted T cells are required for all but the end stages of diabetes development in NOD mice and utilize a prevalent T-cell receptor a chain gene rearrangement. Proc Natl Acad Sci USA 1998;95:12538–12543.

122   Wong FS, Visintin I, Wen L, Flavell RA, Janeway CA: CD8 T cell clones from young nonobese diabetic (NOD) islets can transfer rapid onset of diabetes in NOD mice in the absence of CD4 cells. J Exp Med 1996;183:67–76.

123   Verdaguer J, Schmidt D, Amrani A, Anderson B, Averill N, Santamaria P: Spontaneous autoimmune diabetes in monoclonal T cell nonobese diabetic mice. J Exp Med 1997;186:1663–1676.

124 Maeda T, Sumida T, Kurasawa K, Tomioka H, Itoh I, Yoshida S, et al: T-lymphocyte-receptor repertoire of infiltrating T lymphocytes into NOD mouse pancreas. Diabetes 1991;40:1580–1585.

125 Waters SH, O'Neil JJ, Melican DT, Appel MC: Multiple TCR Vβ gene usage by infiltrates of young NOD mouse islets of Langerhans: A polymerase chain reaction analysis. Diabetes 1992; 41:308–312.

126 Shimizu J, Kanagawa O, Unanue ER: Presentation of β cell antigens to CD4+ and CD8+ T cells of non-obese diabetic mice. J Immunol 1993;151:1723–1730.

127 Daniel D, Gill RG, Schloot N, Wegmann D: Epitope specificity, cytokine production profile and diabetogenic activity of insulin-specific T cell clones isolated from NOD mice. Eur J Immunol 1995;25:1056–1062.

128 Jorgenson JL, Esser U, Fazekas de St. Groth B, Reay PA, Davis MM: Mapping T-cell receptor-peptide contacts by variant peptide immunization of single chain transgenics. Nature 1992; 355:224–230.

129 Kurata A, Berzofsky JA: Analysis of peptide residues interacting with MHC molecule or T cell receptor. Can a peptide bind in more than one way to the same MHC molecule? J Immunol 1990;144:4526–4535.

130 Bhayani H, Patterson Y: Analysis of peptide binding patterns in different MHC/T cell receptor complexes using pigeon cytochrome c-specific hybridomas. J Exp Med 1989;170:1609–1625.

131 Peccound J, Dellabona P, Allen P, Benoist C, Mathis D: Delineation of antigen contact residues on an MHC class II molecule. EMBO J 1990;9:4215–4223.

132 Kaufman DL, Clare-Salzler M, Tian J, Forsthuber T, Ting GSP, Robinson P, et al: Spontaneous loss of T-cell tolerance to glutamic acid decarboxylase in murine insulin-dependent diabetes. Nature 1993;366:69–72.

133 Tisch R, Yang XD, Singer SM, Liblau RS, Fugger L, McDevitt HO: Immune response to glutamic acid decarboxylase correlates with insulitis in non-obese diabetic mice. Nature 1993;366:72–75.

134 Bergman B, Haskins K: Islet-specific T-cell clones from the NOD mouse respond to β-granule antigen. Diabetes 1994;43:197–203.

135 Dallas-Pedretti A, McDuffie M, Haskins K: A diabetes-associated T-cell autoantigen maps to a teleomeric locus on mouse chromosome 6. Proc Natl Acad Sci USA 1995;92:1386–1390.

136 Zekzer D, Wong FS, Ayalon O, Millet I, Altieri M, Shintani S, et al: GAD-reactive CD4+ Th1 cells induce diabetes in NOD/SCID mice. J Clin Invest 1998;101:68–73.

137 Serreze DV, Fleming SA, Chapman HD, Richard SD, Leiter EH, Tisch RM: B-lymphocytes are critical antigen-presenting cells for the initiation of T-cell-mediated autoimmune diabetes in NOD mice. J Immunol 1998;161:3912–3918.

138 Falcone M, Lee J, Patstone G, Yeung B, Sarvetnick N: B lymphocytes are crucial antigen-presenting cells in the pathogenic autoimmune response to GAD65 antigen in nonobese diabetic mice. J Immunol 1998;161:1163–1168.

139 Noorchashm H, Lieu YK, Noorchashm N, Rostami SY, Greely SS, Schlachterman A, et al: I-A^{g7}-mediated antigen presentation by B lymphocytes is critical in overcoming a checkpoint in T-cell tolerance to islet β cells of nonobese diabetic mice. J Immunol 1999;163:743–750.

140 Serreze DV, Chapman HD, Varnum DS, Hanson MS, Reifsnyder PC, Richard SD, et al: B lymphocytes are essential for the initiation of T-cell-mediated autoimmune diabetes: Analysis of a new 'speed congenic' stock of NOD.Igμ^{null} mice. J Exp Med 1996;184:2049–2053.

141 Akashi T, Nagafuchi S, Anzai K, Kondo S, Kitamura D, Wakana S, et al: Direct evidence for the contribution of B cells to the progression of insulitis and the development of diabetes in non-obese diabetic mice. Int Immunol 1997;9:1159–1164.

142 Wong FS, Visintin I, Wen L, Granata J, Flavell R, Janeway CA: The role of lymphocyte subsets in accelerated diabetes in nonobese diabetic-rat insulin promoter-B7-1 (NOD-RIP-B7-1) mice. J Exp Med 1998;187:1985–1993.

143 Wegmann DR, Gill RG, Norbury-Glaser M, Schloot N, Daniel D: Analysis of the spontaneous T cell response to insulin in NOD mice. J Autoimmun 1994;7:833–843.

144 Aguilar-Diosdado M, Parkinson D, Corbett JA, Kwon G, Marshall CA, Gingerich RL, et al: Potential autoantigens in IDDM: Expression of carboxypeptidase-H and insulin but not glutamate decarboxylase on the β-cell surface. Diabetes 1994;43:418–425.

145 Gurlo T, Kawamura K, von Grafenstein H: Role of inflammatory infiltrate in activation and effector function of cloned islet reactive nonobese diabetic CD8+ T cells: Involvement of a nitric oxide-dependent pathway. J Immunol 1999;163:5770–5780.

146 Nagata M, Santamaria P, Kawamura T, Utsugi T, Yoon JW: Evidence for the role of CD8+ cytotoxic T cells in the destruction of pancreatic β-cells in nonobese diabetic mice. J Immunol 1994;152:2042–2050.

147 Santamaria P, Utsugi T, Park BJ, Averill N, Kawazu S, Yoon JW: Beta-cell-cytotoxic CD8+ T cells from nonobese diabetic mice use highly homologous T cell receptor α chain CDR3 sequences. J Immunol 1995;154:2494–2503.

148 Anderson B, Park BJ, Verdaguer J, Amrani A, Santamaria P: Prevalent CD8+ T cell responses against one peptide/MHC complex in autoimmune diabetes. Proc Natl Acad Sci USA 1999; 96:9311–9316.

149 Serreze DV, Johnson EA, Chapman HD, Graser RT, Marron MP, Yoshimura Y, et al: Autoreactive T cells in NOD mice can efficiently expand from a greatly reduced precursor pool. Diabetes 2001; in press.

150 Markees TG, Serreze DV, Phillips NE, Sorli CH, Gordon EJ, Shultz LD, et al: NOD mice have a generalized defect in their response to transplantation tolerance induction. Diabetes 1999;48:967–974.

151 Wicker LS, Miller BJ, Fischer PA, Pressey A, Peterson LB: Genetic control of diabetes and insulitis in the nonobese diabetic mouse. Pedigree analysis of a diabetic H-2nod/b heterozygote. J Immunol 1989;142:781–784.

152 Wicker LS, Appel MC, Dotta F, Pressey A, Miller BJ, DeLarto NH, et al: Autoimmune syndromes in major histocompatibility complex (MHC) congenic strains of nonobese diabetic (NOD) mice. The NOD MHC is dominant for insulitis and cyclophosphamide-induced diabetes. J Exp Med 1992;176:67–77.

153 Carrasco-Marin E, Shimizu J, Kanagawa O, Unanue ER: The class II MHC class I-A^{g7} molecules from non-obese diabetic mice are poor peptide binders. J Immunol 1996;156:450–458.

154 Corper AL, Stratmann T, Apostolopoulos V, Scott CA, Garcia KC, Kang AS, et al: A structural framework for deciphering the link between I-A$^{(g7)}$ and autoimmune diabetes. Science 2000;288:505–511.

155 Schmidt D, Verdaguer J, Averill N, Santamaria P: A mechanism for the major histocompatibility complex-linked resistance to autoimmunity. J Exp Med 1997;186:1059–1075.

156 Schmidt D, Amrani A, Verdaguer J, Bou S, Santamaria P: Autoantigen-independent deletion of diabetogenic CD4+ thymocytes by protective MHC class II molecules. J Immunol 1999;162: 4627–4636.

157 Slattery RM, Miller JFAP, Heath WR, Charlton B: Failure of a protective major histocompatability complex class II molecule to delete autoreactive T cells in autoimmune diabetes. Proc Natl Acad Sci USA 1993;90:10808–10810

158 Singer SM, Tisch R, Yang XD, McDevitt HO: An Abd transgene prevents diabetes in nonobese diabetic mice by inducing regulatory T cells. Proc Natl Acad Sci USA 1993;90:9566–9570.

159 Serreze DV, Leiter EH: Defective activation of T suppressor cell function in nonobese diabetic mice. Potential relation to cytokine deficiencies. J Immunol 1988;140:3801–3807.

160 Serreze DV, Leiter EH: Defective suppressor T cell activation in NOD/Lt mice is associated with pathogenesis and homozygous expression of H2^{g7}. Diabetes 1991;40(suppl 1):49A.

161 Serreze DV, Hamaguchi K, Leiter EH: Immunostimulation circumvents diabetes in NOD/Lt mice. J Autoimmun 1990;2:759–776.

162 Reich EP, Scaringe D, Yagi J, Sherwin RS, Janeway CA: Prevention of diabetes in NOD mice by injection of autoreactive T lymphocytes. Diabetes 1989;38:1647–1651.

163 Choisch N, Harrison LC: Suppression of diabetes mellitus in the non-obese diabetic mouse by an autoreactive (anti-I-A^{g7}) islet-derived CD4+ T-cell line. Diabetologia 1993;36:716–721.

164 Akhtar I, Gold JP, Pan LY, Ferrara JLM, Yang XD, Kim JI, et al: CD4+ β islet cell-reactive T cell clones that suppress autoimmune diabetes in nonobese diabetic mice. J Exp Med 1995;182:87–97.

165 Serreze DV, Leiter EH: Development of diabetogenic T cells from NOD/Lt marrow is blocked when an allo-H-2 haplotype is expressed on cells of hematopoietic origin, but not on thymic epithelium. J Immunol 1991;147:1222–1229.

166 Dahlen E, Hedlund G, Dawe K: Low CD86 expression in the nonobese diabetic mouse results in the impairment of both T cell activation and CTLA-4 up-regulation. J Immunol 2000;164: 2444–2456.

167 Jacob CO, Aiso S, Michie SA, McDevitt HO, Acha-Orbea H: Prevention of diabetes in nonobese diabetic mice by tumor necrosis factor (TNF): Similarities between TNF-α and interleukin-1. Proc Natl Acad Sci USA 1990;87:968–972.

168 Serreze DV, Gaskins HR, Leiter EH: Defects in the differentiation and function of antigen presenting cells in NOD/Lt mice. J Immunol 1993;150:2534–2543.

169 Langmuir P, Bridgett M, Bothwell A, Crispe I: Bone marrow abnormalities in the non-obese diabetic mouse. Int Immunol 1993;5:165–177.

170 Serreze DV, Gaedeke JW, Leiter EH: Hematopoietic stem cell defects underlying abnormal macrophage development and maturation in NOD/Lt mice: Defective regulation of cytokine receptors and protein kinase C. Proc Natl Acad Sci USA 1993;90:9625–9629.

171 Mathieu C, Waer M, Casteels K, Laureys J, Bouillon R: Prevention of type I diabetes in NOD mice by nonhypercalcemic doses of a new structural analog of 1,25-dihydroxyvitamin D_3, KH1060. Endocrinology 1995;136:866–872.

172 Piganelli JD, Martin T, Haskins K: Splenic macrophages from the NOD mouse are defective in ability to present antigen. Diabetes 1998;47:1212–1218.

173 Verdaguer J, Amrani A, Anderson B, Schmidt D, Santamaria P: Two mechanisms for the non-MHC-linked resistance to spontaneous autoimmunity. J Immunol 1999;162:4614–4626.

174 Clare-Salzler M. The immunopathogenic role of antigen presenting cells in the NOD mouse; in Leiter EH, Atkinson MA (eds): NOD Mice and Related Strains: Research Applications in Diabetes, AIDS, Cancer and Other Diseases. Austin, Landes Bioscience Publishers, 1998, pp 101–119.

175 Dinarello CA: Biology of interleukin 1. FASEB J 1988;2:108–115.

176 Cheung DL, Hamilton JA: Regulation of human monocyte DNA synthesis by colony stimulating-factors, cytokines and cyclic adenosine monophosphate. Blood 1992;79:1972–1981.

177 Bakouche O, Moreau JL, Lachman LB: Secretion of IL-1: Role of protein kinase C. J Immunol 1992;148:84–91.

178 Shapira L, Takashiba S, Champagne C, Amar S, Van Dyke TE: Involvement of protein kinase C and protein tyrosine kinase in lipopolysaccharide-induced TNF-α and IL-1β production by human monocytes. J Immunol 1994;153:1818.

179 Goetzl EJ, An S, Zeng L: Specific suppression by prostaglandin E_2 of activation induced apoptosis of human CD4+8+ T lymphoblasts. J Immunol 1995;154:1041–1047.

180 Gerling IC, Serreze DV, Christianson SW, Leiter EH: Intrathymic islet cell transplantation reduces beta-cell autoimmunity and prevents diabetes in NOD/Lt mice. Diabetes 1992;41:1672–1676.

181 Clare-Salzler MJ, Brooks J, Chai A, Herle KV, Anderson C: Prevention of diabetes in nonobese diabetic mice by dendritic cell transfer. J Clin Invest 1992;90:741–748.

182 Lepault F, Faveeuw C, Luan JJ, Gagnerault MC: Lymph node T-cells do not optimally transfer diabetes in NOD mice. Diabetes 1993;42:1823–1828.

183 Zipris D, Lazarus AH, Crow AR, Hadazija M, Delovitch TL: Defective thymic T cell activation by concanavalin A and anti-CD3 in autoimmune nonobese diabetic mice: Evidence of thymic T cell anergy that correlates with the onset of insulitis. J Immunol 1991;146:3763–3771.

184 Rapoport MJ, Lazurus AH, Jaramillo A, Speck E, Delovitch TL: Thymic T cell anergy in autoimmune nonobese diabetic mice is mediated by deficient T cell receptor regulation in the pathway of p21ras activation. J Exp Med 1993;177:1221–1226.

185 Gill BM, Jaramillo A, Ma L, Laupland KB, Delovitch TL: Genetic linkage of thymic T-cell proliferative unresponsiveness to mouse chromosome 11 in NOD mice. Diabetes 1995;44:614–619.

186 Rulifson IC, Sperling AI, Fields PE, Fitch FW, Bluestone JA: CD28 costimulation promotes the production of Th2 cytokines. J Immunol 1997;158:658–665.

187 Tisch R, Wang B, Serreze DV: Induction of GAD65-specific Th2 cells and suppression of autoimmune diabetes at late stages of disease is epitope-dependent. J Immunol 1999;163:1178–1187.

188 Pakala SV, Kurrer MO, Katz JD: T helper 2 (Th2) T cells induce acute pancreatitis and diabetes in immune-compromised nonobese diabetic mice. J Exp Med 1997;186:299–306.

189 Zekzer D, Wong FS, Wen L, Altieri M, Gurlo T, von Grafenstein H, et al: Inhibition of diabetes by an insulin-reactive CD4 T-cell clone in the nonobese diabetic mouse. Diabetes 1997;46:1124–1132.

190 Poulin M, Haskins K: Induction of diabetes in nonobese diabetic mice by Th2 T cell clones from a TCR transgenic mouse. J Immunol 2000;164:3072–3078.

191 Hultgren B, Huang X, Dybdal N, Stewart TA: Genetic absence of γ-interferon delays but does not prevent diabetes in NOD mice. Diabetes 1996;45:812–817.

192 Serreze DV, Post CM, Chapman HD, Johnson EJ, Lu B, Rothman PB. Interferon-γ receptor signaling is dispensable in the development of autoimmune type 1 diabetes in NOD mice. Diabetes 2000;49:2007–2011.

193 Gombert JM, Herbelin A, Tancrede-Bohin E, Dy M, Carnaud C, Bach JF: Early quantitative and functional deficiency of NK1+-like thymocytes in the NOD mouse. Eur J Immunol 1996;26:2989–2998.

194 Baxter AG, Kinder SJ, Hammond KJL, Scollay R, Godfrey DI: Association between αβTCR+ CD4− CD8− T-cell deficiency and IDDM in NOD/Lt mice. Diabetes 1997;46:572–582.

195 Bendelac A, Rivera MN, Park SH, Roark JH: Mouse CD1-specific NK1 T cells: Development, specificity, and function. Annu Rev Immunol 1997;15:535–562.

196 Gombert JM, Tancredebohin E, Hameg A, Leitedemoraes MD, Vicari A, Bach JF: IL-7 reverses NK1(+) T cell-defective IL-4 production in the non-obese diabetic mouse. Int Immunol 1996;8:1751–1758.

197 Hammond KJL, Poulton LD, Palmisano L, Silveira P, Godfrey DI, Baxter AG: α/β T cell receptor CD4-CD8- (NKT) thymocytes prevent insulin-dependent diabetes mellitus in non-obese diabetic mice by influence of interleukin (IL)-4 and/or IL-10. J Exp Med 1998;187:1047–1056.

198 Arase H, Arase N, Saito T. Interferon-γ production by natural killer (NK) cells and NK1.1+ T cells upon NKR-P1 cross-linking. J Exp Med 1996;183:2391–2396.

199 Chen H, Paul WE: Cultured NK1.1+ CD4+ T cells produce large amounts of IL-4 and IFN-γ upon activation by anti-CD3 or CD1. J Immunol 1997;159:2240–2249.

200 Falcone M, Yeung B, Tucker L, Rodriguez E, Sarvetnick N: A defect in interleukin-12-induced activation and interferon-γ secretion of peripheral natural killer T cells in nonobese diabetic mice suggests new pathogenic mechanisms for insulin-dependent diabetes mellitus. J Exp Med 1999;190:963–972.

201 Serreze DV, Chapman HD, Post CM, Johnson EA, Suarez-Pinzon WL, Rabinovitch A: Th1 to Th2 cytokine shifts in NOD mice: Sometimes an outcome, rather than the cause of diabetes resistance elicited by immunostimulation. J Immunol 2001;166:1352–1359.

202 Varadhachary AS, Perdow SN, Hu C, Ramanarayanan M, Salgami P: Differential ability of T cell subsets to undergo activation-induced cell death. Proc Natl Acad Sci USA 1997;94:5778–5783.

203 Zhang X, Brunner T, Carter L, Dutton RW, Rogers P, Bradley L, et al: Unequal death in T helper (Th)1 and Th2 effectors: Th1, but not Th2, effectors undergo rapid Fas/FasL-mediated apoptosis. J Exp Med 1997;185:1837–1849.

Dr. Edward H. Leiter, The Jackson Laboratory,
600 Main St., Bar Harbor, ME 04609 (USA)
Tel. +1 207 288 6370, Fax +1 207 288 6077, E-Mail ehl@jax.org

von Herrath MG (ed): Molecular Pathology of Type 1 Diabetes mellitus.
Curr Dir Autoimmun. Basel, Karger, 2001, vol 4, pp 68–90

The Mucosal Interface between 'Self' and 'Non-Self' Determines the Impact of Environment on Autoimmune Diabetes

Natasha R. Solly, Margo C. Honeyman, Leonard C. Harrison

The Walter & Eliza Hall Institute of Medical Research, Royal Melbourne Hospital, Parkville, Vic., Australia

Introduction: The Mucosa

The primary role of the immune system is defence against pathogens, within the context of maintaining homeostasis between 'self' and 'non-self'. The mucosal surfaces, especially of the gastrointestinal, nasorespiratory and genitourinary tracts represent critical physical and functional interfaces between internal 'self' and external 'non-self'. At these sites, the mucosal immune system plays a seminal role in maintaining the delicate balance between defence against pathogens (immunity) and accommodation of nonpathogenic resident bacteria and a host of potentially immunogenic dietary or inhaled proteins (mucosal tolerance). Given this gatekeeper function of the mucosa at the interface between 'self' and 'non-self', the role of environmental factors in predisposing to or triggering autoimmune diabetes must be considered within the context of mucosal physiology.

Internal 'self' and external 'non-self' are separated by a single layer of epithelial cells covering mucosal surfaces. In addition to being a physical barrier, mucosal epithelial cells are actively involved in mucosal immunity. Intestinal epithelial cells (IEC) constitutively express major histocompatability complex (MHC) class II [1] as well as the nonclassical MHC class I-like molecules CD1d [2] and thymus leukaemia antigen (TLA) [3], and interact directly with intraepithelial lymphocytes (IEL) via cadherin E – $\alpha E\beta7$ integrin, respectively [4]. In vitro, IEC can process and present dietary antigen to primed CD4 T cells [5]. Although they do not form a discrete, organized lymphoid tissue,

IEL are distributed between and at the basement of IEC in number equivalent to that of all T cells present in the spleen and lymph nodes [6]. IEL are the first lymphoid cells to contact external 'non-self', are constitutively cytotoxic, and have a primary role in mucosal immune responses. In mice, half the IEL express Thy-1, $\alpha\beta$ T-cell receptor (TCR) and CD8$\alpha\beta$ heterodimer; the others express $\gamma\delta$($\sim$40%) or $\alpha\beta$ TCR ($\sim$10%) and CD8$\alpha\alpha$ homodimer, and are unique in having an extrathymic ontogeny [reviewed in 7]. In humans, $\gamma\delta$ T cells constitute a lesser proportion of small intestinal IEL, although this increases in the large intestine. In addition to IEL, the mucosal immune system comprises the loosely organized lamina propria lymphocytes located directly beneath the epithelium, lymphoid nodules called Peyer's patches and mesenteric lymph nodes that interface with the systemic immune system.

Mucosal Immunity Is Required to Prevent Autoimmune Diabetes

Paradoxically, the mucosa would be a place of great danger, if it was not so dirty. By dirty we mean colonization by bacteria, which is necessary for the development of the mucosal immune system [8]. The critical role of normal mucosae in regulating autoimmunity is aptly illustrated by the effects of germ-free versus dirty environments on diabetes incidence in the autoimmune nonobese diabetic (NOD) mouse, the model of human type 1 diabetes (T1D). The incidence of spontaneous diabetes in NOD mice differs greatly in colonies around the world and appears to be inversely correlated with exposure to microbial infection [9]. The high incidence of diabetes in NOD mice housed under specific pathogen-free conditions is reduced by conventional conditions of housing and feeding [10]. Under such conventional 'dirty' conditions, bacterial colonization of the intestine is accompanied by an increase in the number of IEL, particularly CD8$\alpha\alpha$ $\alpha\beta$ T cells [11], and by maturation of mucosal immune function [12]. Without natural bacterial colonization and consequent development of mucosal immunity, animals have defective systemic immune tolerance and develop autoimmune disease.

Mechanisms of Immune Tolerance

Discrimination between 'self' and 'non-self' is an essential property of the immune system. For T cells, this is first achieved during their development in the thymus by clonal deletion of cells that recognize self-peptide presented by MHC molecules. However, this editing mechanism is imperfect. T cells with low avidity for self, or with specificity for self-antigens not expressed in the thymus, can

escape this central tolerance mechanism and mature to be potentially autoreactive. Indeed, a diverse T-cell repertoire and therefore an effective immune system
requires that 'self' and 'non-self' overlap at this level. The price paid, however,
is potential for autoimmune disease – if peripheral regulatory mechanisms fail.

(Pro)insulin is the only β-cell-specific autoantigen in T1D in humans and
in NOD mice. In the human thymus decreased expression of proinsulin mRNA
is associated with a risk allele for T1D that maps to a VNTR regulatory region
5′ of the insulin gene [13]. It is likely that reduced expression of proinsulin
epitopes in the thymus leads to failure of central tolerance of proinsulin-reactive
T cells and therefore susceptibility to autoimmune-mediated destruction of
insulin-producing β cells. Prevention of diabetes in the NOD mouse by expression of proinsulin as a transgene in antigen-presenting cells (APC) [14] lends
support to this view.

Lymphocytes that recognize islet antigens are found in some healthy individuals, as well as at higher frequency in individuals with autoimmune diabetes,
implying that their activation in the periphery is normally regulated. Peripheral
tolerance of 'self' has several explanations. First, it may be due simply to
passive 'ignorance', whereby self-antigens remain cryptic or sequestered or their
presentation by MHC molecules remains below a critical threshold for T-cell
activation. This can be overcome by tissue damage or local 'danger' resulting in
release and/or enhanced presentation of self-antigens, usually with upregulation
of co-stimulatory molecules (CD80, CD86 and CD40) on APC. Nonrecognition
of self may also be overcome by immune cross-reactivity with non-self (molecular mimicry). Second, depending on the quality or quantity of antigen presentation, responding T cells may be anergized or deleted by apoptosis. Third,
tolerance can be active and mediated by regulatory T cells whose secreted anti-
inflammatory products (e.g. IL-4, TGF-β, IL-10) or competition for antigen
presentation can antagonize autoreactive T cells.

Peripheral tolerance mechanisms are especially important because they are
potentially inducible for the prevention and treatment of autoimmune disease.
While immune response genes such as those within the MHC determine the
susceptibility to and the outcome of autoreactivity, autoimmune disease susceptibility is not simply 'deterministic'. Family and twin studies in T1D
[reviewed in 15] reveal that the genetic component contributes less than half of
the lifetime risk of the disease; in discordant monozygotic twins the probability
of the second twin developing diabetes is no greater than 36% after 20 years'
follow-up [16]. Thus, susceptibility to T1D must be shaped by environmental
factors that impinge on peripheral tolerance mechanisms.

Mucosal immunity is a significant generator of peripheral tolerance.
Feeding a soluble protein antigen suppresses subsequent systemic priming
to the antigen ('oral tolerance'); a similar phenomenon occurs after antigen

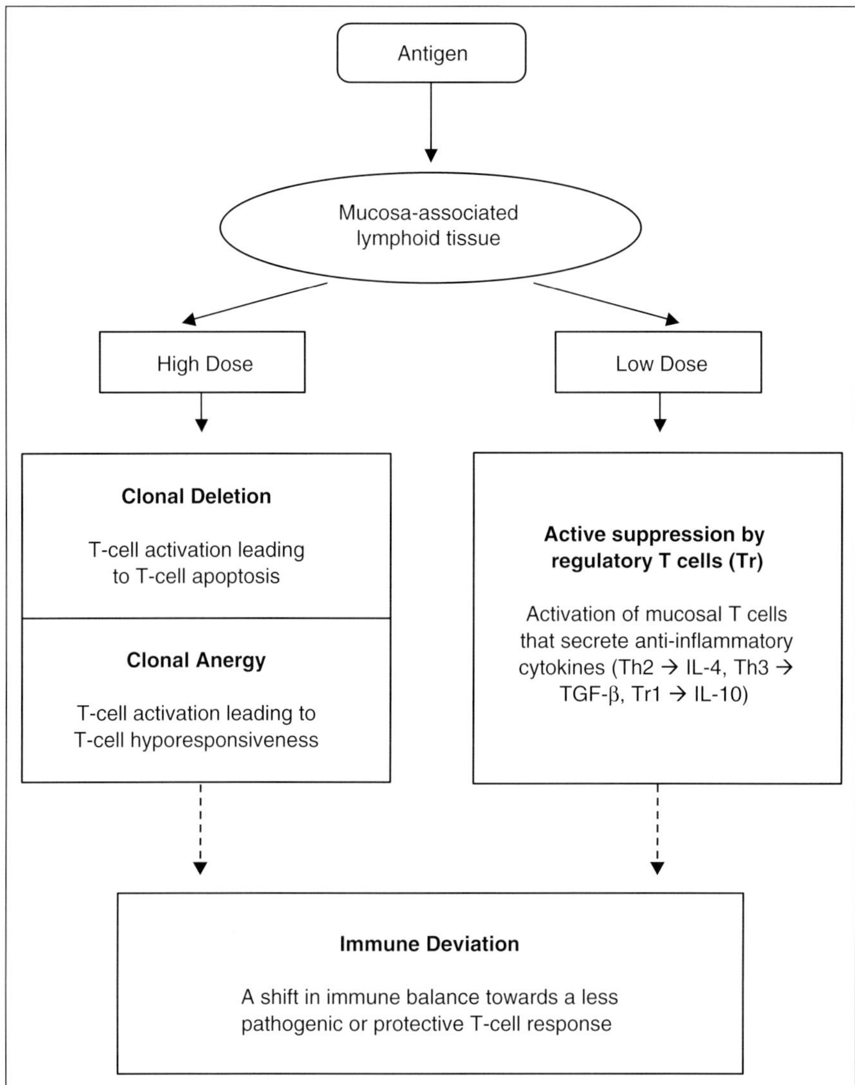

Fig. 1. Mechanisms of mucosal tolerance.

delivery to the nasorespiratory tract and other mucosae [reviewed in 17, 18]. Mucosal tolerance is attributed to several mechanisms (fig. 1) that are not mutually exclusive but may overlap, one or other predominating depending on conditions such as antigen dose, physical form and route of delivery. Cells that survive antigen-induced activation and apoptosis may be anergic and/or exhibit properties of regulatory T cells (Tr) [reviewed in 18]. Tr are triggered in an antigen-specific manner but exert antigen nonspecific 'bystander suppression' in response to recognition of specific antigen locally in the tissues or draining lymph nodes. Bystander suppression due to anti-inflammatory cytokines (IL-4, TGF-β, IL-10) can act as a brake on autoimmunity to multiple self-antigens often seen in human autoimmune disease.

Regulatory T Cells

A balance between immune pathogenic and regulatory mechanisms in autoimmune diabetes was first suggested by the effect of cyclophosphamide [19] or of irradiation of recipients of diabetogenic T cells [20] to accelerate diabetes onset in young NOD mice. A regulatory role for CD4 T cells in young NOD mice was implied by the finding that they prevented diabetes when co-transferred with spleen cells from older diabetic mice to young, irradiated mice [21, 22]. Antidiabetogenic T-cell lines and clones have been isolated from the islets of pre-diabetic NOD mice [e.g. 23]. Lymphocytic infiltration of islets (insulitis) is present for months (mice) to years (humans) before the near-total destruction of β cells leads to diabetes. This protracted preclinical stage is consistent with immunoregulatory mechanisms that hold β-cell destruction in check.

Specific subsets of regulatory CD4 T cells [24–31], CD8 γδ cells [32] induced by mucosal administration of islet autoantigen protein or peptide, as well as CD4 NKT cells [33], have been shown to prevent diabetes in NOD mice. The potential importance of the mucosa in generating Tr that protect against diabetogenic environmental agents is illustrated by an elegant study of Homann et al. [34] in which IL-4- and IL-10-secreting CD4 T cells generated by oral insulin prevented diabetes triggered by lymphocytic choriomeningitis virus (LCMV) in mice expressing an LCMV antigen transgenically in β cells.

γδ IEL – First Line of Mucosal Defence

γδ TCR are present on less than 1% of peripheral T cells, compared to ~30% and ~15% of IEL in rodents and humans, respectively. Their selective localization and relative abundance in the mucosa suggests a key role in the regulation of responses to environmental antigens. Several lines of evidence indicate that γδ IEL regulate autoimmune and other inflammatory disorders. First, γδ T cells have been demonstrated to suppress experimental autoimmune myocarditis [35], airway hyperreactivity to inhaled antigen [36], Lyme arthritis [37], contact sensitivity [38], T1D [32], uveitis [39] and orchitis [40]. Second, as discussed above, germ-free NOD mice have an accelerated onset of diabetes, which is reduced by conventional 'dirty' conditions of housing and feeding that lead to bacterial colonization of the intestine and an associated increase in numbers of IEL. Third, neonatal (3-day) thymectomy (NTX) of mice induces organ-specific autoimmune diseases such as gastritis, insulitis, thyroiditis and oophoritis/orchitis [reviewed in 41] in association with failure to develop IEL [42]. In NOD mice, NTX significantly accelerates the onset and increases the incidence of diabetes in both male and female mice [Solly and Harrison, manuscript submitted]. Fourth, TCR δ-/- mice and treatment with anti-γδ TCR antibody have been used to demonstrate that low-dose oral tolerance is dependent on CD8 γδ T cells [43, 44]. Finally, γδ-/- mice have deficient mucosal IgA synthesis [45] indicating that γδ T cells determine the nature of the mucosal B-cell response.

Holt et al. [46] were the first to show that γδ Tr could be induced by mucosal antigen. Administration of OVA to the nasorespiratory tract before systemic immunization suppressed subsequent airway hyper-responsiveness to inhaled OVA and OVA-specific IgE and IL-4 responses [47]. Notably, these effects were transferable to untreated mice by small numbers of CD8 γδ T cells. In NOD mice, aerosol delivery of insulin to the nasorespiratory mucosa delayed the onset of diabetes and induced CD8 γδ Tr that blocked adoptive transfer of diabetes [32]. Induction of these CD8 γδ Tr requires recognition of intact (but not necessarily hormonally active) insulin [48]. Although classical class I MHC molecules are not involved, how insulin is presented to these CD8 γδ T cells is unknown. Aerosol insulin treatment is followed by the appearance of γδ T cells producing IL-10 specifically in pancreatic lymph nodes [48], indicating that they function as Tr, at least in part via the anti-inflammatory effects of IL-10, analogous to Tr1 cells that suppress experimental inflammatory bowel disease [49].

In summary, several independent studies have implicated γδ T cells as mediators of cellular and humoral mucosal tolerance. Environmental influences on the generation and function of these cells may therefore have a major impact on autoimmune disease risk. Furthermore, IEL cannot be considered without reference to the IEC to which they are intimately physically and functionally related. The integrity of IEC is dependent on IEL [50–53]. Thus, if IEL development or function is impaired, the mucosal permeability barrier is breached, exposing the body to exogenous pathogens and antigens.

Impaired Mucosal Function and Development of Autoimmune Diabetes

Genetic or environmental factors that influence mucosal immune function may predispose to autoimmune disease. Under conventional but not germ-free conditions, IL-2-, IL-10- or TGF-β-deficient mice develop chronic intestinal inflammation resembling human inflammatory bowel disease [54–57]. Thus, in the absence of anti-inflammatory cytokines characteristic of mucosal immune responses [reviewed in 17], normal intestinal microflora evoke pathological responses in the underlying lamina propria. Patients with T1D may show evidence of chronic intestinal inflammation [58, reviewed in 59] indicating that abnormalities in mucosal function or intestinal permeability may contribute to the development of diabetes. Meddings et al. [60] have shown that diabetes-prone biobreeding (BB) rats have increased gastrointestinal permeability compared with diabetes-resistant BB rats. Abnormal gastrointestinal permeability occurred before the development of insulitis and was not induced by dietary diabetogens indicating that impaired exclusion of dietary and bacterial antigens is an inherent defect in these animals which could allow the environment to influence the development of disease.

Coeliac disease, associated with increased gut permeability in the acute phase [61, 62] is strongly linked with the human leukocyte antigen (HLA) DQB1*0201 (DQ2) allele that is also associated with risk for T1D. Approximately one third of T1D patients homozygous for the DQ2 risk allele have evidence of underlying coeliac disease, compared with less than 2% of patients lacking DQ2 [63]. Many investigators report enhanced immune responses to dietary proteins such as gluten [64, 65] and the cows' milk proteins bovine serum albumin (BSA) [66], β-casein [67, 68], and β-lactoglobulin [69, 70] in individuals with or at risk for T1D. However, analysis reveals that this association is with the predisposing HLA haplotype A1-B8-DR3-DQ2 and not necessarily the disease itself [71]. This haplotype is associated not only with coeliac disease and T1D, but with common variable IgA deficiency [72]. Thus, genes on this haplotype may influence the maturation or function of the mucosa, and enhanced responses to dietary antigens could reflect impaired mucosal immune function.

Candidate Diabetogenic Agents and the Mucosa

Dietary Components

Cows' Milk Proteins

Cows' milk is usually the first source of dietary xenogeneic antigens to which the human infant is exposed, at a stage when the mucosal immune system may not have fully matured. Several investigators have noted a high correlation between per capita consumption of cows' milk and the prevalence of T1D between [73, 74] and within [75] countries. However, this observation relates to milk consumption across all ages not just in infancy, and correlations at least as high are reported for coffee and sugar consumption [76]. Sources of cows' milk protein in infancy include dairy products that end up in maternal breast milk, hydrolyzed cows' milk protein in infant formulae and supplements, dairy products such as custard, cheese and yogurt, and cows' milk itself. Cows' milk contains five principal proteins: caseins (70–80%), β-lactoglobulin (10%) which is not present in human milk, α-lactalbumin (5%), γ-globulin (2%) and BSA (1%). IgG antibodies to cows' milk proteins are present in virtually all infants exposed to cows' milk [77, 78] and have even been considered physiological [78]. The significance of increased immune responses to cows' milk proteins in recently-diagnosed patients with T1D has been reviewed and debated [71, 79] and, as discussed, may be associated with impaired oral tolerance to dietary antigens associated with particular HLA haplotypes, such as A1-B8-DR3-DQ2 and with the absence of the protective DQB1*0301 allele, rather than with disease itself [68, 80].

Early introduction of cows' milk to the infant diet was first suggested as an etiological agent in T1D by Borch-Johnsen et al. [81] in 1984, who reported an inverse relationship between breast-feeding frequency/duration and T1D prevalence. This study heralded a rash of over 20 similar studies, all but three strictly retrospective. In a meta-analysis of the first 13 studies, Gerstein [82] concluded that there was only a small protective effect of breast-feeding, lack of which or exposure to cows' milk resulted in a relative risk no greater than 1.5. This was subsequently confirmed in a larger meta-analysis by Norris and Scott [83], who concluded that the apparent weak association could be explained by recall bias in retrospective studies or by disparate control groups. From an immunogenetic perspective, two of the studies analyzed are interesting. Kostraba et al. [76] and Perez-Bravo et al. [84] reported that the relative risks were higher, 11.3 and 13.1, respectively, in children with HLA susceptibility genes for T1D, who had early exposure to cows' milk or shorter periods of breast-feeding. The question of cows' milk and T1D was then addressed by prospective studies of individuals at highest genetic risk.

Norris et al. [85], in the Denver-based Diabetes Autoimmunity Study in the Young (DAISY), retrospectively analyzed infant feeding patterns during the first 6 months of age in relation to the development of islet autoantibodies, markers of T1D, up to 7 years of age. They found no significant associations. In the Australian BabyDiab Study, Couper et al. [86] prospectively analyzed infant feeding patterns and the development of islet autoimmunity in high-risk infants. They looked at the duration of exclusive and total breast-feeding as well as the times at which infant formula, dairy products or cows' milk itself were introduced. Newborns with a first-degree relative with T1D were followed for a median of 29 (9–73) months. Home diaries recorded infant feeding, but no systematic feeding advice was given. Islet cell antibodies (ICA), insulin autoantibodies (IAA), glutamic acid decarboxylase (GAD) antibodies and tyrosine phosphatase IA2 antibodies were measured 6-monthly. Cox proportional hazards survival analysis revealed no association between infant feeding and detection of a single antibody once, a single antibody repeatedly, or two or more antibodies. The same lack of association was also found in a preliminary report from the German Baby-Diab Study [87] and a prospective study of at-risk infants in Finland [88] found that early introduction of cow's milk protein was not a significant risk factor for diabetes. However, while most epidemiological studies have focused only on diet in infancy, Virtanen et al. [88] also investigated childhood diet. Their results suggest that long-term exposure to cows' milk is required to increase the risk of diabetes; the relative risk of developing diabetes associated with high consumption of cows' milk during childhood (>3 glasses per day) was 5.4 in HLA-DQB1-matched children. The result of an ongoing trial of nutritional intervention in Finland in which cows' milk is omitted from the diet of at-risk infants is awaited. At this time, we conclude that there is insufficient evidence for the induction of islet autoimmunity by cows' milk in genetically-susceptible infants.

The development of spontaneous diabetes in either NOD mice or BB rats is not dependent on exposure to cows' milk proteins [89, 90]. Despite the fact that in the BB rat and NOD mouse, synthetic amino acid and casein hydrolysate diets were associated with a lower incidence of diabetes than standard intact casein-containing diets [91, 92], the addition of 25% casein as the only protein source, or BSA or whole cows' milk protein, did not reverse this protection in the BB rat [90]. Feeding a semipurified diet containing 10% skim milk powder (a source of BSA) and 20% casein to NOD mice inhibited development of diabetes [93]. Therefore, in neither rodent model of T1D is there support for the hypothesis that exposure to cows' milk proteins is involved in the pathogenesis of diabetes.

While the evidence does not support a role for cows' milk in the pathogenesis of T1D, it is difficult to separate early exposure to bovine antigens from a lack of breast milk. Breast milk contains a host of growth factors and cytokines, mostly species-specific, many of which appear to have a role in the maturation of intestinal mucosal tissues [94, 95]. Thus, early withdrawal of breast milk could impair development of mucosa-mediated tolerance and promote immunity to dietary antigens (including islet antigens such as insulin). In addition, a lack of passively-transferred immunity via breast milk (including lactadherin, IgA, IgG, IgM antibodies, cytokines such as TGF-β and lymphocytes) may predispose the infant to potentially diabetogenic enteric infections (see below). Shorter duration of breast-feeding may be a surrogate marker of the time of introduction and amount of weaning foods (which may contain other diabetogens).

Increased immunity to cows' milk proteins may occur in certain individuals predisposed to T1D, but is likely to reflect impaired function of mucosa-associated lymphoid tissue associated with specific genotypes [71]. It follows, therefore that cows' milk is not unique, but simply the first dietary antigen encountered, and that predisposed individuals would also exhibit increased immunity to substitutes such as goat and soy milk.

Wheat Proteins

The incidence of T1D is highest in Scandinavia and Finland and lowest in oriental countries such as Japan [96], where dietary wheat flour is replaced by rice. Dietary plant components dramatically affect the incidence of diabetes in the BB rat and NOD mouse. Proteins of plants, specifically those from wheat and soya bean, appear to be the major dietary diabetogens in the BB rat and exert an effect even when animals are first exposed after weaning [92, 97, 98]. Diabetes-prone BB rats are protected from diabetes when fed alternate amino acid sources such as casein, hydrolyzed casein, hydrolyzed soy protein or fish meal [98]. Li et al. [91] showed that plant-derived diabetogens induce over-expression of MHC class I molecules on murine β cells as early as 24 days of age, about 10 days after pups begin to nibble on solid food. No increase in MHC class I expression was seen in diabetes-resistant BB rats fed diabetogenic diets. This may relate to the increased gastrointestinal permeability of diabetes-prone compared with diabetes-resistant BB rats [60]. Although diabetogens may enhance β-cell antigenicity very early on, this does not appear to be sufficient to induce diabetes because long-term exposure to diabetogenic diets are required. Studies in the BB rat by Scott et al. [99], in which a diabetogenic diet was delayed until either 50 or 100 days or switched to a nondiabetogenic diet at 50 days of age, have shown that exposure to food diabetogens at age 50–100 days, corresponding to the period of early puberty to late adolescence, is critical for the dietary modulation of diabetes development.

Gluten, a major protein in wheat flour, and the environmental agent that induces coeliac disease, has been implicated in the development of diabetes. When added to a basic semisynthetic diet at weaning, gluten increased the incidence of diabetes from 15 to 35% in BB rats [100]. The early introduction of a gluten-free diet to NOD mice significantly delayed the onset of diabetes and decreased the incidence of diabetes from 64 to 15% [101]. While there is an increased prevalence of antigliadin antibodies and coeliac disease in patients with T1D [63, 102], there is little evidence that gluten is diabetogenic in humans. Weak peripheral blood T-cell responses to gluten are detectable in only a low percentage of recently-diagnosed T1D patients [64] and a prospective trial in patients indicated no effect of a gluten-free diet on the control of diabetes [103].

NADH ubiquinone reductase in both wheat and soya beans has an identical sequence [104] to a dominant T-cell epitope in the islet antigen, tyrosine phosphatase IA-2 [104, 105], raising the possibility of molecular mimicry. Antibodies from recently-diagnosed T1D patients are able to recognize peptides with similarities to ubiquinone reductases [J. Davies, pers. commun.] lending support to this concept.

Insulin

Immunoreactive insulin is present in human breast milk at concentrations of up to 5 ng/ml [Harrison, unpubl. data]. It is conceivable that tolerance could be generated to insulin (in breast milk) as part of the normal developmental process and that variation in the level of insulin produced in maternal breast milk may affect induction of such tolerance. Also, if maturation of mucosa-associated lymphoid tissue was impaired or delayed, this could lead not only to failure to develop tolerance but to active immunization. In neonatal mice, the ontogeny of mucosal (oral) tolerance is strain-dependent with a clear temporal profile. Intragastric administration of antigen before the first 7–10 days of life does not generally elicit oral tolerance and may in fact prime for systemic immunity [106, 107]. It would be of interest to evaluate breast milk insulin levels and parameters of mucosal function in NOD mice compared to nondiabetes-prone genetically-similar NOR mice. Vaarala et al. [108, 109] proposed that bovine insulin in cows' milk could generate cross-reactive immunity (to human insulin) and demonstrated that IgG antibodies to bovine insulin cross-reacted with human insulin. Although bovine and human insulin differ by only three amino acids, bovine insulin is known to be immunogenic in humans [110].

Viruses

Viruses could theoretically initiate autoimmunity to islets in multiple ways, by direct infection of the target tissue or by a range of indirect means [111–115]. Enteroviruses (Coxsackie, Polio and Echoviruses, all single-stranded RNA) and more recently rotaviruses (double-stranded RNA), that infect intestinal mucosa, have been associated with T1D. However, direct evidence is sparse due to the difficulty of isolating RNA, particularly double-stranded RNA, from the pancreas. Signs of persisting infection are the presence of interferon-α (IFN-α) in the pancreas [116], or less directly, in the blood of newly-diagnosed children [117], and increased levels of 5′ oligoadenylate synthase, a marker of IFN-αβ, in blood mononuclear cells of individuals with T1D [118, 119]. Single, recurrent or chronic infection of the pancreas or islet cells could either be directly cytolytic, or could induce expression of normally sequestered or nonexpressed self-antigens and upregulate MHC and co-stimulator molecules, initiating bystander inflammatory responses. Enteric viruses might also be transported to the pancreas by infected mucosal lymphocytes. Enteric viruses localized to the gut could exert indirect effects via molecular mimicry, production of super-antigens or polyclonal B-cell activators, and alteration of mucosal immune function or permeability.

Enteroviruses

Enteroviruses have a high rate of mutation and recombination, e.g. $>10^4$ variants of coxsackie B4, which accounts for multiple infections despite the development of cross-reactive antibody responses, the persistence of the viruses in nature, and different clinical syndromes associated with infection [120]. Epidemics occur in summer, but there is rapid inactivation of enteroviruses by heat, drying and ultraviolet light. Virus survival in the environment is thus greatest in cold areas of the world where there is a high incidence of T1D compared to warmer areas [121]. Furthermore, in Finland, Norway and Poland, the diagnosis of T1D does not have the marked summer trough in incidence seen in more southerly countries such as France, the UK and Mediterranean countries [122].

Coxsackie viruses enter the intestine via the mouth and replicate in the cervical lymph nodes and Peyer's patches of the pharynx and gut, causing lymphoid hyperplasia and inflammation [120]. Following a 1- to 2-week incubation period in lymphoid tissue, the virus may spread in the blood to organs such as spinal cord and brain, meninges, myocardium and skin, and can be detected in the faeces for several weeks [120]. Infections occur early in life. Coxsackie B viruses may be acquired transplacentally, but more commonly infection in a neonate occurs with contact in the newborn nursery [120]. Twelve percent of infants acquire nonpolio enteroviruses in the first month of life [123] with isolation rates for all enteroviruses highest among infants aged 1–2 months [124]. Infections at this age are, however, largely asymptomatic [123] due to neutralizing IgG transplacental antibodies and/or IgA antibodies in breast milk [123]. Antibody epitopes are located on the VP1, 2 and 3 structural proteins, while T-cell epitopes are within the nonstructural proteins, which include the p2C protein [125]. In coxsackie B3, 4, 5 and A9, p2C contains a sequence (amino acids (aa) 32–47) with strong similarity to aa 250–265 of the T1D autoantigen, GAD65.

Coxsackie B viruses have been associated with both the onset of clinical diabetes and the initiation of islet autoimmunity (see chapter by Manns), and seroepidemiologically with rises in islet autoantibodies [126], but this evidence is mainly circumstantial. Molecular mimicry has been suggested between p2C and GAD65 [127] where virus p2C-reactive T cells could traffic from the gut and recognize the similar GAD peptide in the pancreas or pancreatic lymph node, initiating islet autoimmunity leading to T1D. Although the similar sequences in p2C and GAD bind to the diabetes susceptibility HLA-DR3 [128] and are thus both potentially T-cell epitopes, there is little evidence for these sequences being naturally processed and presented T-cell epitopes [71, 129]. On the other hand, systemic and organ-specific infection by enteroviruses is seen after inoculation of mice with coxsackie B4 virus; infection of the pancreas leads to T1D [130]. Enteroviral RNA has been found in the blood of up to a half of children with newly-diagnosed diabetes [117, 131, 132] and in a quarter of those who subsequently developed diabetes [133, 134]. On the other hand, coxsackie B4 viral RNA was detected in the pancreata of children with recently-diagnosed T1D [135], but this was not subsequently confirmed [136]. Coxsackie B infections have also been shown not to be associated with diabetes in children [137] and to generate only transient immune responses to GAD that do not lead to T1D [138]. In NOD mice, while coxsackie B4 infection accelerates the onset of T1D, it does not initiate islet autoimmunity and depends on a pre-existing critical mass of autoreactive T cells around the islets [139]. This study also suggested that infection with coxsackie B virus prior to islet auto-immunity could block the development of diabetes. Intriguingly, Finns have a sharply rising incidence of diabetes in the very young but very little enteroviral infection, whereas their genetically-similar Estonian neighbors have a strikingly lower incidence of diabetes and high levels of enteroviral infections in the very young [140]. If anything, this implies that enterovirus infection, in humans at this age, is protective. Could Finns and Estonians be the human counterparts of germ-free and dirty NOD mice, respectively? Bjorksten et al. [141] have shown that gut flora differ quantitatively and qualitatively in Scandinavians and Estonians, and have suggested that the increase in number and diversity of gut microbial flora drives maturation of the immune system. Thus the immunological effects of a coxsackie B virus infection could be modified by the age at which the infection occurs in the gut and by the flora populating the gut at the time. The nature of gut flora also reflects whether or not the child is being breast-fed [142].

Rotavirus

Rotavirus is a genus of double-stranded RNA viruses of the Family Reoviridae, which includes reovirus. It has been extensively studied as the single most important cause of diarrhoea in infants and young children world-wide and also as a model for enteric viral infections. Rotavirus is ubiquitous, most children being regularly infected predominantly in winter epidemics [143]. The seasonality of diabetes diagnosis with the summer trough is again worthy of note in this regard [122]. Clinical symptoms are rare after the age of 5, by which time sufficient IgA antibodies have developed to neutralize the virus in the gut. Rotavirus is transmitted from faeces to the mouth, but is not infectious until activated by trypsin, a product of the exocrine pancreas. Thus the virus is activated in the duodenum, just outside the pancreatic and bile ducts, after which it infects IEC of the small intestine at the mature villus tip [144]. Rotavirus double-stranded RNA induces the IEC to secrete the chemokines IL-8, growth-related peptide α and RANTES [145], which are potent attractors of $\alpha4\beta7$ integrin[+] B cells and CD4 and CD8 T cells [146]. The major site of action of rotavirus is the gut, with little evidence of systemic infection. However, there are reports of pancreatitis [147, 148] and biliary atresia [149] associated with rotavirus infection, possibly due to infection of macrophages, B cells and dendritic cells, in which rotavirus RNA has been found [150]. Furthermore, as integrins $\alpha2$ and $\alpha4$ can mediate viral attachment and entry into cells [151], cells in or close to the gut mucosa bearing these integrins, e.g. IEL ($\alpha4$) or pancreatic exocrine cells ($\alpha2$), have the potential to carry infection beyond the IEC or provide additional infective sites.

Rotavirus was associated with T1D by the finding of strong peptide sequence similarities between VP7, the most prevalent rotavirus coat protein, and T-cell epitopes in both GAD65 and tyrosine phosphatase IA-2 [104]. Analysis of sera from children at risk of T1D followed from birth revealed that islet antibodies first appeared or increased temporally with rotavirus infection [152, 153]. However, whether the mechanism is molecular mimicry, direct infection of the pancreas, or both, is not yet determined. The implications for rotavirus vaccines differ depending on the mechanism: if only mimicry is involved, live vaccines could have the potential to trigger islet autoimmunity via the gut, but if the virus enters and infects the pancreas without mimicry, then vaccines could be protective against T1D.

Rotavirus infection may also influence other candidate environmental agents, in particular dietary components, by increasing intestinal permeability [154–156]. Breast milk, which contains both neutralizing IgA and IgG antibodies and lactadherin that binds and inactivates rotavirus, affords symptomatic protection [157]. In the gut of diabetes-susceptible children, low IgA due to cessation of breast-feeding and/or IgA deficiency linked to the diabetes-predisposing HLA-A1-B8-DR3-DQ2 haplotype, could promote rotaviral infectivity, gut permeability and mucosal (non-IgA) immunity to dietary proteins.

In the last decade, hospitals have instituted changes to prevent the previously high level of rotavirus infection in neonates, in whom high-titer anti-rotavirus IgG transplacentally transmitted to the serum is not protective [158].

However, infection is now rife in daycare centers and pre-schools. At the same time, the incidence of T1D in Western Europe and Australia has increased sharply in under 5-year-olds. Daycare attendance, before the age of 3 (with increased exposure to enteric viruses) has been cited as predisposing to T1D [159], but before the age of 1 has been shown to be protective against T1D [160]. Exposure to viruses may elicit different mucosal immune responses depending on age. It is critical to determine the age-related patterns of exposure of susceptible children to enteric viruses, and the consequences.

Concluding Remarks

The mucosa is the gatekeeper between 'self' and infectious as well as non-infectious 'non-self', the interface between the genetic and environmental components of T1D susceptibility. The immune and nonimmune barrier functions of the mucosa actively and passively prevent the development of autoimmune disease. It is highly likely that by developing tools to investigate the very early development of mucosal immunity and function in humans we will gain a far deeper insight into how the prenatal thymus-selected immune repertoire is conditioned to maintain homeostasis of self-reactivity and thereby avert diseases such as T1D.

References

1   Gorvel JP, Sarles J, Maroux S, Olive D, Mawas C: Cellular localization of class I (HLA-A, B, C) and class II (HLA-DR and DQ) MHC antigens on the epithelial cells of normal human jejunum. Biol Cell 1984;52:249–252.
2   Blumberg RS, Terhorst C, Bleicher P, McDermott FV, Allan CH, Landau SB, Trier JS, Balk SP: Expression of a nonpolymorphic MHC class I-like molecule, CD1D, by human intestinal epithelial cells. J Immunol 1991;147:2518–2524.
3   Wu M, van Kaer L, Itohara S, Tonegawa S: Highly restricted expression of the thymus leukemia antigens on intestinal epithelial cells. J Exp Med 1991;174:213–218.
4   Cepek KL, Shaw SK, Parker CM, Russell GJ, Morrow JS, Rimm DL, Brenner MB: Adhesion between epithelial cells and T lymphocytes mediated by E-cadherin and the alpha E beta 7 integrin. Nature 1994;372:190–193.
5   Kaiserlian D: Epithelial cells in antigen sampling and presentation in mucosal tissues. Curr Top Microbiol Immunol 1999;236:55–78.
6   Rocha B, Vassalli P, Guy-Grand D: The V beta repertoire of mouse gut homodimeric alpha CD8+ intraepithelial T cell receptor alpha/beta + lymphocytes reveals a major extrathymic pathway of T cell differentiation. J Exp Med 1991;173:483–486.
7   Lefrancois L, Puddington L: Extrathymic T cell development: Virtual reality? Immunol Today 1995;16:16–21.
8   Cebra JJ, Periwal SB, Lee G, Lee F, Shroff KE: Development and maintenance of the gut-associated lymphoid tissue (GALT): The roles of enteric bacteria and viruses. Dev Immunol 1998;6:13–18.

9 Pozzilli P, Signore A, Williams AJ, Beales PE: NOD mouse colonies around the world – Recent facts and figures. Immunol Today 1993;14:193–196.

10 Suzuki T: Diabetogenic effects of lymphocyte transfusion on the NOD or NOD nude mouse; in Rygaard J (ed): Immune Deficient Animals in Biomedical Research. Basel, Karger, 1987, pp 112–116.

11 Imaoka A, Matsumoto S, Setoyama H, Okada Y, Umesaki Y: Proliferative recruitment of intestinal intraepithelial lymphocytes after microbial colonization of germ-free mice. Eur J Immunol 1996; 26:945–948.

12 Kawaguchi-Miyashita M, Shimizu K, Nanno M, Shimada S, Watanabe T, Koga Y, Matsuoka Y, Ishikawa H, Hashimoto K, Ohwaki M: Development and cytolytic function of intestinal intra-epithelial T lymphocytes in antigen-minimized mice. Immunology 1996;89:268–273.

13 Pugliese A, Zeller M, Fernandez A Jr, Zalcberg LJ, Bartlett RJ, Ricordi C, Pietropaolo M, Eisenbarth GS, Bennett ST, Patel DD: The insulin gene is transcribed in the human thymus and transcription levels correlated with allelic variation at the INS VNTR-IDDM2 susceptibility locus for type 1 diabetes. Nat Genet 1997;15:293–297.

14 French MB, Allison J, Cram DS, Thomas HE, Dempsey-Collier M, Silva A, Georgiou HM, Kay TW, Harrison LC, Lew AM: Transgenic expression of mouse proinsulin II prevents diabetes in nonobese diabetic mice [published erratum appears in Diabetes 1997;46:924]. Diabetes 1997; 46:34–39.

15 Harrison LC, Colman PG, Honeyman MC, Kay TWH: Type 1 diabetes – From pathogenesis to prevention; in Turtle JR, Kaneko T, Osata S (eds): Diabetes in the New Millennium. Sydney, Pot Still Press, 1999, pp 85–100.

16 Redondo MJ, Rewers M, Yu L, Garg S, Pilcher CC, Elliott RB, Eisenbarth GS: Genetic determination of islet cell autoimmunity in monozygotic twin, dizygotic twin, and non-twin siblings of patients with type 1 diabetes: Prospective twin study. BMJ 1999;318:698–702.

17 Faria AM, Weiner HL: Oral tolerance: Mechanisms and therapeutic applications. Adv Immunol 1999;73:153–264.

18 Harrison LC, Hafler DA: Antigen-specific therapy for autoimmune disease. Current Opin Immunol 2000;12:704–711.

19 Harada M, Makino S: Promotion of spontaneous diabetes in non-obese diabetes-prone mice by cyclophosphamide. Diabetologia 1984;27:604–606.

20 Bendelac A, Carnaud C, Boitard C, Bach JF: Syngeneic transfer of autoimmune diabetes from diabetic NOD mice to healthy neonates. Requirement for both L3T4+ and Lyt-2+ T cells. J Exp Med 1987;166:823–832.

21 Boitard C, Yasunami R, Dardenne M, Bach JF: T cell-mediated inhibition of the transfer of autoimmune diabetes in NOD mice. J Exp Med 1989;169:1669–1680.

22 Hutchings PR, Cooke A: The transfer of autoimmune diabetes in NOD mice can be inhibited or accelerated by distinct cell populations present in normal splenocytes taken from young males. J Autoimmun 1990;3:175–185.

23 Chosich N, Harrison LC: Suppression of diabetes mellitus in the non-obese diabetic (NOD) mouse by an autoreactive (anti-I-Ag7) islet-derived CD4+ T-cell line. Diabetologia 1993;36:716–721.

24 Bergerot I, Fabien N, Maguer V, Thivolet C: Oral administration of human insulin to NOD mice generates CD4+ T cells that suppress adoptive transfer of diabetes. J Autoimmun 1994;7:655–663.

25 Daniel D, Wegmann DR: Protection of nonobese diabetic mice from diabetes by intranasal or subcutaneous administration of insulin peptide B-(9-23). Proc Natl Acad Sci USA 1996;93:956–960.

26 Tian J, Clare-Salzler M, Herschenfeld A, Middleton B, Newman D, Mueller R, Arita S, Evans C, Atkinson MA, Mullen Y, Sarvetnick N, Tobin AJ, Lehmann PV, Kaufman DL: Modulating autoimmune responses to GAD inhibits disease progression and prolongs islet graft survival in diabetes-prone mice. Nat Med 1996;2:1348–1353.

27 Von Herrath MG, Dyrberg T, Oldstone MB: Oral insulin treatment suppresses virus-induced antigen-specific destruction of beta cells and prevents autoimmune diabetes in transgenic mice. J Clin Invest 1996;98:1324–1331.

28 Han HS, Jun HS, Utsugi T, Yoon JW: Molecular role of TGF-beta, secreted from a new type of CD4+ suppressor T cell, NY4.2, in the prevention of autoimmune IDDM in NOD mice. J Autoimmun 1997;10:299–307.

29 Ploix C, Bergerot I, Fabien N, Perche S, Moulin V, Thivolet C: Protection against autoimmune diabetes with oral insulin is associated with the presence of IL-4 type 2 T-cells in the pancreas and pancreatic lymph nodes. Diabetes 1998;47:39–44.

30 Maron R, Melican NS, Weiner HL: Regulatory Th2-type T cell lines against insulin and GAD peptides derived from orally- and nasally-treated NOD mice suppress diabetes. J Autoimmun 1999; 12:251–258.

31 Lepault F, Gagnerault MC: Characterization of peripheral regulatory CD4+ T cells that prevent diabetes onset in nonobese diabetic mice. J Immunol 2000;164:240–247.

32 Harrison LC, Dempsey-Collier M, Kramer DR, Takahashi K: Aerosol insulin induces regulatory CD8 gamma delta T cells that prevent murine insulin-dependent diabetes. J Exp Med 1996; 184:2167–2174.

33 Baxter AG, Kinder SJ, Hammond KJ, Scollay R, Godfrey DI: Association between abTCR$^+$CD4$^-$CD8$^-$ T-cell deficiency and IDDM in NOD/Lt mice. Diabetes 1997;46:572–582.

34 Homann D, Holz A, Bot A, Coon B, Wolfe T, Petersen J, Dyrberg TP, Grusby MJ, von Herrath MG: Autoreactive CD4+ T cells protect from autoimmune diabetes via bystander suppression using the IL-4/Stat6 pathway. Immunity 1999;11:463–472.

35 Cardillo F, Falcao RP, Rossi MA, Mengel J: An age-related gamma delta T cell suppressor activity correlates with the outcome of autoimmunity in experimental *Trypanosoma cruzi* infection. Eur J Immunol 1993;23:2597–2605.

36 McMenamin C, Holt PG: The natural immune response to inhaled soluble protein antigens involves major histocompatibility complex (MHC) class I-restricted CD8+ T cell-mediated but MHC class II-restricted CD4+ T cell-dependent immune deviation resulting in selective suppression of immunoglobulin E production. J Exp Med 1993;178:889–899.

37 Vincent MS, Roessner K, Lynch D, Wilson D, Cooper SM, Tschopp J, Sigal LH, Budd RC: Apoptosis of Fas[high] CD4$^+$ synovial T cells by Borrelia-reactive Fas-ligand[high] gamma delta T cells in Lyme arthritis. J Exp Med 1996;184:2109–2117.

38 Szczepanik M, Anderson LR, Ushio H, Ptak W, Owen MJ, Hayday AC, Askenase PW: Gamma delta T cells from tolerized alpha beta T cell receptor (TCR)-deficient mice inhibit contact sensitivity-effector T cells in vivo, and their interferon-gamma production in vitro. J Exp Med 1996;184:2129–2139.

39 Wildner G, Hunig T, Thurau SR: Orally induced, peptide-specific gamma/delta TCR+ cells suppress experimental autoimmune uveitis. Eur J Immunol 1996;26:2140–2148.

40 Mukasa A, Yoshida H, Kobayashi N, Matsuzaki G, Nomoto K: Gamma delta T cells in infection-induced and autoimmune-induced testicular inflammation. Immunology 1998;95:395–401.

41 Bonomo A, Kehn PJ, Shevach EM: Post-thymectomy autoimmunity: Abnormal T-cell homeostasis. Immunol Today 1995;16:61–67.

42 Lin T, Matsuzaki G, Kenai H, Nakamura T, Nomoto K: Thymus influences the development of extrathymically derived intestinal intraepithelial lymphocytes. Eur J Immunol 1993;23:1968–1974.

43 Mengel J, Cardillo F, Aroeira LS, Williams O, Russo M, Vaz NM: Anti-gamma delta T cell antibody blocks the induction and maintenance of oral tolerance to ovalbumin in mice. Immunol Lett 1995;48:97–102.

44 Ke Y, Pearce K, Lake JP, Ziegler HK, Kapp JA: Gamma delta T lymphocytes regulate the induction and maintenance of oral tolerance. J Immunol 1997;158:3610–3618.

45 Fujihashi K, McGhee JR, Kweon MN, Cooper MD, Tonegawa S, Takahashi I, Hiroi T, Mestecky J, Kiyono H: Gamma/delta T cell-deficient mice have impaired mucosal immunoglobulin A responses. J Exp Med 1996;183:1929–1935.

46 Holt PG, Britten D, Sedgwick JD: Suppression of IgE responses by antigen inhalation: Studies on the role of genetic and environmental factors. Immunology 1987;60:97–102.

47 McMenamin C, Pimm C, McKersey M, Holt PG: Regulation of IgE responses to inhaled antigen in mice by antigen-specific gamma delta T cells. Science 1994;265:1869–1871.

48 Hanninen A, Harrison LC: Gamma delta T cells as mediators of mucosal tolerance: The autoimmune diabetes model. Immunol Rev 2000;173:109–119.

49 Groux H, O'Garra A, Bigler M, Rouleau M, Antonenko S, de Vries JE, Roncarolo MG: A CD4+ T-cell subset inhibits antigen-specific T-cell responses and prevents colitis. Nature 1997;389:737–742.

50 Boismenu R, Havran WL: Modulation of epithelial cell growth by intraepithelial gd T cells. Science 1994;266:1253–1255.

51 Komano H, Fujiura Y, Kawaguchi M, Matsumoto S, Hashimoto Y, Obana S, Mombaerts P, Tonegawa S, Yamamoto H, Itohara S, et al: Homeostatic regulation of intestinal epithelia by intraepithelial gamma delta T cells. Proc Natl Acad Sci USA 1995;92:6147–6151.

52 Taylor CT, Murphy A, Kelleher D, Baird AW: Changes in barrier function of a model intestinal epithelium by intraepithelial lymphocytes require new protein synthesis by epithelial cells. Gut 1997;40:634–640.

53 Matsumoto S, Nanno M, Watanabe N, Miyashita M, Amasaki H, Suzuki K, Umesaki Y: Physiological roles of gamma delta T-cell receptor intraepithelial lymphocytes in cytoproliferation and differentiation of mouse intestinal epithelial cells. Immunology 1999;97:18–25.

54 Shull MM, Ormsby I, Kier AB, Pawlowski S, Diebold RJ, Yin M, Allen R, Sidman C, Proetzel G, Calvin D, et al: Targeted disruption of the mouse transforming growth factor-beta-1 gene results in multifocal inflammatory disease. Nature 1992;359:693–699.

55 Kuhn R, Lohler J, Rennick D, Rajewsky K, Muller W: Interleukin-10-deficient mice develop chronic enterocolitis. Cell 1993;75:263–274.

56 Sadlack B, Merz H, Schorle H, Schimpl A, Feller AC, Horak I: Ulcerative colitis-like disease in mice with a disrupted interleukin-2 gene. Cell 1993;75:253–261.

57 Berg DJ, Lynch NA, Lynch RG, Lauricella DM: Rapid development of severe hyperplastic gastritis with gastric epithelial dedifferentiation in *Helicobacter felis*-infected IL-10(-/-) mice. Am J Pathol 1998;152:1377–1386.

58 Savilahti E, Ormala T, Saukkonen T, Sandini-Pohjavuori U, Kantele JM, Arato A, Ilonen J, Akerblom HK: Jejuna of patients with insulin-dependent diabetes mellitus (IDDM) show signs of immune activation. Clin Exp Immunol 1999;116:70–77.

59 Vaarala O: Gut and the induction of immune tolerance in type 1 diabetes. Diabetes Metab Res Rev 1999;15:353–361.

60 Meddings JB, Jarand J, Urbanski SJ, Hardin J, Gall DG: Increased gastrointestinal permeability is an early lesion in the spontaneously diabetic BB rat. Am J Physiol 1999;276:G951–G957.

61 Vogelsang H, Schwarzenhofer M, Oberhuber G: Changes in gastrointestinal permeability in celiac disease. Dig Dis 1998;16:333–336.

62 Smecuol E, Bai JC, Vazquez H, Kogan Z, Cabanne A, Niveloni S, Pedreira S, Boerr L, Maurino E, Meddings JB: Gastrointestinal permeability in celiac disease. Gastroenterology 1997;112:1129–1136.

63 Bao F, Yu L, Babu S, Wang T, Hoffenberg EJ, Rewers M, Eisenbarth GS: One third of HLA DQ2 homozygous patients with type 1 diabetes express celiac disease-associated transglutaminase autoantibodies. J Autoimmun 1999;13:143–148.

64 Klemetti P, Savilahti E, Ilonen J, Akerblom HK, Vaarala O: T-cell reactivity to wheat gluten in patients with insulin-dependent diabetes mellitus. Scand J Immunol 1998;47:48–53.

65 Catassi C, Guerrieri A, Bartolotta E, Coppa GV, Giorgi PL: Antigliadin antibodies at onset of diabetes in children. Lancet 1987;ii:158.

66 Karjalainen J, Martin JM, Knip M, Ilonen J, Robinson BH, Savilahti E, Akerblom HK, Dosch HM: A bovine albumin peptide as a possible trigger of insulin-dependent diabetes mellitus [published erratum appears in N Engl J Med 1992;327:1252]. N Engl J Med 1992;327:302–307.

67 Cavallo MG, Fava D, Monetini L, Barone F, Pozzilli P: Cell-mediated immune response to beta casein in recent-onset insulin-dependent diabetes: Implications for disease pathogenesis [published erratum appears in Lancet 1996;348:1392]. Lancet 1996;348:926–928.

68 Ellis TM, Ottendorfer E, Jodoin E, Salisbury PJ, She JX, Schatz DA, Atkinson MA: Cellular immune responses to beta casein: Elevated in but not specific for individuals with type I diabetes mellitus. Diabetologia 1998;41:731–735.

69 Vaarala O, Klemetti P, Savilahti E, Reijonen H, Ilonen J, Akerblom HK: Cellular immune response to cow's milk beta-lactoglobulin in patients with newly diagnosed IDDM. Diabetes 1996;45:178–182.

70 Savilahti E, Saukkonen TT, Virtala ET, Tuomilehto J, Akerblom HK: Increased levels of cow's milk and β-lactoglobulin antibodies in young children with newly diagnosed IDDM. The Childhood Diabetes in Finland Study Group. Diabetes Care 1993;16:984–989.

71 Harrison LC, Honeyman MC: Cow's milk and type 1 diabetes: The real debate is about mucosal immune function. Diabetes 1999;48:1501–1507.

72 Ambrus M, Hernadi E, Bajtai G: Prevalence of HLA-A1 and HLA-B8 antigens in selective IgA deficiency. Clin Immunol Immunopathol 1977;7:311–314.

73 Scott FW: Cow milk and insulin-dependent diabetes mellitus: Is there a relationship? Am J Clin Nutr 1990;51:489–491.

74 Dahl-Jorgensen K, Joner G, Hanssen KF: Relationship between cows' milk consumption and incidence of IDDM in childhood. Diabetes Care 1991;14:1081–1083.

75 Fava D, Leslie RD, Pozzilli P: Relationship between dairy product consumption and incidence of IDDM in childhood in Italy. Diabetes Care 1994;17:1488–1490.

76 Kostraba JN, Cruickshanks KJ, Lawler-Heavner J, Jobim LF, Rewers MJ, Gay EC, Chase HP, Klingensmith G, Hamman RF: Early exposure to cow's milk and solid foods in infancy, genetic predisposition, and risk of IDDM. Diabetes 1993;42:288–295.

77 Ferguson A: Immunogenicity of cows' milk in man. Influence of age and of disease on serum antibodies to five cows' milk proteins. Ric Clin Lab 1977;7:211–219.

78 Keller KM, Burgin-Wolff A, Lippold R, Wirth S, Lentze MJ: The diagnostic significance of IgG cow's milk protein antibodies re-evaluated. Eur J Pediatr 1996;155:331–337.

79 Kolb H, Pozzilli P: Cow's milk and type I diabetes: The gut immune system deserves attention. Immunol Today 1999;20:108–110.

80 Saukkonen T, Virtanen SM, Karppinen M, Reijonen H, Ilonen J, Rasanen L, Akerblom HK, Savilahti E: Significance of cow's milk protein antibodies as risk factor for childhood IDDM: Interactions with dietary cow's milk intake and HLA-DQB1 genotype. Childhood Diabetes in Finland Study Group. Diabetologia 1998;41:72–78.

81 Borch-Johnsen K, Joner G, Mandrup-Poulsen T, Christy M, Zachau-Christiansen B, Kastrup K, Nerup J: Relation between breast-feeding and incidence rates of insulin-dependent diabetes mellitus. A hypothesis. Lancet 1984;ii:1083–1086.

82 Gerstein H: Cow's milk exposure and type 1 diabetes mellitus. Diabetes Care 1994;17:13–19.

83 Norris JM, Scott FW: A meta-analysis of infant diet and insulin-dependent diabetes mellitus: Do biases play a role? Epidemiology 1996;7:87–92.

84 Perez-Bravo F, Carrasco E, Gutierrez-Lopez MD, Martinez MT, Lopez G, de los Rios MG: Genetic predisposition and environmental factors leading to the development of insulin-dependent diabetes mellitus in Chilean children. J Mol Med 1996;74:105–109.

85 Norris JM, Beaty B, Klingensmith G, Yu L, Hoffman M, Chase HP, Erlich HA, Hamman RF, Eisenbarth GS, Rewers M: Lack of association between early exposure to cow's milk protein and β-cell autoimmunity. Diabetes Autoimmunity Study in the Young (DAISY). JAMA 1996;276: 609–614.

86 Couper JJ, Steele C, Beresford S, Powell T, McCaul K, Pollard A, Gellert S, Tait B, Harrison LC, Colman PG: Lack of association between duration of breast-feeding or introduction of cow's milk and development of islet autoimmunity. Diabetes 1999;48:2145–2149.

87 Hummel M, Schenker M, Ziegler AB, Group TB-DS: Appearance of diabetes-associated antibodies in offspring of parents with type 1 diabetes is independent from environmental factors (abstract). Diabetologia 1998;41(supp 1):A91.

88 Virtanen SM, Laara E, Hypponen E, Reijonen H, Rasanen L, Aro A, Knip M, Ilonen J, Akerblom HK: Cow's milk consumption, HLA-DQB1 genotype, and type 1 diabetes: A nested case-control study of siblings of children with diabetes. Childhood diabetes in Finland study group. Diabetes 2000;49:912–917.

89 Paxson JA, Weber JG, Kulczycki A Jr: Cow's milk-free diet does not prevent diabetes in NOD mice [published erratum appears in Diabetes 1998;47:144]. Diabetes 1997;46:1711–1717.

90 Malkani S, Nompleggi D, Hansen JW, Greiner DL, Mordes JP, Rossini AA: Dietary cow's milk protein does not alter the frequency of diabetes in the BB rat. Diabetes 1997;46:1133–1140.

91 Li XB, Scott FW, Park YH, Yoon JW: Low incidence of autoimmune type I diabetes in BB rats fed a hydrolysed casein-based diet associated with early inhibition of non-macrophage-dependent hyperexpression of MHC class I molecules on β cells. Diabetologia 1995;38:1138–1147.

92 Hoorfar J, Buschard K, Dagnaes-Hansen F: Prophylactic nutritional modification of the incidence of diabetes in autoimmune non-obese diabetic (NOD) mice. Br J Nutr 1993;69:597–607.

93 Coleman DL, Kuzava JE, Leiter EH: Effect of diet on incidence of diabetes in nonobese diabetic mice. Diabetes 1990;39:432–463.

94 Xanthou M, Bines J, Walker WA: Human milk and intestinal host defenses in newborns: An update. Adv Paediat 1995;42:171–208.

95 Srivastava MD, Srivastava A, Brouhard B, Saneto R, Groh-Wargo S, Kubit J: Cytokines in human milk. Res Commun Mol Pathol Pharmacol 1996;93:263–287.

96 Onkamo P, Vaananen S, Karvonen M, Tuomilehto J: Worldwide increase in incidence of type I diabetes – the analysis of the data on published incidence trends. Diabetologia 1999;42:1395–1403.

97 Atkinson MA, Winter WE, Skordis N, Beppu H, Riley WM, Maclaren NK: Dietary protein restriction reduces the frequency and delays the onset of insulin-dependent diabetes in BB rats. Autoimmunity 1988;2:11–19.

98 Scott FW: Food-induced type 1 diabetes in the BB rat. Diabetes Metab Rev 1996;12:341–359.

99 Scott FW, Cloutier HE, Kleemann R, Woerz-Pagenstert U, Rowsell P, Modler HW, Kolb H: Potential mechanisms by which certain foods promote or inhibit the development of spontaneous diabetes in BB rats: Dose, timing, early effect on islet area, and switch in infiltrate from Th1 to Th2 cells. Diabetes 1997;46:589–598.

100 Elliott RB, Martin JM: Dietary protein: A trigger of insulin-dependent diabetes in the BB rat? Diabetologia 1984;26:297–299.

101 Funda DP, Kaas A, Bock T, Tlaskalova-Hogenova H, Buschard K: Gluten-free diet prevents diabetes in NOD mice. Diabetes Metab Res Rev 1999;15:323–327.

102 Vitoria JC, Castano L, Rica I, Bilbao JR, Arrieta A, Garcia-Masdevall MD: Association of insulin-dependent diabetes mellitus and celiac disease: A study based on serologic markers. J Pediatr Gastroenterol Nutr 1998;27:47–52.

103 Kaukinen K, Salmi J, Lahtela J, Siljamaki-Ojansuu U, Koivisto AM, Oksa H, Collin P: No effect of gluten-free diet on the metabolic control of type 1 diabetes in patients with diabetes and celiac disease. Retrospective and controlled prospective survey. Diabetes Care 1999;22:1747–1748.

104 Honeyman MC, Stone NL, Harrison LC: T-cell epitopes in type 1 diabetes autoantigen tyrosine phosphatase IA-2: Potential for mimicry with rotavirus and other environmental agents. Mol Med 1998;4:231–239.

105 Hawkes CJ, Schloot NC, Marks J, Willemen SJ, Drijfhout JW, Mayer EK, Christie MR, Roep BO: T-cell lines reactive to an immunodominant epitope of the tyrosine phosphatase-like autoantigen IA-2 in type 1 diabetes. Diabetes 2000;49:356–366.

106 Strobel S, Mowat AM: Immune responses to dietary antigens: Oral tolerance. Immunol Today 1998;19:173–181.

107 Miller A, Lider O, Abramsky O, Weiner HL: Orally administered myelin basic protein in neonates primes for immune responses and enhances experimental autoimmune encephalomyelitis in adult animals. Eur J Immunol 1994;24:1026–1032.

108 Vaarala O, Paronen J, Otonkoski T, Akerblom HK: Cow milk feeding induces antibodies to insulin in children – a link between cow milk and insulin-dependent diabetes mellitus? Scand J Immunol 1998;47:131–135.

109 Vaarala O, Knip M, Paronen J, Hamalainen AM, Muona P, Vaatainen M, Ilonen J, Simell O, Akerblom HK: Cow's milk formula feeding induces primary immunization to insulin in infants at genetic risk for type 1 diabetes. Diabetes 1999;48:1389–1394.

110 Reeves WG, Kelly U: Insulin antibodies induced by bovine insulin therapy. Clin Exp Immunol 1982;50:163–170.

111 Menser MA, Forrest JM, Bransby RD: Rubella infection and diabetes mellitus. Lancet 1978; i:57–60.

112 Harrison LC, McColl G: Infection and autoimmune disease; in Rose IMN (ed): The Autoimmune Diseases. London, Academic Press, 1998, pp 127–140.

113 Tlaskalova-Hogenova H, Stepankova R, Tuckova L, Farre MA, Funda DP, Verdu EF, Sinkora J, Hudcovic T, Rehakova Z, Cukrowska B, Kozakova H, Prokesova L: Autoimmunity, immunodeficiency and mucosal infections: Chronic intestinal inflammation as a sensitive indicator of immunoregulatory defects in response to normal luminal microflora. Folia Microbiol (Praha) 1998;43:545–550.

114 Moller E: Mechanisms for induction of autoimmunity in humans. Acta Paediatr Suppl 1998; 424:16–20.

115 Rose NR: The role of infection in the pathogenesis of autoimmune disease. Semin Immunol 1998; 10:5–13.

116 Foulis AK, Farquharson MA, Meager A: Immunoreactive alpha-interferon in insulin-secreting beta cells in type 1 diabetes mellitus. Lancet 1987;ii:1423–1427.

117 Chehadeh W, Weill J, Vantyghem MC, Alm G, Lefebvre J, Wattre P, Hober D: Increased level of interferon-alpha in blood of patients with insulin-dependent diabetes mellitus: Relationship with coxsackievirus B infection. J Infect Dis 2000;181:1929–1939.

118 Petrovsky N, Harrison LC, Kyvik KO, Beck-Nielsen H, Green A, Bonnevie-Nielsen V: Evidence for the viral aetiology of IDDM. Autoimmunity 1997;25:251–252.

119 Bonnevie-Nielsen V, Martensen PM, Justesen J, Kyvik KO, Kristensen B, Levin K, Beck-Nielsen H, Worsaa A, Dyrberg T: The antiviral 2′,5′-oligoadenylate synthetase is persistently activated in type 1 diabetes. Clin Immunol 2000;96:11–18.

120 Melnick J: Enteroviruses: Polioviruses, coxsackieviruses, echoviruses, and newer enteroviruses; in Fields KD, Howley PM (eds): Fields Virology. Philodelphia, Lippincott-Raven, 1996, pp 655–712.

121 Wendorf MA: Diabetes and enterovirus autoimmunity in glacial Europe. Med Hypoth 1999; 52:423–429.

122 Levy-Marchal C, Patterson C, Green A: Variation by age group and seasonality at diagnosis of childhood IDDM in Europe. The EURODIAB ACE Study Group. Diabetologia 1995;38:823–830.

123 Jenista JA, Powell KR, Menegus MA: Epidemiology of neonatal enterovirus infection. J Pediatr 1984;104:685–690.

124 Maguire HC, Atkinson P, Sharland M, Bendig J: Enterovirus infections in England and Wales: Laboratory surveillance data: 1975–1994. Commun Dis Public Health 1999;2:122–125.

125 Juhela S, Hyoty H, Hinkkanen A, Elliott J, Roivainen M, Kulmala P, Rahko J, Knip M, Ilonen J: T-cell responses to enterovirus antigens and to beta-cell autoantigens in unaffected children positive for IDDM-associated autoantibodies. J Autoimmun 1999;12:269–278.

126 Hiltunen M, Hyoty H, Knip M, Ilonen J, Reijonen H, Vahasalo P, Roivainen M, Lonnrot M, Leinikki P, Hovi T, Akerblom HK: Islet cell antibody seroconversion in children is temporally associated with enterovirus infections. Childhood Diabetes in Finland (DiMe) Study Group. J Infect Dis 1997;175:554–560.

127 Atkinson MA, Bowman MA, Campbell L, Darrow BL, Kaufman DL, Maclaren NK: Cellular immunity to a determinant common to glutamate decarboxylase and coxsackie virus in insulin-dependent diabetes. J Clin Invest 1994;94:2125–2129.

128 Vreugdenhil GR, Geluk A, Ottenhoff TH, Melchers WJ, Roep BO, Galama JM: Molecular mimicry in diabetes mellitus: The homologous domain in coxsackie B virus protein 2C and islet autoantigen GAD65 is highly conserved in the coxsackie B-like enteroviruses and binds to the diabetes associated HLA-DR3 molecule. Diabetologia 1998;41:40–46.

129 Endl J, Otto H, Jung G, Dreisbusch B, Donie F, Stahl P, Elbracht R, Schmitz G, Meinl E, Hummel M, Ziegler AG, Wank R, Schendel DJ: Identification of naturally processed T cell epitopes from glutamic acid decarboxylase presented in the context of HLA-DR alleles by T lymphocytes of recent onset IDDM patients. J Clin Invest 1997;99:2405–2415.

130 Horwitz MS, Bradley LM, Harbertson J, Krahl T, Lee J, Sarvetnick N: Diabetes induced by Coxsackie virus: Initiation by bystander damage and not molecular mimicry. Nat Med 1998;4:781–785.

131 Nairn C, Galbraith DN, Taylor KW, Clements GB: Enterovirus variants in the serum of children at the onset of type 1 diabetes mellitus. Diabet Med 1999;16:509–513.

132 Lonnrot M, Salminen K, Knip M, Savola K, Kulmala P, Leinikki P, Hyypia T, Akerblom HK, Hyoty H: Enterovirus RNA in serum is a risk factor for beta-cell autoimmunity and clinical type 1 diabetes: A prospective study. Childhood Diabetes in Finland (DiMe) Study Group. J Med Virol 2000;61:214–220.

133 Andreoletti L, Hober D, Hober-Vandenberghe C, Fajardy I, Belaich S, Lambert V, Vantyghem MC, Lefebvre J, Wattre P: Coxsackie B virus infection and beta cell autoantibodies in newly diagnosed IDDM adult patients. Clin Diagn Virol 1998;9:125–133.

134 Hyoty H, Hiltunen M, Knip M, Laakkonen M, Vahasalo P, Karjalainen J, Koskela P, Roivainen M, Leinikki P, Hovi T, et al: A prospective study of the role of coxsackie B and other enterovirus infections in the pathogenesis of IDDM. Childhood Diabetes in Finland (DiMe) Study Group. Diabetes 1995;44:652–657.
135 Szopa TM, Titchener PA, Portwood ND, Taylor KW: Diabetes mellitus due to viruses – some recent developments. Diabetologia 1993;36:687–695.
136 Foulis AK, McGill M, Farquharson MA, Hilton DA: A search for evidence of viral infection in pancreases of newly diagnosed patients with IDDM. Diabetologia 1997;40:53–61.
137 Di Pietro C, Del Guercio MJ, Paolino GP, Barbi M, Ferrante P, Chiumello G: Type 1 diabetes and Coxsackie virus infection. Helv Paediatr Acta 1979;34:557–561.
138 Cainelli F, Manzaroli D, Renzini C, Casali F, Concia E, Vento S: Coxsackie B virus-induced autoimmunity to GAD does not lead to type 1 diabetes. Diabetes Care 2000;23:1021–1022.
139 Serreze DV, Ottendorfer EW, Ellis TM, Gauntt CW, Atkinson MA: Acceleration of type 1 diabetes by a coxsackievirus infection requires a pre-existing critical mass of autoreactive T-cells in pancreatic islets. Diabetes 2000;49:708–711.
140 Lonnrot M, Knip M, Marciulionyte D, Rahko J, Urbonaite B, Moore WP, Vilja P, Hyoty H: Enterovirus antibodies in relation to islet cell antibodies in two populations with high and low incidence of type 1 diabetes. Diabetes Care 1999;22:2086–2088.
141 Bjorksten B: Environment and infant immunity. Proc Nutr Soc 1999;58:729–732.
142 Bjorksten B: The intrauterine and postnatal environments. J Allergy Clin Immunol 1999;104:1119–1127.
143 Coulson BS, Grimwood K, Masendycz PJ, Lund JS, Mermelstein N, Bishop RF, Barnes GL: Comparison of rotavirus immunoglobulin A coproconversion with other indices of rotavirus infection in a longitudinal study in childhood. J Clin Microbiol 1990;28:1367–1374.
144 Rott LS, Rose JR, Bass D, Williams MB, Greenberg HB, Butcher EC: Expression of mucosal homing receptor alpha4beta7 by circulating CD4+ cells with memory for intestinal rotavirus. J Clin Invest 1997;100:1204–1208.
145 Gromkowski SH, Mama K, Yagi J, Sen R, Rath S: Double-stranded RNA and bacterial lipopolysaccharide enhance sensitivity to TNF-alpha-mediated cell death. Int Immunol 1990;2:903–908.
146 Ebert EC: Human intestinal intraepithelial lymphocytes have potent chemotactic activity. Gastroenterology 1995;109:1154–1159.
147 De la Rubia L, Herrera MI, Cebrero M, De Jong JC: Acute pancreatitis associated with rotavirus infection. Pancreas 1996;12:98–99.
148 Nigro G: Pancreatitis with hypoglycemia-associated convulsions following rotavirus gastroenteritis. J Pediatr Gastroenterol Nutr 1991;12:280–282.
149 Gilger MA, Matson DO, Conner ME, Rosenblatt HM, Finegold MJ, Estes MK: Extraintestinal rotavirus infections in children with immunodeficiency. J Pediatr 1992;120:912–917.
150 Brown KA, Offit PA: Rotavirus-specific proteins are detected in murine macrophages in both intestinal and extraintestinal lymphoid tissues. Microb Pathog 1998;24:327–331.
151 Hewish MJ, Takada Y, Coulson BS: Integrins a2b1 and a4b1 can mediate SA11 rotavirus attachment and entry into cells. J Virol 2000;74:228–236.
152 Honeyman MC, Coulson BS, Stone NL, Gellert SA, Goldwater PN, Steele CE, Couper JJ, Tait BD, Colman PG, Harrison LC: Association between rotavirus infection and pancreatic islet autoimmunity in children at risk of developing type 1 diabetes. Diabetes 2000;49:1319–1324.
153 Honeyman MC, Coulson BS, Harrison LC: A novel subtype of type 1 diabetes mellitus. N Engl J Med 2000;342:1835–1837.
154 Jalonen T, Isolauri E, Heyman M, Crain-Denoyelle AM, Sillanaukee P, Koivula T: Increased beta-lactoglobulin absorption during rotavirus enteritis in infants: Relationship to sugar permeability. Pediatr Res 1991;30:290–293.
155 Johansen K, Stintzing G, Magnusson KE, Sundqvist T, Jalil F, Murtaza A, Khan SR, Lindblad BS, Mollby R, Orusild E, et al: Intestinal permeability assessed with polyethylene glycols in children with diarrhea due to rotavirus and common bacterial pathogens in a developing community. J Pediatr Gastroenterol Nutr 1989;9:307–313.

156 Obert G, Peiffer I, Servin AL: Rotavirus-induced structural and functional alterations in tight junctions of polarized intestinal Caco-2 cell monolayers. J Virol 2000;74:4645–4651.
157 Newburg DS, Peterson JA, Ruiz-Palacios GM, Matson DO, Morrow AL, Shults J, Guerrero ML, Chaturvedi P, Newburg SO, Scallan CD, Taylor MR, Ceriani RL, Pickering LK: Role of human-milk lactadherin in protection against symptomatic rotavirus infection. Lancet 1998;351: 1160–1164.
158 Bishop RF, Cameron DJ, Veenstra AA, Barnes GL: Diarrhea and rotavirus infection associated with differing regimens for postnatal care of newborn babies. J Clin Microbiol 1979;9:525–529.
159 Verge CF, Howard NJ, Irwig L, Simpson JM, Mackerras D, Silink M: Environmental factors in childhood IDDM. A population-based, case-control study. Diabetes Care 1994;17:1381–1389.
160 McKinney PA, Okasha M, Parslow RC, Law GR, Gurney KA, Williams R, Bodansky HJ: Early social mixing and childhood Type 1 diabetes mellitus: A case-control study in Yorkshire, UK. Diabet Med 2000;17:236–242.

Leonard C. Harrison, MD, DSc, The Walter & Eliza Hall Institute of Medical Research,
P.O. Royal Melbourne Hospital, Parkville, Vic 3050 (Australia)
Tel. +61 3 9345 2460, Fax +61 3 9347 0852, E-Mail harrison@wehi.edu.au

von Herrath MG (ed): Molecular Pathology of Type 1 Diabetes mellitus.
Curr Dir Autoimmun. Basel, Karger, 2001, vol 4, pp 91–122

Is Activation of Autoreactive Lymphocytes Always Detrimental? Viral Infections and Regulatory Circuits in Autoimmunity

Matthias G. von Herrath, Michael Oldstone, Dirk Homann, Urs Christen

Division of Virology, Department of Neuropharmacology & Immunology,
The Scripps Research Institute, La Jolla, Calif., USA

Introduction: The Concept of Virus-Mediated Endocrine Diseases

Viral infections are implicated in the pathogenesis of autoimmune disorders through several mechanisms [1–12]. Many of these rely on the fact that potentially autoreactive but resting lymphocytes, which 'escaped' thymic negative selection, are present in the periphery of most individuals [13–16]. There are multiple 'built-in' safety mechanisms to prevent the activation of these cells under normal circumstances. However, an external inflammatory insult such as a viral infection leading to immune activation could disturb the carefully established equilibrium of self-tolerance [6, 17, 18]. Various virus-induced mechanisms that break 'tolerance' or 'unresponsiveness' to self are discussed in this chapter. However, direct evidence for viral infections causing autoimmune diseases has been difficult to obtain *in humans*. One reason is that humans are exposed to a multitude of infections during lifetime and therefore a direct causal association between a particular virus and a disease is hard to establish. The second reason is that viruses have the ability to mutate frequently and one strain can contain multiple different sequences ('quasi-species') that can differ in their diabetogenicity. Last, some viral infections disrupt autoimmune processes and provide, at least in experimental models, a cure from disease. Several animal models have been generated to create conditions under which viruses induce or dampen autoimmunity and to analyze the principles by which this can occur [19, 20]. These models serve as valuable tools for understanding basic pathogenetic concepts and also for planning and using treatment strategies. The RIP-LCMV transgenic model [19, 21] of virus-induced autoimmune diabetes (T1D) has been of great value in this regard and will be described and discussed in more depth. This will provide the reader with several recent concepts of how antiviral responses affect autoimmunity and that activation of autoreactive lymphocytes does not always enhance disease, but can have beneficial effects.

How Can Viruses Cause or Enhance Diabetes? Direct Viral Non-Autoimmune Effects versus Induction of Autoimmunity

A prerequisite for the development of T- or B-cell-mediated autoimmune diseases is the ability of potentially autoreactive lymphocytes to 'escape' thymic negative selection [13–15, 22]. Thymic selection of lymphocytes is dependent on the affinity/avidity of the T-cell receptor (TcR)/peptide/MHC complex [23]. Higher affinity interactions are thought to lead to deletion of lymphocytes, a process termed negative selection. This mechanism is probably responsible for the elimination of most autoreactive lymphocytes, in particular those with high affinity for 'self'. In contrast, lower affinity interactions do not result in complete elimination of lymphocytes and consequently the escaping cells appear as naïve CD4, CD8 or B lymphocytes in the periphery [21, 24–26]. It is possible that the receptors of these lymphocytes have lower affinity for 'self' or recognize 'cryptic' epitopes [27], but at the same time can recognize foreign antigens with an avidity that is equal or higher than that of the regular T- or B-cell repertoire [28]. When this occurs through the infection by a microbe like a virus, the scenario is termed 'molecular mimicry'. It describes the ability of T or B lymphocytes to cross-react with self and foreign MHC peptide complexes or antigens from disparate genes, that leads to T- or B-cell activation [7, 29–32]. Similar conformations or sequences between viral and self-peptides are candidates for inducing this cross-reactivity, but the final determining factor is the shared conformation of the antigen or MHC/peptide complex [17, 33, 34]. Infection with the 'cross-reactive virus' leads to both activation of the antiviral lymphocyte repertoire and activation of a certain number of the self-reactive cells depending on the viral strain [35] and presence of costimulatory signals [36]. The result is elimination of the viral infection ('hit-and-run' event) and a varied degree of auto-reactivity. When the numbers and affinities of self-reactive lymphocytes are high enough, a sufficient amount of tissues or cells will be destroyed to initiate an inflammatory reaction that can, over time, lead to destruction of enough tissue and thus to autoimmune disease. The lag period between the triggering viral infection and the time, when disease is detected, may vary considerably from a few weeks to several years. For example, in human auto-immune diabetes, autoantibodies to islet cell antigens precede the clinical onset of disease by up to 7 years [10, 37–43]. Factors such as MHC background of the host, the type of virus infection, the number and affinities of autoreactive lymphocytes and the extent of cross-reactivity determine the time period. Because of these multiple factors in human autoimmune disease it is difficult to establish a causative association between viral infection(s) and the autoimmune disease per se. During the interim period of time, each individual may have several unrelated viral infections, any of which could be associated with the disease and some of which might even suppress the autoimmune process. Thus, antibodies to the virus or T cells might not be present at the time, when auto-immune disease becomes evident. Further, for molecular mimicry to 'function', it is not required that the infectious agent remains present in the tissue targeted by the autoimmune disease. Consequently, analysis of tissues might not establish a connection with an infectious agent, because viral antigens are not found in the target organ after they have been cleared by the host's immune response. Multiple linear sequence homologies between viruses and self-ligands can be detected through extensive data bank searches. However, since conformational homo-logies cannot be precisely predicted [33, 34, 44–46], putative cross-reactive epitopes need to be tested in vitro and in vivo [6, 17, 47, 48]. An example implicat-ing molecular mimicry in human autoimmunity comes from work performed by Wucherpfennig and Strominger [48]. They identified multiple T-cell clones from peripheral blood of multiple sclerosis (MS) patients that showed cross-reactivity with viral peptides restricted by MHC class II alleles in vivo. Further, cross-reactivity between herpes simplex virus antigens and 'self' might cause autoimmune uveitis [49]. Other studies along these lines should bring more clarity to this problem and its approach. In the long run this will provide a tool to identify potential cross-reactive ligands and epitopes. It is important to bear in mind that 'mimicry' might not only enhance, but could also abrogate or mod-ulate the disease process. Given the multitude of different microbial and self-ligands it is likely that some of the cross-reactive interactions will have the effect attributed and described for altered peptide ligands (APL) [50]. Such effects can include partial activation anergy or a modified cytokine profile of the responder T cell [50]. Indeed, recent evidence from an experimental model strongly supports this notion [51].

The second mechanism for how viral infections could trigger autoimmune diseases involving molecular mimicry or bystander (inflammatory) effects are depicted in figure 1 and involves several sequential infections with related or unrelated viruses. This mechanism was recently termed the 'fertile field' hypo-thesis by our colleague Lindsay Whitton at The Scripps Research Institute and postulates that viral infections only cause autoimmunity when pre-existing

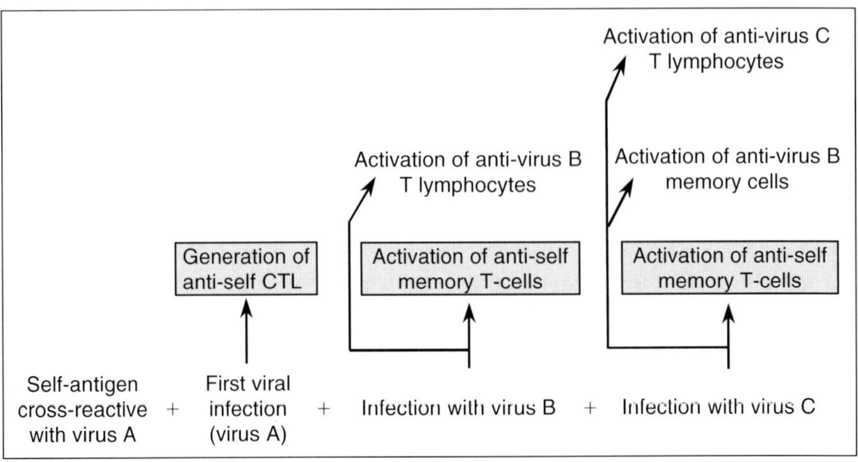

Fig. 1. A hypothesis for the mechanism how sequential related or unrelated viral infections can induce autoimmune disease.

damage of the target organ exists or, conversely, provide an inflammatory lesion out of which autoimmune disease can develop. It could also involve memory lymphocytes generated in response to a previous viral infection can be reactivated during a subsequent related or unrelated infection either directly through cross-reactivity [52–54] or indirectly via bystander activation involving cytokines and other inflammatory factors [55]. When these lymphocytes react at the same time to a self-antigen, their chronic restimulation by repeated viral infections could contribute to the maintenance and expansion of a potential local or systemic autoimmune process. Figure 1 cartoons how this mechanism could lead to disease. Indeed, observations by Evans and her colleagues [56] using a transgenic mouse model of autoimmune disease of the central nervous system (CNS) have demonstrated increased infiltration of the CNS after repeated infections with related or unrelated viruses. However, very recent observations show that infection with a second virus that shares sequence homologies can also reduce disease by increasing apoptosis of autoreactive lymphocytes [von Herrath and Oldstone, unpubl. data].

There are other ways, in addition to molecular mimicry, by which viruses can directly attack target cells or tissues associated with autoimmune diseases. Viral infection of a specific cell, for example the β cell of the islets of Langerhans, can induce the presentation of viral antigen on the cell's surface. This can result in destruction, either through direct cytopathicity of the virus or attack of the infected cell by antiviral cytotoxic T lymphocytes (CTL) [21, 57, 58]. In addition, if persistent infection by a virus occurs, alteration of differentiated cellular functions (i.e. production of insulin) of the infected cell [59, 60] or induction of cytokines that can enhance expression of MHC molecules or activate resting anti-self T or B lymphocytes can result [55]. Consequently, direct viral infection of targeted cells can have three different outcomes: The first is that cells are directly destroyed by the viral infection or antiviral immune response and disease will follow immediately. In this case, a direct association between the disease and the viral infection is most evident. Examples would be acute-onset T1D following rubella virus, Coxsackie B4 virus or mumps virus infections. These viruses and their products have been detected in the pancreas of children with acute-onset T1D [9, 45, 61–64]. The second process is associated with a lag period. When immune mediated death of only few target cells or activation of limited anti-self lymphocytes occurs, the numbers of effector cells may be initially insufficient to cause disease unless either more effectors are generated or more self-antigens are released leading to recruitment of additional autoreactive cells. The immune system will contain enough cells reactive to some of the self-antigens, because they are not expressed in the thymus or lower avidity cells escaped negative selection. Activation of autoreactive lymphocyte-sis achieved by presentation of the self-antigens taken up in the area of the primary viral lesion by a 'professional' antigen-presenting cell (APC) such as a macrophage or dendritic cell, capable of providing the required costimulatory signals [65]. Once a number of autoreactive cells are activated, a chronic attack of the target cell or tissue occurs that leads to autoimmune disease. The viral infection may be eliminated at this time, but autoreactive cells persist. Most elegant proof for this situation involving 'secondary' autoimmunity after a viral trigger was recently provided by Steve Miller's laboratory [66, 67]. In their model, infection of the CNS with Theiler's virus lead to secondary presentation of self-antigens by resident APCs that were then able to activate autoreactive lymphocytes. APCs thus play a central role in regulating autoimmunity. Very interesting work by Kurts et al. [68, 69] shows that cross-presentation of β-cell antigens in the pancreatic draining node plays a fundamental role in determining the degree of diabetes in RIP-mOva transgenic mice.

Another, and third mechanism, involves the persistence of a virus or viral antigen. This can result in disruption of the cells differentiated functions (i.e. production of hormones or neurotransmitters), without destroying the cell [59]. One example is the disruption of the transcription of growth hormone observed in C3H/St mice persistently infected with LCMV [70]. Another is virus persistence in β cells disrupting the synthesis of insulin [59]. In such cases, viral persistence occurs without causing any functional or structural damage to a cell. Accelerated disease can develop when the host encounters the same virus or a cross-reactive virus again (fig. 1). Then cells are activated that recognize the persisting viral antigen on the target cell and, dependent on the extent and nature of the infection, will lead to destruction of enough cells to cause disease by an autoimmune mechanism. It is important to keep in mind that not only direct activation of autoreactive lymphocytes but also indirect 'bystander' effects involving inflammatory or regulatory cytokines can have a profound effect on the progression of an ongoing autoimmune process. For example, bystander activation after Coxsackie virus infection was found to enhance autoimmune diabetes in BDC 2.5 TcR transgenic NOD mice [71], as well as regular NODs during a susceptible phase [72]. However, secondary spreading of the autoreactive process to cryptic [73, 74] or antigenetically distinct self-determinants can induce regulatory self-reactive cells [75] able to prevent disease [76]. Thus, as discussed later in this chapter, the precise cytokine profile secreted by primary or secondary autoreactive cells or induced by a local viral infection can have a strong influence on the progression or abrogation of an ongoing autoimmune process.

Viruses and Type 1 Diabetes mellitus (T1D)

The association of a wide variety of viruses with T1D has been the subject of considerable interest. Human epidemiological and genetic studies as well as investigations in animals over the past half century produced an intriguing picture of likely viral involvement in the pathogenesis of T1D as reviewed elsewhere [77–87]. Studies have suggested a role for Coxsackie B4 virus infection by way of molecular mimicry with glutamic acid decarboxylase (GAD) [47, 88–90]. However, more recent studies have shown that bystander activation potentially involving inflammatory cytokines might play a major role in pancreatic Coxsackie virus infections [71]. In 1997 the group of Bernhard Mach [91] reported a retroviral superantigen as a candidate autoimmune glue for human type 1 diabetes. However, several other groups could not confirm this finding to date and its implications are very controversial [79, 92–97]. More promising associations have been identified with reoviruses and human diabetes [98]. The most intriguing association has been reported recently [99] showing a timewise close association of Rotavirus infection in young children and conversion to autoantibody positivity. The underlying mechanism might be molecular mimicry. Viruses reportedly associated with T1D are listed in table 1.

How Viruses Can Abrogate an Ongoing Autoimmune Process

Several autoimmune diseases, among them type 1 diabetes and MS, occur with higher frequency in countries of the northern hemisphere. It has been

Table 1. Viruses and T1D

Virus	Reported association with T1D	Year of publication	Ref.
FMD virus	Diabetogenic in cattle	1966	176
BVD-MD	Diabetogenic in cattle	1992	177
KRV	Diabetogenic in diabetes-resistant BB rats	1991	178
EMC virus	Diabetogenic in mice	1968	179
	Diabetes induction dependent on mouse strain and substrains	1981, 86	1, 9
	Diabetes induction dependent on virus substrain (Ala$_{776}$ confers T1D whereas Thr$_{776}$ is found in all nondiabetic variants)	1980	180
Reovirus	Diabetogenic in male suckling SJL/J mice	1978	181
	In vitro infection of human β cells	1981	182
Rotavirus	Association with rotavirus infections and human T1D	1999	98
	Association with rotavirus infections in young children with type 1 diabetes	2000	99
Mengo virus	Diabetogenic in mice	1984	183
LCMV	LCMV infection of mice transgenic for LCMV NP or GP in β cells leads to T1D (discussed in detail in the following chapters)	1991	19, 20
	Persistent infection of β cells by LCMV causes abnormal glucose tolerance test	1984	59
	LCMV abrogates diabetes in NODs	1986	100
Poliovirus	One patient developing T1D after infection with poliovirus	1934	184
Mumps virus	One patient developing T1D after mumps infection	1898	185
	Association of T1D with antecedent mumps infection	1949	186
	Increased incidence of T1D after mumps outbreak	1975	146
	In vitro infection of human β cells with mumps virus	1978	187
	Mumps preceding T1D in children	1980	188
	Islet cell antibodies in children with mumps	1980	189

Virus	Reported association with T1D	Year of publication	Ref.
	In vitro upregulation of HLA I molecules after infection of β cells with mumps	1992	190
	Induction of IL-1 and IL-6 release after mumps infection of human insulinoma cell line	1992	191
Rubella	Higher incidence of T1D in rubella patients with congenital rubella syndrome (CRS)	1971–86	63, 145, 147, 187, 192–194
	Prospective study (242 patients): 10–20% of CRS patients develop T1D in 5–20 years	1986	194
	50-year follow-up study: CRS patients have higher incidence of T1D [CRS reviewed in 195]	1992	196
CMV	Infection of pancreatic islets in children with fatal CMV	1980	5
	20% of T1D patients have CMV genomic DNA in pancreatic islets	1988	62
	CMV induces autoantibody reactive with pancreatic islets	1990	82, 197
EBV	Abnormalities in EBV-specific immune responses in children with new-onset T1D	1991	198
	Homology between EBV BOLF1 molecule with HLA-DQw8	1991	199
	5′-Amino acid sequence in HLA-DQ β (near Asp_{57}) is successively repeated 6 times in EBV-BERF4, 2 patients producing epitope-specific EBV antibody subsequently developed T1D	1994	200
HAV	Three cases of T1D following 2–3 weeks after HAV infection	1992	201
HCV	High prevalence of hepatitis C infection in patients with T1D	1996	202
Echovirus	Echovirus 4: effects on human and mouse β cells	1987, 92	203, 204
	Echovirus 2: raised IgM levels in diabetic patients	1992	205

Virus	Reported association with T1D	Year of publication	Ref.
Retrovirus	Type C retrovirus production in pancreatic β cells of C3H-db/db ('diabetes') mice	1985	206
	β-Cell-specific expression of endogenous retrovirus in NOD mice; involvement of macrophages suggested	1988	207, 208
	HTLV-1 gag-related sequences in leukocyte DNA of patients with polyendocrinopathies (Basedow-Graves' disease and T1D)	1991	209
	Insulin autoantibodies from T1D patients cross-react with retroviral antigen p73	1993	210
	Retroviral etiology for human T1D was proposed but could not be confirmed by other groups	1998–99	91, 93–97
VEV	Decreased glucose tolerance in Syrian hamsters	1981	211
CBV	*Contrasting results possibly due to inability to distinguish between Coxsackie B4 variants by routine ELISA testing in the following 5 citations:*		
	Islet cell degeneration, no inflammation	1959	212
	Association of high CBV-specific antibody titers and T1D	1969–92	209, 213–228
	No correlation between CBV infection and T1D	1981–93	229–234
	Association of high CBV-specific antibody titers and nondiabetic controls	1982, 89	235, 236
	Isolation of 13 variants of Coxsackie B4 virus virus	1982	237
	Isolation of CBV or presence of CBV viral antigen in several case reports [reviewed in 104]	1976–82	43, 238, 239
	Expression of 64 kD autoantigen (GAD) is increased after Coxsackie B4 infection of SJL/J and CD1 mice before onset of hyperglycemia	1988	240
	Identification of GAD as an autoantigen in T1D	1990	88

Table 1 (continued)

Virus	Reported association with T1D	Year of publication	Ref.
	Molecular mimicry proposed for antibodies cross-reacting with GAD and Coxsackie B4 non-capsid protein P2-C	1992	241
	Spontaneous loss of T-cell tolerance to GAD in murine T1D	1993	242
	Association of Coxsackie A virus with T1D	1992, 93	80, 205
	Bystander activation effects cause increased T1D in TcR transgenic NODs	1997	71
	Enhancement of pre-existing autoimmunity in NOD mice by Coxsackie B3 infection	2000	72

Based in part on reviews [77–79]. FMD = Foot-and-mouth disease; BVD-MD = bovine viral diarrhea mucosal disease; KRV = Kilham's rat virus; EMC = encephalomyocarditis virus; LCM = lymphocytic choriomeningitis virus; CMV = cytomegalovirus; EBV = Epstein-Barr virus; HAV = hepatitis A virus; HCV = hepatitis C virus; CBV = Coxsackie B virus; GAD = glutamic acid decarboxylase; VEV = Venezuelan encephalitis virus.

proposed that the lack of some infectious diseases predisposes to autoimmunity. Indeed, this inverse correlation has been validated in several animal models, where viral infections can abrogate an ongoing autoimmune process. For example, lymphocytic choriomeningitis virus (LCMV) infection can abrogate diabetes in NOD mice [100]. This occurs through local modulation of the pancreatic microenvironment. Most recently, we have shown [von Herrath et al. unpubl. data] that high-dose infection with some LCMV strains can abolish diabetes development in RIP-LCMV mice. In this case, overstimulation of the autoreactive response leads to apoptosis of autoreactive lymphocytes. Thus, a certain baseline stimulation of the immune system might be beneficial in maintaining tolerance. Additionally, overstimulation of an ongoing autoimmune process can reveal its inherent 'fragility' and disrupt it when too many autoaggressive lymphocytes undergo apoptosis. Further, there is some yet unpublished evidence that short-term induction of inflammatory cytokines can result in a similar disruption of autoimmunity. Therefore, some localized, but not systemic, autoimmune reactions, such as the ones causing diabetes or MS, might be better viewed as a fragile chronic inflammation that respond to more 'gentle' immune interventions than, for example, an antiviral response would (see also Conclusions and fig. 4).

The RIP-LCMV Model to Probe Pathogenesis, Prevention and Treatment of T1D

Rationale to Establish the Model and General Description
The RIP-LCMV transgenic mice express LCMV viral antigens (glycoprotein, GP, or nucleoprotein, NP) under control of the RIP in their pancreatic β cells. The model was independently described by the laboratories of Oldstone [19] and Zinkernagel [20] in 1991. Expression of the viral transgene neither leads to islet pathology and alteration of islet function nor autoimmune response against the β cells. However, the transgenic mice are not 'tolerant' [13] to the transgene, since in all established lines (over 15), LCMV-reactive CTL 'escape' negative selection in the thymus and are present in the periphery. This is true for both lines devoid of thymic expression of the LCMV ('self') transgene and lines with thymic expression [21]. The reason for RIP-LCMV mice not developing spontaneous autoimmunity is that the transgene is not expressed on or (cross-) presented by 'professional' APCs and therefore T cells are *unresponsive* towards the viral β-cell proteins [21]. The evidence for this point is 2-fold. First, when APCs (*Drosophila* cells transfected with and expressing the appropriate MHC class I allele and coated with the respective viral peptide) are incubated with peripheral T cells from RIP-LCMV transgenic mice, the T cells become activated and kill LCMV-infected targets in a MHC-restricted antigen-specific manner [101]. Second, in vivo LCMV infection or repeated immunization of viral-peptide expressing dendritic cells [102] generates MHC-restricted CTL in RIP-LCMV mice [19, 21] and results in activation of peripheral naïve LCMV-specific T lymphocytes, because LCMV (self) proteins are now presented by infected 'professional' APCs to T cells. The consequence is not only elimination of the virus and virus-infected cells, but also a direct attack of the β cells expressing the viral transgene. As a result, β cells are destroyed and T1D develops in over 90% of RIP-LCMV transgenic mice. The main lymphocyte class required for destruction of β cells and T1D in RIP-LCMV mice is the MHC class I-restricted CD8+ CTL [19, 21, 103]. T1D does not occur in MHC class I-deficient or CD8-depleted mice [21]. Current investigations show that, although β cells can be directly lysed by CTL [103–105], indirect islet cell damage occurs through secretion of inflammatory mediators such as interferon-γ [104, 106] and is a main course for their death in vivo [107].

A major advantage of the RIP-LCMV model is that a known, quantifiable cellular autoimmune response can be induced by a viral infection to a clearly defined self-antigen expressed constitutively on β cells (fig. 2). This allows fine dissection of the sequence of events leading to T1D as well as the design of novel immune-therapeutic approaches [108], which are discussed below.

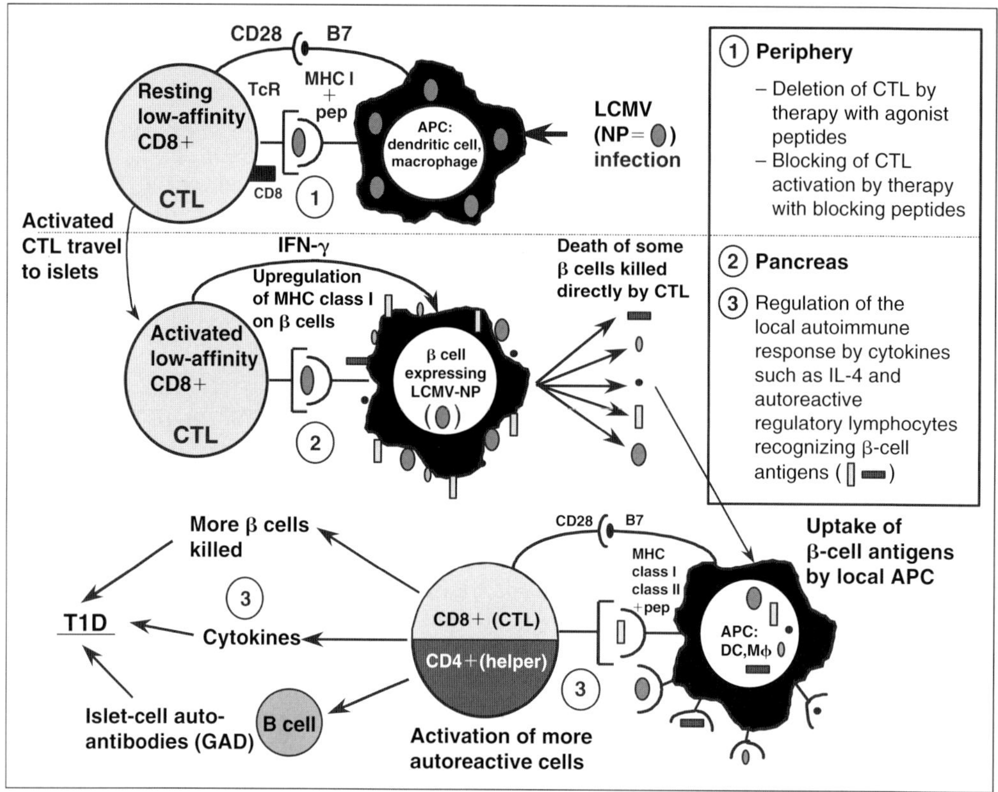

Fig. 2. Development of slow-onset T1D in RIP-LCMV mice. Different stages (1–3) during the pathogenetic process were identified, where immunotherapies can be effective in preventing T1D. See the text for a description of the different interventions tested.

How Virus Can Induce Rapid- and Slow-Onset T1D: Role of the Thymus in Determining Affinities and Numbers of Autoreactive CTL

Analysis of several different RIP-LCMV transgenic mouse lines resulted in an interesting discovery. Some lines developed rapid-onset T1D that occurred 10–18 days postinfection with LCMV. In contrast, other lines developed a slower-onset T1D occurring 4–6 months post-LCMV inoculation [21]. One line from the rapid and one from the slow onset were selected for further study. Both were bred to the MHC H-2^b (C57Bl6/J) genetic background. Direct comparison showed that, whereas both lines expressed equivalent levels of the transgene in the pancreas, the slow-onset RIP-NP mice also exhibited expression of the viral transgene in their thymus [21]. Further analysis demonstrated that thymic expression of the transgene lead to deletion of most LCMV-specific (anti-self) high-affinity CTL, while lower affinity CTL escaped negative selection and were detectable in the periphery [25]. These low-affinity CTL were able to trigger T1D. However, the development was slower and required the participation of CD4+ helper lymphocytes. In contrast, rapid-onset T1D did not depend on the participation of CD4+ help [109]. Thus, thymic expression of a self-antigen did not lead to elimination of all autoreactive lymphocytes and the CTL escaping negative selection were able to cause slowly progressive autoimmune disease. Thus, the slow-onset RIP-LCMV model reproduces a scenario similar to human T1D and, because of the lag phase between viral exposure and onset of disease, offers a suitable time frame to test therapeutic interventions. Other evidence [110] suggests that the autoimmune attack that is initially induced to the LCMV transgene, is associated with autoimmune response to other islet-cell antigens prior to onset of T1D (see fig. 2). This is reflected in the generation of antibodies to GAD and insulin in RIP-LCMV slow-onset mice [76] and parallels the findings of islet cell antibodies preceding human T1D [111, 112].

Role of MHC and Non-MHC-Linked Polymorphism in Virus-Induced T1D

To test the influence of the MHC background on the development of virus-induced T1D in RIP-LCMV transgenic mice, RIP-NP (slow-onset) mice were back-crossed onto the MHC H-2^b, H-2^d and H-2^k backgrounds. T1D occurred within 4–6 months in the H-2^b mice (incidence >90%), by 1–2 months in H-2^d mice (incidence >90%) and significantly slower (>8 months) and less frequent in H-2^k mice (40%) [21, 113], or not at all in SV129 crosses [von Herrath et al., unpubl. data, 2000]. The incidence of T1D on the different MHC backgrounds correlated directly with the magnitude of the LCMV-specific CTL response to the transgene protein as assessed by numbers and/or affinities of autoreactive CTL. RIP-NP H-2^d mice generated more CTL than RIP-NP H-2^b mice, which generated more CTL than H-2^k mice. Our studies showed that the increased numbers of CTL in RIP-NP H-2^d mice as compared to RIP-NP H-2^b mice were due to the higher affinity of the H-2^b NP epitope, which resulted in more efficient negative selection in the thymus. Other studies showed an involvement of MHC genes in the susceptibility of human T1D [114–116]. Thus, the RIP-LCMV model illustrates and incorporates this finding and offers a possible mechanism. Since most responses to pathogens, in particular viruses, are controlled in a MHC-restricted manner, the MHC linkage to T1D may also be related to interaction with different viruses.

Non-MHC-linked genes can influence the incidence of T1D in animal models. For example, a multitude of T1D susceptibility genes has been identified in the NOD mouse [117]. Further, studies with RIP-influenza hemagglutinin transgenic mice have shown that incidence of T1D varies significantly between transgenic mice bred to different mouse backgrounds that do not differ in their MHC [118]. The incidence of rapid- and slow-onset T1D in the RIP-LCMV model was drastically reduced (90%) in F_1 mice obtained when H-2^b transgenic mice were crossed to the 129 H-2^b background for one generation [119]. This occurred even though immune responses to LCMV and clearance of viral infection were similar between these strains and the F_1 hybrid. It is currently under study whether the 129 background provides T1D-protective genes and/or β-cell regeneration occurs [von Herrath et al., submitted, 2000].

Role of Cytokines in Virus-Induced T1D

Cytokines are important regulatory molecules that constitute a 'cross-talk' network that connects lymphocytes, APCs and other cells of the immune system. In the recent past, the so-called 'Th$_1$/Th$_2$ paradigm' has gained increasing importance as well as controversy in describing the regulation of immune processes [120–123]. This scheme divides cytokines into two groups: Th$_1$ or T$_1$ cytokines. Th$_1$ cytokines include interferon-γ (IFN-γ), TNF-α, IL-12 and others and are classified as inflammatory cytokines. Secretion of these cytokines results in upregulation of MHC molecules on APCs, enhancement of CTL killing, increased activation-induced cell death and proliferation and recruitment of other cells that secrete Th$_1$ cytokines. The main immunoglobulin subclass induced by a Th$_1$ response is IgG2a. In contrast, the T$_2$ group of cytokines includes IL-4, IL-5, IL-6 and IL-10, which are thought to have mainly regulatory functions [124] including the dampening of Th$_1$ processes, but also activation of B cells to secrete IgE and IgG1, enhancement of cellular proliferation and increment of the life-spans of naïve lymphocytes [125]. Thus, the 'Th$_1$/Th$_2$ paradigm' is not an entirely 'black and white' but a 'gray zone', where Th$_2$ cytokines might enhance or maintain an inflammatory process and Th$_1$ cytokines might conversely lead to the termination of an inflammation through induction of more activation-dependent cell death. For example, Marrack and her colleagues [126] recently discovered that IL-4 significantly increase the life-span of naïve lymphocytes in vitro. Evaluation of cytokine-dependent effects are necessary for each model analyzed as they vary between different autoimmune diseases and depend on the organ affected. In the NOD mouse, Th$_1$ and Th$_2$ clones have been isolated from the pancreas and in some instances Th$_2$ clones conferred protection [120], whereas in other studies they exhibited diabetogenic potential [121, 127].

With this in mind, we explored the role of various cytokines in the RIP-LCMV model of virus-induced T1D. Two approaches were chosen. First, various cytokines were expressed over basal levels locally in the β cells by generating RIP cytokines expressing transgenic mice that were then crossed with the RIP-LCMV transgenic mice. Second, cytokine 'knock-out' mice were back-crossed to the RIP-LCMV transgenic mice. Incidence of spontaneous and LCMV-induced T1D was then monitored in both groups of mice and compared.

When IFN-γ was overexpressed in β cells, T1D occurred spontaneously without LCMV infection in RIP-LCMV × RIP-IFN-γ double transgenic mice. Inflammation and upregulation of MHC class I molecules was seen and anti-LCMV CTL were activated spontaneously without LCMV infection. These antivirus (self)-specific cells were found in the islets [128, 129]. In contrast, T1D did not occur when RIP-LCMV IFN-γ-deficient mice were infected with LCMV [106], despite generation of good levels of anti-LCMV (self) CTL that infiltrated the pancreas. MHC class I upregulation on β cells, usually seen in RIP-LCMV transgenic mice after LCMV infection, was not detectable. From these observations we concluded that (a) IFN-γ has a strong inflammatory effect by upregulation of MHC molecules on β cells [107] and that (b) IFN-γ is a prerequisite for β-cell destruction, since CTL directed against a self-antigen on β cells do not successfully destroy β cells in the absence of IFN-γ [107]. These studies also illustrate the 'built-in' safeties an organism has to prevent autoimmune disease: Basal MHC class I levels on β cells are insufficient to support β cell destruction by autoreactive CTL [107].

Interleukin-2 was evaluated by us in collaboration with Miller's and Allison's laboratories using a double transgenic RIP-LCMV × RIP-IL-2 model [113]. In contrast to IFN-γ, spontaneous T1D did not occur despite the presence of local insulitis in the absence of LCMV infection. Further spontaneous generation of anti-LMCV CTL did not occur. However, enhanced T1D developed in these double transgenic mice, when infected with LCMV. Thus, although IL-2 induces lymphocyte proliferation, it is not able to break unresponsiveness to self. Possible reasons are that not enough lymphocytes are activated in the pancreas through the local presence of IL-2 or alternatively lymphocytes were not fully activated by IL-2.

Several Th₂ cytokines were evaluated. Unexpectedly, *IL-10* promoted T1D, showing that the Th$_{1/2}$ paradigm is clearly not always applicable [130]. *IL-4* prevented T1D. These findings are similar to those described using NOD mice. IL-4 was shown to prevent islet destruction under certain, but not all conditions [131], whereas IL-10 did not prevent T1D. Thus, the regulation of the local milieu to favor or dampen autoimmune disease development is complex. In addition to the amount, the timing of cytokine production in relation to the stage of the autoimmune process appears to be important.

When direct immunohistochemical or ELISPOT analysis of cytokines was performed on the islets, an interesting scheme emerged. Mice that were prediabetic or had received successful immune-intervention therapy [108] had more IL-4 than IFN-γ in their islets. In contrast, mice with T1D showed a predominance of IFN-γ [101]. This indicates a regulatory or islet-protective role for cytokines such as IL-4 and provides a mechanism how immune interventive approaches could influence the local inflammatory process in the islets [76].

Activation of Autoaggressive Lymphocytes Alone Is Not Sufficient for Autoimmune Disease – Immunopathologic Changes in the Target Organ Are a Prerequisite for Their Ability to Cause Disease in vivo

An important question is whether sufficient quantities of activated, autoreactive lymphocytes are present and able to cause disease, or if other additional factors are necessary. Recently, several factors were found to be required to drive the inflammatory process that destroys the islets to cause T1D [104]. Adoptive transfer of large numbers of highly activated LCMV (self-transgene) specific CTL into RIP-LCMV transgenic uninfected recipients led to insulitis but not T1D (blood glucose <350 mg/dl, pancreatic insulin <10 μg/mg of pancreatic tissue) [132]. These findings prompted us to investigate whether unique changes in islets of LCMV-infected transgenic mice preceded infiltration and the accumulation of autoreactive LCMV-specific CTL. Noted was the upregulation of MHC class II and activation of macrophages in and around the islets by 2–4 days postinfection before CTL and CD4 lymphocytes homed to the islets, usually by day 7 [132]. Infectious virus (LCMV) is found in the pancreas and occasionally in the islets (10% of studied transgenic animals) as early as 2–4 days postinfection. Thus, when the LCMV-activated lymphocytes reach the target organ, antigen presentation of viral antigens occurs and might allow local expansion of effector lymphocytes (CTL). The viral (self) transgene expressed on B cells is then recognized by these lymphocytes, which are able to directly or indirectly kill β cells. The role played by cross-presentation of β-cell antigens including the LCMV transgenes in the continuation of the autoimmune process is presently not known. Later, when T1D develops, most β cells have been destroyed, islet infiltration is organized similar to a 'mini-lymph node' around a network of dendritic cells, and activated macrophages are only in the peripheral areas surrounding islets [21]. Similar findings have been reported with the RIP influenza-HA transgenic mouse model. The detection of islet cell antigen such as GAD-65 prior to the onset of clinical T1D in RIP-LCMV mice adds weight to the concept of cross-presentation of β-cell antigens by APC in the islets.

Therapeutic Approaches for T1D

T1D is thought to be caused by autoaggressive lymphocytes that enter the islets of Langerhans, where they destroy β cells [11, 39, 111, 127, 133–139]. Activation of such cells is probably multifactorial involving a genetic predisposition

[114, 117, 140, 141], environmental triggers [142–144] such as viruses and maybe local damage to the pancreas/islets [43, 72, 80, 83, 87, 98, 99, 145–148], for example caused by a local proinflammatory reaction. Since the autoaggressive process is usually fairly advanced when prediabetic human individuals are identified by screening for islet-cell antibodies [149–151], one can assume that aggressive responses to more than one islet-antigen will be ongoing during this stage of the disease [74, 152, 153]. It is, therefore, not necessarily practical to attempt to anergize or delete all of the autoaggressive lymphocytes using antigen-specific immunotherapy [154, 155]. Furthermore, strong forms of non-specific systemic immunosuppression are usually not acceptable, since diabetes frequently affects young individuals and insulin substitution affords a reasonable quality of life. However, insulin cannot prevent all of the late complications of diabetes and life expectancy is usually reduced by 10–15 years. Therefore, a curative immune-based intervention with low side effects is very desirable [156].

Not all components of an ongoing autoreactive process are necessarily damaging to the targeted organ. Indeed, autoaggressive and autoreactive regulatory responses have been described in many experimental models for autoimmune diseases [75, 76, 108, 152, 157–165]. These coexist in a relatively fragile equilibrium, which usually shifts in favor of the aggressive response before clinically manifested autoimmunity (in diabetes associated with destruction of more than 90% of all pancreatic β cells) develops [101, 152]. Enhancing the regulatory component by external immunization with the 'regulatory' protein autoantigens via the oral and nasal routes [166, 167] or intramuscular immunization with plasmid DNA-encoded antigens [164, 168] has been successful in preventing diabetes in several animal models (including our systems [76, 108, 161]). In these experiments the curative effect extended over the life of the animal without requiring continuous immunizations. However, in order to bring this strategy closer to a potential application in humans, several obstacles still have to be overcome. It is not precisely known what effector mechanisms are used by autoreactive regulatory cells, how to safely choose the appropriate autoantigen suited for immune intervention, or how to optimize the efficacy of the immunotherapy.

The RIP-LCMV transgenic model is useful for testing novel approaches of antigen specific immunotherapy [169, 170], because the induction of T1D can be controlled by viral infection and a precise prediabetic phase can be experimentally defined, which is suited to test interventions. Further, the immune response to the viral (self) transgene is well known and has been dissected on the molecular level [171, 172]. As shown in figures 3 and 4, antigen-specific interventions are successful when used at different stages during development of T1D. Two successful approaches, oral antigen administration [108, 173, 174], and DNA vaccines will be discussed in more detail.

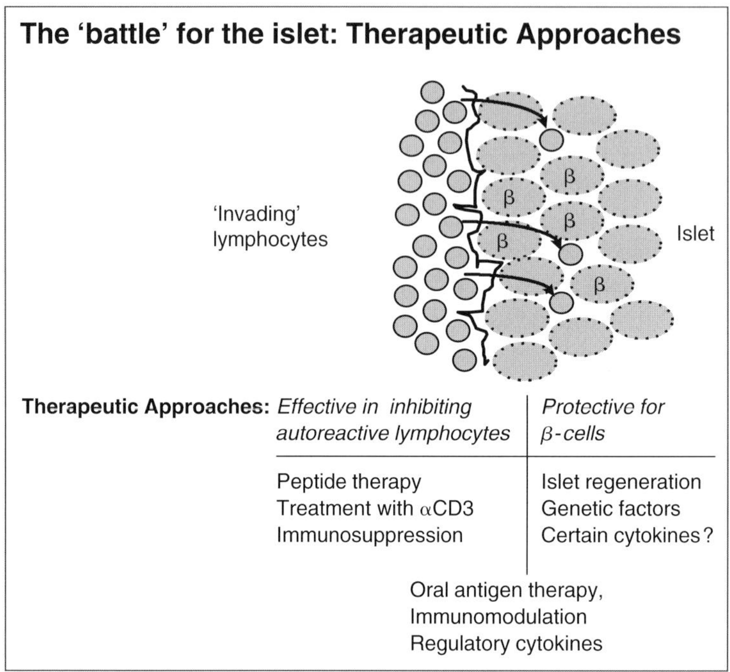

Fig. 3. Therapeutic approaches to prevent autoimmune diabetes. Infiltration and destruction of islets by autoreactive lymphocytes and APCs is a complex process that can be modulated by various mechanisms even after initiation. It is not clear whether 'attack' of an islet results in increased regeneration of β cells and whether in this way the islets 'fight back' to maintain homeostasis. Islet destruction should be viewed as a dynamic process that depends on the degree of local inflammatory mediators and the balance between numbers of destructive versus protective lymphocytes.

Oral Antigens

When RIP-LCMV tg mice that develop slow-onset T1D receive insulin orally twice per week starting 10 days after LCMV infection until 6 weeks postinfection, T1D is prevented in 40–60% of tg mice (observation time 6 months) [108]. Anti-self (viral) CTL are generated at equal levels in either insulin-treated mice with and without T1D. However, insulitis is markedly reduced in insulin-treated tg mice without T1D. Infiltrating lymphocytes are found around but do not enter the islets, and B cells are not destroyed. Pancreas from insulin-treated tg mice without T1D contains 6-fold more IL-4, IL-10 or TGF-β and 2-fold less IL-2 or IFN-γ producing lymphocytes than the pancreas from insulin-treated mice with T1D. Thus, oral administration of insulin as a potential self-antigen, that is not involved in the primary induction of T1D in

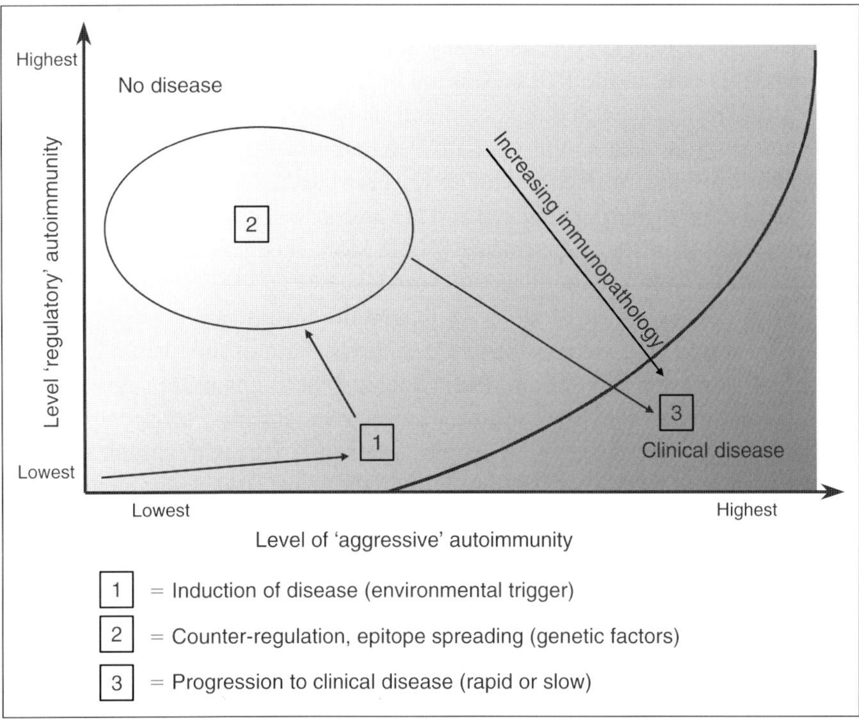

Fig. 4. Autoimmune diabetes results as a consequence of immune dysregulation. The goal of immune-based interventions is to preserve stage 2 and prevent its progressing towards the clinical stage 3. Molecules known to be instrumental in this decision are inflammatory and regulatory cytokines, chemokines, adhesion molecules, and the activation profile of autoreactive lymphocytes as well as APCs. Many of these molecules can have beneficial or detrimental effects based on the time and level of expression in relation to the ongoing disease process. Due to the complexity of this situation, it has therefore been very difficult to make good predictions about the safety and efficacy of a given approach. Importantly, many of these molecules can be assessed as markers in the peripheral blood and could potentially be used as a basis for a predictive model that would allow rating the success of immune interventions.

RIP-LCMV tg mice, prevents autoimmune disease. Our recent findings demonstrate that a *'bystander effect' of regulatory T cells* changing the ratio of Th_2/Th_1 type cytokines in the target tissue is the key factor in preventing and treating T1D in this situation. Autoreactive regulatory lymphocytes specific for the insulin B chain are activated after oral insulin administration. These cells are found in most lymphoid organs, but only proliferate in the pancreatic draining lymph node (PDLN), where they recognize their coguate antigen (insulin) [76]. Insulin peptides are likely presented by these only after islet destruction has been initiated and professional APCs will take up insulin fragments. As a consequence, these regulatory cells locally suppress LCMV-specific autoaggressive responses only in the PDLN, but not other lymphoid organs, a 'bystander suppression' process that requires IL-4 and likely an effect on local APCs [76]. In studies using the NOD model for T1D and other animal models for different autoimmune disease, Weiner and his group [174] described a similar protective effect for insulin, myelin or collagen and the presence of regulatory lymphocytes secreting TGF-β in the target tissue [108]. More recently, insulin-B-specific regulatory cells were found in the NOD model as well [157].

The protective effect observed with porcine insulin was abrogated when one or two amino acids were substituted in the β chain of the insulin molecule [160]. Oral administration of an insulin analog (Phe/β chain 25(aa25) $\rightarrow$ Asn; $AlaB_{30} \rightarrow$ Thr) with over a 100-fold reduction in the intrinsic activity of insulin and oral administration of human insulin ($AlaB_{30} \rightarrow$ Thr) with similar intrinsic hormonal activity to porcine insulin did not abort the T1D. These results suggest first that the efficacy of insulin in preventing T1D is not due to a direct hormonal effect of the insulin. Second, a one amino acid substitution can completely abrogate the ability of insulin to induce oral tolerance. Thus, the precise sequence of the autoantigens used for immunotherapy has to be evaluated carefully. Further, T1D can be potentiated by selecting the 'wrong' antigen [175], and therefore use of oral antigens requires a carefully evaluation and a reliable marker prior to human therapeutic trials.

DNA Vaccines

For the second strategy we reasoned that insulin B-chain-specific regulatory lymphocytes induced by oral insulin should also be activated by other means of immunization. We therefore immunized prediabetic RIP-LCMV mice with a DNA vaccine construct expressing the insulin-B chain [168]. As a result, diabetes was prevented to a similar degree as we previously demonstrated for oral insulin (50–60% reduction). It is noteworthy that vaccination with plasmids expressing LCMV self-antigens was not successful in preventing T1D in RIP-LCMV mice [168]. The conclusion is that islet autoantigens for the induction of autoreactive regulatory lymphocytes have to be chosen based on the ongoing auto-reactive response. T cells already committed to an aggressive phenotype, such as anti-LCMV lymphocytes in RIP-LCMV mice after LCMV infection, will not be turned around to become 'regulatory'. In contrast, secondary autoreactive responses, such as the one to insulin, might be 'weaker' in nature, for example of lower affinity, and more prone to mature into a 'regulatory' phenotype. A reliable marker will be required to assess the role of a given autoantigen in an autoreactive process and safely predict the outcome of vaccines with such an antigen.

Conclusions and Future Directions

Many autoimmune diseases are associated with a recent exposure to an infectious agent. Despite the lack of direct mechanistic proof in humans, there are several scenarios in animal models showing how this can occur (reviewed in this chapter). Studies like those pioneered by Strominger's laboratory [17] could ultimately yield direct evidence for cross-reactive T cells that recognize viral and self-ligands. This approach will be more straightforward when additional information about conformational requirements becomes available. However, there is a clear possibility that the association of infectious diseases with autoimmune disorders is multifactorial and many different scenarios can lead to the same outcome, i.e., destruction of >90% of all β cells. Assuming this, the possible dual role of infectious diseases in autoimmunity would become more clear. Some infectious processes will stimulate the autoaggressive response directly or indirectly in 'just the right way' thereby enhancing disease development. In contrast, overstimulation of aggressive lymphocytes might disrupt an ongoing autoimmune process and 'cure' from disease. This argumentation could possibly explain both the reduced incidence of T1D in equatorial countries, as well as its association with recent infections. Regulation of the aggressive response is not the only factor determining, whether clinical disease will develop. Based on more recent evidence one can postulate that each auto-immune process also has a regulatory, autoreactive component. This will coexist with the aggressive counterpart in a fragile equilibrium and suitable external immunizations with such 'regulatory' autoantigens can augment it and in this way prevent progression to disease. Animal models provided and will continue to supply important leads for understanding the pathogenetic scenarios involved in virus-induced autoimmune disease(s). From such insights, intervention strategies can be devised and tested.

Acknowledgements

This is publication 10837-NP from the Department of Neuropharmacology, Division of Virology, The Scripps Research Institute. This work was supported by NIH grants DK 51091 and AI44451 to M. von Herrath and grants AG04342 and DL 995005 to M. Oldstone. M. von Herrath is also the recipient of a Career Development Award by the Juvenile Diabetes Foundation International. D. Homann was supported by a Fellowship from the DAAD and Deutsche Studienstiftung and a Fellowship by the Juvenile Diabetes Foundation International. We thank Diana Frye for excellent assistance with the manuscript preparation.

References

1   Babu PG, Huber SA, Craighead JE: Contrasting features of T-lymphocyte-mediated diabetes in encephalomyocarditis virus-infected Balb/cBy and Balb/cCum mice. Am J Pathol 1986;124:193–198.

2   Baum H: Mitochondrial antigens, molecular mimicry and autoimmune disease. Biochim Biophys Acta 1995;1271:111–121.

3   Gianani R, Sarvetnick N: Viruses, cytokines, antigens, and autoimmunity. Proc Natl Acad Sci USA 1996;93:2257–2259.

4   Foulis AK, McGill M, Farquharson MA, Hilton DA: A search for evidence of viral infection in pancreases of newly diagnosed patients with IDDM. Diabetologia 1997;40:53–61.

5   Jenson AB, Rosenberg HS, Notkins AL: Pancreatic islet cell damage in children with fatal viral infections. Lancet 1980;ii:354–358.

6   Shrinivasappa J, Saegusa J, Prabhakar B, Gentry M, Buchmeier M, Wiktor T, Koprowski H, Oldstone M, Notkins A: Frequency of reactivity of monoclonal antiviral antibodies with normal tissues. J Virol 1986;57:397–401.

7   Oldstone MBA: Molecular mimicry and autoimmune disease. Cell 1987;50:819–820.

8   Oldstone MBA: Molecular mimicry as a mechanism for the cause and as a probe uncovering etiologic agent(s) of autoimmune disease. Curr Top Microbiol Immunol 1989;145:127–135.

9   Notkins AL, Yoon JW: Virus-induced diabetes mellitus; in Notkins AL, Oldstone MBA (eds): Concepts in Viral Pathogenesis. New York, Springer, 1984, pp 241–247.

10 Bach JF: Predictive medicine in autoimmune diseases: From the identification of genetic predisposition and environmental influence to precocious immunotherapy. Clin Immunol Immunopathol 1994;72:156–161.

11 Bach JF: Insulin-dependent diabetes mellitus as an autoimmune disease. Endocr Rev 1994;15: 516–542.

12 Barnett LA, Fujinami RS: Molecular mimicry: A mechanism for autoimmune injury. FASEB J 1992;6:840–844.

13 Miller JFAP: Role of the thymus in immunity – 30 years of progress. Immunologist 1993;1:9–13.

14 Sprent J, Webb S: Intrathymic and extrathymic deletion of T-cells. Curr Opin Immunol 1995;7: 196–205.

15 Genain C, Lee-Parritz D, Nguyen M, Massacesi L, Joshi N, Ferrante R, Hoffman K, Moseley K, Letvin N, Hauser S: In healthy primates, circulating autoreactive T-cells mediate autoimmune disease. J Clin Invest 1994;94:1339–1345.

16 Lohse AW, Dinkelmann M, Kimmig M, Herkel J, Meyer zum Buschenfelde KH: Estimation of the frequency of self-reactive T cells in health and inflammatory diseases by limiting dilution analysis and single cell cloning. J Autoimmun 1996;9:667–675.

17 Wucherpfennig KW, Strominger JL: Molecular mimicry in T-cell mediated autoimmunity: Viral peptides activate human T-cell clones specific for myelin basic protein. Cell 1995;80:695–705.

18 Oldstone MBA: Infectious agents as etiologic triggers of autoimmune disease; in Oldstone MBA (ed): CTMI. New York, Springer, 1989, pp 1–4.

19 Oldstone MBA, Nerenberg M, Southern P, Price J, Lewicki H: Virus infection triggers insulin-dependent diabetes mellitus in a transgenic model: Role of anti-self (virus) immune response. Cell 1991;65:319–331.

20 Ohashi P, Oehen S, Buerki K, Pircher H, Ohashi C, Odermatt B, Malissen B, Zinkernagel R, Hengartner H: Ablation of 'tolerance' and induction of diabetes by virus infection in viral antigen transgenic mice. Cell 1991;65:305–317.

21 Von Herrath MG, Dockter J, Oldstone MBA: How virus induces a rapid or slow onset insulin-dependent diabetes mellitus in a transgenic model. Immunity 1994;1:231–242.

22 Steinman L: A few autoreactive cells in an autoimmune infiltrate control a vast population of nonspecific cells: A tale of smart bombs and the infantry. Proc Natl Acad Sci USA 1996;93: 2253–2256.

23 Ashton-Rickardt PG, Bandeira A, Delaney J, Van Kaer L, Pircher HP, Zinkernagel RM, Tonegawa S: Evidence for a differential avidity model of T-cell selection in the thymus. Cell 1994;76:651–663.

24 Cerasoli DM, McGrath J, Carding SR, Shih FF, Knowles BB, Caton AJ: Low avidity recognition of a class II-restricted neo-self peptide by virus-specific T cells. Int Immunol 1995;7:935–945.

25 Von Herrath MG, Dockter J, Nerenberg M, Gairin JE, Oldstone MBA: Thymic selection and adaptability of cytotoxic T lymphocyte responses in transgenic mice expressing a viral protein in the thymus. J Exp Med 1994;180:1901–1910.

26 Heath WR, Karamalis F, Donoghue J, Miller JF: Autoimmunity caused by ignorant CD8+ T cells is transient and depends on avidity. J Immunol 1995;155:2339–2349.

27 Moudgil KD, Sercarz EE: The T cell repertoire against cryptic self determinants and its involvement in autoimmunity and cancer. Clin Immunol Immunopathol 1994:283–289.

28 Haggerty DT, Allen PM: Identification and analysis of two cross-reactive T cell epitopes within a single protein. J Immunol 1995;155:2993–3001.

29 Garza KM, Tung KS: Frequency of molecular mimicry among T cell peptides as the basis for autoimmune disease and autoantibody induction. J Immunol 1995;155:5444–5448.

30 Silvestris F, Williams RC Jr, Dammacco F: Autoreactivity in HIV-1 infection: The role of molecular mimicry. Clin Immunol Immunopathol 1995;75:197–205.

31 Norazmi MN, Peakman M, Vergani D, Baum H: Shared amino acid sequences between glutamic acid decarboxylase 65 and 67 and alpha-2-macroglobulin. A focus for cross-reactive autoantibodies? Diabetologia 1995;38:874–875.

32 Lawson CM, O'Donoghue HL, Reed WD: Mouse cytomegalovirus infection induces antibodies which cross-react with virus and cardiac myosin: A model for the study of molecular mimicry in the pathogenesis of viral myocarditis. Immunology 1992;75:513–519.

33 Wildner G, Thurau SR: Database screening for molecular mimicry. Immunol Today 1997;18:252.

34 Roudier J: Response to Wildner et al and Baum et al. Immunol Today 1997;18:253.

35 Dutko FJ, Oldstone MBA: Genomic and biological variation among commonly used lymphocytic choriomeningitis virus strains. J Gen Virol 1983;64:1689–1698.

36 Croft M: Activation of naive, memory and effector T cells. Curr Opin Immunol 1994;6:431–437.

37 Bach JF: Organ-specific autoimmunity. Immunol Today 1995;16:353–355.

38 Berdanier CD: Diet, autoimmunity, and insulin-dependent diabetes mellitus: A controversy. Proc Soc Exp Biol Med 1995;209:223–230.

39 Eisenbarth GA: Type 1 diabetes mellitus. A chronic autoimmune disease. N Engl J Med 1986;314:1360–1368.

40 Hiltunen M, Hyoty H, Knip M, Ilonen J, Reijonen H, Vahasalo P, Roivainen M, Lonnrot M, Leinikki P, Hovi T, Akerblom HK: Islet cell antibody seroconversion in children is temporally associated with enterovirus infections. Childhood Diabetes in Finland (DiMe) Study Group. J Infect Dis 1997;175:554–560.

41 Peterson JS, Dyrberg T, Karlsen AE, Molvig J, Michelsen B, Nerup J, Mandrup-Poulsen T: Glutamic acid decarboxylase (GAD$_{65}$) autoantibodies in prediction of β-cell function and remission in recent onset IDDM after cyclosporin treatment. Diabetes 1994;43:1291–1296.

42 Peterson JS, Hejnaes KR, Moody A, Karlsen AE, Marshall MO, Hoier-Madsen M, Boel E, Michelsen BK, Dyrberg T: Detection of GAD$_{65}$ antibodies in diabetes and other autoimmune diseases using a simple radioligand assay. Diabetes 1994;43:459–467.

43 Yoon JW, Austin M, Onodera T, Notkins AL: Virus-induced diabetes mellitus: Isolation of a virus from the pancreas of a child with diabetic ketoacidosis. N Engl J Med 1979;300:1173–1179.

44 Akamine H, Takasu N, Komiya I, Ishikawa K, Shinjyo T, Nakachi K, Masuda M: Association of HTLV-1 with autoimmune thyroiditis in patients with adult T-cell leukemia (ATL) and in HTLV-I carriers. Clin Endocrinol (Oxf) 1996;45:461–466.

45 Richter W, Mertens T, School B, Muir P, Ritzkowsky A, Scherbaum WA, Boehm BO: Sequence homology of the diabetes-associated autoantigen glutamate decarboxylase with coxsackie B4-2C protein and heat shock protein 60 mediates no molecular mimicry of autoantibodies. J Exp Med 1994;180:721–726.

46 Roudier C, Auger I, Roudier J: Molecular mimicry reflected through database screening: Serendipity or survival strategy? Immunol Today 1996;17:357–358.

47 Tian J, Lehmann PV, Kaufman DL: T cell cross-reactivity between coxsackie virus and glutamate decarboxylase is associated with a murine diabetes susceptibility allele. J Exp Med 1994;180:1979–1984.

48 Wucherpfennig KW, Yu B, Bhol K, Monos DS, Argyris E, Karr RW, Ahmed AR, Strominger JL: Structural basis for major histocompatibility complex (MHC)-linked susceptibility to autoimmunity: Charged residues of a single MHC binding pocket confer selective presentation of self-peptides in pemphigus vulgaris. Proc Natl Acad Sci USA 1995;92:11935–11939.

49 Zhao ZS, Granucci F, Yeh L, Schaffer PA, Cantor H: Molecular mimicry by herpes simplex virus-type 1: Autoimmune disease after viral infection. Science 1998;279:1344–1347.

50 Sloan-Lancaster J, Allen PM: Altered peptide ligand-induced partial T cell activation: Molecular mechanisms and role in T cell biology. Annu Rev Immunol 1996;14:1–27.

51 Ruiz PJ, Garren H, Hirschberg DL, Langer-Gould AM, Levite M, Karpuj MV, Southwood S, Sette A, Conlon P, Steinman L: Microbial epitopes act as altered peptide ligands to prevent experimental autoimmune encephalomyelitis. J Exp Med 1999;189:1275–1283.

52 Selin L, Nahill S, Welsh R: Cross-reactivities in memory CTL recognition of heterologous viruses. J Exp Med 1994;179:1933–1943.

53 Selin LK, Vergilis K, Welsh RM, Nahill SR: Reduction of otherwise remarkably stable virus-specific cytotoxic T lymphocyte memory by heterologous viral infections. J Exp Med 1996;183:2489–2499.

54 Ahmed R: Tickling memory T cells. Science 1996;272:1904.

55 Tough DF, Borrow P, Sprent J: Induction of bystander T cell proliferation by viruses and type 1 interferon in vivo. Science 1996;272:1947–1950.

56 Evans CF, Horwitz MS, Hobbs MV, Oldstone MBA: Viral infection of transgenic mice expressing a viral protein in oligodendrocytes leads to chronic central nervous system autoimmune disease. J Exp Med 1996;184:2371–2384.

57 Panina-Bordignon P, Lang R, van Endert PM, Benazzi E, Felix AM, Pastore RM, Spinas GA, Sinigaglia F: Cytotoxic T cells specific for glutamic acid decarboxylase in autoimmune diabetes. J Exp Med 1995;181:1923–1927.

58 Parenti DM, Steinberg W, Kang P: Infectious causes of acute pancreatitis. Pancreas 1996;13: 356–371.

59 Oldstone MBA, Southern P, Rodriguez M, Lampert P: Virus persists in beta cells of islets of Langerhans and is associated with chemical manifestations of diabetes. Science 1984;224: 1440–1443.

60 Tonietti G, Oldstone MBA, Dixon FJ: The effect of induced chronic viral infections of the immunologic diseases of New Zealand mice. J Exp Med 1970;132:89–109.

61 King ML, Shaikh A, Bidwell D, Voller A, Banatvala JE: Coxsackie B virus specific IgM responses in children with insulin-dependent (juvenile-onset, type1) diabetes mellitus. Lancet 1983;i:1397–1399.

62 Pak CY, Eun HM, McArthur RG, Yoon JW: Association of cytomegalovirus infection with autoimmune type 1 diabetes. Lancet 1988;ii:1–4.

63 Patterson K, Chandra RS, Jenson AB: Congenital rubella, insulitis, and diabetes mellitus in an infant. Lancet 1981;i:1048–1049.

64 Dahlquist G, Frisk G, Ivarsson SA, Svanberg L, Forsgren M, Diderholm H: Indications that maternal coxsackie B virus infection during pregnancy is a risk factor for childhood-onset IDDM. Diabetologia 1995;38:1371–1373.

65 Matzinger P: Tolerance, danger, and the extended family. Annu Rev Immunol 1994;12:991–1045.

66 Miller DM, Sedmak DD: Viral effects on antigen processing. Curr Opin Immunol 1999;11:94–99.

67 Katz-Levy Y, Neville KL, Girvin AM, Vanderlugt CL, Pope JG, Tan LJ, Miller SD: Endogenous presentation of self myelin epitopes by CNS-resident APCs in Theiler's virus-infected mice. J Clin Invest 1999;104:599–610.

68 Kurts C, Kosaka H, Carbone FR, Miller JFAP, Heath WR: Class I-restricted cross-presentation of exogenous self-antigens leads to deletion of autoreactive CD8[+] T cells. J Exp Med 1997;186: 239–245.

69 Kurts C, Carbone FR, Barnden M, Blanas E, Allison J, Heath WR, Miller FAP: CD4[+] T cell help impairs CD8[+] T cell deletion induced by cross-presentation of *self* antigens and favors autoimmunity. J Exp Med 1997;186:2057–2062.

70 De la Torre JC, Oldstone MBA: Selective disruption of growth hormone transcription machinery by viral infection. Proc Natl Acad Sci USA 1992;89:9939–9943.

71 Horwitz MS, Bradley LM, Harbertson J, Krahl T, Lee J, Sarvetnick N: Diabetes induced by Coxsackie virus: Initiation by bystander damage and not molecular mimicry. Nat Med 1998;4:781–785.

72 Serreze DV, Ottendorfer EW, Ellis TM, Gauntt CJ, Atkinson MA: Acceleration of type 1 diabetes by a coxsackievirus infection requires a preexisting critical mass of autoreactive T-cells in pancreatic islets. Diabetes 2000;49:708–711.

73 Moudgil KD, Grewal IS, Jensen PE, Sercarz EE: Unresponsiveness to a self-peptide of mouse lysozyme owing to hindrance of T cell receptor-major histocompatibility complex/peptide interaction caused by flanking epitopic residues. J Exp Med 1996;183:535–546.

74 Moudgil KD, Southwood S, Ametani A, Kim K, Sette A, Sercarz EE: The self-directed T cell repertoire against mouse lysozyme reflects the influence of the hierarchy of its own determinants and can be engaged by a foreign lysozyme. J Immunol 1999;163:4232–4237.

75 Tian J, Kaufman DL: Attenuation of inducible Th2 immunity with autoimmune disease progression. J Immunol 1998;161:5399–5403.

76 Homann D, Holz A, Bot A, Coon B, Wolfe T, Petersen J, Dyrberg TP, Grusby MJ, von Herrath MG: Autoreactive CD4+ lymphocytes protect from autoimmune diabetes via bystander suppression using the IL-4/STAT6 pathway. Immunity 1999;11:463–472.

77 Tomer Y, Davies TF: Infections and autoimmune endocrine disease. Baillieres Clin Endocrinol Metab 1995;9:47–70.

78 Szopa TM, Titchener PA, Portwood ND, Taylor KW: Diabetes mellitus due to viruses – Some recent developments. Diabetologia 1993;36:687–695.

79 Yoon JW: A new look at viruses in type 1 diabetes. Diabetes Metab Rev 1995;11:83–107.

80 Fohlman J, Friman G: Is juvenile diabetes a viral disease? Ann Med 1993;25:569–574.

81 Yoon JW: Induction and prevention of type 1 diabetes mellitus by viruses. Diabète Metab 1992;18:378–386.

82 Yoon JW: The role of viruses and environmental factors in the induction of diabetes. Curr Top Microbiol Immunol 1990;164:95–123.

83 Yoon JW, Ihm SH, Kim KW: Viruses as a triggering factor of type 1 diabetes and genetic markers related to the susceptibility to the virus-associated diabetes. Diabetes Res Clin Pract 1989;1: S47–S58.

84 Yoon JW, Eun HM, Essani K, Roncari DA, Bryan LE: Possible mechanisms in the pathogenesis of virus-induced diabetes mellitus. Clin Invest Med 1987;10:450–456.

85 Leiter EH, Hamaguchi K: Viruses and diabetes: Diabetogenic role for endogenous retroviruses in NOD mice? J Autoimmun 1990;1:31–40.

86 Patrick AW, Collier A: An infectious aetiology of insulin-dependent diabetes mellitus? FEMS Microbiol Immunol 1989;1:411–416.

87 Falke D, Muntefering H, Stallmach D: Is type I diabetes a virus-induced disease? Wien Klin Wochenschr 1988;100:422–430.

88 Baekkeskov S, Anastoot HJ, Christgau S, et al: Identification of the 64K autoantigen in insulin-dependent diabetes as the GABA-synthesizing enzyme glutamic acid decarboxylase. Nature 1990;347:151–156.

89 Jones DB, McLaughlin PJ, Armstrong N, et al: Does Coxsackie B4 virus infection induce GAD and HSP 65 autoreactivity in type I diabetes? Diabetic Med 1992;9:524.

90 Hou J, Sheikh S, Martin DL, Chatterjee NK: Coxsackievirus B4 alters pancreatic glutamate decarboxylase expression in mice soon after infection. J Autoimm 1993;6:529–542.

91 Conrad B, Weissmahr RN, Boni J, Arcari R, Schupbach J, Mach B: A human endogenous retroviral superantigen as candidate autoimmune gene in type I diabetes. Cell 1997;90:303–313.

92 Manns MP, Obermayer-Straub P: Viral induction of autoimmunity: Mechanisms and examples in hapatology. J Viral Hepat 1997;4:42–47.

93 Knerr I, Repp R, Dotsch J, Gratzki N, Hanze J, Kapellen T, Rascher W: Quantitation of gene expression by real-time PCR disproves a 'retroviral hypothesis' for childhood-onset diabetes mellitus. Pediatr Res 1999;46:57–60.

94 Kim A, Jun HS, Wong L, Stephure D, Pacaud D, Trussell RA, Yoon JW: Human endogenous retrovirus with a high genomic sequence homology with IDDMK(1, 2)22 is not specific for type I (insulin-dependent) diabetic patients but ubiquitous. Diabetologia 1999;42:413–418.

95 Muir A, Ruan QG, Marron MP, She JX: The IDDMK(1, 2)22 retrovirus is not detectable in either mRNA or genomic DNA from patients with type 1 diabetes. Diabetes 1999;48:219–222.

96 Badenhoop K, Donner H, Neumann J, Herwig J, Kurth R, Usadel KH, Tonjes RR: IDDM patients neither show humoral reactivities against endogenous retroviral envelope protein nor do they differ in retroviral mRNA expression from healthy relatives or normal individuals. Diabetes 1999;48: 215–218.

97 Jaeckel E, Heringlake S, Berger D, Brabant G, Hunsmann G, Manns MP: No evidence for association between IDDMK(1, 2)22, a novel isolated retrovirus, and IDDM. Diabetes 1999;48:209–214.

98 Honeyman MC, Stone NL, Harrison LC: T-cell epitopes in type 1 diabetes autoantigen tyrosine phosphatase IA-2: Potential for mimicry with rotavirus and other environmental agents. Mol Med 1998;4:231–239.

99 Honeyman MC, Coulson BS, Stone NL, Gellert SA, Goldwater PN, Steele CE, Couper JJ, Tait BD, Colman PG, Harrison LC: Association between rotavirus infection and pancreatic islet autoimmunity in children at risk of developing type 1 diabetes. Diabetes 2000;49:1319–1324.

100 Oldstone MB: Prevention of type I diabetes in nonobese diabetic mice by virus infection. Science 1988;239:500–502.

101 Von Herrath MG, Guerder S, Lewicki H, Flavell R, Oldstone MBA: Coexpression of B7.1 and viral (self) transgenes in pancreatic β-cells can break peripheral ignorance and lead to spontaneous autoimmune diabetes. Immunity 1995;3:727–738.

102 Ludewig B, Odermatt B, Landmann S, Hengartner H, Zinkernagel RM: Dendritic cells induce autoimmune diabetes and maintain disease via de novo formation of local lymphoid tissue. J Exp Med 1998;188:1493–1501.

103 Kagi D, Odermatt B, Ohashi PS, Zinkernagel RM, Hengartner H: Development of insulitis without diabetes in transgenic mice lacking perforin-dependent cytotoxicity. J Exp Med 1996;183:2143–2152.

104 Homo-Delarche F, Boitard C: Autoimmune diabetes: The role of the islets of Langerhans. Immunol Today 1996;17:456–460.

105 Kagi D, Odermatt B, Seiler P, Zinkernagel RM, Mak TW, Hengartner H: Reduced incidence and delayed onset of diabetes in perforin-deficient nonobese diabetic mice. J Exp Med 1997;7:989–997.

106 Von Herrath MG, Oldstone MBA: IFN-gamma is essential for beta-cell destruction by CTL. J Exp Med 1997;185:531–539.

107 Seewaldt S, Thomas HE, Ejrnaes M, Christen U, Wolfe T, Rodrigo E, Coon B, Michelsen B, Kay TWH, von Herrath MG: Virus-induced autoimmune diabetes: Most beta-cells die through inflammatory cytokines and not perforin from autoreactive (anti-viral) CTL. Diabetes 2000;49:1801–1809.

108 Von Herrath MG, Dyrberg T, Oldstone MBA: Oral insulin treatment suppresses virus-induced antigen-specific destruction of beta cells and prevents autoimmune diabetes in transgenic mice. J Clin Invest 1996;98:1324–1331.

109 Laufer TM, von Herrath MG, Grusby MJ, Oldstone MBA, Glimcher LH: Autoimmune diabetes can be induced in transgenic major histocompatibility complex class II-deficient mice. J Exp Med 1993;178:589–596.

110 Holz A, Dyrberg T, Hagopian W, Homann D, von Herrath M, Oldstone MBA: Neither B-lymphocytes nor antibodies directed against self antigens of the islets of Langerhans are required for development of virus-induced autoimmune diabetes. J Immunol 2000;165:5945–5953.

111 McDevitt HO: Autoimmune diabetes and its antigenic triggers. Hosp Pract 1995;30:55–62.

112 Dahlquist G, Palmer J, Lernmark A: Glutamate decarboxylase, insulin, and islet cell antibodies and HLA typing to detect diabetes in a general population-based study of Swedish children. J Clin Invest 1995;95:1505–1511.

113 Von Herrath MG, Allison J, Miller JF, Oldstone MBA: Focal expression of interleukin-2 does not break unresponsiveness to 'self' (viral) antigen expressed in beta cells but enhances development of autoimmune disease (diabetes) after initiation of an anti-self immune response. J Clin Invest 1995;95:477–485.

114 Vyse TJ, Todd JA: Genetic analysis of autoimmune disease. Cell 1996;85:311–318.

115 Todd JA, Acha-Orbea H, Bell JI, Chao N, Fronek Z, Jacobs CO, McDermott M, Sinha AA, Timmerman L, Steinman L, McDevitt HO: A molecular basis for MHC-associated autoimmunity. Science 1988;240:1003–1009.

116 Green A: The role of genetic factors in development of IDDM. Curr Top Microbiol Immunol 1990;164:3–17.

117 Wicker LS, Todd JA, Peterson LB: Genetic control of autoimmune diabetes in the NOD mouse. Annu Rev Immunol 1995;13:179–200.

118 Scott B, Liblau R, Degermann S, Marconi LA, Ogata L, Caton AJ, McDevitt HO, Lo D: A role for non-MHC genetic polymorphism in susceptibility to spontaneous autoimmunity. Immunity 1994;1:73–82.

119 Von Herrath M, Coon B, Homann D, Wolfe T, Guidotti LG: Thymic tolerance to only one viral protein reduces lymphocytic choriomeningitis virus-induced immunopathology and increases survival in perforin-deficient mice. J Virol 1999;73:5918–5925.

120 Katz J, Benoist C, Mathis DX: T helper cell subsets in IDDM. Science 1995;268:1185–1188.

121 Healey D, Ozegbe P, Arden S, Chandler P, Hutton J, Cooke A: In vivo activity and in vitro specificity of CD4$^+$ Th1 and Th2 cells derived from the spleens of diabetic NOD mice. J Clin Invest 1995;95:2979–2985.

122 Liblau RS, Singer SM, McDevitt H: Th1 and Th2 CD4$^+$ T cells in the pathogenesis of organ-specific autoimmune diseases. Immunol Today 1995;16:34–38.

123 Swain SL: Generation and in vivo persistence of polarized Th1 and Th2 memory cells. Immunity 1994;1:543–552.

124 Modlin R, Nutman TB: Type 2 cytokines and negative immune regulation in human infections. Curr Opin Immunol 1993;5:511–517.

125 Vella AT, Mitchell T, Groth B, Linsley PS, Green JM, Thompson CB, Kappler JW, Marrack P: CD28 engagement and proinflammatory cytokines contribute to T cell expansion and long-term survival in vivo. J Immunol 1997;158:4714–4720.

126 Vella A, Teague TK, Ihle J, Kappler J, Marrack P: Interleukin (IL)-4 or IL-7 prevents the death of resting T cells: Stat6 is probably not required for the effect of IL-4. J Exp Med 1997; 186:325–330.

127 Haskins K, Wegmann D: Diabetogenic T-cell clones. Diabetes 1996;45:1299–1305.

128 Lee MS, von Herrath MG, Reiser H, Oldstone MBA, Sarvetnick N: Sensitization to self antigens by in situ expression of interferon-γ. J Clin Invest 1995;95:486–492.

129 Sarvetnick N, Shizuru J, Liggitt D, Martin L, McIntyre B, Gregory A, Parslow T, Stewart T: Loss of pancreatic islet tolerance induced by B-cell expression of interferon-γ. Nature 1990;346:844–847.

130 Lee MS, Mueller R, Wicker LS, Peterson LB, Sarvetnick N: IL-10 is necessary and sufficient for autoimmune diabetes in conjunction with NOD MHC homozygosity. J Exp Med 1996;183: 2663–2668.

131 Mueller RT, Sarvetnick N: Pancreatic expression of IL-4 abrogates insulitis and diabetes in NOD mice. J Exp Med 1996;184:1093–1099.

132 Von Herrath MG, Holz A: Pathological changes in the islet milieu precede infiltration of islets and destruction of beta-cells by autoreactive lymphocytes in a transgenic model of virus-induced IDDM. J Autoimmun 1997;10:231–238.

133 Atkinson MA, Maclaren NK: The pathogenesis of insulin-dependent diabetes mellitus. N Engl J Med 1994;331:1428–1436.

134 Brooks-Worrell BM, Juneja R, Minokadeh A, Greenbaum CJ, Palmer JP: Cellular immune responses to human islet proteins in antibody-positive type 2 diabetic patients. Diabetes 1999;49:983–988.

135 Boyton RJ, Lohmann T, Londei M, Kalbacher H, Halder T, Frater AJ, Douek DC, Leslie DG, Flavell RA, Altmann DM: Glutamic acid decarboxylase T lymphocyte responses associated with susceptibility or resistance to type I diabetes: Analysis in disease discordant human twins, non-obese diabetic mice and HLA-DQ transgenic mice. Int Immunol 1998;10:1765–1776.

136 Wegmann DR, Norbury-Glaser M, Daniel D: Insulin-specific T cells are a predominant component of islet infiltrates in pre-diabetic NOD mice. Eur J Immunol 1994;24:1853–1857.

137 Durinovic-Bello I, Hummel M, Ziegler AG: Cellular immune response to diverse islet cell antigens in IDDM. Diabetes 1996;45:795–800.

138 Hawkes CJ, Schloot NC, Marks J, Willemen JM, Drijfhout JW, Mayer EK, Christie MR, Roep BO: T-cell lines reactive to an immunodominant epitope of the tyrosine phosphatase-like autoantigen IA-2 in type 1 diabetes. Diabetes 2000;49:356–366.

139 Lernmark A, Baekkeskov S, Gerling I, Kastern W, Knutson C, Michelsen B: Immunological aspects of type 1 and 2 diabetes mellitus. Adv Exp Med Biol 1985;189:107–127.

140 Merriman TR, Todd JA: Genetics of autoimmune disease. Curr Opin Immunol 1995;7:786–792.

141 Redondo MJ, Rewers M, Yu L, Garg S, Pilcher CC, Elliott RB, Eisenbarth GS: Genetic determination of islet cell autoimmunity in monozygotic twin, dizygotic twin, and non-twin siblings of patients with type 1 diabetes: Prospective twin study. BMJ 1999;318:698–702.

142 Harrison LC, Honeyman MC: Cow's milk and type 1 diabetes. Diabetes 1999;48:1501–1507.

143 Hypponen E, Kenward MG, Virtanen SM, Piitulainen A, Virta-Autio P, Tuomilehto J, Knip M, Akerblom HK: Infant feeding, early weight gain, and risk of type 1 diabetes. Diabetes Care 1999; 22:1961.

144 Kolb H, Pozzilli P: Cow's milk and type I diabetes: The gut immune system deserves attention. Immunol Today 1999;20:108–110.

145 Forrest J, Menser MA, Burgess J: High frequency of diabetes mellitus in young adults with congenital rubella. Lancet 1991;i:332–334.

146 Sultz HA, Hart BA, Zielezny M, et al: Is mumps virus an aetiological factor in juvenile diabetes mellitus? J Pediatr 1975;86:654–656.

147 Smithells R, Sheppard S, Marshall W, Peckham C: Congenital rubella and diabetes mellitus. Lancet 1978;i:439.

148 Von Herrath MG: Obstacles to identifying viruses that cause autoimmune disease. J Neuroimmunol 2000;107:154–160.

149 Bonifacio E, Lampasona V, Bernasconi L, Ziegler AG: Maturation of the humoral autoimmune response to epitopes of GAD in preclinical childhood type 1 diabetes. Diabetes 2000;49:202–208.

150 Fuchtenbusch M, Kredel K, Bonifacio E, Schnell O, Ziegler AG: Exposure to exogenous insulin promotes IgG1 and the T-helper 2-associated IgG4 responses to insulin but not to other islet autoantigens. Diabetes 2000;49:918–925.

151 Eisenbarth GS, Gianani R, Yu L, Pietropaolo M, Harrison C, Jackson R: Dual-parameter model for prediction of type I diabetes mellitus. Proc Assoc Am Physicians 1998;110:126–135.

152 Tian J, Lehmann PV, Kaufman DL: Determinant spreading of T helper cell 2 (Th2) responses to pancreatic islet autoantigens. J Exp Med 1997;186:2039–2043.

153 Lehmann PV, Forsthuber T, Miller A, Sercarz EE: Spreading of T-cell autoimmunity to cryptic determinants of an autoantigen. Nature 1992;358:155–157.

154 Tisch R, McDevitt HO: Antigen-specific immunotherapy: Is it a real possibility to combat T-cell-mediated autoimmunity? Proc Natl Acad Sci USA 1994;91:437–438.

155 Liblau R, Tisch R, Bercovici N, McDevitt HO: Systemic antigen in the treatment of T-cell-mediated autoimmune diseases. Immunol Today 1997;18:599–604.

156 Von Herrath MG: Selective immunotherapy of IDDM: A discussion based on new findings from the RIP-LCMV model for autoimmune diabetes. Transplant Proc 1998;30:4115–4121.

157 Bergerot I, Arreaza GA, Cameron MJ, Burdick MD, Strieter RM, Chensue SW, Chakrabarti S, Delovitch TL: Insulin B-chain reactive CD4+ regulatory T-cells induced by oral insulin treatment protect from type 1 diabetes by blocking the cytokine secretion and pancreatic infiltration of diabetogenic effector T-cells. Diabetes 1999;48:1720–1729.

158 Holz A, Bot A, Coon B, Wolfe T, Grusby MJ, von Herrath MG: Disruption of the STAT4 signaling pathway protects from autoimmune diabetes while retaining anti-viral immune competence. J Immunol 1999;163:5374–5382.

159 Von Herrath MG: Bystander suppression induced by oral tolerance. Res Immunol 1997;148:541–554.

160 Homann D, Dyrberg T, Petersen J, Oldstone MBA, von Herrath MG: Insulin in oral immune 'tolerance': A one-amino acid change in the B chain makes the difference. J Immunol 1999;163: 1833–1838.

161 Von Herrath MG, Homann D, Gairin JE, Oldstone MBA: Pathogenesis and treatment of virus-induced autoimmune diabetes: Novel insights gained from the RIP-LCMV transgenic mouse model. Biochem Soc Trans 1997;25:630–635.

162 Inobe J, Slavin AJ, Komagata Y, Chen Y, Liu L, Weiner HL: IL-4 is a differentiation factor for transforming growth factor-beta secreting Th3 cells and oral administration of IL-4 enhances oral tolerance in experimental allergic encephalomyelitis. Eur J Immunol 1998;9:2780–2790.

163 Weiner HL: Oral tolerance for the treatment of autoimmune diseases. Annu Rev Med 1997;48: 341–351.

164 Tisch R, Wang B, Serreze DV: Induction of glutamic acid decarboxylase 65-specific Th2 cells and suppression of autoimmune diabetes at late stages of disease is epitope dependent. J Immunol 1999;163:1178–1187.

165 Zhang ZJ, Davidson L, Eisenbarth G, Weiner HL: Suppression of diabetes in nonobese diabetic mice by oral administration of porcine insulin. Proc Natl Acad Sci USA 1991;88:10252–10256.

166 Hanninen A, Harrison LC: $\gamma\delta$ T cells as mediators of mucosal tolerance: The autoimmune diabetes model. Immunol Rev 2000;173:109–119.

167 Harrison LC, Dempsey-Collier M, Kramer DR, Takahashi K: Aerosol insulin induces regulatory CD8 gamma delta T cells that prevent murine insulin-dependent diabetes. J Exp Med 1996;184: 2167–2174.

168 Coon B, An LL, Whitton JL, von Herrath MG: DNA immunization to prevent autoimmune diabetes. J Clin Invest 1999;104:189–194.

169 Tisch R, McDevitt H: Antigen-specific immunotherapy: Is it a real possibility to combat T-cell medicated autoimmunity? Proc Natl Acad Sci USA 1994;91:437–438.

170 Lamont A: Are we closer to selective immunotherapy for autoimmune diseases? Immunol Today 1994;15:45–47.

171 Oldstone MBA, Lewicki H, Borrow P, Hudrisier D, Gairin JE: Discriminated selection among viral peptides with the appropriate anchor residues: Implications for the size of the cytotoxic T-lymphocyte repertoire and control of viral infection. J Virol 1995;69:7423–7429.

172 Hudrisier D, Mazarguil H, Laval F, Oldstone MBA, Gairin JE: Binding of viral antigens to major histocompatibility complex class I H-2D^b molecules is controlled by dominant negative elements at peptide non-anchor residues. J Biol Chem 1996;271:17829–17836.

173 Zang JA, Davidson L, Eisenbarth G, Weiner H: Suppression of diabetes in NOD mice by oral administration of porcine insulin. Proc Natl Acad Sci USA 1991;88:10252–10256.

174 Weiner HL, Friedman A, Miller A, Khoury SJ, al-Sabbagh A, Santos L, Sayegh M, Nussenblatt RB, Trentham DE, Hafler DA: Oral tolerance: Immunologic mechanisms and treatment of animal and human organ-specific autoimmune diseases by oral administration of autoantigens. Annu Rev Immunol 1994;12:809–837.

175 Blanas E, Carbone FR, Allison J, Miller JF, Heath WR: Induction of autoimmune diabetes by oral administration of autoantigen. Science 1996;274:1707–1709.

176 Barboni E, Mannochio I, Asdurbaki G: Observations on diabetes mellitus associated with experimental foot and mouth disease in cattle. Vet Ital 1966;17:362–368.

177 Tajima M, Yazawa T, Hagiwara K, Kurosawa T, Takahashi K: Diabetes mellitus in cattle infected with bovine viral diarrhea mucosal disease virus. J Vet Med Assoc 1992;39:616–620.

178 Guberski DL, Thomas VA, Shek WR, Like AA, Handler ES, Rossini AA, Wallace JE, Welsh RM: Induction of type I diabetes by Kilham's rat virus in diabetes-resistant BB/Wor rats. Science 1991;254:1010–1013.

179 Craighead JE, McLane MF: Diabetes mellitus: Introduction in mice by encephalomyocarditis virus. Science 1968;162:913–914.

180 Yoon JW, McClintock PR, Onodera T, Notkins AL: Virus-induced diabetes mellitus (XVIII). Inhibition by a nondiabetogenic variant of encephalomyocarditis virus. J Exp Med 1980;152: 878–892.

181 Onodera T, Jenson AB, Yoon JW, Notkins AL: Virus-induced diabetes mellitus: Reovirus infection of pancreatic beta cells in mice. Science 1978;301:529–531.

182 Yoon JW, Selvaggio S, Onodera T, Wheeler J, Jenson AB: Infection of cultured human pancreatic β cell with reovirus type 3. Diabetologia 1981;20:462–467.

183 Yoon JW, Morishima T, McClintock PR, Austin M, Notkins AL: Virus-induced diabetes mellitus: Mengovirus infects pancreatic beta cells in strains of mice resistant to the diabetogenic effects of encephalomyocarditis virus. J Virol 1984;50:684–690.

184 John HJ: The diabetic child. Etiologic factors. Ann Intern Med 1934;8:198–213.

185 Harris HF: A case of diabetes mellitus quickly following mumps on the pathological alterations of salivary glands, closely resembling those found in pancreas, in a case of diabetes mellitus. Boston Med Surg J 1898;CXL:465.

186 John HJ: Diabetes mellitus in children. J Pediatr 1949;39:723–744.

187 Prince G, Jenson AB, Billups L, Notkins AL: Infection of human pancreatic beta cell cultures with mumps virus. Nature 1978;27:158–161.

188 Gamble DR: Relation of antecedent illness to development of diabetes in children. Br Med J 1980;ii:99–101.

189 Helmke K, Otten A, Willems W: Islet cell antibodies in children with mumps infection. Lancet 1980;ii:211–222.

190 Parkkonen P, Hyoty H, Koskinen L, Leinikki P: Mumps virus infects beta cells in human fetal islet cell cultures upregulating the expression of HLA class I molecules. Diabetologia 1992;35:63–69.

191 Cavallo MG, Baroni MG, Toto A, Gearing AJ, Forsey T, Andreani D, Thorpe R, Pozzilli P: Viral infection induces cytokine release by beta islet cells. Immunology 1992;75:664–668.

192 Schopfer K, Matter L, Flueler U, Werder E: Diabetes mellitus, endocrine autoantibodies and prenatal rubella infection. Lancet 1982;ii:159.

193 Menser MA, Forrest J, Bransby R: Rubella infection and diabetes mellitus. Lancet 1978;i:57–60.

194 Rabinowe S, George K, Loughlin R, Soeldner J: Predisposition to type I diabetes and organ-specific autoimmunity in congenital rubella: T cell abnormalities in young adults. Diabetes 1986;35:187A.

195 Freij BJ, South MA, Sever JL: Maternal rubella and the congenital rubella syndrome. Clin Perinatol 1988;15:247–257.

196 McIntosh EDG, Menser MA: A fifty-year follow-up of congenital rubella. Lancet 1992;340:414–415.

197 Pak CY, Cha CY, Rajotte RV, McArthur RG, Yoon JW: Human pancreatic islet cell-specific 38 kDa autoantigen identified by cytomegalovirus-induced monoclonal islet cell autoantibody. Diabetologia 1990;33:569–572.

198 Hyoty H, Rasanen L, Hiltunen M, Lehtinen M, Huupponen T, Leinikki P: Decreased antibody reactivity to Epstein-Barr virus capsid antigen in type 1 (insulin-dependent) diabetes mellitus. APMIS 1991;99:359–363.

199 Sairenji T, Daibata M, Sorli CH, Qvistback H, Humphreys RE, Ludvigsson J, Palmer J, Landin-Olsson M, Sundkvist G, Michelsen B, Lernmark A, Dyrberg T: Relating homology between the Epstein-Barr virus BOLF1 molecule and HLA-DQw8 beta chain to recent onset type 1 (insulin-dependent) diabetes mellitus. Diabetologia 1991;34:33–39.

200 Parkkonen P, Hyoty H, Ilonen J, Reijonen H, Yla-Herttuala S, Leinikki P: Antibody reactivity to an Epstein-Barr virus BERF4-encoded epitope occurring also in Asp-57 region of HLA-DQ8 β chain. Clin Exp Immunol 1994;95:287–293.

201 Makeen AM: The association of infective hepatitis type A (HAV) and diabetes mellitus. Trop Geogr Med 1992;44:362–364.

202 Simo R, Hernandez C, Genesca J, Jardi R, Mesa J: High prevalence of hepatitis C virus infection in diabetic patients. Diabetes Care 1996;19:998–1000.

203 Uriarte A, Cabrere E, Ventura R, Vargas J: Islet cell antibodies and echo 4 virus infection. Diabetologia 1987;30:590.

204 Szopa TM, Ward T, Taylor KW: Disturbance of mouse pancreatic β-cell function following echo 4 virus infection. Biochem Soc Trans 1992;20:3155.

205 Frisk G, Nilsson E, Tuvemo T, Friman G, Diderholm H: The possible role of coxsackie A and echo virus in the pathogenesis of type I diabetes mellitus studied by IgM analysis. J Infect 1992;24:13–22.

206 Leiter EH: Type C retrovirus production by pancreatic beta cells. Associated with accelerated pathogenesis in C3H-db/db (diabetes) mice. Am J Pathol 1985;119:22–32.

207 Suenaga K, Yoon JW: Association of beta cell-specific expression of endogenous retrovirus with the development of insulitis and diabetes in NOD mice. Diabetes 1988;37:1722–1726.

208 Lee KU, Amano K, Yoon JW: Evidence for initial involvement of macrophages in development of insulitis in NOD mice. Diabetes 1988;37:1989–1991.

209 Lagaye S, Vexiau P, Morozov V, Saal F, Gony J, Poorters AM, Loosfeld-Poirot Y, Hors J, Cathelineau G, Emanoil-Ravier R, Peries J: Detection of HTLV-1 gag-related sequences in leukocyte DNA from patients with polyendocrinopathies (Basendow-Graves' disease and insulin-dependent diabetes). C R Acad Sci III 1991;312:309–315.

210 Hao W, Serreze DV, McCulloch DK, Neifing JL, Palmer JP: Insulin (auto) antibodies from human IDDM cross-react with retroviral antigen p73. J Autoimmun 1993;6:787–798.

211 Rayfield EJ, Seto Y, Walsh S, McEvoy RC: Virus-induced alterations in insulin release in hamster islets of Langerhans. J Clin Invest 1981;68:1172–1181.

212 Sussman ML, Strauss L, Hodes HL: Fatal Coxsackie group B virus infection in the newborn. Am J Dis Child 1959;97:483–492.

213 Friman G, Fohlman J, Frisk G, Diderholm H, Ewald U, Kobbah M, Tuvemo T: An incidence peak of juvenile diabetes. Relation to Coxsackie B virus immune response. Acta Pediatr Scand 1985; 320:14–19.

214 Banatvala J, Schernthaner G, Schober E, DeSilva L, Bryant J, Borkenstein M, Brown D, DeSilva LM, Menser MA, Silink M: Coxsackie B, mumps, rubella and cytomegalovirus specific IgM responses in patients with junvenile-onset insulin-dependent diabetes mellitus in Britain, Austria and Australia. Lancet 1985;i:1409–1412.

215 King M, Shaikh A, Bidwell D, Voller A, Banatvala J: Coxsackie B-virus specific IgM responses in children with insulin-dependent diabetes mellitus. Lancet 1983;i:1397–1399.

216 Mertens T, Gruneklee D, Eggers HJ: Neutralizing antibodies against Coxsackie B viruses in patients with recent onset of type I diabetes. Eur J Pediatr 1983;140:293–294.

217 Mirkovic RR, Varma KS, Yoon JW: Incidence of Coxsackievirus B type 4 (CB4) infection concomitant with onset of insulin-dependent diabetes mellitus. J Med Virol 1984;14:9–16.

218 Buschard K, Madsbad SA: Longitudinal study of virus antibodies in patients with newly diagnosed type 1 (insulin-dependent) diabetes mellitus. J Clin Lab Immunol 1984;13:65–70.

219 Frisk G, Fohlman J, Kobbah M, Ewald U, Tuvemo T, Diderholm H, Friman G: High frequency of Coxsackie-B-virus-specific IgM in children developing during a period of high diabetes morbidity. J Med Virol 1985;17:219–227.

220 Alberti AM, Amato C, Candela A, Constantino F, Grandolfo ME, Lombardi F, Novello F, Orsini M, Santoro R: Serum antibodies against Coxsackie B1-6 viruses in type 1 diabetics. Acta Diabetol Lat 1985;22:33–38.

221 Fohlman J, Bohme J, Rask L, Frisk G, Diderholm H, Friman G, Tuvemo T: Matching of host genotype and serotypes of Coxsackie B virus in the development of juvenile diabetes. Scand J Immunol 1987;26:110.

222 Michlakova D, Petrovicova A, Kolar J, Jancova E, Silesova J: The role of Coxsackie virus infection in the development of type 1 diabetes mellitus and its effect on the postinitial course of the disease in childhood. Cesk Pediatr 1989;44:257–262.

223 Michalkova D, Kostal M, Rajcani J, Petrovicova A, Bircak J, Barak L, Hrabcakova D: Cytoplasmic islet cell antibodies in children with type I diabetes. Bratisl Lek Listy 1989;90:159–167.

224 Verma IC: The challenge of childhood diabetes mellitus in India. Indian J Pediatr 1989;56:S33–S38.

225 Tuskiewicz-Misztal E: Epidemiologic factors and serum antibody titer to Coxsackie B-4 virus in patients with type I diabetes. Kinderärztl Prax 1991;59:88–91.

226 Frisk G, Friman G, Tuvemo T, Fohlman J, Diderholm H: Coxsackie B virus IgM in children at onset of type 1 (insulin-dependent) diabetes mellitus: Evidence for IgM induction by a recent or current infection. Diabetologia 1992;35:249–253.

227 Gamble DR, Taylor KW, Cumming H: Coxsackie viruses and diabetes mellitus. Br Med J 1973;iv: 260–262.

228 Wagenknecht L, Roseman J, Herman W: Increased incidence of insulin-dependent diabetes mellitus following an epidemic of Coxsackie virus B5. Am J Epidemiol 1991;133:1024–1031.

229 Palmer JP, Cooney MK, Crossley JR, Hollander PH, Asplin CM: Antibodies to viruses and to pancreatic islets in nondiabetic and insulin-dependent diabetic patients. Diabetes Care 1981;4:525–528.

230 Pagano G, Cavallo-Perin P, Cavallo TF, Dall'Omo AM, Mascioloa P, Surinani R, Amoroso A, Curtoni SE, Borelli I, Lenti G: Genetic, immunologic, and environmental herterogeneity of IDDM. Incidence and 12-month follow-up of an Italian population. Diabetes 1987;36:859–863.

231 Emekdas G, Rota S, Kusitmur S, Kocabeyoglu O: Antibody levels against Coxsackie B viruses in patients with type 1 diabetes mellitus. Mikrobiyol Bul 1992;26:116–120.

232 Gunczler P, Lanes R, Layrisse Z, Esparza B, Salas R, Hernandez L, Arnaiz-Villena A: Epidemiology and immunogenetics in recently diagnosed Venezuelan children with insulin-dependent diabetes mellitus. J Pediatr Endocrinol 1993;6:165–171.

233 Ajuwon ZA, Olaleye OD, Omilabu SA, Baba SS: Complement fixing antibodies against selected viruses in diabetic patients and non-diabetic control subjects in Ibadan Nigeria. Rev Roum Virol 1992;43:3–5.

234 Pato E, Cour MI, Gonzalez-Cuadrado S, Gonzalez-Gomez C, Munoz JJ, Figueredo A: Coxsackie B4 and cytomegalovirus in patients with insulin-dependent diabetes. Anal Med Interna 1992;9:30–32.

235 Palmer JP, Cooney MK, Ward RH, Jansen JA, Brodsky JB, Ray CG, Crossley JR, Asplin CM, Williams RH: Reduced Coxsackie antibody titres in type 1 (insulin-dependent) diabetic patients presenting during an outbreak of Coxsackie B3 and B4 infection. Diabetologia 1982;22:426–429.

236 Tuvemo T, Dahlquist G, Frisk G, Blom L, Friman G, Landin-Olsson M, Diderholm H: The Swedish childhood diabetes study. III. IgM against Coxsackie B viruses in newly diagnosed type 1 (insulin-dependent) diabetic children – No evidence of increased antibody frequency. Diabetologia 1989;32:745–747.

237 Prabhaker BS, Haspel MV, McClintock PR, Notkins AL: High frequency of antigenic variants among naturally occurring human Coxsackie B4 virus isolates identified by monoclonal antibodies. Nature 1982;300:374–376.

238 Gladisch R, Hoffmann W, Waldherr R: Myocarditis and insulitis in Coxsackie virus infection. Z Kardiol 1976;65:873–881.

239 Champsaur H, Bottazzo G, Bertrams J, Assan R, Bach C: Virologic, immunologic and genetic factors in insulin-dependent diabetes mellitus. J Pediatr 1982;100:15–20.

240 Gerling I, Nejman C, Chatterjee NK: Effect of Coxsackievirus B4 infection in mice on expression of 64,000 Mr autoantigen and glucose sensitivity of islets before development of hyperglycemia. Diabetes 1988;37:1419–1425.

241 Kaufman DL, Erlander MG, Clare-Salzler MJ, Atkinson MA, Maclaren NK, Tobin AJ: Autoimmunity to two forms of glutamate decarboxylase in insulin-dependent diabetes mellitus. J Clin Invest 1992;89:283–292.

242 Kaufman DL, Clare-Salzler MG, Tian J, Forsthuber T, Ting G, Robinson P, Atkinson MA, Sercarz EE, Tobin AJ, Lehmann PV: Spontaneous loss of T-cell tolerance to glutamic acid decarboxylase in murine insulin-dependent diabetes. Nature 1993;366:69–72.

Matthias G. von Herrath, Division of Virology, IMM6,
Department of Neuropharmacology & Immunology, The Scripps Research Institute,
10550 North Torrey Pines Road, La Jolla, CA 92037 (USA)
Tel. +1 858 784 9602, Fax +1 858 784 9981, E-Mail matthias@scripps.edu

von Herrath MG (ed): Molecular Pathology of Type 1 Diabetes mellitus.
Curr Dir Autoimmun. Basel, Karger, 2001, vol 4, pp 123–143

Defects in Deletional Tolerance of CD8+ T Cells in Autoimmune Diabetes

Huub T.C. Kreuwel, Linda A. Sherman

Department of Immunology, The Scripps Research Institute, La Jolla, Calif., USA

Type 1 diabetes mellitus (T1D) is a T-cell-mediated autoimmune disease, characterized by the destruction of insulin-producing pancreatic β cells. In humans, this disease occurs in children and young adults and is often referred to as juvenile diabetes. One of the best rodent models for T1D is the nonobese diabetic mouse (NOD) [1–4]. In diabetic patients and NOD mice, immuno-histological studies of pancreata reveal the presence of both CD4+ and CD8+ T cells, suggesting a role for CD8+ T cells in disease [5–10]. Studies of disease progression in the NOD mouse model have confirmed that CD8+ T cells play an important role [11–17]. The elimination of CD8+ T cells either through the use of in vivo depleting antibodies, or by producing NOD mice deficient in MHC class I (β_2-microglobulin knockout mice) that lack functional CD8+ T cells, results in the inhibition of autoimmune diabetes [18–22]. Also, β-islet-specific CD8+ T cells have been isolated from NOD mice that have a direct pathogenic effect upon adoptive transfer into nondiabetic recipients [17, 23]. Here, we will consider the possibility that defects in the deletion of high-affinity CD8+ T cells specific for β-cell antigens may contribute to autoimmune diabetes. Before considering this issue, it is helpful to review what is currently known about how CD8+ T-cell tolerance to peripheral antigens is normally established and its potential role in averting autoimmunity.

CD8+ T-Cell Tolerance in the Thymus

The primary mechanism of T-cell tolerance is thymic deletion. Such intrathymic clonal deletion was first described in normal mice expressing specific endogenous viral superantigens and substantiated using T-cell receptor (TCR) transgenic mice, that recognize a known self-antigen [24–26]. Most developing T cells never migrate to the periphery due to the rigorous selection processes they must first undergo in the thymus. The fate of a thymocyte is determined on the basis of the strength of the signal received through its clonotypic TCR [27–29]. These receptors focus on recognition of MHC molecules which contain self-peptide ligands [30, 31]. The strength of this signal appears to depend not only on the affinity of the TCR for the MHC-peptide complex, but also on the duration of this recognition event [32, 33]. In the complete absence of recognition of self-MHC-peptide, the thymocyte does not receive the signals necessary to survive and complete maturation. Weak recognition of self-antigen signals cell survival. Parenthetically, the signal delivered to the T cell on the basis of its recognition of self-MHC-peptide is required not only for thymocyte maturation, but also for continued survival of mature T cells in the periphery [34–38]. However, if the strength of this signal exceeds a certain threshold, the thymocyte is triggered to undergo apoptosis [39]. Antigens are most efficiently presented by bone-marrow-derived antigen-presenting cells (APC), either through their direct synthesis or by cross-presentation. This latter mechanism occurs when antigen produced by a non-APC is acquired and re-presented by the MHC molecules of an APC [40]. Most strong signals are delivered by such professional APCs and these cells are primarily responsible for deletion of potentially autoreactive T cells [41–44].

Recent data suggests that thymic deletion may also eliminate autoreactive T cells that recognize peripheral antigens that are not thought to be normally expressed in the thymus. Indeed, studies from a number of laboratories have demonstrated that low level expression of tissue-specific proteins, including pancreatic proteins and proteins previously thought to be unique to the nervous system, can occur within the thymus [45–51]. An elegant example is the work of Klein et al. [51], in which an inverse correlation was established between thymic expression of antigenic epitopes of proteolipid protein, a protein expressed in the central nervous system, and susceptibility to experimental autoimmune encephalomyelitis (EAE). The simplest explanation for expression of peripheral antigens in the thymus is that it facilitates self-tolerance and prevents otherwise lethal autoimmune disorders. This raises the question of whether mechanisms of peripheral tolerance are actually necessary to prevent autoimmunity. If peripheral antigens are expressed in the thymus, and available for tolerance induction, why do we need a mechanism of peripheral tolerance? First, not all peripheral antigens are expressed in the thymus, and second, deletion is most efficient for T cells that are strongly signaled by the presence of a stable MHC-peptide ligand. As discussed below, there are many exceptions to these stringent requirements, which leaves room for the escape to the periphery of large numbers of T cells with specificity for self-antigens.

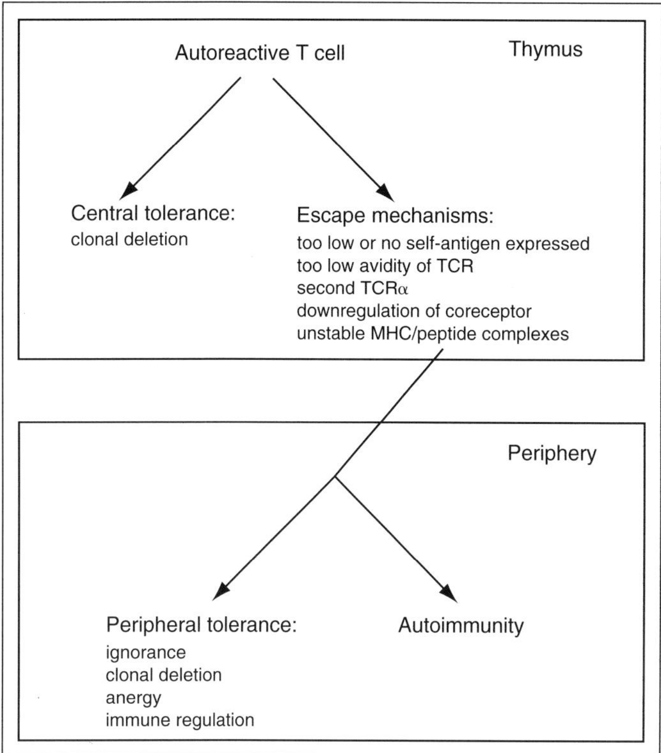

Fig. 1. The fate of an autoreactive T cell. Potentially autoreactive T cells can escape thymic deletion through several different mechanisms that have been described in recent years. These cells escape into the periphery, where they are dealt with by other tolerogenic mechanisms, such as ignorance, clonal deletion, anergy and immune regulation. Why these mechanisms are not operative during autoimmune diseases is still poorly understood.

Several mechanisms have been described that demonstrate how potentially autoreactive T cells escape thymic deletion (fig. 1). First, not all peripheral antigens are expressed in the thymus, or expressed at levels that are sufficient to facilitate deletion of all potentially autoreactive T cells. There is an inverse correlation between the amount of antigen available for recognition and the affinity of the T cells that are deleted. If very little antigen is available, only those thymocytes with the highest affinity receptors will be deleted. This could leave behind many T cells capable of recognition of this same antigen in the periphery [52–54]. Second, some thymocytes escape deletion by strategies that lower the strength of their interaction with the APC. An important example is the expression of more than one TCRα chain. Since the total amount of CD3 limits the amount of TCR expressed on the surface of T cells, a T cell that expresses two different TCRα chains, one of which is specific for a self-antigen, reduces the signal received by the cell and may avert apoptosis. A number of reports using TCR transgenic mice specific for thymic expressed antigen have demonstrated, that expression of two TCR can result in the escape to the periphery of potentially autoreactive T cells [55–59]. Similarly, it has been observed that some thymocytes undergoing negative selection can escape deletion through downregulation of the CD8 co-receptor [60–62].

The duration of signal through the TCR is an important consideration in the success of signaling. Accordingly, another mechanism that can account for the escape of self-specific cells from thymic deletion is based on the stability of the MHC-peptide complex. There is evidence that an unstable MHC-peptide complex can result in a deficiency in thymic deletion of autoreactive T cells specific for proteins from the central nervous system and pancreatic islets [63–67]. Finally, T cells can escape negative selection through a process called tuning, which actively reduces T-cell responsiveness to self-antigen encountered in the thymus by dampening signal transduction [68–70].

CD8+ T-Cell Tolerance in the Periphery

Considering that many potentially autoreactive T cells do make it out to the periphery, and yet autoimmunity is a relatively rare event, it is clear that these T cells must be dealt with in the periphery. Strategies that maintain self-tolerance include ignorance, clonal deletion, anergy and immune regulation (fig. 1). Ignorance and clonal deletion appear to be the major pathways for peripheral tolerance of CD8+ T cells. The important distinguishing feature is that ignorance can be overcome through some forms of antigenic stimulation, resulting in the emergence of autoreactive T cells, whereas deletion, by definition, cannot. Much progress has been made in recent years, elucidating the mechanism of CD8+ T-cell peripheral tolerance, using transgenic mice that express a defined model self-antigen and TCR transgenic mice that recognize this antigen [71–77].

Early studies by two different groups using the glycoprotein or nucleoprotein from lymphocytic choriomeningitis virus (LCMV-GP or NP), expressed as a self-antigen in the pancreatic β islets under the rat insulin promoter (RIP), revealed the ability of T cells specific for a peripheral antigen to remain in a naïve state of ignorance [71, 72]. Despite the expression of this self-antigen, these mice harbor GP/NP-specific CD8+ T cells that seem to ignore the cognate self-antigen in the periphery, either because the self-antigen is not presented by APCs, or because the T cells that are present have too low an avidity to detect the antigen. These autoreactive T cells could be activated by infection of the host with LCMV that contains the same GP or NP protein, subsequently leading to autoimmune diabetes. The same ignorance was observed when GP-specific CD8+ TCR transgenic mice were crossed with RIP-GP transgenic mice. Diabetes does not occur in these double transgenic mice, unless they are immunized with LCMV [71].

Ignorance also appears to be the modus operandi of many T cells specific for naturally occurring peripheral antigens. A number of autoimmune situations are initiated only after self-antigen is introduced concurrent with a strong adjuvant. One of the best understood models is EAE. Such experimentally-induced disease often resolves, however, it does indicate that potentially harmful autoreactive T cells are present, yet they simply ignore the antigen, when in their naïve state [78].

However, not all transgenic mice expressing model self-antigens in their β-islet cells demonstrate ignorance. More recently, Heath et al. [76, 77] have identified a mechanism of CD8+ T-cell peripheral tolerance using transgenic mice that express ovalbumin under the control of the rat insulin promoter (RIP-OVA). Upon adoptive transfer of CD8+ TCR transgenic cells that can recognize ovalbumin-peptide, OT-I, peripheral clonal deletion of these autoreactive CD8+ T cells occurs. This process relies upon the acquisition and presentation of self-antigen by professional APCs. These APCs acquire antigen in the islets and then migrate to the regional draining lymph nodes where it is presented to naïve T cells. The interaction between professional APCs and naïve autoreactive T cells leads to T-cell activation and division. However, in the absence of an inflammatory environment [79–84] or activated CD4+ T cells [85], the outcome of this activation is deletion of the T cells, rather than memory cell generation [77].

Although these studies indicated that a mechanism of peripheral deletion of autoreactive T cell exists, they could not demonstrate that this mechanism was actually employed to eliminate potentially autoreactive T cells. Indeed, it was not used as a mechanism of tolerance by mice expressing the RIP-OVA transgene. When mice expressing the OVA-specific TCR were mated to RIP-OVA mice, significant thymic deletion was observed due to expression of OVA in the thymus [76]. There seemed to be no need for peripheral tolerance in this model, since central tolerance already eliminated these potentially autoreactive CD8+ T cells. Furthermore, adoptive transfer of OT-I T cells (from mice that did not express RIP-OVA) into RIP-OVA mice leads to insulitis and diabetes, rather than peripheral tolerance. In order to circumvent this autoimmune attack, it was necessary to construct allogeneic bone marrow chimeras, in which the β-islet cells did not present the MHC molecules recognized by the OT-I cells, yet the bone-marrow-derived APCs could present the antigen to the T cells. Only then was it possible to observe deletion of OT-1 cells [77].

However, evidence that this mechanism of deletional peripheral tolerance was actually used to eliminate potentially autoreactive CD8+ T cells was later obtained in a different transgenic model studied in our laboratory, the InsHA mouse. These transgenic mice express the influenza hemagglutinin (HA) on pancreatic β-islet cells under the control of the rat insulin promoter [73]. In InsHA transgenic mice, peripheral expression of HA is sufficient to result in elimination of HA-specific CD8+ T cells. This was determined by experiments in which InsHA mice were thymectomized, irradiated and reconstituted with bone marrow and thymus from normal nontransgenic mice. These chimeras demonstrated tolerance to HA, even after immunization with influenza virus that contained the same HA molecule [73]. Examination of the few HA-specific CD8+ T cells that could be obtained from these mice, revealed that the HA-specific T-cell repertoire was functionally altered, such that CD8+ T cells with high avidity for this model self-antigen were deleted from the repertoire [73, 87, 88].

The mechanism responsible for this peripheral clonal deletion was revealed in studies using a CD8+ TCR transgenic mouse that can recognize a dominant HA epitope, Clone-4 TCR [74]. In contrast to the situation observed in RIP-OVA mice, Clone-4 TCR cells were not negatively selected in the thymus of InsHA transgenic mice [88]. Upon adoptive transfer of naïve Clone-4 TCR cells into InsHA mice, no diabetes was observed. By tracking the cells with the fluorescent dye, CFSE, it was revealed that they became activated in the pancreatic lymph nodes and underwent several rounds of division, yet the pancreas of InsHA recipient mice remained completely free of infiltrates. Rather, the T cells appeared to undergo deletion [74]. This was similar to the type of peripheral tolerance mechanism that had been described by Heath and co-workers [77], in which naïve T cells became activated in the pancreatic lymph nodes and were subsequently eliminated. An important difference between these two models is the fact that the T cells activated by HA were eliminated before they became effectors [J. Hernandez and L.A. Sherman manuscript in preparation].

Further evidence of peripheral deletion in this model came from studies in neonatal InsHA mice. Although adult InsHA mice demonstrate profound tolerance of HA, to our surprise it was observed that during the first few weeks after birth, immunization of InsHA mice with influenza virus resulted in the development of insulitis and diabetes. With increasing age, the ability to induce diabetes in the neonate gradually waned and tolerance was achieved by 4–8 weeks of age [74]. By adoptive transfer of CFSE-labeled Clone-4 TCR cells into InsHA neonates, it was found that activation of HA-specific CD8+ T cells in the pancreatic lymph nodes did not occur until sometime between 2 and 4 weeks of age. Considering that activation through cross-presentation is required for peripheral deletion, the inability to observe stimulation of Clone-4 TCR cells in the neonate correlated with the lack of deletion of the high-affinity HA-specific CD8+ T cells. This data suggests that induction of peripheral tolerance in this transgenic model is dependent upon the ability of the T cells to become activated in the pancreatic lymph nodes.

It is unknown why cross-presentation of HA does not occur in the neonate. It is possible that either the self-antigen or the relevant APC is limiting in quantity. However, the fact that deletional tolerance does not occur until the mice become mature presents the intriguing possibility that autoimmune diabetes occurs primarily in children and young adults because a greater number of CD8+ T cells are present in this juvenile period that have high-affinity for islet antigens. They have not yet been peripherally deleted, and are therefore available for participation in β-cell destruction.

In addition to the onset of presentation of self-antigen, several other parameters were found to affect the rate of peripheral deletion of CD8+ T cells, such as the amount of the cross-presented self-antigen [75, 89]. Indeed, it was directly demonstrated in the RIP-OVA model that the difference between ignorance and deletional tolerance was the amount of antigen available for cross-presentation. Using the InsHA mice, it was found that peripheral deletion of Clone-4 TCR cells occurred more quickly if there was a higher concentration of cross-presented HA-antigen. When InsHA$^{+/-}$ and InsHA$^{+/+}$ transgenic mice were compared with respect to their ability to peripherally tolerize naïve Clone-4 TCR cells, a higher proportion of the adoptively transferred cells were activated and proliferated in homozygous recipients, as compared to hemizygous recipients. This resulted in a more rapid elimination of Clone-4 TCR cells in the homozygous recipient mice [75].

Another important parameter that determines the rate of peripheral deletion is the actual number of HA-specific precursor T cells that are present. At any given time, only a small number of the HA-specific T cells became activated. As a consequence, the larger the number of naïve HA-specific T cells, the longer it took for all of the T cells to become activated and subsequently, eliminated. This would pose little problem within a conventional repertoire, as the number of T cells specific for any one epitope is exceedingly small [75]. It is unclear why there is a limitation of the number of T cells that become activated, however it may reflect a limited number of APCs that present sufficient amounts of antigen to activate these cells.

Despite the occurrence of peripheral deletion in this model, HA-specific CD8+ T cells do exist in InsHA animals [73, 86, 87]. After immunization of InsHA mice with influenza virus, a small proportion of pancreatic islets (<10%) demonstrate peri-insulitis that includes CD8+ T cells. Although these cells surround the islets, no destruction is apparent. These CD8+ T cells can be

Table 1. Comparison of the three major transgenic models for CD8+ T-cell tolerance

	LCMV [71, 72]	InsHA [73–75]	OVA [76, 77]
Antigen expressed under RIP	LCMV-nucleoprotein (NP) glycoprotein (GP)	Influenza hemagglutinin (HA)	Membrane ovalbumin (OVA)
Viral infection leads to	Diabetes	Tolerance	ND
Amount of antigen expressed	+	++	+++
Cross-presentation of self-antigen	Not detectable	Yes	Yes
CD8+ TCR transgenic	TCR-GP	Clone-4 TCR	OT-I
Thymic deletion	No	No	Yes
Spontaneous diabetes	No	Yes	No
Model for	Ignorance	Peripheral tolerance	Central tolerance

Three different transgenic models expressing model self-antigens under control of the rat insulin promoter have revealed different mechanisms of CD8+ T-cell tolerance. Depending on for instance the concentration of self-antigen tolerance is induced either by thymic deletion (OVA), peripheral deletion (InsHA) or leads to ignorance (LCMV).

retrieved from InsHA mice by stimulation in vitro with very high concentration of HA-peptide. Also, as compared to HA-specific CTL from nontransgenic mice, much more peptide is required to trigger their effector functions. Characterization of these HA-reactive CD8+ T cells on the basis of their ability to stably bind HA-containing MHC-tetramers, further confirmed that these CTL have low avidity for HA-antigen [87]. Presumably these cells were of sufficiently low avidity for the HA epitope that they escaped both thymic and peripheral deletion.

In conclusion, different transgenic models have revealed that the immune system uses different strategies to prevent autoimmunity (table 1). The InsHA transgenic model has been very helpful in understanding one of the strategies used to achieve CD8+ T-cell peripheral tolerance. In this model, naïve HA-specific CD8+ T cells, that have escaped negative selection in the thymus, circulate through the lymphoid tissue, until they encounter HA-antigen cross-presented by professional APCs in the pancreatic draining lymph nodes. Here, interaction leads to activation, division and deletion of the potentially autoreactive CD8+ T cells. This peripheral tolerance is developmentally regulated and over time results in an alteration of the CD8+ T-cell repertoire, leaving only HA-specific CD8+ T cells, that are of low avidity and form no threat to the immune system. The frequency with which this type of peripheral deletion mechanism is used to maintain tolerance to naturally-expressed self-antigens remains to be determined.

Tolerance to Islet Antigens in NOD Mice

Despite the tolerance mechanisms previously described, the immune system sometimes fails to maintain tolerance and autoimmunity occurs. A clear example is T1D, where T-cell-mediated destruction of pancreatic β cells leads to hyperglycemia and eventually death [90, 91]. Susceptibility to diabetes is determined by both environmental and genetic factors. Genetic predisposition is polygenic, with as many as 15–20 genes involved in progression to type I diabetes in mice and man. These observations suggest that autoimmmune disease represents a situation in which numerous events conspire to override multiple checkpoints that normally maintain self-tolerance. One of the best rodent models for autoimmune diabetes is the NOD. The strongest genetic determinant of autoimmune diabetes, both in mice and man, is the MHC genotype. The presence of certain MHC class II molecules, HLA-DQ8 in humans and I-A^{g7} in NOD mice, is strongly associated with disease susceptibility [92, 93]. These molecules share the same non-aspartic acid substitution at residue 57 in the β chain. As a consequence, the role of CD4+ T cells and the MHC class II molecules they recognize has been a major focus of research in diabetes. Previous data have demonstrated a generalized defect in self-tolerance within the CD4 compartment to certain self-antigens, such as GAD, carboxypeptidase, insulin, hsp60 and peripherin [94–96]. This defect results in spontaneous proliferation to self-antigens when presented by MHC class II I-A^{g7}. For this autoproliferation to occur, expression of this unusual MHC class II gene is required, but not sufficient, since I-A^{g7}-positive B10 mice do not exhibit autoproliferation, nor do NOD mice that lack I-A^{g7} molecules [97, 98]. This suggests that multiple NOD genes are required in order for CD4+ T-cell autoproliferation to occur.

Progression towards spontaneous autoimmune diabetes in NOD mice requires the presence of both CD4+ and CD8+ T cells [11–17]. There is ample evidence that CD4+ T cells in NOD mice are unusual in that they respond vigorously to self-antigens. However, much less is known concerning the status of self-tolerance within the CD8+ T-cell repertoire. To address this issue, we have employed the InsHA model described above. In other murine strains that express this transgene, such as BALB/c and B10.D2, there are no CD8+ T cells that exhibit high affinity for HA [73, 86, 87]. The HA-specific CD8+ T cells that are present in these mice demonstrate low avidity by a number of criteria, including dose response to cognate antigen, incomplete signal transduction in response to HA [Murtaza and Sherman, unpubl. results], and weak binding of K^dHA tetramers. Most importantly, they are unable to initiate islet destruction [87]. In contrast, examination of the immune response to HA in NOD-InsHA mice suggests that some autoreactive CD8+ T cells have escaped deletional tolerance and can be retrieved from NOD-InsHA mice. Characterization of these CTL showed a high avidity towards self-antigens as measured by efficient binding of K^dHA tetramers and responsiveness to low-dose antigen. More importantly, upon adoptive transfer into nondiabetic NOD-InsHA recipient mice, these HA-specific CD8+ T cells are able to initiate diabetes. Taken together, these data demonstrate that CD8+ T-cell peripheral tolerance to islet antigens is less successful in NOD mice than other strains [Kreuwel et al., submitted]. Whether CD8+ T-cell tolerance towards naturally-expressed islet antigens is defective as well, is unknown. Of interest, Wong et al. [99], have demonstrated the presence in NOD mice of pro-insulin-specific CD8+ T cells. However, it is not know if the presence of such T cells is unique to these diabetes-prone mice. Such T cells may normally be present in all strains and persist due to ignorance. If this is the case, than it is their spontaneous activation and survival which is unique to NOD, rather than their lack of deletion. Recent data, which demonstrates a weak interaction between the pro-insulin peptide in question and the K^d molecule which presents this epitope, would be compatible with the persistence of CD8+ T cells that ignore the antigen [100].

Disruption of any one of a number of steps, that constitute the normal chain of events that leads to CD8+ T tolerance, may result in incomplete CD8+ T-cell tolerance to HA in NOD-InsHA mice. Included among these would be the presence of autoreactive CD4+ T cells, intrinsic defects in CD8+ T-cell signaling or apoptosis, or defective antigen presentation by professional APCs. These possibilities are discussed below.

The Role of CD4+ T Cells

One of the factors that plays an important role in preventing tolerance of CD8+ T cells is the presence of CD4+ T cells [85, 101–103]. Recent data by Heath and co-workers [85] demonstrated that, in RIP-OVA mice, the presence of ovalbumin-specific CD4+ T cells impaired the clonal deletion of ovalbumin-specific CD8+ T cells. As a result, CD8+ T-cell tolerance was prevented and a large proportion of mice developed autoimmune diabetes. Normally, in the absence of CD4+ T-cell help, these autoreactive CD8+ T cells would undergo

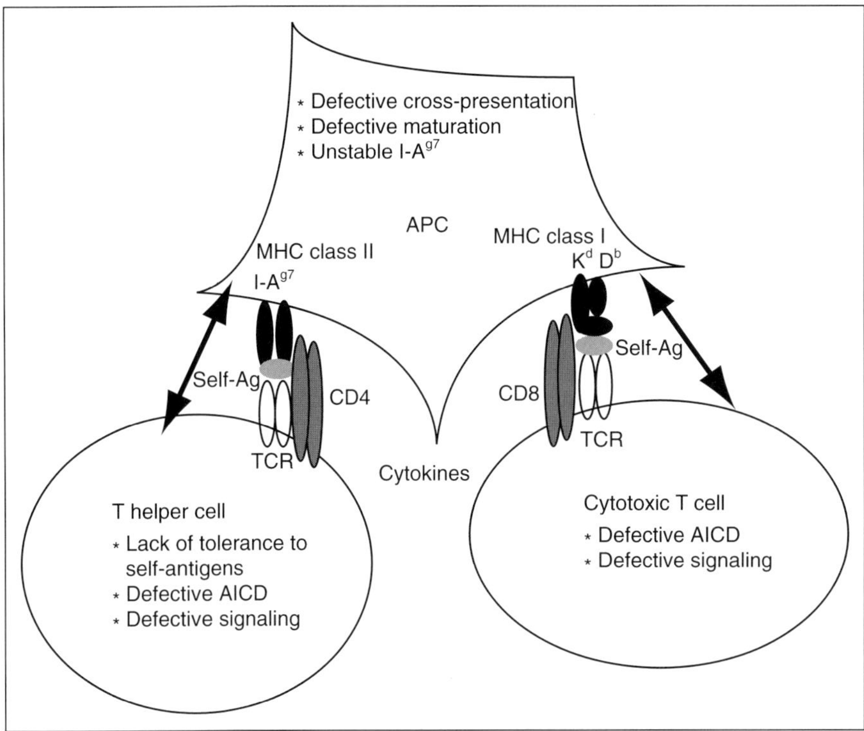

Fig. 2. Parameters that could prevent CD8+ T cell tolerance from occurring in the NOD. The defect in NOD CD8+ T cell tolerance could be attributable to intrinsic defects in the CD8+ T cell, such as resistance to activation-induced cell death (AICD) or defective T-cell signaling. Other cell types that could play a role in preventing CD8+ T-cell tolerance are CD4+ T-cell help and APCs.

peripheral deletion (as described above). The mechanism through which CD4+ T cells can prevent this deletion is unknown, but in this OVA model their presence seemed to increase the lifespan of the activated CD8+ T cells, rather than directly increasing the amount of clonal expansion.

NOD mice contain CD4+ T cells that spontaneously proliferate to self-antigens [94–96]. Therefore, it is possible that these autoreactive CD4+ T cells prevent the deletion of CD8+ T cells specific for islet antigens, as was observed in the RIP-OVA model. Autoreactive CD4+ T cells may prevent the normal induction of CD8+ T-cell peripheral tolerance either directly through production of cytokines or indirectly, through their effects on APCs [104–106]. We are currently investigating these possibilities in the NOD-InsHA model.

Intrinsic Defects in CD8+ T-Cell Signaling

Another factor, that could contribute to defects in T-cell tolerance, is an intrinsic defect in signal transduction through the TCR. This possibility is suggested by results from Delovitch and colleagues [107, 108]. They have observed that, as compared to thymocytes from other strains, both CD4+ and CD8+ SP thymocytes from NOD mice respond weakly to a variety of TCR stimuli including conA, anti-CD3 or anti-TCR antibodies as assessed on the basis of proliferative responses. This hyporesponsiveness was also observed in mature T cells in the periphery, when splenic T cells where cross-linked with anti-CD3 antibodies [107, 108]. This T-cell hyporesponsiveness is independent of APCs, suggesting it reflects an intrinsic defect in T-cell function. NOD T cells demonstrate diminished production of IL-2 and IL-4 which could be overcome by administration of exogenous IL-4 [108].

It is possible that defective signaling alters the threshold of antigen required for deletion of autoreactive thymocytes and peripheral T cells. Such reduced strength of signal could result in incomplete tolerance in the NOD T-cell repertoire.

Intrinsic Defects in T-Cell Apoptosis

Another way an intrinsic defect in T-cell function may contribute to incomplete T-cell tolerance involves molecules that act downstream of TCR signal transduction to regulate cellular apoptosis. Defects in the regulation of apoptosis could lead to loss of self-tolerance and immune homeostasis, resulting in autoimmunity.

Using several different apoptosis-inducing agents, such as corticosteroids, γ-irradiation, dexamethasone and cyclophosphamide, it was shown that both NOD thymocytes and mature T cells are more resistant to apoptosis than lymphocytes from other strains of mice [109–112]. Numerous molecules have been identified in recent years, that play a role in regulating apoptosis and autoimmunity, in particular members of the bcl-2 and TNF superfamily [113–119].

Several candidate molecules have been implicated in this defect in apoptosis. The apoptosis defect has been linked to Idd5, located on chromosome 1. Genes that have been mapped near this region include bcl-2 [111], which can prevent apoptosis, and the co-stimulatory molecules, CD28 and CTLA-4 [112]. These molecules are important in regulating IL-2 expression, which can regulate T-cell survival. To date, no difference in bcl-2 expression or sequence has been found between NOD and normal strains [110, 111]. However, decreased expression of both CD28 and CTLA-4 on CD4+ and CD8+ peripheral T cells after anti-CD3 cross-linking was found. Expression of CTLA-4 is necessary for downregulating the T-cell response. This downregulation could impair the control of T-cell activation, leading to autoimmunity [112]. Furthermore, CD28-deficient NOD mice demonstrate an increased incidence of diabetes [120].

Another interesting candidate that can regulate T-cell apoptosis is IL-2 [121]. IL-2 has pleiotropic effects during an immune response. It is an important growth and survival factor for T cells and it can also sensitize T lymphocytes to undergo Fas-mediated activation-induced cell death [122, 123]. In accordance, mice that lack the IL-2 gene, or a functional IL-2 receptor, accumulate T cells with an activated phenotype and develop autoimmunity [124–128]. Also the common IL-2Rγ chain has been implied in regulating IL-2-dependent activation-induced cell death in CD8+ T cells. Signaling through the IL-2Rγ chain can prevent apoptosis by inducing the anti-apoptosis genes bcl-2 and bcl-xl [129, 130]. Interestingly, it has been shown in NOD mice that the IL-2 gene closely maps to Idd3 [131, 132]. Alteration in IL-2 expression could contribute to defective T-cell apoptosis and autoimmunity.

More recently, three members of the TNF superfamily (CD30, TNFRII and CD137), located on chromosome 4 (Idd9) were identified as candidate genes controlling the development of both insulitis and diabetes in NOD mice [133]. Although the exact role of these molecules is still not clear, both CD30 and TNFRII have been implicated in regulating CD8+ T-cell apoptosis [134, 135].

The Role of APCs

APCs play an important role in tolerizing CD8+ T cells, both in the thymus and in the periphery. During T-cell development, recognition of self-peptides presented by bone marrow derived dendritic cells is responsible for deletion of autoreactive T cells [41–44]. In addition, APCs play an important role in the periphery by inducing CD8+ T-cell peripheral tolerance, as previously described [74, 76, 77]. Since APCs play such an important role in T-cell tolerance, any defects in their function could directly contribute to a deficiency of CD8+ T cell tolerance.

Indeed, certain defects in APCs have been reported both in diabetic patients and NOD mice. In diabetic patients, these differences include cell numbers, maturation and function of dendritic cells [136, 137]. These APCs express significantly lower amounts of B7-1 and B7-2 co-stimulatory molecules and have a reduced ability to stimulate autologous and allogeneic CD4+ T cells [137]. They also have a reduced ability to form clusters with T cells [136]. In NOD mice, similar defects in differentiation and function of bone-marrow-derived APCs were found. Suboptimal numbers of dendritic cells in NOD mice may contribute to the onset of diabetes. This is supported by the fact that adoptive transfer of dendritic cells into NOD mice can prevent diabetes in vivo [138, 139].

In summary, the observed defect in CD8+ T-cell tolerance in NOD mice could be attributable to disruption of any of one of a number of steps that are normally required for induction of CD8+ T-cell tolerance. Future studies will identify the cellular and genetic basis for the defect in CD8+ T-cell tolerance, which may lead to a better understanding of why autoimmune diabetes develops.

Acknowledgments

We would like to thank members of the Sherman laboratory for helpful discussion. This work was supported by National Institutes of Health grants DK 50824 and DK 57644.

References

1 Makino S, Kunimoto K, Muraoka Y, Mizushima Y, Katagiri K, Tochino Y: Breeding of a non-obese, diabetic strain of mice. Exp Anim 1980;29:1–13.
2 Tochino Y, Kanaya T, Makino S: Genetics of NOD mice; in Mimura G, Bab S, Goto Y, Kobberling J, (eds): Clinico-Genetic Genesis of Diabetes mellitus. Amsterdam, Excerpta Medica, 1981, pp 285–291.
3 Fujita T, Yui R, Kusumoto Y, Serizawa Y, Makino S, Tochino Y: Lymphocytic insulitis in a 'non-obese diabetic (NOD)' strain of mice: An immunohistochemical and electron microscope investigation. Biomed Res 1982;3:429–437.
4 Kanazawa Y, Komeda K, Sato S, Mori S, Akanuma K, Takaku F: Nonobese-diabetic mice: Immune mechanisms of pancreatic beta-cell destruction. Diabetologia 1984;27:113–115.
5 Bottazzo GF, Dean BM, McNally JM, MacHay EH, Swift PGF, Gamble DR: In situ characterization of autoimmune phenomena and expression of HLA molecules in the pancreas in diabetic insulitis. N Engl J Med 1985;313:353–360.
6 Hanninen A, Jalkanen S, Salmi M, Tiokkanen S, Nikolakaros G, Simell O: Macrophages, T cell receptor usage and endothelial cell activation in the pancreas at the onset of insulin-dependent diabetes mellitus. J Clin Invest 1992;90:1901–1910.
7 Itoh N, Hanafusa T, Miyazaki A, Miyagawa J, Yamagata K, Yamamoto K, Waguri M, Imagawa A, Tamura S, Inada M, et al: Mononuclear cell infiltration and its relation to the expression of major histocompatibility complex antigens and adhesion molecules in pancreas biopsy specimens from newly diagnosed insulin-dependent diabetes mellitus patients. J Clin Invest 1993;92:2313–2322.
8 Sibley RK, Sutherland DER, Goetz FC, Michael AF: Recurrent diabetes mellitus in the pancreas iso- and allograft: A light and electron microscopic and immunohistochemical analysis of four cases. Lab Invest 1985;53:132–144.
9 Koike T, Itoh Y, Ishii T, Ito I, Takabayashi K, Maruyama N, Tomioka H, Yoshida S: Preventive effect of monoclonal anti-L3T4 antibody on development of diabetes in NOD mice. Diabetes 1987;36:539–541.
10 Miyazaki A, Hanafusa T, Yamada K, Miyagawa J, Fujino-Kurihara H, Nakajima H, Nonaka K, Tarui S: Predominance of T lymphocytes in pancreatic islets and spleen of pre-diabetic non-obese diabetic (NOD) mice: A longitudinal study. Clin Exp Immunol 1985;60:622–630.
11 Hanafusa T, Sugihara S, Fujino-Kurrihara H, Miyagawa JI, Miyazaki A, Yoshioka T, Yamada K, Nakajima H, Asakawa H, Kono N, Fujiwara H, Hamaoka T, Tarui S. Induction of insulitis by adoptive transfer with L3T4$^+$ and Lyt2$^+$ T-lymphocytes in T-lymphocyte depleted NOD mice. Diabetes 1988;37:204–208.

12 Miller BJ, Appel MC, O'Neil JJ, Wicker LS: Both the Lyt-2$^+$ and L3T4$^+$ T cell subsets are required for the transfer of diabetes in nonobese diabetic mice. J Immunol 1988;140:52–58.

13 Bendelac A, Carnaud C, Boitard C, Bach JF: Syngeneic transfer of autoimmune diabetes from diabetic NOD mice to healthy neonates: Requirement for both L3T4$^+$ and Lyt2$^+$ T cells. J Exp Med 1987;166:823–832.

14 O'Reilly LA, Hutchings PR, Crocker PR, Simpson E, Lund T, Kioussis D, Takei F, Baird J, Cooke A: Characterization of pancreatic islet cell infiltrates in NOD mice: Effect of cell transfer and transgene expression. Eur J Immunol 1991;21:1171–1180.

15 Reich EP, Sherwin RS, Kanagawa O, Janeway C Jr: An explanation for the protective effect of the MHC class II I-E molecule in murine diabetes. Nature (Lond) 1989;341:326–328.

16 Shimizu J, Kanagawa O, Unanue ER: Presentation of β-cell antigens to CD4$^+$ and CD8$^+$ T cells of nonobese diabetic mice. J Immunol 1993;151:1723–1730.

17 Nagata M, Santamaria P, Kawamura T, Utsugi T, Yoon JW: Evidence for the role of CD8+ cytotoxic T cells in the destruction of pancreatic β cells in nonobese diabetic mice. J Immunol 1994;152:2042–2050.

18 Katz J, Benoist C, Mathis D: Major histocompatibility complex class I molecules are required for the generation of insulitis in non-obese diabetic mice. Eur J Immunol 1993;23:3358–3360.

19 Wicker LS, Leiter EH, Todd JA, Renjilian RJ, Peterson E, Fischer PA, Podolin PL, Zijlstra M, Jaenisch R, Peterson LB: β$_2$-Microglobulin-deficient NOD mice do not develop insulitis or diabetes. Diabetes 1994;43:500–504.

20 Serreze DV, Leiter EH, Christianson GJ, Greiner D, Roopenian DC: Major histocompatibility complex class I-deficient NOD-β$_2$m^{null} mice are diabetes and insulitis resistant. Diabetes 1994;4:505–509.

21 Sumida T, Furukawa M, Sakamoto A, Namekawa T, Maeda T, Zijlstra M, Iwamoto I, Koike T, Yoshida S, Tomioka H, Taniguchi M: Prevention of insulitis and diabetes in β$_2$-microglobulin-deficient non-obese diabetic mice. Int Immunol 1994;6:1445–1449.

22 Wang B, Gonzalez C, Benoist C, Mathis D: The role of CD8$^+$ T cells in initiation of insulin-dependent diabetes mellitus. Eur J Immunol 1996;26:1762–1769.

23 Wong SF, Visintin I, Wen L, Flavell RA, Janeway CA Jr: CD8 T cell clones from young nonobese diabetic (NOD) islets can transfer rapid onset of diabetes in NOD mice in the absence of CD4 cells. J Exp Med 1996;183:67–76.

24 Kappler JW, Roehm N, Marrack P: T cell tolerance by clonal elimination in the thymus. Cell 1987;49:273–280.

25 Kisielow P, Bluthmann H, Staerz UD, Steinmetz M, von Boehmer H: Tolerance in T cell receptor transgenic mice involves deletion of non-mature CD4+8+ thymocytes. Nature 1988;333:742–746.

26 MacDonald HR, Schneider R, Lees RK, Howe RC, Acha-Orbea H, Festenstein H, Zinkernagel RM, Hengartner H: T cell receptor Vb use predicts reactivity and tolerance to MlsA-encoded antigens. Nature 1988;322:40–45.

27 Sprent J, Webb SR: Function and specificity of T cell subsets in the mouse. Adv Immunol 1987;41:39–133.

28 Sebzda E, Wallace VA, Mayer J, Yeung RS, Mak TW, Ohashi PS: Positive and negative thymocyte selection induced by different concentrations of a single peptide. Science 1994;263:1615–1618.

29 Ashton-Rickardt PG, Bandeira A, Delaney JR, van Kaer L, Pircher HP, Zinkernagel RM, Tonegawa S: Evidence for a differential avidity model of T cell selection in the thymus. Cell 1994;76:651–663.

30 Bevan MJ: In a radiation chimera, host H-2 antigens determine immune responsiveness of donor cytotoxic cells. Nature 1977;269:417–418.

31 Zinkernagel RM, Callahan GN, Klein J, Dennert G: Cytotoxic T cells learn specificity for self-H-2 during differentiation in the thymus. Nature 1987;271:251–253.

32 Lyons DS, Lieberman SA, Hampl J, Boniface JJ, Chien Y, Berg LJ, Davis MM: A TCR binds to antagonist ligands with lower affinities and faster dissociation rates than to agonists. Immunity 1996;5:53–61.

33 Alam SM, Tanvers PJ, Wung JL, Nasholds W, Redpath S, Jameson SC, Gascoigne NRJ: T-cell-receptor affinity and thymocyte positive selection. Nature 1996;381:616–620.

34 Ernst B, Lee DS, Chang JM, Sprent J, Surh CD: The peptide ligands mediating positive selection in the thymus control T cell survival and homeostatic proliferation in the periphery. Immunity 1999;11:173–181.

35 Markiewicz MA, Girao C, Opferman JT, Sun J, Hu Q, Agulnik AA, Bishop CE, Thompson CB , Ashton-Rickardt PG: Long-term T cell memory requires the surface expression of self-peptide/major histocompatibility complex molecules. Proc Natl Acad Sci USA 1998;95: 3065–3070.

36 Kirberg J, Berns A, von Boehmer H: Peripheral T cell survival requires continual ligation of the T cell receptor to major histocompatibility complex-encoded molecules. J Exp Med 1997;186: 1269–1275.

37 Brocker T: Survival of mature CD4 T lymphocytes is dependent on major histocompatability complex class II-expressing dendritic cells. J Exp Med 1997;186:1223–1232.

38 Tanchot C, Lemonnier FA, Perarnau B, Freitas AA, Rocha B: Differential requirements for survival and proliferation of CD8 naïve or memory T cells. Science 1997;276:2057–2062.

39 Surh CD, Sprent J: T-cell apoptosis detected in situ during positive and negative selection. Nature 1994;372:100–103.

40 Bevan MJ: Cross-priming for a secondary cytotoxic response to minor H antigens with H-2 congenic cells which do not cross-react in the cytotoxic assay. J Exp Med 1976;143:1283–1288.

41 Brocker T, Riedinger M, Karjalainen K: Targeted expression of major histocompatability complex (MHC) class II moleculed demonstrates that dendritic cells can induce negative but not positive selection of thymocytes in vivo. J Exp Med 1997;185:541–550.

42 Matzinger P, Guerder S: Does T-cell tolerance require a dedicated antigen-presenting cell? Nature 1989;338:74–76.

43 Marrack P, Lo D, Brinster R, Palmiter L, Burkly R, Flavell H, Kappler J: The effect of thymus environment on T cell development and tolerance. Cell 1988;53:627–634.

44 Sprent J, Lo D, Gao EK, Ron Y: T cell selection in the thymus. Immunol Rev 1988;101:173–190.

45 Kojima K, Reindl M, Lassmann H, Wekerle H, Linington C: The thymus and self-tolerance: Co-existence of encephalitogenic S100β-specific T cells and their nominal autoantigen in the normal adult rat thymus. Int Immunol 1997;8:897–904.

46 Hanahan D: Peripheral-antigen-expressing cells in thymic medulla: Factors in self-tolerance and autoimmunity. Curr Opin Immunol 1998;10:656–662.

47 Klein L, Klein T, Rüther U, Kyewski B: CD4 T-cell tolerance to human C-reactive protein, an inducible serum protein, is mediated by medullary thymic epithelium. J Exp Med 1998;188:5–16.

48 Heath VL, Moore NC, Parnell SM, Mason DW: Intrathymic expression of genes involved in organ-specific autoimmune disease. J Autoimmun 1998;11:309–318.

49 Sospedra M, Ferrer-Francesch X, Dominguez O, Juan M, Foz-Sala M, Pujol-Borrell R: Transcription of a broad range of self-antigens in human thymus suggests a role for central mechanisms in tolerance toward peripheral antigens. J Immunol 1998;161:5918–5929.

50 Anderson AC, Nicholson LB, Legge KL, Turchin V, Zaghouani H, Kuchroo V: High frequency of autoreactive myelin proteolipid protein-specific T cells in the periphery of naïve mice: Mechanisms of selection of the self-reactive repertoire. J Exp Med 2000;191:761–770.

51 Klein L, Klugmann M, Nave KA, Tuohy V, Kyewski B: Shaping of the autoreactive T-cell repertoire by a splice variant of self-protein expressed in thymic epithelial cells. Nat Med 2000;6:56–61.

52 Sandberg JK, Franksson L, Sundbäck J, Michaelsson J, Petersson M, Achour A, Wallin RPA, Sherman NE, Bergman T, Jörnvall H, Hunt DF, Kiessling R, Kärre K: T cell tolerance based on avidity thresholds rather than complete deletion allows maintenance of maximal repertoire diversity. J Immunol 2000;165:25–33.

53 Oehen SU, Ohashi PS, Burki K, Hengartner H, Zinkernagel RM, Aichele P: Escape of thymocytes and mature T cells from clonal deletion due to limiting tolerogen expression levels. Cell Immunol 1994;158:342–352.

54 Liu GY, Fairchild PJ, Smith RM, Prowle JR, Kioussis D, Wraith DC: Low avidity recognition of self-antigen by T cells permits escape from central tolerance. Immunity 1995;4:407–415.

55 Padovan E, Casorali G, Dellatona P, Meyer S, Brockhaus M, Lanzavecchia A: Expression of two T cell receptor α chains: Dual receptor T cells. Science 1993;264:422–424.

56 Heath WR, Carbone FR, Bertolino P, Kelly J, Cose S, Miller JFAP: Expression of two T cell receptor α chains on the surface of normal murine T cells. Eur J Immunol 1995;25:1617–1623.

57 Blichfeldt E, Munthe LA, Rotnes JS, Bogen B: Dual T cell receptor T cells have a decreased sensitivity to physiological ligands due to reduced density of each T cell receptor. Eur J Immunol 1995;26:2876–2884.

58 Zal T, Weiss S, Mellor A, Stockinger B: Expression of a second receptor rescues self-specific T cells from thymic deletion and allows activation of autoreactive effector function. Proc Natl Acad Sci USA 1996;93:9102–9107.

59 Sarukhan A, Garcia C, Lanoue A, von Boehmer H: Allelic inclusion of T cell receptor alpha genes poses an autoimmune hazard due to low-level expression of autospecific receptors. Immunity 1998;8:563–570.

60 Teh HS, Kishi H, Scott B, von Boehmer H: Deletion of autospecific T cells in T cell receptor (TCR) transgenic mice spares cells with normal TCR levels and low levels of CD8 molecule. J Exp Med 1989;169:795–806.

61 Sha WC, Nelson CA, Newberry RD, Pullen JK, Pease LR, Russell JH, Loh DY: Positive selection of transgenic receptor-bearing thymocytes by K^b antigen is altered by K^b mutations that involve peptide binding. Proc Natl Acad Sci USA 1990;87:6186–6190.

62 Barnden MJ, Heath WR, Carbone FR: Down-modulation of CD8 β-chain in response to an altered peptide ligand enables developing thymocytes to escape negative selection. Cell Immunol 1997; 175:111–119.

63 Reich EP, von Grafenstein H, Barlow A, Swenson KE, Williams K, Janeway CA Jr: Self-peptides isolated from MHC glycoproteins of non-obese diabetic mice. J Immunol 1994;152:2279–2288.

64 Carrasco-Marin E, Shimizu J, Kanagawa O, Unanue E: The class II MHC I-A^{g7} molecules from non-obese diabetic mice are poor peptide binders. J Immunol 1996;156:450–458.

65 Harrington CJ, Paez A, Hunkapiller T, Mannikko V, Brabb T, Ahearn M, Beeson C, Goverman J: Differential tolerance is induced in T cells recognizing distinct epitopes of myelin basic protein. Immunity 1998;8:571–580.

66 Mason K, Denney DW Jr, McConnell HM: Kinetics of the reaction of a myelin basic protein peptide with solube Iau. Biochemistry 1995;34:14874–14878.

67 Fairchild PJ, Wildgoose R, Atherton E, Webb S, Wraith DC: An autoantigenic T cell epitope forms unstable complexes with class II MHC: A novel route for escape from tolerance induction. Int Immunol 1993;5:1151–1158.

68 Grossman Z, Singer A: Tuning of activation thresholds explains flexibility in the selection and development of T cells in the thymus. Proc Natl Acad Sci USA 1996;93:14747–14752.

69 Sebzda E, Kündig TM, Thompson CT, Aoki K, Mak SY, Mayer JP, Zamborelli T, Nathenson SG, Ohashi PS: Mature T cell reactivity altered by peptide agonist that induces positive selection. J Exp Med 1996;183:1093–1104.

70 Kawai K, Ohashi PS: Immunological function of a defined T cell population tolerized to low-affinity self-antigens. Nature 1995;374:68–69.

71 Ohashi PS, Oehen S, Buerki K, Pircher H, Ohashi CT, Odermatt B, Malissen B, Zinkernagel RM, Hentgartner H: Ablation of 'tolerance' and induction of diabetes by virus infection in viral antigen transgenic mice. Cell 1991;65:305–317.

72 Oldstone MB, Nerenberg M, Southern P, Price J, Lewicki H: Virus infection triggers insulin-dependent diabetes mellitus in a transgenic model: Role of anti-self (virus) immune response. Cell 1991;65:319–331.

73 Lo D, Freedman J, Hesse S, Palmiter RD, Brinster RL, Sherman LA: Peripheral tolerance to an islet cell-specific hemagglutinin transgene affects both CD4$^+$ and CD8$^+$ T cells. Eur J Immunol 1992;22:1013–1022.

74 Morgan DJ, Kurts C, Kreuwel HTC, Holst K, Heath WR, Sherman LA: Ontogeny of T cell tolerance to peripherally expressed antigens. Proc Natl Acad Sci USA 1999;96:3854–3858.

75 Morgan DJ, Kreuwel HTC, Sherman LA: Antigen concentration and precursor frequency determine the rate of CD8$^+$ T cell tolerance to peripherally expressed antigens. J Immunol 1999;163: 723–727.

76 Kurts C, Heath WR, Carbone FR, Allison J, Miller JFAP, Kosaka H: Constitutive Class I-restricted exogenous presentation of self-antigens in vivo. J Exp Med 1996;184:923–930.

77 Kurts C, Kosaka H, Carbone FR, Miller JFAP, Heath WR: Class I-restricted cross-presentation of exogenous self-antigens leads to deletion of autoreactive CD8+ T cells. J Exp Med 1997;186: 239–245.

78 Mokhtarian F, McFarlin DE, Raine CS: Adoptive transfer of myelin basic protein-sensitized T cells produces chronic relapsing demyelinating disease in mice. Nature 1984;309:356–358

79 Ehl S, Hombach J, Aichele P, Rülicke T, Odermatt B, Hengartner H, Zinkernagel R, Pircher H: Viral and bacterial infections interfere with peripheral tolerance induction and activate CD8+ T cells to cause immunopathology. J Exp Med 1998;187:763–774.

80 Marrack P, Kappler J, Mitchell T: Type I interferons keep activated T cells alive. J Exp Med 1999;189:521–529.

81 Mitchell T, Kappler J, Marrack P: Bystander virus infection prolongs activated T cell survival. J Immunol 1999;162:4527–4535.

82 Tough DF, Borrow P, Sprent J: Induction of bystander T cell proliferation by viruses and type I interferon in vivo. Science 1996;272:1947–1950.

83 Rocken M, Urban JF, Shevach EM: Infection breaks T-cell tolerance. Nature 1992;359:79–82.

84 Aoki Y, Hiromatsu K, Usami J, Makino M, Igarashi H, Ogasawara J, Nagata S, Yoshikai Y: Clonal expansion but lack of subsequent clonal deletion of bacterial superantigen-reactive T cells in murine retroviral infection J Immunol 1994;153:3611–3621.

85 Kurts C, Carbone FR, Barnden M, Blanas E, Allison J, Heath WR, Miller JFAP: CD4$^+$ T cell help impairs CD8$^+$ T cell deletion induced by cross-presentation of self-antigens and favors auto-immunity. J Exp Med 1997;186:2057–2062.

86 Morgan DJ, Kreuwel HTC, Fleck S, Levitsky HI, Pardoll DM, Sherman LA: Activation of low avidity CTL specific for a self-epitope results in tumor rejection but not autoimmunity. J Immunol 1998;160:643–651.

87 Nugent CT, Morgan DJ, Biggs JA, Ko A, Pilip IM, Pamer EG, Sherman LA: Characterization of CD8$^+$ T lymphocytes which persist after peripheral tolerance to a self-antigen expressed in the pancreas. J Immunol 2000;164:191–200.

88 Morgan DJ, Liblau R, Scott B, Fleck S, McDevitt HO, Lo D, Sherman LA: CD8$^+$ T cell mediated spontaneous diabetes in neonatal mice. J Immunol 1996;157:978–983.

89 Kurts C, Sutherland RM, Davey G, Li M, Lew AM, Blanas E, Carbone FR, Miller JFAP, Heath WR: CD8 T cell ignorance or tolerance to islet antigens depends on antigen dose. Proc Natl Acad Sci USA 1999;22:12703–12707.

90 Tisch R, McDevitt H: Insulin-dependent diabetes mellitus. Cell 1996;85:291–297.

91 Bach JF: Insulin-dependent diabetes mellitus as an autoimmune disease. Endocr Rev 1994;15: 516–542.

92 Acha-Orbea H, McDevitt H: The first external domain of the nonobese diabetic mouse class II I-A beta chain is unique. Proc Natl Acad Sci USA 1997;84:2435–2439.

93 Todd JA, Bell JI, McDevitt H: HLA-DQ beta gene contributes to susceptibility and resistance to insulin-dependent diabetes mellitus. Nature 1987;329:599–604.

94 Kaufman DL, Clare-Salzler M, Tian J, Forsthuber T, Ting GSP, Robinson P, Atkinson MA, Sercarz EE, Tobin AJ, Lehmann PV: Spontaneous loss of T-cell tolerance to glutamic acid decarbo-xylase in murine insulin-dependent diabetes. Nature 1993;366:69–72.

95 Tisch R, Yang XD, Singer SM, Liblau RS, Fugger L, McDevitt H: Immune response to glutamic acid decarboxylase correlates with insulitis in non-obese diabetic mice. Nature 1993;366: 72–75.

96 Ridgway WM, Fasso M, Lanctot A, Garvey C, Fathman CG: Breaking self-tolerance in nonobese diabetic mice. J Exp Med 1996;183:1657–1662.

97 Wicker LS, Miller BJ, Coker LZ, McNally SE, Scott S, Mullen Y, Appel MC: Genetic control of diabetes and insulitits in the nonobese diabetic (NOD) mouse. J Exp Med 1987;165:1639–1654.

98 Ridgway WM, Ito H, Fasso M, Yu C, Fathman CG: Analysis of the role of variation of major histocompatibility complex class II expression on nonobese diabetic (NOD) peripheral T cell response. J Exp Med 1998;188:2267–2275.

99   Wong SF, Karttunen J, Dumont D, Wen L, Visintin I, Pilip IM, Shastri N, Pamer EG, Janeway CA Jr: Identification of an MHC class I-restricted autoantigen in type I diabetes by screening an organ-specific cDNA library. Nat Med 1999;5:1026–1031.

100 Wong FS, Wen L, Visintin I, Janeway CA Jr: The peptide recognized by highly cytotoxic CD8 T cells in type 1 diabetes binds poorly to the MHC; in Mechanisms of Immunologic Tolerance and its Breakdown. Steamboat Springs, Colo, Keystone Symposia, 2000.

101 Guerder S, Matzinger P: A fail-safe mechanism for maintaining self-tolerance. J Exp Med 1992; 176:553–564.

102 Rees MA, Rosenberg AS, Munitz TI, Singer A: In vivo induction of antigen-specific transplantation tolerance to Qa1a by exposure to alloantigen in the absence of T-cell help. Proc Natl Acad Sci USA 1990;87:2765–2769.

103 Kirberg J, Bruno L, von Boehmer H: CD4$^+$ 8$^-$ help prevents rapid deletion of CD8$^+$ cells after a transient response to antigen. Eur J Immunol 1993;23:1963–1967.

104 Ridge JP, Di Rosa F, Matzinger P: A conditional dendritic cell can be a temporal bridge between a CD4$^+$ T-helper and a T-killer cell. Nature 1998;393:474–478.

105 Bennett SMR, Carbone FR, Karamalis F, Miller JFAP, Heath WR: Help for inducing cytotoxic T cell responses by cross-priming is mediated via CD40 signaling. Nature 1998;393:478–480.

106 Schoenberger SP, Toes RE, van der Voort IE, Offringa R, Melief CJ: T cell help for cytotoxic T lymphocytes is mediated by CD40-CD40L interactions. Nature 1998;393:480–483.

107 Zipris D, Lazarus AH, Crow AR, Hadzija M, Delovitch TL: Defective thymic T cell activation by concanavalin A and anti-CD3 in autoimmune nonobese diabetic mice. Evidence for thymic T cell anergy that correlates with the onset of insulitis. J Immunol 1991;146:3763–3771.

108 Rapaport M, Jaramillo JA, Zipris D, Lazarus AH, Serreze DV, Leiter EH, Cyopick P, Danska JS, Delovitch TL: Interleukin-4 reverses T cell proliferative unresponsiveness and prevents the onset of diabetes in nonobese diabetic mice. J Exp Med 1993;178:87–99.

109 Lamhamedi-Cherradi SE, Luan JJ, Eloy L, Fluteau G, Bach JF, Garchon HJ: Resistance of T cells to apoptosis in autoimmune diabetic (NOD) mice is increased early in life and is associated with dysregulated expression of Bcl-x. Diabetologia 1998;41:178–184.

110 Leijon K, Hammarström B, Holmberg D: Non-obese diabetic (NOD) mice display enhanced immune responses and prolonged survival of lymphoid cells. Int Immunol 1994;6:339–345.

111 Garchon H, Luan JJ, Eloy L, Bédossa P, Bach JF: Genetic analysis of immune dysfunction in non-obese diabetic (NOD) mice: Mapping of a susceptibility locus close to the Bcl-2 gene correlates with increased resistance of NOD T cells to apoptosis induction. Eur J Immunol 1994;24: 380–384.

112 Colucci F, Bergman ML, Penha-Gonçalves C, Cilio CM, Holmberg D: Apoptosis resistance of nonobese diabetic peripheral lymphocytes linked to the Idd5 diabetes susceptibility region. Proc Natl Acad Sci USA 1997;94:8670–8674.

113 Watanabe-Fukunaga R, Brannan CI, Copeland NG, Jenkins NA, Nagata S: Lymphoproliferation disorder in mice explained by defects in Fas antigen that mediates apoptosis. Nature 1992;356: 314–317.

114 Takahashi T, Tanaka M, Brannan DI, Jenkins NA, Copeland NG, Suda T, Nagata S: Generalized lymphoproliferative disease in mice, caused by a point mutation in the Fas ligand. Cell 1994;76: 969–976.

115 Suzuki H, Kündig TM, Furlonger C, Wakeham A, Timms E, Matsuyama T, Schmits R, Simard JJL, Ohashi PS, Griesser H, Taniguchi T, Paige CJ, Mak TW: Deregulated T cell activation and autoimmunity in mice lacking interleukin-2 receptor β. Science 1995;268:1472–1476.

116 Willerford DM, Chen J, Ferry JA, Davidson L, Ma A, Alt FW: Interleukin-2 receptor (chain regulated the size and content of the peripheral lymphoid compartment. Immunity 1995;3: 521–530.

117 Di Christofano A, Kotsi P, Peng YF, Cordon-Cardo C, Elkon KB, Pandolfi PP: Impaired Fas response and autoimmunity in Pten $+/-$ mice. Science 1999;285:2122–2125.

118 Bouillet P, Metcalf D, Huang DC, Tarlington DM, Kay TW, Kontgen F, Adams JM, Strasser A: Proapoptotic Bcl-2 relative Bim required for certain apoptotic responses, leukocyte homeostasis, and to preclude autoimmunity. Science 1999;286:1735–1738.

Defects in Deletional Tolerance of CD8+ T Cells in Autoimmune Diabetes

119 Van Parijs L, Refaeli Y, Abbas AK, Baltimore D: Autoimmunity as a consequence of retrovirus-mediated expression of C-FLIP in lymphocytes. Immunity 1999;11:763–770.

120 Lenschow DJ, Herold KC, Rhee L, Patel B, Koons A, Qin HY, Fuchs E, Singh B, Thompson CB, Bluestone JA: CD28/B7 regulation of Th1 and Th2 subsets in the development of autoimmune diabetes. Immunity 1996;5:285–293.

121 Lenardo MJ: Interleukin-2 programs mouse $\alpha\beta$ T lymphocytes for apoptosis. Nature 1991;353:858–861.

122 Refaeli Y, Van Parijs L, London CA, Tschopp J, Abbas AK: Biochemical mechanisms of IL-2-regulated Fas-mediated T cell apoptosis. Immunity 1998;8:615–623.

123 Kneitz B, Herrmann T, Yonehara S, Schimpl A: Normal clonal expansion but impaired Fas-mediated cell death and anergy induction in interleukin-2-deficient mice. Eur J Immunol 1995;25:2572–2577.

124 Sadlack B, Merz H, Schorle H, Schimpl A, Feller AC, Horak I: Ulcerative colitis-like disease in mice a disrupted interleukin-2 gene. Cell 1993;75:253–261.

125 Suzuki H, Kundig TM, Furlonger C, Wakeman A, Timms E, Matsuyama T, Schmitts T, Simard J, Ohashi PS, Greisser H, et al: Deregulated T cell activation and autoimmunity in mice lacking IL-2 receptor beta. Science 1995;268:1472–1476.

126 Willerford DM, Chen J, Ferry JA, Davidson L, Ma A, Alt FW: Interleukin-2 receptor alpha chain regulates the size and content of the peripheral lymphoid compartment. Immunity 1995;3:521–530.

127 Cao X, Shores EW, Hu-Li J, Anver MR, Kelsall BL, Russell SM, Drago J, Noguchi M, Grinberg A, Bloom ET: Defective lymphoid development in mice lacking expression of the common cytokine receptor γ chain. Immunity 1995;2:223–238.

128 DiSanto JP, Muller W, Guy-Grand D, Fischer A, Rajewsky K: Lymphoid development in mice with a targeted deletion of the interleukin-2 receptor gamma chain. Proc Natl Acad Sci USA 1995;92:377–381.

129 Dai Z, Arakelov A, Wagener M, Konieczny BT, Lakkis FG: The role of the common cytokine receptor γ-chain in regulating IL-2-dependent, activation-induced CD8[+] T cell death. J Immunol 1999;163:3131–3137.

130 Akbar AN, Borthwick NJ, Wickremasinghe RG, Panayiotidis P, Pilling D, Bofill M, Krajewski S, Reed JC, Salmon M: Interleukin-2 receptor common γ chain signaling cytokines regulate activated T cell apoptosis in response to growth factor withdrawal: Selective induction of anti-apoptotic (bcl-2, bcl-x_l) but not pro-apoptotic (bax, bcl-x_s) gene expression. Eur J Immunol 1996;26:294–299.

131 Wicker LS, Todd JA, Prins JB, Podolin PL, Renjilian RJ, Peterson LB: Resistance alleles at two non-major histocompatibility complex-linked insulin-dependent diabetes loci on chromosome 3, $Idd3$ and $Idd10$, protect nonobese diabetic mice from diabetes. J Exp Med 1994;180:1705–1713.

132 Denny P, Lord CJ, Hill NJ, Goy JV, Levy ER, Podolin PL, Peterson LB, Wicker LS, Todd JA, Lyons PA: Mapping of the IDDM locus, $Idd3$ to a 0.35-cM interval containing the $Interleukin-2$ gene. Diabetes 1997;46:695–700.

133 Lyons PA, Hancock WW, Denny P, Lord CJ, Hill NJ, Armitage N, Siegmund T, Todd JA, Philips MS, Hess JF, Chen SL, Fischer PA, Peterson LB, Wicker LS: The NOD Idd9 genetic interval influences the pathogenicity of insulitis and contains molecular variants of Cd30, Tnfr2 and Cd137. Immunity 2000;13:107–115.

134 Zheng L, Fisher G, Miller RE, Peschon J, Lynch DH, Lenardo MJ: Induction of apoptosis in mature T cells by tumour necrosis factor. Nature 1995;377:348–351.

135 Telford WG, Nam SY, Podack ER, Miller RA: CD30-regulated apoptosis in murine CD8 T cells after cessation of TCR signals. Cell Immunol 1997;182:125–136.

136 Jansen A, van Hagen M, Drexhage HA: Defective maturation and function of antigen-presenting cells in type 1 diabetes. Lancet 1995;345:491–492.

137 Takahashi K, Honeyman MC, Harrison LC: Impaired yield, phenotype and function of monocyte-derived dendritic cells in humans at risk for insulin-dependent diabetes. J Immunol 1998;161:2629–2635.

138 Feili-Hariri M, Dong X, Alber SM, Watkins SC, Salter RD, Morel PA: Immunotherapy of NOD
     mice with bone marrow-derived dendritic cells. Diabetes 1999;48:2300–2308.
139 Clare-Salzler MJ, Brooks J, Chai A, Van Herle K, Anderson C: Prevention of diabetes in nonobese
     diabetic mice by dendritic cell transfer. J Clin Invest 1992;90:741–748.

Linda A. Sherman, PhD, Department of Immunology, The Scripps Research Institute,
10550 North Torrey Pines Road, IMM-15, La Jolla, CA 92037 (USA)
Tel. +1 858 784 8052, Fax +1 858 784 8298, E-Mail lsherman@scripps.edu

von Herrath MG (ed.): Molecular Pathology of Type 1 Diabetes mellitus.
Curr Dir Autoimmun. Basel, Karger, 2001, vol 4, pp 144–170

....

How Beta Cells Die in Type 1 Diabetes

Helen E. Thomas, Thomas W.H. Kay

The Walter and Eliza Hall Institute of Medical Research,
Royal Melbourne Hospital, Parkville, Vic., Australia

Pancreatic β cells secrete insulin in response to glucose to maintain normoglycemia. Type 1 (insulin-dependent) diabetes results from T-cell-mediated destruction of pancreatic β cells. There is a considerable amount of research investigating the mechanisms by which this selective β-cell destruction occurs. This review will focus on the way in which β cells can be killed both in experimental systems in vitro and in vivo during progression to diabetes.

Many studies on β-cell destruction have been performed using animal models of diabetes such as the nonobese diabetic (NOD) mouse, as well as different transgenic models. This is mainly due to the limited availability of human tissue to study, and the ability to manipulate animal models to study the effects of transgenes or deletion of genes on diabetes development. In vitro experiments have also been performed with both rodent and human islets. However, caution must be taken when extrapolating results from animal models to humans.

Because of the similarities with the human disease, the NOD mouse is a very useful tool to study the mechanisms of β-cell destruction. Transgenes can now be introduced into NOD mice by direct microinjection of NOD embryos [1–4], and recently NOD embryonic stem cells have been developed [5], which will facilitate the analysis of gene-targeted NOD mice because extensive backcrossing will not be required. Large numbers of genes have already been analyzed in backcrossed transgenic and knockout NOD mice (table 1). In addition, the tools for studying β cells in vitro have improved, as has our knowledge about the molecular processes of cell death. Together these advances will enable specific characterization of the mechanisms of β-cell destruction.

Table 1. Transgenic and knockout models in the NOD mouse

Gene	Effect on insulitis	Effect on diabetes	Ref.
IFNγ−/−	Unchanged	Unchanged	6
IFNγR−/−	Unchanged	Unchanged	7, 8
RIP-ΔγR transgenic	Unchanged	Unchanged	9
iNOS−/−	Not known	Unchanged	10
TNFR1−/−	Unchanged	Reduced	11
RIP-TNF (early onset)	Increased	Accelerated	12
RIP-TNF (adult onset)	Increased	Reduced	13
lpr/lpr	Absent	Absent	14–16
gld/+	Unchanged	Absent	17
Perforin−/−	Unchanged	Reduced	18, 19
RIP-Bcl2	Unchanged	Unchanged	20
RIP-thioredoxin	Unchanged	Absent	21
RIP-FasL	Variable	Variable	14

Apoptosis

Morphological Features That Distinguish Apoptosis and Necrosis

Apoptosis is a mechanism of cell death required for normal physiological processes such as morphogenesis, homeostatic maintenance of tissues and removal of harmful cells [22, 23]. Apoptosis can also occur at pathological sites. Necrosis is the result of direct toxic or physical injury to tissue.

In the process of apoptosis, dying cells become detached from neighboring cells and shrink in size, together with condensation of the nucleus and cytoplasm. The mitochondria release cytochrome c into the cytoplasm, the nuclear envelope breaks apart and there is condensation of chromatin, which becomes cleaved intranucleosomally into approximately 180-bp fragments. The plasma membrane then forms blebs which divide into smaller apoptotic bodies and are removed by phagocytes. This apoptotic process occurs within 24 h and there is no leakage of cellular contents as the plasma membrane remains intact (fig. 1).

In necrosis the cells take up water and swell, the plasma membrane bursts and the contents of the cytoplasm are released. There is DNA degradation but it occurs later and is more random than that of apoptosis, thus the characteristic DNA laddering observed in apoptosis is not seen. ATP is essential for apoptosis but not required for necrosis [24].

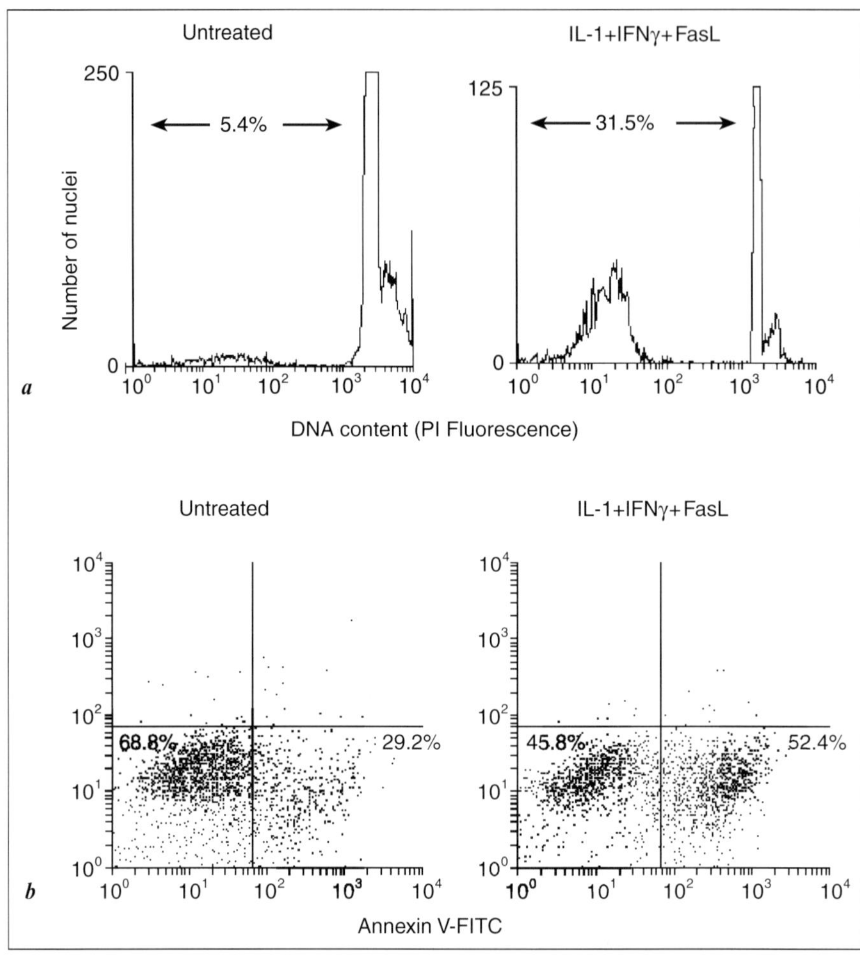

Fig. 1. Two methods for detection of FasL-induced apoptosis of primary mouse β cells. Islets from NOD mice were incubated 4 days in culture with IL-1 + IFNγ to upregulate Fas expression and soluble FasL to induce apoptosis. *a* Islets were trypsinized into single cells and analyzed by flow cytometry for fragmented nuclei with propidium iodide. Fragmented nuclei are observed on the left of the profile. *b* Islet cells were stained with Annexin V-FITC to detect early apoptotic events (Annexin V-positive cells).

Mechanisms of Apoptosis

Apoptotic signaling pathways are highly conserved through evolution. Most of the molecules which signal apoptosis in the nematode worm, *Caenorhabditis elegans*, have mammalian homologues.

The process of cell death occurs by activation of a group of cysteine proteases called caspases [25]. Caspases exist in cells as inactive proenzymes and are activated by proteolysis within seconds of an apoptotic signal. Caspases cleave specific cellular substrates after aspartate residues, and this eventually leads to the characteristics of apoptosis including DNA fragmentation and nuclear condensation.

Different death signals activate specific caspases in different cell types. Caspase 8 is required for apoptosis induced by FasL, TNFα and Apo3L [26], whereas caspase 9 is activated in thymocytes by γ-irradiation and release of cytochrome c from mitochondria [27, 28]. Other caspases, such as caspase 3, are activated downstream of the specific molecules, and act as effector caspases [29].

Mitochondria-Dependent Apoptosis

The mitochondria release cytochrome c into the cytoplasm after an apoptotic signal such as exposure to dexamethasone, etoposide or γ-irradiation [reviewed in 30, 31]. Cytochrome c is believed to bind the molecule Apaf-1, that in turn associates with caspase 9 causing its cleavage and activation. Activated caspase 9 then activates downstream caspases that cleave cellular substrates leading to cell death. This cascade is inhibited by the antiapoptotic molecules Bcl2 and Bcl-X_L, that function upstream of Apaf-1 (fig. 2).

Cell Death Receptors

The tumor necrosis factor (TNF) family is a group of structurally related proteins which signal cell growth and death via a common pathway [reviewed in 32]. Members of this family include Fas ligand (FasL/CD95L), TNFα, lymphotoxin (LT)-α and -β, TNF-related apoptosis-inducing ligand (TRAIL/ Apo2L) and Apo3L (TWEAK). Ligands of the TNF family usually act as membrane-bound homotrimers, with the exception of LTβ, which forms trimers with LTα. FasL and TNFα function as both soluble and membrane-bound ligands, although they are more efficient in the membrane-bound form [33–35]. Receptors for this family which promote cell death include Fas (CD95/Apo1), TNFR1 (CD120a), DR3 (Apo3), DR4 (TRAIL-R), DR5 (TRAIL-R2) and DR6. Some members of the TNFR family (including CD27, CD30, CD40, TNFR2 (CD120b), Ox40, 4-1BB, LIGHTR and the p75 NGFR) oppose cell death and instead promote cell survival. Others act as decoy receptors (including decoy receptor (DcR)1 (TRAIL-R3), DcR2 (TRAIL-R4), DcR3 and osteoprotegerin (OPG)), and function as inhibitors of death signaling [reviewed in 36, 37].

In addition to cell death via caspase activation, other signaling pathways induced by the TNFR family include activation of the transcription factors

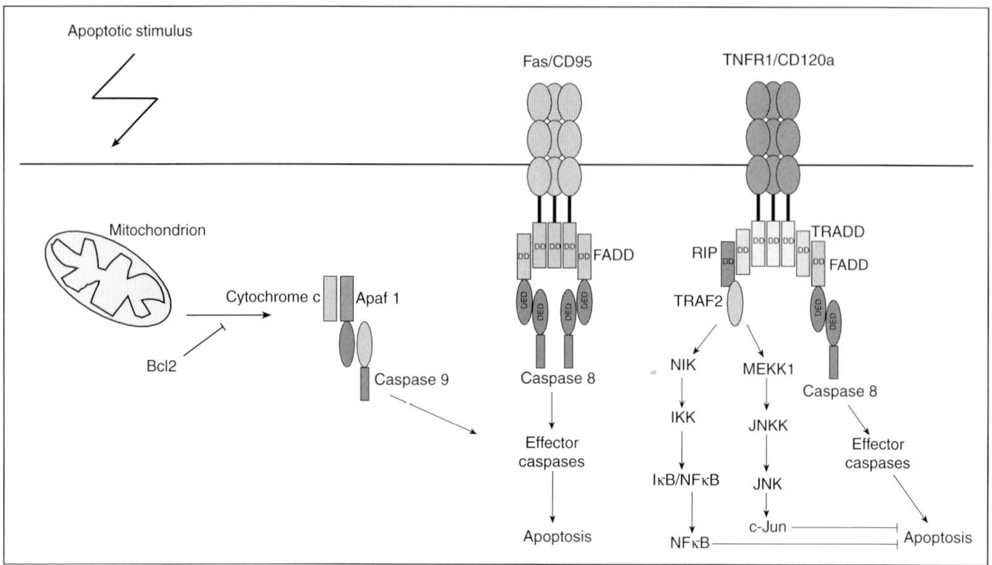

Fig. 2. Caspase-dependent cell death effector mechanisms. Fas (CD95) and TNFR1 (CD120a) form homodimers upon ligand binding. With both Fas and TNFR1, this can lead to association of Fas-associated death domain (FADD) and TNFR-associated death domain (TRADD) via a death domain (DD), and recruitment of caspase 8 via a death effector domain (DED), resulting in activation of downstream effector caspases and apoptosis. In the case of TNFR1, ligand binding can also lead to activation of the transcription factors AP-1 and NF-κB through association of receptor interacting protein (RIP) and TNFR-associated factor-2 (TRAF2) with the receptor, followed by activation of MAPK and NIK/IKK kinase cascades. Activation of these pathways is thought to protect cells from TNF-mediated apoptosis by inducing expression of anti-apoptotic genes. Stimuli such as γ-irradiation or chemotherapeutic drugs induce apoptosis through activation of cytochrome c in the mitochondria. Cytochrome c then interacts with Apaf 1 and caspase 9 is recruited, thus activating downstream caspases and apoptosis of the cell. This process is inhibited by anti-apoptotic members of the Bcl2 family.

NF-κB and AP-1 [reviewed in 32]. NF-κB and AP-1 are believed to play a role in signaling cell survival [38].

Death Receptor Signaling

The signaling events which occur through TNFR1 and Fas have been well characterized [37] TNF, produced by activated macrophages and T cells, normally induces proinflammatory gene expression [39]. However, in the right conditions, such as the absence of protein synthesis, TNF and FasL induce apoptosis. Signaling through TNFR1 and Fas is outlined in figure 2.

β-Cell Apoptosis

There is now considerable evidence that β cells are destroyed by apoptotic mechanisms. In the NOD mouse, apoptotic cells have been detected using TUNEL in tissue sections. While many TUNEL-positive cells appear in infiltrated islets, and the number increases with severity of disease, very few apoptotic cells are insulin-positive (0–1 apoptotic β cells per 100 islets) [40, 41]. In accelerated diabetes models [42, 43] TUNEL-positive β cells can be more readily identified. β-Cell-specific diabetogenic BDC2.5 CD4+ T cells were used to transfer rapid diabetes to NOD*scid* mice [42]. This transfer model was used to detect about 75 apoptotic β cells per mouse. Calculations were performed to determine the clearance rate of apoptotic β cells to be 1.7 min explaining why detection of apoptosis in NOD β cells is so difficult.

In biopsies taken from recent-onset type 1 diabetic patients [44] Fas expression was detected on β cells from islets with infiltration, and FasL was found on infiltrating cells suggesting that Fas-mediated apoptosis of β cells may occur. TUNEL-positive cells were not detected, however this may be due to the rapid clearance rate of apoptotic cells and the long time course of disease in humans.

It is not known whether the β cells are gradually destroyed during the long preclinical phase, or if they are rapidly destroyed immediately prior to the onset of hyperglycemia. In humans, the 'continuous model' suggests that diabetes results from a slow, progressive destruction of β cells [45]. In the NOD mouse, Lafferty and co-workers [46, 47] have suggested that a late change from a benign to malignant insulitis leads to massive β-cell loss and hyperglycemia. Using morphometric methods to assess β-cell mass in NOD mice over time, evidence supporting both models has been reported [48–50]. A compensatory increase in β-cell proliferation [50] may occur. β-Cell function may progressively decline before β-cell destruction [51]. Knowledge of the time course of β-Cell destruction and loss of function is important for determining when therapy is required in the preclinical phase of diabetes. Devising more quantitative techniques to study this would be useful.

β-Cell mass in the rat pancreas was found to increase linearly from birth to 100 days of age, with the exception of a plateau between 5 and 20 days of age [52]. The plateau suggested that a 'wave' of β-cell death must occur at this time and more apoptotic β-cells were found then. This time frame corresponds to the time of weaning in rats, and a similar phenomenon was observed in NOD mice just before weaning [53]. This is the time at which the first infiltrating cells are detected in the islets of NOD mice. This rise in apoptotic β cells may lead to priming of autoreactive T cells in susceptible individuals [53]. Apoptosis of β cells was also thought to be responsible for initiation of diabetes in mice transgenically expressing TNFα (RIP-TNF) in β cells [12].

β-Cell Destruction in vitro

Interleukin-1 (IL-1) and Nitric Oxide (NO)

IL-1 was found to be the active compound in mitogen-activated PBMC supernatants that inhibited insulin secretion [54, 55], and affected insulin synthesis in a dose-dependent manner [56, 57]. IL-1 has since been shown to induce DNA damage [58], reduction of DNA content [59] and cell death [60] of rodent and human islets, as well as β-cell lines [61–63].

The inhibitory action of IL-1 involves inducible nitric oxide synthase (iNOS) upregulation and NO production [64–66]. NO mediates destruction of the iron-sulfur clusters in iron-containing mitochondrial enzymes such as aconitase [67] leading to inhibition of mitochondrial function, a decrease in cellular ATP concentration and inhibition of D-glucose oxidation to CO_2 [68].

IL-1-induced islet damage can be blocked by inhibiting NO production with the arginine analog compounds N^G-monomethyl-L-arginine (NMMA) or aminoguanidine (AG) [64, 69–71]. Addition of IL-1 receptor antagonist (IL-1Ra) to cultures also prevents NO production and NO-mediated β-cell damage [72] NO is not responsible for all IL-1-mediated effects, however, as inhibition of NO in human islet cultures does not affect IL-1 induced DNA fragmentation [73–75].

Treatment of islets with TNFα, LPS and IFNγ also leads to intra-islet NO production and inhibition of islet function [76–78]. Islet damage induced with TNFα, LPS and IFNγ can be inhibited with IL-1Ra or anti-IL-1R Ab [78]. These cytokines are therefore believed to induce IL-1 secretion by resident islet macrophages, thus stimulating NO production and β-cell damage in an indirect manner [79].

Reactive Oxygen Species

While most in vitro studies of the effects of cytokines on β cells have focussed on NO, oxygen-reactive species may also induce β-cell damage. Peroxynitrite is produced by a reaction between superoxide and NO [80], and is toxic to human islets in vitro, causing DNA strand breaks, reduced glucose oxidation and necrosis [81]. Peroxynitrite is also found in the majority of β cells from NOD mice at the onset of diabetes [82]. Hydrogen peroxide and lipid peroxidation are also able to induce necrosis of β cells in vitro [83, 84].

Interferon-γ (IFNγ)

IFNγ, together with IL-1 and/or TNFα, induces β-cell damage by transcriptionally regulating iNOS expression and production of NO. While IFNγ is normally not thought to be able to damage β cells on its own, there is some in vitro evidence to suggest that IFNγ may be directly toxic to β cells. High concentrations of IFNγ inhibit glucose-stimulated insulin secretion by primary mouse islets [77, 85] and β-cell lines [62, 86], and incubation of rat islets with IFNγ leads to a decrease in insulin accumulation in tissue culture medium [76].

Cell Death Ligands

Fas Ligand

Fas expression on β cells, and Fas-mediated lysis of β cells in vitro have been extensively studied as a means of determining whether this is a potential mechanism of β-cell destruction. Fas expression can be upregulated by cytokines on primary β cells and β-cell lines [87–91], and Fas-expressing β cells are sensitive to apoptosis induced by soluble FasL or agonist anti-Fas Ab in vitro [88, 89, 91, 92]. We found that even though IL-1 is able to upregulate low levels of Fas on mouse β cells, a signal from IFNγ is also required for sensitivity to FasL-induced death [91]. It may be possible that the low level of Fas induced by IL-1 alone is insufficient for activated cell death, and that the high level induced by both cytokines is required. Additionally IFNγ is known to regulate caspase expression [93, 94] and may also activate other intracellular cell death signaling molecules.

While it has been suggested that FasL is expressed on β cells [92], we found no functional evidence that β-cell 'suicide' occurs due to constitutive expression of FasL on β cells and upregulation of Fas by cytokines [91]. One study suggested that NO is able to induce Fas expression on human β cells in vitro, and thus prime them for Fas-mediated destruction [89]. This does not appear to be the case in mouse islets [91].

Tumor Necrosis Factor

While the direct effects of TNFα on β cells in the progression to diabetes remain unclear, there is evidence that TNFα is able to induce β-cell death. TNFα potentiates NO-dependent β-cell damage mediated by IL-1 [95–97]. On its own, TNFα inhibits glucose-stimulated insulin secretion of isolated islets, an effect which is enhanced by addition of IFNγ [77, 85]. This is thought to be an NO-dependent process, however, there is some evidence that with high concentrations of TNFα, NO-independent mechanisms also exist [77]. In the NIT-1 β-cell line, caspase-dependent apoptosis was induced by TNFα [98]. Primary mouse β cells express low levels of TNFR1, indicating that direct action of TNFα on β cells is possible [98].

Other Toxic Stimuli

β Cells appear to be relatively resistant to cell death induced by agents which induce apoptosis in other cell types such as γ-irradiation or anoxia [H.T., unpubl. data]. This resistance to cell death may be due to expression of protective molecules such as Bcl2 family members. The broad-spectrum protein kinase inhibitor staurosporine is able to induce β-cell apoptosis in vitro [20], although even this highly toxic drug was used at much greater concentrations to induce β-cell death (5 μM) than those required for other cell types, suggesting a relative resistance of β cells to toxic stimuli.

β-Cell Destruction in vivo

The precise mechanisms by which β cells are killed in vivo remains largely undefined in humans. In animal models such as the NOD mouse, the in vivo mechanisms of β-cell destruction have been studied using transgenic and knockout mice, as well as diabetogenic T-cell clones. The main candidates for β-cell destruction appear to be perforin, FasL, and inflammatory cytokines such as IL-1, TNF and IFNγ.

Autoimmune diabetes is T-cell-mediated. Both CD4+ and CD8+ T-cell subsets are required for development of diabetes in the NOD mouse [reviewed in 99]. CD8+ T cells are predominant in the insulitis of most recent-onset diabetic patients [100–102], although others show a predominance of CD4+ T cells. CD8+ T cells are also observed at the time of islet transplant rejection in recurrent diabetes [103, 104]. Islet-specific T-cell clones have also been isolated from NOD mice, which can be used to clarify T-cell-mediated effector mechanisms of β-cell destruction. These include both CD4+ and CD8+ T-cell clones, with varying ability to transfer insulitis and/or diabetes [105] (table 2).

Perforin

Perforin is the major mechanism of cytotoxicity used by CTLs. CD4+ T cells generally do not utilize perforin for direct cytotoxicity, but employ other mechanisms such as Fas/FasL, TNF or TRAIL [115, 116]. When a CTL interacts with a target cell, it releases the contents of its cytoplasmic granules, containing perforin and granzymes, into the intercellular space between the lymphocyte and the target cell. Perforin is then integrated into the target cell membrane in a Ca^{2+}-dependent manner, and forms pores in the membrane. After pore formation and permeabilization of the target cell membrane, granzymes are able to

Table 2. Mechanism of β-cell destruction utilized by mouse T-cell clones

	Clone name	Postulated mechanism of β-cell destruction	Ref.
CD4+	4.1	FasL	106
	BDC2.5	TNFR1-dependent, unknown mechanism	107, 108
CD8+	8.3	FasL	19
	A14	Unknown/perforin?	109
	G9C8	FasL	14, 110, 111
	Clone-4 (HA-specific)	Perforin	112
	LCMV-GP	Perforin/IFNγ	113, 114

enter the cell and cleave procaspases and downstream caspase substrates to activate apoptosis [117].

In the LCMV model of diabetes, mice express the nucleoprotein (NP) or glycoprotein (GP) from lymphocytic choriomeningitis virus (LCMV) in their pancreatic β cells (RIP-NP/RIP-GP) [118, 119]. Perforin was found to be crucial for β-cell destruction after infection of RIP-GP mice with LCMV [18]. This result has complexities, however, since perforin-deficient mice are unable to clear LCMV infection and die after 2–3 weeks. Therefore, LCMV-specific CTL, derived from TCR transgenic mice, were used to induce diabetes without the need for LCMV infection. These cells were activated in vitro with LCMV-GP recombinant vaccinia virus. Perforin-deficient CTL activated in this way transfer insulitis but not diabetes to RIP-GP recipients. These data suggest that perforin is less important in initiation of disease in this model, but important for β-cell destruction leading to diabetes (table 3).

In another CD8+ T-cell model, activated TCR transgenic T cells (clone-4 TCR) were adoptively transferred into mice expressing the influenza virus hemagglutinin on their β cells (ins-HA) [112]. To assess the relative roles of perforin and FasL in β-cell destruction, HA-specific clone-4 TCR transgenic mice were crossed with either perforin- or FasL (*gld*)-deficient ins-HA mice. The number of *gld* TCR transgenic cells required to induce diabetes was equivalent to that of wild-type cells, while a 30-fold increase in cell number from perforin-deficient mice was required, suggesting that perforin rather than FasL is important in β-cell destruction in this model.

In the NOD mouse, perforin-mediated lysis of β cells is also a major mechanism of β-cell destruction. Perforin-deficient NOD mice develop insulitis, but diabetes incidence is reduced (16% in perforin−/− vs. 77% in +/+) and delayed (median onset 39.5 weeks in perforin−/− vs. 19 weeks in +/+) [113].

Table 3. Effect on diabetes of transgenic and knockout models in the RIP-LCMV system

Transgene or knockout	Effect on diabetes	Ref.
RIP-TNF	Increased	120
TNFR1−/−	Unchanged	121
RIP-IFNγ	Increased	122
IFNγ−/−	Absent	123
RIP-ΔγR	Reduced	114
Perforin−/−	Reduced	18
RIP-AdE3	Absent	124

Thus, initiation of disease does not require perforin-dependent β-cell lysis, but perforin is important as an effector mechanism of β-cell destruction by CD8+ T cells in the NOD mouse.

CD4+ T-cell clones have been isolated from infiltrated islets which can effectively induce diabetes in the absence of CD8+ T cells [105]. This, together with the fact that diabetes can occur in some perforin-deficient NOD mice suggests that there are alternative mechanisms utilized by CD4+ T cells.

Fas Ligand

Despite the large amount of literature on the role of Fas in the progression to diabetes, its role remains controversial. Many groups have made use of Fas-deficient NOD*lpr* and FasL-deficient NOD*gld* mice to study the role of Fas. Additionally, expression studies have been performed by immunostaining of pancreas sections or islet cell preparations. In humans, Fas and FasL expression have been detected in pancreas sections but there is little functional evidence for β-cell FasL expression. Fas was detected on β cells from infiltrated islets of recent-onset diabetic patients, and FasL was detected on the infiltrating T cells and macrophages [44]. It remains uncertain, however, whether this Fas and FasL expression is involved in β-cell destruction in human diabetes.

The most striking evidence that Fas plays a role in development of diabetes in the NOD mouse came from two studies using NOD*lpr* mice [14, 15]. These mice do not develop insulitis or diabetes either spontaneously or after adoptive transfer of diabetic NOD splenocytes or the CD8+ islet-specific T-cell clone G9C8 [14, 110]. The conclusion from these studies was that Fas-dependent lysis is a predominant mechanism for T-cell-mediated β-cell damage. However, subsequent studies have challenged this conclusion.

Fas-deficient NOD*lpr* islet grafts were transplanted into diabetic wild-type NOD mice to assess the ability of Fas-deficient β cells to survive autoimmune attack in an immunologically normal setting [16, 125]. The NOD*lpr* grafts were unable to restore normoglycemia of the recipient diabetic mice, and they were not protected from autoimmune destruction. Additionally, when wild-type NOD islets were grafted into NOD*lpr* mice, even wild-type islets remained healthy and were not destroyed after adoptive transfer of diabetogenic splenocytes [16]. This result would not be expected if Fas-mediated lysis was important in β-cell destruction.

Studies with NOD*gld* mice, however, suggest that Fas-mediated lysis of β cells may be important in diabetes in the NOD mouse [17]. Diabetes was absent in heterozygous NOD*gld/*+ mice (reduced functional FasL on T cells with no lymphadenopathy), although the mice developed insulitis suggesting that an islet-specific immune response was generated. This result supports a role for FasL in β-cell destruction. Also, NOD*scid.lpr* mice, which have no T-cell-associated FasL-mediated lytic activity, had a reduced and delayed incidence of diabetes after adoptive transfer of diabetic NOD splenocytes [17]. These authors were also unable to restore normoglycemia of diabetic NOD mice with NOD*lpr* islet grafts, however, suggesting that different mechanisms exist in the different experimental systems [17].

To more specifically address the issue of β-cell Fas expression in the NOD mouse, we used flow cytometry to directly identify Fas-expressing β cells during the progression to diabetes. Even very close to the onset of hyperglycemia in NOD mice, we were only able to detect 1–5% of β cells expressing Fas [91]. Also, it appears that these Fas-positive cells may be CD45+ cells (e.g. macrophages) contaminating the β-cell population gated on the FACS [R. Darwiche and T.W. Kay, unpubl. data]. The fact that we only found few, if any, Fas-expressing β cells in islets suggests that at least in the spontaneous disease in the NOD mouse this may not be an important mechanism of β-cell destruction. This does not exclude the possibility that FasL is able to destroy β cells in other models of autoimmune diabetes.

The anti-insulin CD8+ T-cell clone G9C8 was isolated from the islet infiltrate of a 7-week-old NOD female, and therefore may be representative of T cells involved early in the disease process [110]. These cells destroy β cells by Fas-mediated lysis [14], however, the experiments to prove this are confounded by the fact that it is difficult to transfer diabetes into NOD*lpr* mice due to their high numbers of FasL-bearing T cells.

The CD8+ T-cell clone 8.3 [19] has been used to generate 8.3 TCR transgenic NOD mice which develop diabetes by 21 days of age. T cells from these mice were able to kill antigen-pulsed non-β-cell targets via either perforin or Fas-dependent mechanisms. However, while NOD islets could be killed by perforin-deficient 8.3 T cells in vitro, NOD*lpr* islets could not, indicating that NOD β cells are only killed by 8.3 CD8+ T cells in vitro via Fas. Additionally, perforin-deficient 8.3 NOD mice developed diabetes at the same rate as wild-type 8.3 NOD mice, suggesting that perforin is not an important mechanism of killing used by these cells. The authors suggest that this may represent the way in which early infiltrating T cells initiate β-cell destruction in the NOD mouse.

The same group has generated 4.1 NOD mice which bear a diabetogenic I-Ag⁷-restricted, β-cell-specific TCR [106]. Like 8.3 CD8+ T cells, CD4+ T cells from these mice can develop into both perforin and FasL-expressing CTLs in vivo. RAG2-deficient 4.1 mice cannot rearrange endogenous TCRs and therefore have a monoclonal 4.1 TCR with no CD8+ T cells. RAG2-deficient 4.1 NOD*lpr* mice had a markedly reduced incidence of diabetes, while RAG2-deficient/perforin-deficient 4.1 NOD mice developed normal diabetes, suggesting that 4.1 T cells only kill β cells in a Fas-dependent manner. This Fas-mediated killing of β cells depends upon exposure of β cells to macrophage-derived cytokines (IL-1 or TNFα) which upregulate Fas expression and thus 'mark' β cells for CD4+ T-cell cytotoxicity. These results with 8.3 and 4.1 TCR transgenic mice suggest that there may be Fas-mediated lysis of β cells during progression to diabetes.

The idea that macrophage-derived cytokines such as IL-1 upregulate β-cell Fas expression, and that FasL-expressing T cells are involved in killing β cells, is attractive as a mechanism of β-cell destruction, given that IL-1 mRNA is observed in the insulitis lesion [126], FasL can kill β cells in vitro [91], and that T-cell clones isolated from NOD mice kill β cells in a Fas-dependent manner [19, 106]. It is possible that this is a more important mechanism early on in the disease process [14, 19], and this theory remains relatively untested. It has been suggested that the Fas/FasL pathway of cytotoxicity induced by CTLs may be more important than perforin when antigen is limiting and weaker TCR signals are provided, which may be the case in the early stages of diabetes [112, 127].

Interleukin-1 and Nitric Oxide

NO can kill β cells in vitro but there is little direct evidence that this is the case in vivo. Expression of iNOS is greater in 16-week-old female NOD islets compared with those from 5-week-old mice [128]. Evidence for iNOS expression is also found in BB rat islets [129], and after injection of cyclophosphamide into NOD mice [130, 131].

IL-1 mRNA and protein expression are also detected in the infiltrate of islets from NOD mice [126] and Bio-Breeding (BB) rats [132]. Injection of recombinant IL-1 into normal rats induces transient hyperglycemia [133], and a high dose of IL-1 administered to BB rats accelerates the onset of diabetes. However, low doses of IL-1 are protective when given to BB rats or NOD mice [134, 135]. Neutralization of IL-1 by prolonged treatment with IL-1Ra (BB rats) or soluble IL-1R (cyclophosphamide-treated NOD mice) delays and reduces the incidence of diabetes [136].

Diabetes induced by multiple low-dose streptozotocin (STZ) is reduced in iNOS-deficient C57Bl/6 mice [137]. Inhibition of NOS in vivo by treatment with $N\omega$-nitro-L-arginine methyl ester (L-NAME) leads to a modest decrease in diabetes incidence in BB rats [138], and selective inhibition of iNOS with AG in normal rats prevents hyperglycemia induced by injection of IL-1β [133]. In NOD mice, treatment of recipient mice with AG delays the onset of diabetes after adoptive transfer of diabetogenic splenocytes, and blocks islet NO production [71].

In the BDC2.5 CD4+ T-cell model, iNOS-deficient islets are destroyed at the same rate as wild-type islets [108], suggesting that islets themselves do not need to produce NO to undergo destruction. Spontaneous diabetes is still observed in iNOS-deficient NOD mice [10], although information about insulitis and adoptive transfer in these mice has not been reported.

IL-1 receptors are expressed on islet cells [139]. Many islet cell types, including β cells, are capable of producing NO in response to IL-1 [78, 140–142], but it appears that the cellular source of NO important for NO-dependent β-cell damage is the β cell itself. We found that addition of the NO donor compound sodium nitroprusside to islets in culture did not induce damage, despite high concentrations of nitrite detected in the culture supernatant [91]. Additionally, RIP-$\Delta\gamma$R islets appear to be protected from damage induced by IL-1 and IFNγ, even though NO is produced by non-β cells within these islets (such as macrophages or endothelial cells). Also, transgenic overexpression of iNOS in β cells leads to NO-dependent β-cell destruction [143].

Tumor Necrosis Factor

The time of TNF expression is critical in determining its ability to enhance or prevent the onset of diabetes [144]. Transgenic overexpression of TNFα in β cells of adult NOD mice leads to rapid infiltration in the islets without loss of β cells or diabetes [13]. TNFα impairs the development of autoreactive T cells, as there is no proliferative response to islet antigens, and spleen cells from these mice are unable to transfer disease into NOD*scid* recipients. When TNFα is expressed in β cells of neonatal NOD mice diabetes occurs rapidly in all mice [12]. TNF is thought to activate APCs to present islet antigens to CD4+ T cells, which then leads to activation of CD8+ T-cell effectors. In neither of these transgenic models does TNFα induce diabetes by direct β-cell apoptosis.

TNFR1-deficient NOD mice develop normal insulitis but not diabetes [11]. When splenocytes from wild-type diabetic mice are transferred into TNFR1-deficient recipients, diabetes is delayed compared with wild-type recipients, suggesting that TNFα is acting on recipient APCs, endothelium or β cells, not T cells. Perforin-deficient splenocytes are unable to transfer diabetes to TNFR1-deficient recipients, which suggests that both perforin and TNFR1-dependent mechanisms are required for T-cell-mediated β-cell destruction in the NOD mouse.

Activated splenocytes from BDC2.5 TCR transgenic mice [107] induce apoptosis of β cells and diabetes in NOD*scid* recipients [42]. Pakala and co-workers developed a transplant model using BDC2.5 TCR transgenic CD4+ T cells to examine β-cell destruction mediated by these cells. Islets from mice deficient in TNFR1, TNFR2, IFNγR, iNOS and Fas were grafted into STZ-treated NOD*scid* mice [108]. Diabetes was then induced by adoptive transfer of BDC2.5 T cells. The only islets protected from destruction in this system were TNFR1-deficient islets. However, TNFR1-deficient islet grafts were not protected when wild-type islets were also present in the graft. That is, an islet cell response to TNFα is required for CD4+ T-cell activation. Destruction of both wild-type and TNFR1-deficient β cells then proceeds in the absence of direct action of TNFα on islet cells [108].

Together these data suggest that TNFα does not directly induce β-cell apoptosis in vivo, even though it is capable of doing so in vitro [98]. Therefore, the role of TNFα in diabetes in the NOD mouse may be to facilitate islet antigen presentation by APCs and CD4+ T-cell activation. The studies with TNFR1-deficient mice suggest that TNFα action on islet cells (e.g. β cells, endothelial cells or DCs) is important in the development of autoreactive CD4+ T cells and subsequent diabetes.

Interferon-γ

IFNγ is produced by T cells and NK cells. IFNγ production is stimulated by macrophage-derived cytokines including TNFα, IL-12, IL-18, and by IFNγ itself. Although IFNγ was originally defined for its antiviral activity, this cytokine also has immunomodulatory functions, many of which may be involved in the pathogenesis of diabetes. These include regulation of MHC molecule expression, the development of a Th1 cell phenotype, macrophage activation, apoptosis and adhesion.

When IFNγ is expressed in pancreatic β cells using the human insulin promoter (ins-IFNγ), nondiabetes-prone mice develop autoimmune diabetes [145] that is probably not due to direct β-cell toxicity of IFNγ [146]. Treatment of

NOD mice with anti-IFNγ monoclonal Abs or soluble IFNγR [147–150] reduces diabetes and insulitis.

Mice with targeted disruption of IFNγ, however, develop insulitis and diabetes [6]. The fact that diabetes occurs in IFNγ-deficient NOD mice suggests a lesser role for IFNγ in diabetes development than was originally expected, or that alternative cytokine(s) could compensate for loss of IFNγ action in these mice. IFNγR-deficient NOD mice were originally thought to develop very little infiltrate with no diabetes [7]. These studies were performed on NOD IFNγR-deficient mice after 4–8 backcross generations. The discrepancy between the ligand and receptor-deficient animals remained unexplained until recently. After more extensive backcrossing of IFNγR-deficient mice to the NOD genetic background (14 generations), mice developed diabetes, suggesting that there is a diabetes resistance gene(s) linked to the IFNγR locus in the 129 mouse strain [8]. These further backcrossed IFNγR-deficient mice are still resistant to cyclophosphamide-induced diabetes, consistent with the earlier experiments using monoclonal Abs [147].

We produced RIP-ΔγR mice to further clarify which cell type is responding to IFNγ in the development of diabetes [9]. These transgenic mice express a dominant-negative IFNγR on their β cells, making β cells unresponsive to IFNγ in vitro and in vivo. All other cell types, including other islet cells, remain IFNγ-responsive. Using this mouse model, we were able to dissect the precise role of IFNγ action on pancreatic β cells in the development of diabetes in an immunologically normal setting. NOD RIP-ΔγR mice develop diabetes at the same rate as nontransgenic control mice, suggesting that direct IFNγ action on β cells is not required for diabetes in the NOD mouse.

Insulitis was, however, delayed and diabetes absent in IFNγR-deficient BDC2.5 transgenic mice [7]. These experiments suggest that diabetes induced by CD4+ T cells may require IFNγ responsiveness. In this model, IFNγ may exert its effects on recipient cells such as APCs, β cells or endothelial cells [7].

In contrast to the relatively minor effect of IFNγ deficiency on spontaneous diabetes in the NOD mouse, IFNγ-deficient RIP-NP or GP mice did not develop diabetes or islet infiltrate when challenged with LCMV [123]. Class I MHC overexpression was not detected on the IFNγ-deficient RIP-NP β cells compared with the high levels detected on wild-type (RIP-NP) β cells. Diabetes induced in this model requires class I MHC overexpression on β cells for efficient presentation of the LCMV-NP antigen [123]. These results suggest that IFNγ may be required in this model for inducing class I MHC expression on β cells. IFNγ may also be needed for priming of the response in the lymph node and for the breaking of tolerance.

Using RIP-ΔγR transgenic mice we were able to examine the direct effect of IFNγ on β cells in the LCMV model [114]. RIP-ΔγR mice were crossed with

RIP-GP mice, and upregulation of class I MHC as well as development of diabetes was examined. It was expected that RIP-ΔγR β cells would not have elevated class I MHC levels after virus infection due to β-cell unresponsiveness to IFNγ, and that they would be protected from autoimmune attack in this class I MHC, perforin and CD8+ T-cell-dependent model of diabetes. It was found that RIP-ΔγR mice did not develop diabetes after LCMV infection; however, class I MHC was upregulated on β cells and islets were killed by CTL in vitro. More work needs to be done to clarify the exact mechanism, however it seems that IFNγ-responsive β cells are required for β-cell destruction but not for class I MHC upregulation in this model.

Other Mechanisms of β-Cell Death

The cytotoxic drug STZ kills β cells by a mechanism which does not appear to be inhibited by Bcl2 [20]. STZ has a glucose moiety which allows its transport into the β cell via the GLUT2 receptor. STZ is thought to methylate DNA to cause DNA strand breaks, overactivation of the DNA repair enzyme PARP and depletion of NAD [151]. Administration of STZ to mice in high doses (140 mg/kg) leads to β-cell apoptosis within hours [20].

Administration of steroids such as hydrocortisone leads to β-cell degranulation and death in animal models, and steroids have an inhibitory effect on insulin synthesis in humans [152, 153]. The immunosuppressive drugs tacrolimus and cyclosporine also induce β-cell abnormalities such as degranulation and vacuolization in animals. In a study performed on patients receiving pancreas allografts, a correlation was found between islet damage and serum levels of tacrolimus and cyclosporine [154].

Concluding Remarks and Future Perspectives

There are many potential players leading to β-cell destruction. These include factors such as perforin, cytokines (IFNγ, IL-1, FasL, TNFα) and free radicals (fig. 3). Perforin is the only factor definitely implicated in β-cell killing in the NOD mouse. As diabetes can sometimes occur independent of perforin, some of these other factors must play a role. We believe that other, as yet undefined, factors are also involved in β-cell destruction. These may include cell death ligands such as TRAIL or LIGHT. Expression of receptors for these molecules on β cells will be the first step towards determining if they are relevant to diabetes. In the search for as yet unidentified mediators of β-cell destruction, factors may be purified from supernatants of activated T cells or identified

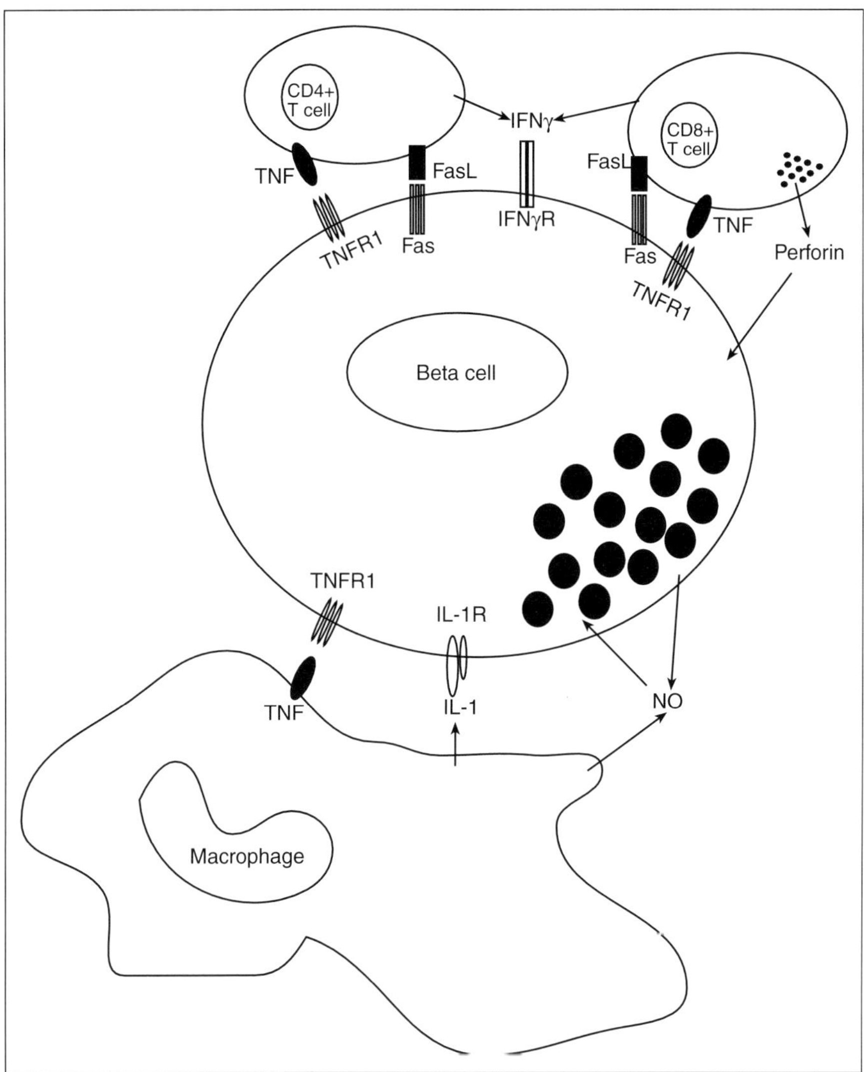

Fig. 3. Cellular and molecular mediators of β-cell destruction in the NOD mouse. CD4+ and CD8+ T cells as well as macrophages are required for development of diabetes. Perforin, produced by CD8+ T cells, is important in β-cell destruction. Cell death ligands including FasL and TNF are able to kill β cells in vitro and may also be important in vivo in the NOD mouse. Additionally, soluble cytokines such as IL-1 and IFNγ may be required for production of NO either by macrophages or by the β cell itself, leading to β-cell death.

in the insulitis lesions of humans and mice. The next few years should be exciting in this field as we begin to discover the precise mechanisms of β-cell destruction.

Transgenic expression of protective molecules in NOD mouse β cells may provide protection from cell death induced by individual factors or groups of factors with similar mechanisms of action. This remains the chief rationale for these experiments. Additionally they may help to clarify the exact role of the different mechanisms of β-cell destruction in diabetes development. However, evidence suggests that to prevent diabetes in the NOD mouse, multiple effector pathways need to be blocked, including perforin, IFNγ-dependent mechanisms such as Fas and NO, and IFNγ-independent mechanisms, some of which are yet to be discovered.

Animal models have been developed to study the precise mechanisms of progression to type 1 diabetes, as the events leading to disease in humans remain largely unknown. This review has focussed on animal models in which a great deal is now known about the mechanisms of β-cell destruction. The events and kinetics that can lead to β-cell destruction in these animal models are variable. Which model best fits the human disease is not yet known, nor is it clear whether the mechanisms are the same in all humans. The study of β-cell destruction in humans, using isolated islets and human pancreas sections in the early, prediabetic stages of disease will help determine which, if any, of the animal models is most relevant.

Acknowledgments

This work was supported by the National Health and Medical Research Council (Regkey 973002) and the Juvenile Diabetes Foundation via a Diabetes Interdisciplinary Research Program Grant and Career Development Award (to T.W.H.K.). We thank Dr. Jan Allison for critical reading.

References

1 Lund T, O'Reilly L, Hutchings P, Kanagawa O, Simpson E, Gravely R, Chandler P, Dyson J, Picard JK, Edwards A, et al: Prevention of insulin-dependent diabetes mellitus in non-obese diabetic mice by transgenes encoding modified I-A beta-chain or normal I-E alpha-chain. Nature 1990;345:727–729.
2 Miyazaki T, Uno M, Uehira M, Kikutani H, Kishimoto T, Kimoto M, Nishimoto H, Miyazaki J, Yamamura K: Direct evidence for the contribution of the unique I-ANOD to the development of insulitis in non-obese diabetic mice. Nature 1990;345:722–724.
3 Slattery RM, Kjer-Nielsen L, Allison J, Charlton B, Mandel TE, Miller JF: Prevention of diabetes in non-obese diabetic I-Ak transgenic mice. Nature 1990;345:724–726.

4 French MB, Allison J, Cram DS, Thomas HE, Dempsey-Collier M, Silva A, Georgiou HM, Kay TW, Harrison LC, Lew AM: Transgenic expression of mouse proinsulin II prevents diabetes in nonobese diabetic mice. Diabetes 1997;46:34–39.

5 Nagafuchi S, Katsuta H, Kogawa K, Akashi T, Kondo S, Sakai Y, Tsukiyama T, Kitamura D, Niho Y, Watanabe T: Establishment of an embryonic stem (ES) cell line derived from a non-obese diabetic (NOD) mouse: In vivo differentiation into lymphocytes and potential for germ line transmission. FEBS Lett 1999;455:101–104.

6 Hultgren B, Huang X, Dybdal N, Stewart TA: Genetic absence of gamma-interferon delays but does not prevent diabetes in NOD mice. Diabetes 1996;45:812–817.

7 Wang B, Andre I, Gonzalez A, Katz JD, Aguet M, Benoist C, Mathis D: Interferon-gamma impacts at multiple points during the progression of autoimmune diabetes. Proc Nat Acad Sci USA 1997; 94:13844–13849.

8 Kanagawa O, Xu G, Tevaarwerk A, Vaupel BA: Protection of nonobese diabetic mice from diabetes by gene(s) closely linked to IFN-gamma receptor loci. J Immunol 2000;164:3919–3923.

9 Thomas HE, Parker JL, Schreiber RD, Kay TW: IFN-gamma action on pancreatic beta cells causes class I MHC upregulation but not diabetes. J Clin Invest 1998;102:1249–1257.

10 Kolb H, Kolb-Bachofen V: Nitric oxide in autoimmune disease: Cytotoxic or regulatory mediator? Immunol Today 1998;19:556–561.

11 Kagi D, Ho A, Odermatt B, Zakarian A, Ohashi PS, Mak TW: TNF receptor 1-dependent beta cell toxicity as an effector pathway in autoimmune diabetes. J Immunol 1999;162:4598–4605.

12 Green EA, Eynon EE, Flavell RA: Local expression of TNF-alpha in neonatal NOD mice promotes diabetes by enhancing presentation of islet antigens. Immunity 1998;9:733–743.

13 Grewal IS, Grewal KD, Wong FS, Picarella DE, Janeway CA Jr, Flavell RA: Local expression of transgene encoded TNF-alpha in islets prevents autoimmune diabetes in nonobese diabetic (NOD) mice by preventing the development of auto-reactive islet-specific T cells. J Exp Med 1996;184: 1963–1974.

14 Chervonsky AV, Wang Y, Wong FS, Visintin I, Flavell RA, Janeway CA Jr, Matis LA: The role of Fas in autoimmune diabetes. Cell 1997;89:17–24.

15 Itoh N, Imagawa A, Hanafusa T, Waguri M, Yamamoto K, Iwahashi H, Moriwaki M, Nakajima H, Miyagawa J, Namba M, Makino S, Nagata S, Kono N, Matsuzawa Y: Requirement of Fas for the development of autoimmune diabetes in nonobese diabetic mice. J Exp Med 1997;186:613–618.

16 Allison J, Strasser A: Mechanisms of beta cell death in diabetes: A minor role for CD95. Proc Natl Acad Sci USA 1998;95:13818–13822.

17 Su X, Hu Q, Kristan JM, Costa C, Shen Y, Gero D, Matis LA, Wang Y: Significant role for Fas in the pathogenesis of autoimmune diabetes. J Immunol 2000;164:2523–2532.

18 Kagi D, Odermatt B, Ohashi PS, Zinkernagel RM, Hengartner H: Development of insulitis without diabetes in transgenic mice lacking perforin-dependent cytotoxicity. J Exp Med 1996;183: 2143–2152.

19 Amrani A, Verdaguer J, Anderson B, Utsugi T, Bou S, Santamaria P: Perforin-independent beta-cell destruction by diabetogenic CD8(+) T lymphocytes in transgenic nonobese diabetic mice. J Clin Invest 1999;103:1201–1209.

20 Allison J, Thomas HE, Beck D, Brady J, Lew AM, Elephanty A, Kosaka H, Kay TWH, Huang DCS, Strasser A: Transgenic over-expression of human Bcl2 in islet beta cells inhibits apoptosis but does not prevent autoimmune destruction. Int Immunol 2000;12:9–17.

21 Hotta M, Tashiro F, Ikegami H, Niwa H, Ogihara T, Yodoi J, Miyazaki J: Pancreatic beta cell-specific expression of thioredoxin, an antioxidative and antiapoptotic protein, prevents autoimmune and streptozotocin-induced diabetes. J Exp Med 1998;188:1445–1451.

22 Wyllie AH, Kerr JF, Currie AR: Cell death: The significance of apoptosis. Int Rev Cytol 1980;68: 251–306.

23 Ellis RE, Yuan JY, Horvitz HR: Mechanisms and functions of cell death. Annu Rev Cell Biol 1991;7:663–698.

24 Tsujimoto Y: Apoptosis and necrosis: Intracellular ATP level as a determinant for cell death modes. Cell Death Differ 1997;4:429–434.

25 Nicholson DW, Thornberry NA: Caspases: Killer proteases. Trends Biochem Sci 1997;22:299–306.

26 Varfolomeev EE, Schuchmann M, Luria V, Chiannilkulchai N, Beckmann JS, Mett IL, Rebrikov D, Brodianski VM, Kemper OC, Kollet O, Lapidot T, Soffer D, Sobe T, Avraham KB, Goncharov T, Holtmann H, Lonai P, Wallach D: Targeted disruption of the mouse caspase 8 gene ablates cell death induction by the TNF receptors, Fas/Apo1, and DR3 and is lethal prenatally. Immunity 1998;9:267–276.
27 Hakem R, Hakem A, Duncan GS, Henderson JT, Woo M, Soengas MS, Elia A, de la Pompa JL, Kagi D, Khoo W, Potter J, Yoshida R, Kaufman SA, Lowe SW, Penninger JM, Mak TW: Differential requirement for caspase 9 in apoptotic pathways in vivo. Cell 1998;94:339–352.
28 Kuida K, Haydar TF, Kuan CY, Gu Y, Taya C, Karasuyama H, Su MS, Rakic P, Flavell RA: Reduced apoptosis and cytochrome c-mediated caspase activation in mice lacking caspase 9. Cell 1998;94:325–337.
29 Cohen GM: Caspases: The executioners of apoptosis. Biochem J 1997;326:1–16.
30 Rathmell JC, Thompson CB: The central effectors of cell death in the immune system. Annu Rev Immunol 1999;17:781–828.
31 Scaffidi C, Kirchhoff S, Krammer PH, Peter ME: Apoptosis signaling in lymphocytes. Curr Opin Immunol 1999;11:277–285.
32 Wallach D, Varfolomeev EE, Malinin NL, Goltsev YV, Kovalenko AV, Boldin MP: Tumor necrosis factor receptor and Fas signaling mechanisms. Annu Rev Immunol 1999;17:331–367.
33 Grell M, Douni E, Wajant H, Lohden M, Clauss M, Maxeiner B, Georgopoulos S, Lesslauer W, Kollias G, Pfizenmaier K, et al: The transmembrane form of tumor necrosis factor is the prime activating ligand of the 80 kDa tumor necrosis factor receptor. Cell 1995;83:793–802.
34 Schneider P, Holler N, Bodmer JL, Hahne M, Frei K, Fontana A, Tschopp J: Conversion of membrane-bound Fas(CD95) ligand to its soluble form is associated with downregulation of its proapoptotic activity and loss of liver toxicity. J Exp Med 1988;187:1205–1213.
35 Tanaka M, Itai T, Adachi M, Nagata S: Downregulation of Fas ligand by shedding. Nat Med 1998; 4:31–36.
36 Smith CA, Farrah T, Goodwin RG: The TNF receptor superfamily of cellular and viral proteins: Activation, costimulation, and death. Cell 1994;76:959–962.
37 Ashkenazi A, Dixit VM: Death receptors: Signaling and modulation. Science 1998;281:1305–1308.
38 Van Antwerp DJ, Martin SJ, Verma IM, Green DR: Inhibition of TNF-induced apoptosis by NF-kappa B. Trends Cell Biol 1998;8:107–111.
39 Vassalli P: The pathophysiology of tumor necrosis factors. Annu Rev Immunol 1992;10:411–452.
40 O'Brien BA, Harmon BV, Cameron DP, Allan DJ: Apoptosis is the mode of beta-cell death responsible for the development of IDDM in the nonobese diabetic (NOD) mouse. Diabetes 1997;46: 750–757.
41 Augstein P, Elefanty AG, Allison J, Harrison LC: Apoptosis and beta-cell destruction in pancreatic islets of NOD mice with spontaneous and cyclophosphamide-accelerated diabetes. Diabetologia 1998;41:1381–1388.
42 Kurrer MO, Pakala SV, Hanson HL, Katz JD: Beta cell apoptosis in T cell-mediated autoimmune diabetes. Proc Natl Acad Sci USA 1997;94:213–218.
43 Augstein P, Stephens LA, Allison J, Elefanty AG, Ekberg M, Kay TW, Harrison LC: Beta-cell apoptosis in an accelerated model of autoimmune diabetes. Mol Med 1998;4:495–501.
44 Moriwaki M, Itoh N, Miyagawa J, Yamamoto K, Imagawa A, Yamagata K, Iwahashi H, Nakajima H, Namba M, Nagata S, Hanafusa T, Matsuzawa Y: Fas and Fas ligand expression in inflamed islets in pancreas sections of patients with recent-onset type I diabetes mellitus. Diabetologia 1999;42: 1332–1340.
45 Eisenbarth GS: Type I diabetes mellitus. A chronic autoimmune disease. N Engl J Med 1986;314: 1360–1368.
46 Gazda LS, Charlton B, Lafferty KJ: Diabetes results from a late change in the autoimmune response of NOD mice. J Autoimmun 1997;10:261–270.
47 Dilts SM, Lafferty KJ: Autoimmune diabetes: The involvement of benign and malignant autoimmunity. J Autoimmun 1999;12:229–232.
48 Debussche X, Lormeau B, Boitard C, Toublanc M, Assan R: Course of pancreatic beta cell destruction in prediabetic NOD mice: A histomorphometric evaluation. Diabète Metab 1994;20:282–290.

49 Signore A, Procaccini E, Toscano AM, Ferretti E, Williams AJ, Beales PE, Cugini P, Pozzilli P: Histological study of pancreatic beta-cell loss in relation to the insulitis process in the non-obese diabetic mouse. Histochemistry 1994;101:263–269.

50 Sreenan S, Pick AJ, Levisetti M, Baldwin AC, Pugh W, Polonsky KS: Increased beta-cell proliferation and reduced mass before diabetes onset in the nonobese diabetic mouse. Diabetes 1999;48: 989–996.

51 Strandell E, Eizirik DL, Sandler S: Reversal of beta-cell suppression in vitro in pancreatic islets isolated from nonobese diabetic mice during the phase preceding insulin-dependent diabetes mellitus. J Clin Invest 1990;85:1944–1950.

52 Finegood DT, Scaglia L, Bonner-Weir S: Dynamics of beta-cell mass in the growing rat pancreas. Estimation with a simple mathematical model. Diabetes 1995;44:249–256.

53 Trudeau JD, Dutz JP, Arany E, Hill DJ, Fieldus WE, Finegood DT: Neonatal beta-cell apoptosis: A trigger for autoimmune diabetes? Diabetes 2000;49:1–7.

54 Mandrup-Poulsen T, Bendtzen K, Nielsen JH, Bendixen G, Nerup J: Cytokines cause functional and structural damage to isolated islets of Langerhans. Allergy 1985;40:424–429.

55 Mandrup-Poulsen T, Bendtzen K, Nerup J, Dinarello CA, Svenson M, Nielsen JH: Affinity-purified human interleukin-1 is cytotoxic to isolated islets of Langerhans. Diabetologia 1986; 29:63–67.

56 Comens PG, Wolf BA, Unanue ER, Lacy PE, McDaniel ML: Interleukin-1 is potent modulator of insulin secretion from isolated rat islets of Langerhans. Diabetes 1987;36:963–970.

57 Spinas GA, Hansen BS, Linde S, Kastern W, Molvig J, Mandrup-Poulsen T, Dinarello CA, Nielsen JH, Nerup J: Interleukin-1 dose-dependently affects the biosynthesis of (pro)insulin in isolated rat islets of Langerhans. Diabetologia 1987;30:474–480.

58 Delaney CA, Green MH, Lowe JE, Green IC: Endogenous nitric oxide induced by interleukin-1-beta in rat islets of Langerhans and HIT-T15 cells causes significant DNA damage as measured by the 'comet' assay. FEBS Lett 1993;333:291–295.

59 Sandler S, Andersson A, Hellerstrom C: Inhibitory effects of interleukin-1 on insulin secretion, insulin biosynthesis, and oxidative metabolism of isolated rat pancreatic islets. Endocrinology 1987;121:1424–1431.

60 Bolaffi JL, Rodd GG, Wang J, Grodsky GM: Interrelationship of changes in islet nicotine adenine dinucleotide, insulin secretion, and cell viability induced by interleukin-1-beta. Endocrinology 1994;134:537–542.

61 Sandler S, Bendtzen K, Eizirik DL, Sjoholm A, Welsh N: Decreased cell replication and polyamine content in insulin-producing cells after exposure to human interleukin-1-beta. Immunol Lett 1989; 22:267–272.

62 Hamaguchi K, Leiter EH: Comparison of cytokine effects on mouse pancreatic alpha-cell and beta-cell lines. Viability, secretory function, and MHC antigen expression. Diabetes 1990;39:415–425.

63 Janjic D, Asfari M: Effects of cytokines on rat insulinoma INS-1 cells. J Endocrinol 1992;132: 67–76.

64 Southern C, Schulster D, Green IC: Inhibition of insulin secretion by interleukin-1-beta and tumour necrosis factor-alpha via an L-arginine-dependent nitric oxide generating mechanism. FEBS Lett 1990;276:42–44.

65 Corbett JA, Lancaster JR Jr, Sweetland MA, McDaniel ML: Interleukin-1 beta-induced formation of EPR-detectable iron-nitrosyl complexes in islets of Langerhans. Role of nitric oxide in inter-leukin-1-beta-induced inhibition of insulin secretion. J Biol Chem 1991;266:21351–21354.

66 Welsh N, Eizirik DL, Bendtzen K, Sandler S: Interleukin-1-beta-induced nitric oxide production in isolated rat pancreatic islets requires gene transcription and may lead to inhibition of the Krebs cycle enzyme aconitase. Endocrinology 1991;129:3167–3173.

67 Corbett JA, Wang JL, Hughes JH, Wolf BA, Sweetland MA, Lancaster JR Jr, McDaniel ML: Nitric oxide and cyclic GMP formation induced by interleukin-1-beta in islets of Langerhans. Evidence for an effector role of nitric oxide in islet dysfunction. Biochem J 1992;287:229–235.

68 Sandler S, Bendtzen K, Borg LA, Eizirik DL, Strandell E, Welsh N: Studies on the mechanisms causing inhibition of insulin secretion in rat pancreatic islets exposed to human interleukin-1-beta indicate a perturbation in the mitochondrial function. Endocrinology 1989;124:1492–1501.

69 Corbett JA, Tilton RG, Chang K, Hasan KS, Ido Y, Wang JL, Sweetland MA, Lancaster JR Jr, Williamson JR, McDaniel ML: Aminoguanidine, a novel inhibitor of nitric oxide formation, prevents diabetic vascular dysfunction. Diabetes 1992;41:552–556.

70 Welsh N, Sandler S: Interleukin-1-beta induces nitric oxide production and inhibits the activity of aconitase without decreasing glucose oxidation rates in isolated mouse pancreatic islets. Biochem Biophys Res Commun 1992;182:333–340.

71 Corbett JA, Mikhael A, Shimizu J, Frederick K, Misko TP, McDaniel ML, Kanagawa O, Unanue ER: Nitric oxide production in islets from nonobese diabetic mice: Aminoguanidine-sensitive and -resistant stages in the immunological diabetic process. Proc Natl Acad Sci USA 1993;90:8992–8995.

72 Eizirik DL, Tracey DE, Bendtzen K, Sandler S: An interleukin-1 receptor antagonist protein protects insulin-producing beta cells against suppressive effects of interleukin-1-beta. Diabetologia 1991;34:445–448.

73 Eizirik DL, Sandler S, Welsh N, Cetkovic-Cvrlje M, Nieman A, Geller DA, Pipeleers DG, Bendtzen K, Hellerstrom C: Cytokines suppress human islet function irrespective of their effects on nitric oxide generation. J Clin Invest 1994;93:1968–1974.

74 Rabinovitch A, Suarez-Pinzon WL, Strynadka K, Schulz R, Lakey JR, Warnock GL, Rajotte RV: Human pancreatic islet beta-cell destruction by cytokines is independent of nitric oxide production. J Clin Endocrinol Metab 1994;79:1058–1062.

75 Delaney CA, Pavlovic D, Hoorens A, Pipeleers DG, Eizirik DL: Cytokines induce deoxyribonucleic acid strand breaks and apoptosis in human pancreatic islet cells. Endocrinology 1997;138:2610–2614.

76 Sternesjo J, Bendtzen K, Sandler S: Effects of prolonged exposure in vitro to interferon-gamma and tumour necrosis factor-alpha on nitric oxide and insulin production of rat pancreatic islets. Autoimmunity 1995;20:185–190.

77 Dunger A, Cunningham JM, Delaney CA, Lowe JE, Green MH, Bone AJ, Green IC: Tumor necrosis factor-alpha and interferon-gamma inhibit insulin secretion and cause DNA damage in unweaned-rat islets. Extent of nitric oxide involvement. Diabetes 1996;45:183–189.

78 Arnush M, Heitmeier MR, Scarim AL, Marino MH, Manning PT, Corbett JA: IL-1 produced and released endogenously within human islets inhibits beta cell function. J Clin Invest 1998;102:516–526.

79 Corbett JA, McDaniel ML: Intraislet release of interleukin-1 inhibits beta cell function by inducing beta cell expression of inducible nitric oxide synthase. J Exp Med 1995;181:559–568.

80 Beckman JS, Beckman TW, Chen J, Marshall PA, Freeman BA: Apparent hydroxyl radical production by peroxynitrite: Implications for endothelial injury from nitric oxide and superoxide. Proc Natl Acad Sci USA 1990;87:1620–1624.

81 Delaney CA, Tyrberg B, Bouwens L, Vaghef H, Hellman B, Eizirik DL: Sensitivity of human pancreatic islets to peroxynitrite-induced cell dysfunction and death. FEBS Lett 1996;394:300–306.

82 Suarez-Pinzon WL, Szabo C, Rabinovitch A: Development of autoimmune diabetes in NOD mice is associated with the formation of peroxynitrite in pancreatic islet beta-cells. Diabetes 1997;46:907–911.

83 Rabinovitch A, Suarez WL, Thomas PD, Strynadka K, Simpson I: Cytotoxic effects of cytokines on rat islets: Evidence for involvement of free radicals and lipid peroxidation. Diabetologia 1992;35:409–413.

84 Delaney CA, Green IC, Lowe JE, Cunningham JM, Butler AR, Renton L, D'Costa I, Green MH: Use of the comet assay to investigate possible interactions of nitric oxide and reactive oxygen species in the induction of DNA damage and inhibition of function in an insulin-secreting cell line. Mutat Res 1997;375:137–146.

85 Campbell IL, Iscaro A, Harrison LC: IFN-gamma and tumor necrosis factor-alpha. Cytotoxicity to murine islets of Langerhans. J Immunol 1988;141:2325–2329.

86 Baldeon ME, Neece DJ, Nandi D, Monaco JJ, Gaskins HR: Interferon-gamma independently activates the MHC class I antigen processing pathway and diminishes glucose responsiveness in pancreatic beta-cell lines. Diabetes 1997;46:770–778.

87 Stassi G, Todaro M, Richiusa P, Giordano M, Mattina A, Sbriglia MS, Lo Monte A, Buscemi G, Galluzzo A, Giordano C: Expression of apoptosis-inducing CD95 (Fas/Apo-1) on human beta-cells sorted by flow cytometry and cultured in vitro. Transplant Proc 1995;27:3271–3275.

88 Yamada K, Takane-Gyotoku N, Yuan X, Ichikawa F, Inada C, Nonaka K: Mouse islet cell lysis mediated by interleukin-1-induced Fas. Diabetologia 1996;39:1306–1312.

89 Stassi G, Maria RD, Trucco G, Rudert W, Testi R, Galluzzo A, Giordano C, Trucco M: Nitric oxide primes pancreatic beta cells for Fas-mediated destruction in insulin-dependent diabetes mellitus. J Exp Med 1997;186:1193–1200.

90 Harrison M, Dunger AM, Berg S, Mabley J, John N, Green MH, Green IC: Growth factor protection against cytokine-induced apoptosis in neonatal rat islets of Langerhans: Role of Fas. FEBS Lett 1998;435:207–210.

91 Thomas HE, Darwiche R, Corbett JA, Kay TW: Evidence that beta cell death in the nonobese diabetic mouse is Fas independent. J Immunol 1999;163:1562–1569.

92 Loweth AC, Williams GT, James RF, Scarpello JH, Morgan NG: Human islets of Langerhans express Fas ligand and undergo apoptosis in response to interleukin-1-beta and Fas ligation. Diabetes 1998;47:727–732.

93 Chin YE, Kitagawa M, Kuida K, Flavell RA, Fu XY: Activation of the STAT signaling pathway can cause expression of caspase 1 and apoptosis. Mol Cell Biol 1997;17:5328–5337.

94 Kumar A, Commane M, Flickinger TW, Horvath CM, Stark GR: Defective TNF-alpha-induced apoptosis in STAT1-null cells due to low constitutive levels of caspases. Science 1997;278: 1630–1632.

95 Mandrup-Poulsen T, Bendtzen K, Dinarello CA, Nerup J: Human tumor necrosis factor potentiates human interleukin-1-mediated rat pancreatic beta-cell cytotoxicity. J Immunol 1987;139: 4077–4082.

96 Cetkovic-Cvrlje M, Eizirik DL: TNF-alpha and IFN-gamma potentiate the deleterious effects of IL-1-beta on mouse pancreatic islets mainly via generation of nitric oxide. Cytokine 1994;6: 399–406.

97 Kaneto H, Fujii J, Seo HG, Suzuki K, Matsuoka T, Nakamura M, Tatsumi H, Yamasaki Y, Kamada T, Taniguchi N: Apoptotic cell death triggered by nitric oxide in pancreatic beta-cells. Diabetes 1995;44:733–738.

98 Stephens LA, Thomas HE, Ming L, Grell M, Darwiche R, Volodin L, Kay TW: Tumor necrosis factor-alpha-activated cell death pathways in NIT-1 insulinoma cells and primary pancreatic beta cells. Endocrinology 1999;140:3219–3227.

99 Bach JF: Insulin-dependent diabetes mellitus as an autoimmune disease. Endocrine Reviews 1994;15:516–542.

100 Bottazzo GF, Dean BM, McNally JM, MacKay EH, Swift PG, Gamble DR: In situ characterization of autoimmune phenomena and expression of HLA molecules in the pancreas in diabetic insulitis. N Engl J Med 1985;313:353–360.

101 Foulis AK, McGill M, Farquharson MA: Insulitis in type 1 (insulin-dependent) diabetes mellitus in man – Macrophages, lymphocytes, and interferon-gamma containing cells. J Pathol 1991;165: 97–103.

102 Somoza N, Vargas F, Roura-Mir C, Vives-Pi M, Fernandez-Figueras MT, Ariza A, Gomis R, Bragado R, Marti M, Jaraquemada D, Pujol-Borrell R: Pancreas in recent onset insulin-dependent diabetes mellitus. Changes in HLA, adhesion molecules and autoantigens, restricted T cell receptor V beta usage, and cytokine profile. J Immunol 1994;153:1360–1377.

103 Sibley RK, Sutherland DE, Goetz F, Michael AF: Recurrent diabetes mellitus in the pancreas iso- and allograft. A light and electron microscopic and immunohistochemical analysis of four cases. Lab Invest 1985;53:132–144.

104 Santamaria P, Nakhleh RE, Sutherland DE, Barbosa JJ: Characterization of T lymphocytes infiltrating human pancreas allograft affected by isletitis and recurrent diabetes. Diabetes 1992;41:53–61.

105 Haskins K, Wegmann D: Diabetogenic T-cell clones. Diabetes 1996;45:1299–1305.

106 Amrani A, Verdaguer J, Thiessen S, Bou S, Santamaria P: IL-1-alpha, IL-1-beta, and IFN-gamma mark beta cells for Fas-dependent destruction by diabetogenic CD4(+) T lymphocytes. J Clin Invest 2000;105:459–468.

107 Katz JD, Wang B, Haskins K, Benoist C, Mathis D: Following a diabetogenic T cell from genesis through pathogenesis. Cell 1993;74:1089–1100.

108 Pakala SV, Chivetta M, Kelly CB, Katz JD: In autoimmune diabetes the transition from benign to pernicious insulitis requires an islet cell response to tumor necrosis factor alpha. J Exp Med 1999;189:1053–1062.

109 DiLorenzo TP, Graser RT, Ono T, Christianson GJ, Chapman HD, Roopenian DC, Nathenson SG, Serreze DV: Major histocompatibility complex class I-restricted T cells are required for all but the end stages of diabetes development in nonobese diabetic mice and use a prevalent T cell receptor alpha chain gene rearrangement. Proc Natl Acad Sci USA 1998;95:12538–12543.

110 Wong FS, Visintin I, Wen L, Flavell RA, Janeway CA Jr: CD8 T cell clones from young nonobese diabetic (NOD) islets can transfer rapid onset of diabetes in NOD mice in the absence of CD4 cells. J Exp Med 1996;183:67–76.

111 Wong FS, Karttunen J, Dumont C, Wen L, Visintin I, Pilip IM, Shastri N, Pamer EG, Janeway CA Jr: Identification of an MHC class I-restricted autoantigen in type 1 diabetes by screening an organ-specific cDNA library. Nat Med 1999;5:1026–1031.

112 Kreuwel HTC, Morgan DJ, Krahl T, Ko A, Sarvetnick N, Sherman LA: Comparing the relative role of perforin/granzyme versus Fas/Fas ligand cytotoxic pathways in CD8+ T cell-mediated insulin-dependent diabetes mellitus. J Immunol 1999;163:4335–4341.

113 Kagi D, Odermatt B, Seiler P, Zinkernagel RM, Mak TW, Hengartner H: Reduced incidence and delayed onset of diabetes in perforin-deficient nonobese diabetic mice. J Exp Med 1997;186: 989–997.

114 Seewaldt S, Thomas HE, Ejrnaes M, Christen U, Wolfe T, Rodrigo E, Coon B, Michelsen B, Kay TWH, von Herrath MG: Virus-induced autoimmune diabetes: Most beta cells die through inflammatory cytokines and not perforin from autoreactive (anti-viral) CTL. Diabetes 2000;49: 1801–1809.

115 Stalder T, Hahn S, Erb P: Fas antigen is the major target molecule for CD4+ T cell-mediated cytotoxicity. J Immunol 1994;152:1127–1133.

116 Oxenius A, Zinkernagel RM, Hengartner H: CD4+ T-cell induction and effector functions: A comparison of immunity against soluble antigens and viral infections. Adv Immunol 1998;70: 313–367.

117 Andrade F, Roy S, Nicholson D, Thornberry N, Rosen A, Casciola-Rosen L: Granzyme B directly and efficiently cleaves several downstream caspase substrates: Implications for CTL-induced apoptosis. Immunity 1998;8:451–460.

118 Ohashi PS, Oehen S, Buerki K, Pircher H, Ohashi CT, Odermatt B, Malissen B, Zinkernagel RM, Hengartner H: Ablation of 'tolerance' and induction of diabetes by virus infection in viral antigen transgenic mice. Cell 1991;65:305–317.

119 Oldstone MB, Nerenberg M, Southern P, Price J, Lewicki H: Virus infection triggers insulin-dependent diabetes mellitus in a transgenic model: Role of anti-self (virus) immune response. Cell 1991;65:319–331.

120 Ohashi PS, Oehen S, Aichele P, Pircher H, Odermatt B, Herrera P, Higuchi Y, Buerki K, Hengartner H, Zinkernagel RM: Induction of diabetes is influenced by the infectious virus and local expression of MHC class I and tumor necrosis factor-alpha. J Immunol 1993;150: 5185–5194.

121 McKall-Faienza KJ, Kawai K, Kundig TM, Odermatt B, Bachmann MF, Zakarian A, Mak TW, Ohashi PS: Absence of TNFRp55 influences virus-induced autoimmunity despite efficient lymphocytic infiltration. Int Immunol 1998;10:405–412.

122 Lee MS, von Herrath M, Reiser H, Oldstone MB, Sarvetnick N: Sensitization to self (virus) antigen by in situ expression of murine interferon-gamma. J Clin Invest 1995;95:486–492.

123 Von Herrath MG, Oldstone MB: Interferon-gamma is essential for destruction of beta cells and development of insulin-dependent diabetes mellitus. J Exp Med 1997;185:531–539.

124 Von Herrath MG, Efrat S, Oldstone MB, Horwitz MS: Expression of adenoviral E3 transgenes in beta cells prevents autoimmune diabetes. Proc Natl Acad Sci USA 1997;94:9808–9813.

125 Kim YH, Kim S, Kim KA, Yagita H, Kayagaki N, Kim KW, Lee MS: Apoptosis of pancreatic beta-cells detected in accelerated diabetes of NOD mice: No role of Fas-Fas ligand interaction in autoimmune diabetes. Eur J Immunol 1999;29:455–465.

126 Rabinovitch A, Suarez-Pinzon WL, Sorensen O, Bleackley RC: Inducible nitric oxide synthase (iNOS) in pancreatic islets of nonobese diabetic mice: Identification of iNOS-expressing cells and relationships to cytokines expressed in the islets. Endocrinology 1996;137:2093–2099.

127 Kessler B, Hudrisier D, Schroeter M, Tschopp J, Cerottini JC, Luescher IF: Peptide modification or blocking of CD8, resulting in weak TCR signaling, can activate CTL for Fas- but not perforin-dependent cytotoxicity or cytokine production. J Immunol 1998;161:6939–6946.

128 Welsh M, Welsh N, Bendtzen K, Mares J, Strandell E, Oberg C, Sandler S: Comparison of mRNA contents of interleukin-1-beta and nitric oxide synthase in pancreatic islets isolated from female and male nonobese diabetic mice. Diabetologia 1995;38:153–160.

129 Kleemann R, Rothe H, Kolb-Bachofen V, Xie QW, Nathan C, Martin S, Kolb H: Transcription and translation of inducible nitric oxide synthase in the pancreas of prediabetic BB rats. FEBS Lett 1993;328:9–12.

130 Rothe H, Faust A, Schade U, Kleemann R, Bosse G, Hibino T, Martin S, Kolb H: Cyclophosphamide treatment of female non-obese diabetic mice causes enhanced expression of inducible nitric oxide synthase and interferon-gamma, but not of interleukin-4. Diabetologia 1994;37: 1154–1158.

131 Reddy S, Kaill S, Poole CA, Ross J: Inducible nitric oxide synthase in pancreatic islets of the non-obese diabetic mouse: A light and confocal microscopical study of its ontogeny, co-localization and up-regulation following cytokine administration. Histochem J 1997;29:53–64.

132 Jiang Z, Woda BA: Cytokine gene expression in the islets of the diabetic Biobreeding/Worcester rat. J Immunol 1991;146:2990–2994.

133 Reimers JI, Bjerre U, Mandrup-Poulsen T, Nerup J: Interleukin-1-beta induces diabetes and fever in normal rats by nitric oxide via induction of different nitric oxide synthases. Cytokine 1994;6: 512–520.

134 Jacob CO, Aiso S, Michie SA, McDevitt HO, Acha-Orbea H: Prevention of diabetes in nonobese diabetic mice by tumor necrosis factor (TNF): Similarities between TNF-alpha and interleukin-1. Proc Natl Acad Sci USA 1990;87:968–972.

135 Wilson CA, Jacobs C, Baker P, Baskin DG, Dower S, Lernmark A, Toivola B, Vertrees S, Wilson D: IL-1-beta modulation of spontaneous autoimmune diabetes and thyroiditis in the BB rat. J Immunol 1990;144:3784–3788.

136 Nicoletti F, Di Marco R, Barcellini W, Magro G, Schorlemmer HU, Kurrle R, Lunetta M, Grasso S, Zaccone P, Meroni P: Protection from experimental autoimmune diabetes in the non-obese diabetic mouse with soluble interleukin-1 receptor. Eur J Immunol 1994;24:1843–1847.

137 Flodstrom M, Tyrberg B, Eizirik DL, Sandler S: Reduced sensitivity of inducible nitric oxide synthase-deficient mice to multiple low-dose streptozotocin-induced diabetes. Diabetes 1999;48: 706–713.

138 Lindsay RM, Smith W, Rossiter SP, McIntyre MA, Williams BC, Baird JD: N-omega-nitro-L-arginine methyl ester reduces the incidence of IDDM in BB/E rats. Diabetes 1995;44:365–368.

139 Jafarian-Tehrani M, Amrani A, Homo-Delarche F, Marquette C, Dardenne M, Haour F: Localization and characterization of interleukin-1 receptors in the islets of Langerhans from control and nonobese diabetic mice. Endocrinology 1995;136:609–613.

140 Wiegand F, Kroncke KD, Kolb Bachofen V: Macrophage generated nitric oxide as cytotoxic factor in destruction of alginate-encapsulated islets. Protection by arginine analogs and/or coencapsulated erythrocytes. Transplantation 1993;56:1206–1212.

141 Steiner L, Kroncke K, Fehsel K, Kolb-Bachofen V: Endothelial cells as cytotoxic effector cells: Cytokine-activated rat islet endothelial cells lyse syngeneic islet cells via nitric oxide. Diabetologia 1997;40:150–155.

142 Pavlovic D, Chen MC, Bouwens L, Eizirik DL, Pipeleers D: Contribution of ductal cells to cytokine responses by human pancreatic islets. Diabetes 1999;48:29–33.

143 Takamura T, Kato I, Kimura N, Nakazawa T, Yonekura H, Takasawa S, Okamoto H: Transgenic mice overexpressing type 2 nitric-oxide synthase in pancreatic beta cells develop insulin-dependent diabetes without insulitis. J Biol Chem 1998;273:2493–2496.

144 Sarvetnick N: Mechanisms of cytokine-mediated localized immunoprotection. J Exp Med 1996; 184:1597–1600.

145 Sarvetnick N, Liggitt D, Pitts SL, Hansen SE, Stewart TA: Insulin-dependent diabetes mellitus induced in transgenic mice by ectopic expression of class II MHC and interferon-gamma. Cell 1988;52:773–782.

146 Sarvetnick N, Shizuru J, Liggitt D, Martin L, McIntyre B, Gregory A, Parslow T, Stewart T: Loss of pancreatic islet tolerance induced by beta-cell expression of interferon-gamma. Nature 1990;346:844–847.

147 Campbell IL, Kay TW, Oxbrow L, Harrison LC: Essential role for interferon-gamma and interleukin-6 in autoimmune insulin-dependent diabetes in NOD/Wehi mice. J Clin Invest 1991;87: 739–742.

148 Debray-Sachs M, Carnaud C, Boitard C, Cohen H, Gresser I, Bedossa P, Bach JF: Prevention of diabetes in NOD mice treated with antibody to murine IFN-gamma. J Autoimmun 1991;4:237–248.

149 Kay TW, Campbell IL, Oxbrow L, Harrison LC: Overexpression of class I major histocompatibility complex accompanies insulitis in the non-obese diabetic mouse and is prevented by anti-interferon-gamma antibody. Diabetologia 1991;34:779–785.

150 Nicoletti F, Zaccone P, Di Marco R, Di Mauro M, Magro G, Grasso S, Mughini L, Meroni P, Garotta G: The effects of a nonimmunogenic form of murine soluble interferon-gamma receptor on the development of autoimmune diabetes in the NOD mouse. Endocrinology 1996;137: 5567–5575.

151 Burkart V, Wang ZQ, Radons J, Heller B, Herceg Z, Stingl L, Wagner EF, Kolb H: Mice lacking the poly(ADP-ribose) polymerase gene are resistant to pancreatic beta-cell destruction and diabetes development induced by streptozocin. Nat Med 1999;5:314–319.

152 Madar I, Mihail N: Age-related effect of treatment by hydrocortisone upon the histological structure of endocrine pancreas and on the blood glucose in normal and diabetic young rats. Endocrinologie 1983;21:37–42.

153 Venkatesan N, Davidson MB, Hutchinson A: Possible role for the glucose-fatty acid cycle in dexamethasone-induced insulin antagonism in rats. Metabolism 1987;36:883–891.

154 Drachenberg CB, Klassen DK, Weir MR, Wiland A, Fink JC, Bartlett ST, Cangro CB, Blahut S, Papadimitriou JC: Islet cell damage associated with tacrolimus and cyclosporine: Morphological features in pancreas allograft biopsies and clinical correlation. Transplantation 1999;68:396–402.

Dr. Thomas W.H. Kay, Autoimmunity and Transplantation Division,
The Walter and Eliza Hall Institute, P.O. Royal Melbourne Hospital,
Parkville, Vic 3050 (Australia)
Tel. +61 3 9345 2457, Fax +61 3 9347 0852, E-Mail kay@wehi.edu.au

von Herrath MG (ed): Molecular Pathology of Type 1 Diabetes mellitus.
Curr Dir Autoimmun. Basel, Karger, 2001, vol 4, pp 171–192

Interactions of Effectors and Regulators Are Decisive in the Manifestations of Type 1 Diabetes in Nonobese Diabetic Mice

Anthony Quinn[a], *Vipin Kumar*[a], *Kent P. Jensen*[a,b], *Eli E. Sercarz*[a,b]

[a] Division of Immune Regulation, La Jolla Institute for Allergy and Immunology,
San Diego, Calif., and
[b] Department of Microbiology, Immunology and Molecular Genetics,
UCLA School of Medicine, Los Angeles, Calif., USA

Type 1 Diabetes mellitus, Nonobese Diabetic Mice and β-Cell Autoantigens

Type 1 diabetes mellitus (T1D) is an autoimmune disease in which the insulin-producing β cells within the islets of Langerhans of the pancreas are selectively destroyed. The nonobese diabetic (NOD) mouse serves as a good model of the human disease and it shares several of its disease characteristics. In NOD mice, the disease is characterized by the infiltration of the islets first by dendritic cells and macrophages, and then by autoreactive T cells (both CD4+ and CD8+) and B cells (which serve in the secretion of autoantibodies as well as antigen-presenting cells (APCs)) [1]. The infiltration of APCs is followed by the slow, destructive recruitment and activation of autoreactive T cells directed against several β-cell-specific islet antigens. These antigens include insulin, glutamic acid decarboxylase (GAD-65 and GAD-67), a tyrosine phosphatase I-A2/phogrin, carboxypeptidase H, and HSP 60/65 [1]. Among these antigens, it has been shown that B-cell and T-cell responses to GAD-65 [2, 3], and insulin [4] are among the first to arise in NOD mice as well as in humans. The initial responses arise in the NOD at about 3–4 weeks of age, a time which precedes both insulitis and diabetes.

T-Cell Responses to GAD-65 in NOD Mice

In 1993, spontaneous T-cell responses to GAD-65 in NOD mice were first described by two groups [2, 3]. Kaufman et al. [2] demonstrated that the immune response to GAD-65 was initially restricted to the C-terminal portion of the molecule (to determinants within p509–528 and p524–543) and that the response then spread to include other regions of GAD-65 (within p78–97, p246–266, p340–356, p479–493, p539–558 and to p570–585) as well as to other autoantigens. Both studies also demonstrated that treatment of young NOD mice with GAD-65, either intravenously or intrathymically, was able to prevent the disease and the spread of T-cell autoimmunity [2, 3]. These data indicated an important role for GAD-65 in the pathogenesis of NOD T1D. Further studies by both of these groups (as well as others) have demonstrated that other routes of GAD-65 administration, as well as its peptides, are effective in the prevention of T1D [5–12]. Finally, it has been demonstrated that prevention of GAD-65 and GAD-67 expression in the islets through the expression of antisense RNA in the β cells was effective in preventing diabetes as well as insulitis [13].

Despite this strong evidence for an important role for GAD-65 in NOD diabetes, the relevance of GAD-65 remains controversial. This is owing in part to the fact that (1) the expression of GAD-65 in the β cells of NOD mice appears to be relatively low when compared to GAD-67 and other autoantigens [14–19], (2) the expression of GAD-65 appears to be short-lived (peaking between 5–8 weeks of age) in NOD mice [16, 18, 19], and (3) the difficulty by some labs to reproduce the spontaneous proliferation data to peptides described by Kaufman et al. [2].

In 1997, a second set of I-A^{g7}-restricted T-cell determinants was described by Chao et al. [20] and shortly thereafter by Zechel et al. [21]. In contrast to the 'spontaneous' T-cell determinants described by Kaufman et al. [2] and Tisch et al. [3], this second set of determinants was originally described as being elicited following immunization with whole GAD-65 (although spontaneous T-cell responses for these determinants have also been described) [21]. We typically refer to members of this second set of GAD-65 T-cell determinants as 'inducible' (owing to the manner in which they were first described) and they include determinants within p78–97, p206–220, p221–235, p286–300, p400–410 and p570–585.

Evidently, some regions of a self-molecule are more readily processed and presented than others, becoming dominant within the self context. Other regions may gain prominence when antigen is presented in a large bolus from external sources. Thus, the 'spontaneous' and 'inducible' sets of determinants are not totally distinct pools.

The existence of so many CD4+, GAD-65-restricted T cells in the NOD brings up some interesting questions. What are the clinically relevant T-cell determinants on GAD-65 in NOD T1D? Are T cells directed against these determinants restricted in their TcR usage? Which determinants of GAD-65 can be processed and presented by NOD APCs on their I-A^{g7} MHC II molecules? What are the CD8+ T-cell determinants of GAD-65 in the NOD mouse? Do they share any geographical features in common with CD4+ GAD-65 T-cell determinants? Which if any of these T-cell determinants are regulatory for NOD T1D? We will attempt to answer some of these questions in the rest of this review.

The Self-Antigen and the Self-Directed Repertoire

The two overriding elements that must be understood in unravelling the mysteries of autoimmunity are the nature of the self-antigenic determinants involved and the quality of the T-cell repertoire engaged in the process. Within the confines of any potential antigen there are several features which are crucial in determining the immunogenic fate of its determinants. How is the molecule initially unfolded, or are there portions of the molecule which are available enough to bind directly to an MHC groove? Beginning at an endopeptidic site of initial attack, what is the subsequent pathway of processing of the antigen? Upon sufficient unfolding of the molecule to reveal areas which have sufficient binding energy for the MHC class II binding site, the most available, high-affinity antigenic determinant for the site will be the winner in competition with other determinants. Concomitantly, several MHC molecules may compete for different determinants along the available stretch of the unfolding antigen. At this point, residues flanking the class II binding site may protrude and cause some interference with interaction, until the flanking sites are trimmed away.

A recurrent theme in determinant choice is the power of proximal determinants. When a single determinant region on a protein molecule has several neighboring or overlapping sets of motifs, this region is much more likely to be preferred in the competition for binding to MHC class II molecules. There are many multideterminant stretches on known antigens where two or three separate determinants overlap, producing a favored, dominant determinant region [22–24]. Within these overlapping sets, there will be competition for dominance, and this will be settled by the relative affinity of the determinants in question for the MHC.

Shaping the Self-Reactive T-Cell Repertoire

In addition to these considerations, for self-molecules there are other issues relating to the extent of tolerance within the self-directed population. For well-expressed determinants, a large portion of the T-cell repertoire may have been nullified owing to tolerance induction during development. GAD-65, with its numerous determinants for which residual T cells abound, represents a self-molecule with limited distribution in the body, but like most self-antigens, it is not clear what proportion of the originally available repertoire was shut down by tolerance mechanisms. We will refer to the GAD-65-specific repertoire in the adult as the 'residual repertoire', and in this chapter describe our studies with four different GAD-65 determinants and the repertoires which they elicit.

The whole question of the extent of the residual repertoire remaining to a self-antigen is very pertinent to this discussion. Although it is clear that high-avidity T cells should be purged by self determinants that are presented at any reasonable level, it still is probable that for the poorly presented determinants, many high-avidity T cells will be spared. The great majority of the rest of the T-cell repertoire, specific for poorly presented (cryptic) determinants, is no doubt fully available. Of course, what remains an issue is the breadth of the potential T-cell repertoire to any one determinant: if the repertoire is narrow, but highly avid in one case, while broad and diverse in another, the residual repertoires may not be representative of the original, potential repertoires.

Relationships among Antigenic Determinants

In the study of T-helper determinants and the B cells which are helped by them, there is a preferential pairing of these determinants on a single antigen molecule [25]. In this case, it isn't clear whether the determinants were proximal, but they needed to occupy a particular relationship to each other for the collaboration between T and B cells to work. This is also true in the work of Watts and Lanzavecchia [26], while in the β-galactosidase (GZ) model, the proximity rule, per se, was invoked by the authors and it appeared necessary for the B-cell determinant (a hapten) to be quite close to the T-cell-helper determinant.

Previously it was shown that not all Th-induced responses to β-galactosidase were suppressor T-cell-sensitive, but only those T_H determinants residing near CD8+ Ts-inducing determinants [27, 28], a finding that implies a regional interaction between MHC class I and class II responses. Overlapping MHC class I and class II determinants have also been described in humans [29–31]. More recently, it was discovered that a diabetogenic CD8+ T-cell clone recognized a site on the insulin B chain [32] which overlaps a well-described Th-inducing determinant B9–23 [33, 34]. The mechanism(s) that controls and mediates this regional response is not understood, though it is tempting to suggest that if there is to be a productive CD8 response, the processing of neighboring MHC class I and class II determinants should occur in the same vicinity. Not all CD8+ T-cell-inducing determinants are located proximal to Th-inducing ones, just as not all CTL responses are predicated on CD40–CD40L interactions [35, 36]. The observation is nonetheless interesting and may provide a useful approach for identifying relevant CD8+ T-cell determinants in infectious disease, cancer and autoimmunity.

Functional Differences among T Cells

It should be stated very clearly that it is not our desire to present any one islet protein molecule as more important than any other with respect to its involvement in T1D. Others have studied insulin/proinsulin or heat-shock protein 65, and it is very evident that a role in disease causation could easily be attributed to these molecules. However, it may be of some importance that one rather than another determinant on one of these molecules *drives* the response and is responsible for propagating it via determinant spreading. There need not be a single driver, nor a single initiator, and these two functions need not be contained within one determinant.

This question bears investigating since it is very relevant to therapy, where it would be most interesting to focus regulation on a dominant T-cell population. It is noteworthy that EAE in the B10.PL mouse, where the disease is self-limiting owing to a regulatory response directed against the TcR of the predominant driver clone, that concomitant with the disappearance of symptoms, there is the disappearance of the driver clone [37]. Accordingly, one possible way to abrogate T1D might be through the induction of a regulatory clone which arises in response to a driver, and then controls the Th1/Th2 milieu balance via bystander suppression.

A framework for consideration in the discussion to follow is that several determinants on a self-antigen may be instigators of a response and lead to determinant spreading. These instigators may or may not be the 'drivers' which provide the energy and fuel the inflammatory response. Maintenance of the Th1 response may be helped by propagators which are more passive, but which are recruitable. However, at this point, the tide may turn. The driver clones, first of all, may harbor the seeds for a supervening regulatory response. Likewise, recruitment of clones which have deviated in a Th2 direction can contribute to downregulating the response, as can certain types of CD8+ regulatory cells (fig. 1). Many influences will serve to tip the balance in one or the other direction, but it appears that once stabilized, it may be difficult to change.

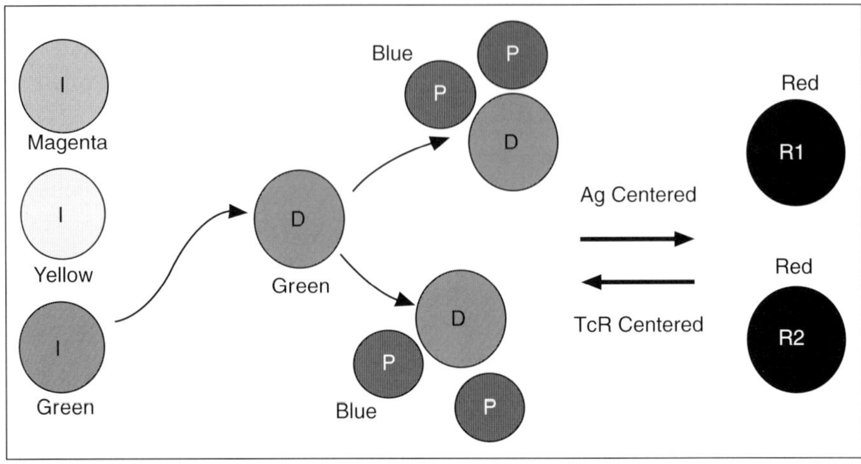

Fig. 1. One of the initiator (I) clones becomes the major driver (D) clone in inducing the disease. Subsequently, perpetuator (P) clones with other antigen specificities get recruited which helps to propagate disease. Regulatory (R) clones that are either antigen-(R1) or TcR-(R2) based can then be engaged in controlling driver clones, and possibly perpetuators.

Differences among I-A^{g7+} Mice in GAD-65 Antigen Processing and Presentation

As we described above, a great deal of effort has been put forth addressing which GAD-65-reactive T cells are important in NOD T1D. However, it is often overlooked that it is the APC and its proteolytic enzymes which are the central players in determining whether an immune response is initiated or tolerance to a self-antigen is induced. We therefore wanted to address the question of which determinants of mGAD-65 can be processed and presented on NOD MHC class II molecules? To address this question, a panel of CD4+ I-A^{g7}-restricted T-cell hybridomas were generated directed against 9 of the 14 known CD4 determinants of mGAD-65. These hybridomas have been used as detectors of GAD-65 antigen processing and presentation in vitro using NOD bulk spleen cells and purified B cells, macrophages and dendritic cells as APCs. In young NOD mice, most of the known CD4 determinants of mGAD-65 are well processed and presented on I-A^{g7} molecules when the protein is provided exogenously (fig. 2).

One of the interesting findings is that young NOD mice were able to process and present determinants within p246–266 at an age prior to the onset of insulitis but do not present this determinant(s) by the time the mice begin to undergo infiltration of immune cells into the pancreas. Further, we found that

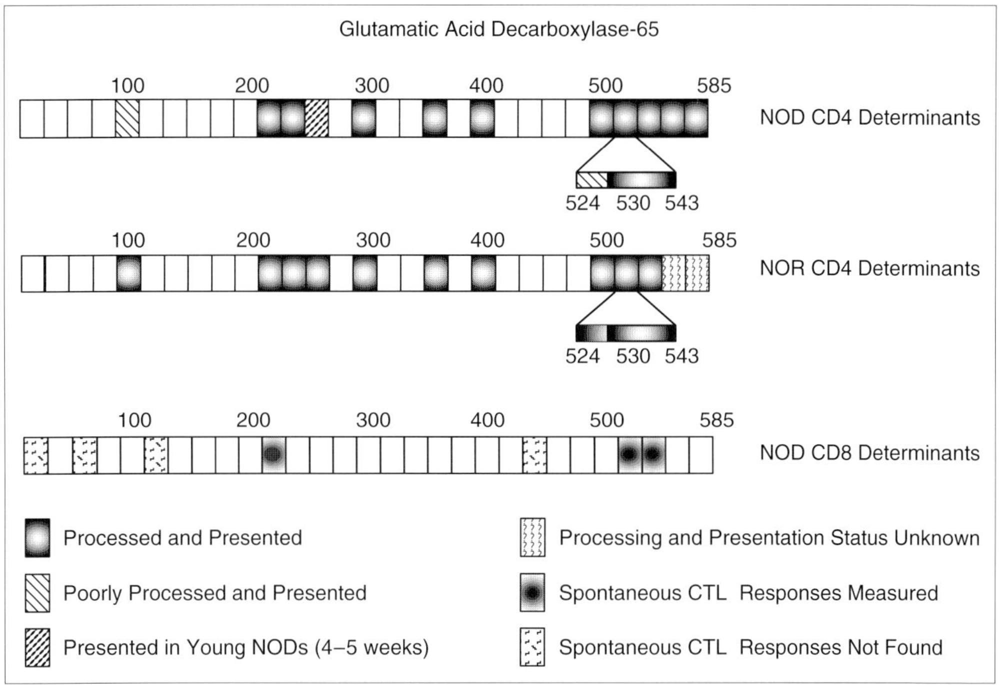

Fig. 2. Summary of differences in the antigen processing and presentation GAD-65 determinants on NOD and NOR I-A^{g7} MHC class II molecules and on NOD K^d MHC class I molecules in vitro. The CD4 determinants are 14-mers to 15-mers and the CD8 determinants are 9-mers.

NOD mice do not typically present determinants within p524–538. This region is of particular interest becuase T cells directed against this determinant are able to regulate spontaneous and cyclophosphamide-induced T1D in NOD mice [38].

The NOR mouse strain is a strain that is very similar to the NOD, but is resistant to T1D. They share the same MHC haplotype and have about 90% of their genome in common with the NOD [39]. This 10% genomic difference permits the NOR mouse to remain completely resistant to spontaneous and cyclophosphamide-induced diabetes [39]. When we examined the processing and presentation of GAD-65 by APCs from NOR mice, it was found that NOR APCs were able to process and present all nine of the determinants that were examined, when GAD-65 was exogenously provided (fig. 2). This presentation included determinants within p78–97, p246–266 and p524–538.

Peptide p524–538 appears to induce Vβ12-bearing regulatory cells which are able to control T1D after adoptive transfer. This determinant is not well presented by NOD APCs and it is possible that these cells only arise in an inflammatory focus owing to some inductive stimulus arising from the activation of p530-specific aggressive cells. Interestingly, NOR mice, bearing the same I-A^{g7} MHC class II molecule, are nevertheless able to present p524 very successfully to either NOR or NOD T cells. Although perhaps too simplistic, this ready ability of the NOR strain to activate the specific p524 regulatory T cell may contribute to the resistance of NOR mice to T1D.

Three Contiguous GAD-65 Determinants and the Different Functions of T Cells Directed Against Each of Them

One of the most interesting determinant appositions to be discussed is the trio of determinants close to the C-terminus of GAD-65. Not only are there the two CD4 overlapping determinants, 524–543 and 530–543, but also the CTL-inducing determinant, p546–554. This region of GAD-65 might be co-processed and in that sense, comprises a functional entity. For example, the determinant that is dominant within GAD-65 as it is handled in the NOD mouse is 530–543, and Vβ4+ T cells specific for this determinant appear to arise at about 2–3 weeks of age and soon afterwards have infiltrated into the islets of Langerhans. The overlapping determinant, 524–538, is poorly presented by NOD I-A^{g7+} APCs but quite well by NOR I-A^{g7+} APCs to NOD T cells: such T cells are potent regulators as will be documented below.

It may be that p530-specific T cells initiate a response in a still unknown way, serving as 'driver clones' in causing disease and also could provide a stimulus to the APC helping to recruit CD8 CTL directed against 546–554, which play a required part in T1D [40]. Although the evidence is circumstantial in linking these responses as we have done above, the fact that a spontaneous response to the CD8 determinant arises only when there is a nearby CD4 determinant is supportive of this view [also see below].

Overlapping Determinants on GAD-65 Induce Distinct T Cells with Divergent Effector Functions

T-cell responses to GAD-65 have been shown to be predictors of T1D in humans and in mice [1]. More specifically, these responses are thought to be pathognomonic in NOD mice, such that therapies designed to block or reverse the immunopathology associated with T1D are often judged by their ability to alter the T-cell response to particular determinants on GAD-65, qualitatively or quantitatively. These distinctive T-cell responses to selected epitopes in T1D, i.e. GAD65 530–543, are suggestive of a selective pressure at the level of the T-cell clone. Therefore, it would be expected that certain TCR structural motifs could also be used to define the T cells involved in the earliest responses to GAD-65 in NOD mice.

Two Distinct T-Cell Determinants within GAD-65 p524–543

We addressed the nature of GAD-65 p524–543-specific CD4+ T cells which spontaneously arise in the NOD mouse [2] and those that are activated as a result of immunization with this 20-mer peptide. We were able to show that two unique CD4+ T-cell repertoires are recruited to this diabetes-associated region within GAD-65, one dominantly primed by intrinsically-provided GAD-65 (against aa 530–543) and the other induced after immunization with GAD-65 p524–543 or the smaller GAD-65 fragment p524–538 [43]. Furthermore, the response to these two overlapping registers within the 20-mer peptide of GAD-65 p524–543 is mediated by two functionally distinct T-cell repertoires. One set of these T cells arises spontaneously and is specific for a GAD-65 determinant within the p530–543 sequence (p530) while the other is induced only after immunization with peptides within GAD-65 p524–543. These GAD-65-reactive T cells utilize distinct TCR Vβ families, and the amino-terminal moiety induces T cells that are capable of playing a regulatory role in T1D. It is also of interest that a T-cell line reactive with GAD-65 has been shown to induce T1D [42], while GAD-65 expression appears to be required for β cell damage to occur spontaneously in NOD mice [13].

Correspondingly, GAD-65 or certain GAD-65 peptides administered to NOD mice in a tolerogenic manner (intravenously, mucosally or intraperitoneally) have the ability to prevent the spontaneous development of T1D [2, 3, 41, 44], a further indication that the two different effector functions of cellular destruction and immune regulation might be dissociable within GAD-65-specific T-cell responses. Strikingly, all p530-specific T-cell hybridomas utilize Vβ4 chains in their antigen receptors. This preferential Vβ usage was also observed in T-cell lines and clones derived from unimmunized prediabetic NOD mice. Perhaps this focus on Vβ4 would be of little note except for the fact that the oft-studied diabetogenic T-cell clone BDC-2.5 also expresses the Vβ4 TcR chain [45]. The BDC-2.5 clone has been shown to be responsive to pancreatic islet cells and possesses diabetogenic activity. A combinatorial peptide library study deconvoluted the specificity of BDC2.5 to a peptide from GAD-65 (526–541) and GAD-67 (538–551) [46]. Accordingly, a link was forged between the earliest appearing spontaneous T-cell clones in the NOD mouse model and a TCR-Tg mouse model of type I diabetes.

The precise role of p530-specific T cells in possible causation of T1D is currently being sought; however, in accord with molecular mimicry ideas, it is possible that a latent provirus, characteristic of the NOD genome, becomes induced early in the life of most NOD mice and displays a p530 cross-reactive determinant, in response to which Vβ4 clones generally arise. As an example, in the deconvolution experiments of Judkowski et al. [46], strongly heteroclitic versions of the native GAD-65 528–539 peptide also stimulate the BDC2.5 clone. Conceivably, the GAD-65 peptide p530–543 itself, an overlapping GAD-65 or GAD-67 peptide, or a ubiquitous environmental peptide, might induce self-reactivity owing to particularly efficient processing and presentation in the young NOD animal. The mimic-peptide and the GAD-65 peptide need not be identical although it is remarkable how closely they do resemble each other. Many reports testify to the vast degeneracy of specificity of a single TcR [e.g., 47–49]. It is apparent that the similarity of the deconvoluted peptide for which BDC2.5 is specific, to p530 is indeed more than sufficient to surpass the cross-reactivity threshold needed to induce disease in transfer experiments [46].

GAD-65 p524–538 Regulatory Cells

The mechanisms by which the p524-reactive T cells control T1D is not yet clear. Numerous cells have been reported to have regulatory functions that can modulate autoimmune disease, including T cells that are characterized by particular cell surface markers (e.g. CD45B, CD25, etc.) as well as those which display apparently unique cytokine secretion patterns (e.g. Th2, Th3, Tr1). The p524-reactive regulatory cells were CD4+ TcR Vβ12+ T cells that were potent producers of IL-2, IL-5 and IFN-γ. Since none produced detectable levels of IL-4, IL-10 or TGF-β, they could not be categorized as TR1 [50] or Th3 type clones [51], leaving IL-5 and IFN-γ as the candidates that might be responsible for the effector function of these T cells. Interestingly, although IL-4 and IL-10 have been viewed as anti-inflammatory cytokines that are antagonistic to Th1 cells, a similar role has not been previously ascribed to IL-5 in autoimmune diabetes. On the other hand, IL-5 has a well-established association with mast cells and the mediation of allergic responses, which may be antagonistic to the diabetogenic process. In addition, while the idea that IFN-γ-producing cells can play a protective role in an inflammatory disease might seem counterintuitive, it was recently shown that a Th1 phenotype was required in generating TCR-specific regulatory T cells in the EAE model [52]. Curiously, a GAD-65 p524–543-reactive line, containing Vβ12+ T cells, has been shown to have diabetogenic potential in NOD.scid mice [42]. It remains to be seen whether the

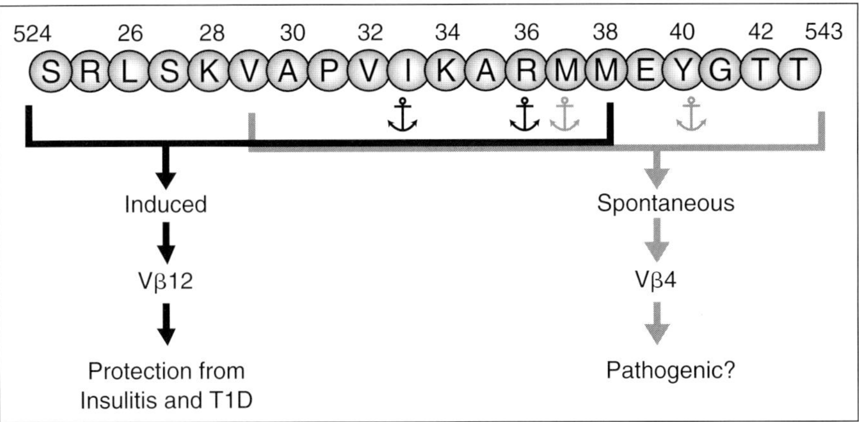

Fig. 3. Summary of the two overlapping CD4+ determinants within the p524–543 region of GAD-65 found in NOD mice. Putative I-A^{g7} P6- and P9-binding residues for each determinant are indicated with an anchor.

Vβ12+ T cells in that line have diabetogenic potential or if they share the same fine specificity for p524 with the previously described Vβ12+ T cells.

The observation that two overlapping determinants reside within the GAD-65 p524–543 sequence demonstrates that the response to an apparently unique determinant in a self-molecule may instead induce responsiveness within at least two functionally distinct T-cell repertoires: one moiety of the determinant spontaneously activates a proliferative, putatively autoaggressive response, while a regulatory response is evoked by the overlapping moiety. Consequently, the fine details of antigen processing could be very influential in determinant choice and thereby, disease outcome. In any case, it may be possible to actively induce antigen-specific regulatory T cells that recognize determinants distinct from those that arise during the natural course of autoimmune disease (fig. 3).

CD8+ T Cells in T1D

Autoimmune diabetes is a manifestation of permanent β-cell loss leading to an inability to maintain glucose homeostasis. Whether the critical threshold of β-cell damage is due to chronic immunological assault or rather to an acute attack precipitated by some yet undescribed event, is a matter of debate [53, 54]. However, one of the central issues in the discussion is the mechanism by which β cells are destroyed. Despite the investigational focus on CD4+ T-cell responses to GAD-65, both CD4+ and CD8+ T cells are required for disease progression in NOD mice [55].

NOD mice deficient in MHC class I or β_2M expression are protected from both T1D and insulitis [56, 57], and adoptive transfer experiments using T-cell clones or spleen cells from diabetic mice have clearly demonstrated the necessity for both CD4+ and CD8+ T cells if severe disease is to develop in the recipients [55, 58]. Moreover, the findings of Serreze et al. [59] suggest that CD8+ T cells are not only necessary for insulitis and T1D, but that MHC class I-restricted T-cell responses may be the early initiators of islet β-cell damage. Thus, while spleno-cytes from diabetic NOD mice could transfer disease to recipients lacking MHC class I molecules (NOD-scid.β2m-null), similar cells from young prediabetic NOD mice could only transfer disease to recipients that expressed MHC class I [59].

Although CD4+ T cells are requisite for spontaneous T1D and some Th clones can transfer disease without the need for donor CD8+ T cells [60], the means by which they perpetuate hyperglycemia is not clear since murine β-cells do not express MHC class II [61, 62]. It is possible that certain Th cells mediate islet damage via the production of cytokines [63] or through Fas-Fas L inter-actions between activated T cells and β cells [56]. Alternatively, CD8+ T cells may initiate early β-cell damage, which then leads to the release of islet antigens and the activation of CD4+ T cells [58, 64]. The islet-reactive Th cells could then home to the pancreas, secrete cytokines/chemokines which attract other inflammatory cells, and escalate the inflammatory process. Accordingly, previously described diabetogenic CD4+ T-cell clones, which transfer T1D without the need for CD8+ T cells, were themselves recovered from mice that possessed a CD8+ T-cell repertoire [60, 65].

Thus, while the role of CD8+ T cells, either as initiators or terminal effectors of β-cell damage in type I diabetes is unresolved, we sought to investigate the activities of GAD-65-specific CD8+ T cells in T1D (given the prominence of GAD-65 as an autoantigen in T1D) [13]. To identify relevant CD8+ T cell-inducing determinants on GAD-65, we utilized information regarding the K^d allele-specific binding motif [66, 67]. To further refine the search, we took into consideration findings of previous reports wherein the dominant CD8- and CD4-inducing specificities on β-galactosidase were defined [27, 28] as well as the nature of the interactions of cells recognizing each determinant. Those reports suggested a proximity relationship between CD4-inducing determinants and CD8-inducing determinants such that the most important feature of the relationship between the two T-cell types was the proximity of the distinct determinants that they recognized. We demonstrated that CTL specific for the K^d-restricted GAD-65 determinants p206–214 (p206) and p546–554 (p546) were detectable in the spleens of prediabetic NOD mice. Most importantly, such CTL responses were found in 3-week-old NOD mice, an age at which insulitis is difficult to detect histologically. Spontaneous responses to other putative CD8 determinants were not observed. We showed that CD8+ T cells specific for GAD-65 peptides p206 and p546 were cytotoxic for GAD-65-expressing cells and produced IFN-γ in response to antigenic stimulation. Such MHC class I-restricted responses may provide the necessary inflammatory focus for the generation of islet-specific CD4+ T cells.

Thus, it would be expected that the CD8+ T-cell response to GAD-65 would have a role in the development of T1D and alterations in these spontaneous responses might alter the course of T1D. It was not surprising then that NOD mice nasally pretreated with p546 and p206 show a delayed onset of spontaneous T1D. Moreover, what was most impressive was the ability of p546 to confer protection from CY-induced T1D, a very aggressive model of T1D. It should be noted that even in the p546-treated mice that succumbed to inducible diabetes, the onset of the disease occurred much later (10–12 days after CY treatment compared to 5–7 days in the saline-treated mice).

The use of MHC class I-restricted peptides to ameliorate T1D in the NOD mouse model represents an important step in our efforts to treat autoimmune disease. Although a great deal of success has been achieved using MHC class II-restricted peptides therapeutically in a number of animal models of autoimmune disease, a comparable level of achievement has been difficult to approach in humans. One of the obstacles to human therapy lies in predicting and designing the appropriate MHC class II-binding peptides for an outbred population. Fortunately, MHC class I alleles typically display a more limited predilection for peptides, forming binding motifs that are easier to characterize. The use of CD8+ T-cell-inducing peptides to treat autoimmunity may represent a more pragmatic approach to therapy. To our knowledge this is the first reported demonstration of MHC class I-restricted self-peptides being effectively used to treat an autoimmune disease. The finding that the CD8+ T-cell-inducing determinants on a protein molecule lie near CD4+ T-cell-inducing determinants of the same molecule is intriguing.

Identification of Potential Diabetogenic Islet Antigen-Reactive T Cells Using TCR-Based Regulation

Finally, it is of interest to discuss the last of the four responses, that of Vβ8.2+ T cells to GAD-65 p206–220, which may be regulated through TcR-centered circuitry, in parallel to the Vβ8.2 regulation found in B10.PL mice. The active determinant in the latter strain is part of the framework 3 region on the B5 TcR peptide (aa Vβ8.2(76–101). B5 is unusual in binding well to a large variety of class II molecules, among them I-A^{g7}, I-A^u and I-A^k. The evidence presented in this review establishes that T1D can be prevented by pretreatment with the B5 peptide and that this only occurs if the Vβ8.2+ T cells are present and functional.

Elucidating normal mechanisms for the maintenance of tolerance to self-antigens is important in understanding the development and ultimately the treatment of autoimmune diseases. Although deletional tolerance in the thymus occurs to some dominant antigenic determinants, T cells reactive to many other determinants on self-antigens remain in healthy individuals. In two different transgenic systems, it has been shown that despite the presence of a huge pathogenic self-reactive (islet antigen or myelin basic protein) T-cell repertoire, animals do not develop spontaneous autoimmune pathology unless they are challenged with antigen or are devoid of potential regulatory cells or subjected to different environmental conditions [68, 69]. This clearly shows that a very large number of competent self-reactive T cells can be present without causing disease [70]. It is not clear what controls the progression from benign ongoing autoimmunity to clinical disease. This may require a qualitative/quantitative change in the autoimmune response as a result of differential stimulation of self-reactive pathogenic T cells or regulatory T cells.

During the course of an autoimmune response, in addition to antigen-reactive disease-causing T cells, regulatory CD4+ and CD8+ populations reactive to distinct TCR determinants also arise. For example, during the autoimmune response to MBP in B10.PL mice, several distinct T-cell populations expand: the effectors mediating EAE which are MBP-reactive, CD4+ and predominantly TCR Vβ8.2+ [71, 72], and at least two regulatory populations [73–76]. One of the latter is comprised of Vβ14+ or Vβ3+ CD4 T cells reactive to a framework region 3 determinant (B5, aa 76–101) on the Vβ8.2 chain in the I-A^u context. A second population is CD8+ and reactive to another Vβ8.2 determinant from the CDR1/2 region.

The combined action of these two regulatory cell types leads to a global skewing of disease-causing effector cells in a Th2 direction, resulting in spontaneous recovery from disease. Thus, adoptive transfer of regulatory CD4 T-cell clones or vaccination of naive mice with TCR peptide B5 leads to significant protection from antigen-induced EAE. Furthermore, inactivation of regulatory T cells using anti-Vβ14 and anti-Vβ3 antibodies results in an increase in the severity and duration of disease [75]. These findings define a physiological role for the TCR-peptide-reactive CD4 T$_{reg}$ population in the control of MBP-reactive encephalitogenic T cells resulting in spontaneous recovery from EAE [75, 76]. Consistent with this, it has been shown that a prior expansion of B5-reactive T cells following immunization with TCR in the form of peptides, recombinant single-chain proteins, plasmid DNA, or vaccinia or adenovirus delivery systems results in immune deviation and prevention of disease [Kumar et al. and Braciak et al., unpubl. data; 73, 74, 77, 78].

A number of experimental observations are consistent with the presence of regulatory CD4+ T cells in prediabetic NOD mice: (a) sublethal irradiation is required to transfer disease by diabetogenic T cells [79]; (b) adoptive transfer of diabetes can be blocked by co-transfer of CD4+ T cells from young nondiabetic males or females [80, 81]; (c) cyclophosphamide treatment induces acute T1D in NOD [82]; (d) thymectomy and CD4 depletion potentiates the development of diabetes in NOD males [83]. These observations suggest that islet infiltrates contain both effector populations that are capable of causing islet-cell destruction as well as regulatory populations capable of protecting them.

The antigen specificity of these regulatory T cells and their physiological role is not known. In a number of antigen-induced autoimmune disease models, autoantigen-reactive disease-inducing T cells have been shown to be oligoclonal with respect to their TCR V-gene usage. It has been difficult to identify such oligoclonality in T cells involved in autoimmune diseases that develop spontaneously, such as diabetes in NOD mice. There could be several reasons for the difficulty in identifying this oligoclonality. For example, the route and form of the antigen depot (in adjuvant) can change the nature of antigen presentation, including processing, co-stimulation, and strength of the signal for T cells. Furthermore, in induced models, T-cell responses are monitored in the lymphoid organs (draining lymph nodes or spleen) following a predictable time range, such as 6–10 days after immunization with the autoantigen.

It is difficult to know the appropriate timing for the detection of initiating T cells into the target organ in a spontaneous disease model. Furthermore, even in EAE, initial Vβ8.2 predominance in the central nervous system (CNS) is only transitory as a result of the dominant regulatory circuit, and it is followed by the infiltration of T cells bearing different Vβs [84]. These infiltrates contain myelin-reactive T cells of different specificities recruited as a result of determinant spreading, as well as nonmyelin-reactive T cells [2, 3]. Consistent with this, islet antigen-specific T cells isolated from the periphery as well as from islets of both prediabetic and diabetic NOD mice, have been shown to recognize several different islet antigens and to use different TCRs [reviewed in 1]. It may be that with a large molecule such as GAD-65, distinct domains may lead to parallel activation of several effector populations. These effectors may each arouse populations of regulators which through a variety of mechanisms, could combine to successfully prevent responsiveness.

The Role of Vβ8.2 T Cells

Some recent findings suggest that Vβ8+ T cells represent one of the important T-cell families in anti-islet responses and their regulation. Fathman's laboratory has shown that a TCR Vβ8 oligoclonal T-cell population, perhaps recognizing a single β-cell autoantigenic determinant, was detectable in islets at 2 weeks of age [85]. By 1 month, the subsequent nonspecific inflammatory response obscured the oligoclonal nature of the initiating event. In a separate study, anti-Vβ8 antibody treatment has been shown to protect NOD mice from insulitis or even from cyclophosphamide-induced T1D [86]. One might then question how mice genetically lacking Vβ8 genes are still susceptible to diabetes [87, 88].

The redundancy in immune response is clear from some of the knockout as well as from the genetic deletion studies. When T lymphocytes bearing preferred TCR genes are not available from the very beginning of development, T cells expressing alternative Vβ genes may be used [89]. Therefore, the occurrence of diabetes in mice lacking Vβ8 genes does not necessarily imply that these cells remain uninvolved when they are present.

Although several islet-derived autoantigens have been implicated in the development of diabetes (GAD-65, insulin and hsp60) [reviewed in 1, 90, 91], the one initiating this process is unclear. Since predominance of Vβ8 T cells among diabetogenic T cells has not been established, we have asked how control of this single population could influence diabetogenic T cells and as a result, spontaneous diabetes. As discussed above, we have recently discovered that the action of TCR peptide-reactive regulatory T cells in the EAE model leads to deviation of the MBP-specific response in a Th2 direction [76]. We have examined whether such a deviated Vβ8.2+ islet-reactive T-cell population can provide an anti-inflammatory environment which leads to deviation of other islet-specific T cells in a similar protective direction (bystander effects) [92].

T-Cell-Receptor-Centered Regulatory Circuitry

We found that a single TCR peptide derived from the Vβ8.2 chain (B5, aa 76–101) binds to I-A^{g7} and induces a proliferative CD4 T-cell response in NOD mice upon challenge. Notably, a single adjuvant-free priming of 2-week-old NOD females with TCR peptide results in reduced insulitis and significant prevention of diabetes. In parallel, identical priming with immunogenic peptides derived from the TCR Vβ17 chain, or with lysozyme, had no effect. The prevention of diabetes is dependent upon the presence of Vβ8+ cells, as priming of NOD-Vβa mice lacking this target population with the TCR peptide does not result in protection from T1D. This does not exclude the idea that other protective mechanisms can be engaged to control T1D.

In a novel inverted approach to defining disease-relevant regions of self-antigens, we have examined whether the cytokine profiles of T cells reactive to any of the dominant determinants from candidate islet antigens are deviated following B5 challenge. Since spontaneous responses to islet antigenic determinants are not easy to detect, we have determined cytokine secretion profiles of immunized islet antigens following priming with TCR peptide B5. We found that the frequency of IFN-γ-secreting T cells decreased whereas the frequency of IL-4-secreting cells increased in response to GAD-65 p206–220 in the B5-treated group. There was no increase in Th2 response to other determinants on GAD-65 p524–543, or to HSP60/65, or insulin. The specific deviation of only the GAD-65 p206–220 response following priming of the regulatory T cells suggests that (a) these cells are among the pathogenic T cells and (b) the TCR Vβ8 may constitute a predominant pathogenic population reactive to this determinant.

Based on our data in experimentally-induced autoimmune disease models, we propose that a TCR peptide B5-reactive, CD4 T_{reg} response of type 1 is required for effective regulation and eventual immune deviation of GAD-65 p206–220-reactive Vβ8.2 T cells. Indeed, in male NOD mice, with their Th2 bias, nasal priming with B5 results in exacerbation of diabetes which can be reversed by priming in the presence of IL-12. Inflammatory cytokines secreted by CD4+ T_{reg} promote efficient recruitment/activation of regulatory CD8 T cells. Regulatory CD8 cells recognize the class I MHC-peptide complex on the surface of an activated Vβ8.2+ CD4+ T-cell population. Thus CD8+ T cells may induce apoptosis or anergy of the initial rapidly expanding, high avidity, disease-causing Vβ8.2 Th1-cell population, enabling low avidity, GAD-65 p206–220-reactive type 2 T cells (which may or may not express Vβ8.2) to expand, resulting in immune deviation. Immune deviation of antigen-specific T cells at the population level may explain how TCR-based regulation directed to a single Vβ chain is able to control disease-inducing, GAD-65 p206–220-reactive T cells that use other TCR V chains, e.g. Vβ6.

Furthermore, such modulation of T-cell responsiveness to the target antigenic determinant may provide a suppressive environment for responses to other antigenic determinants from the GAD-65, as well as from other islet antigens that may arise as a result of determinant spreading. It will be interesting to determine if the regulatory determinant GAD-65 p530–543 or p524–538-reactive T cells become engaged following the skewing of GAD-65 p206–220-specific responses and participate in a cooperative manner to effectively maintain tolerance to islet antigens. A crucial question currently being addressed is whether TCR peptide-reactive T_{reg} are also physiologically involved in the maintenance of peripheral tolerance to islet antigens in NOD mice. Using the sensitive single-cell ELISA spot assay, we have been able to detect the presence of TCR peptide B5-reactive T cells in 2- to 4-week-old NOD females. This would allow us to examine whether there is a qualitative or quantitative alteration in this population during the change from benign to destructive insulitis.

Concluding Remarks

In summary, the various cellular components underlying self-responsiveness in T1D could have been analyzed on one of many diabetogenic islet antigens. We have chosen to consider glutamic acid decarboxylase-65, a large antigen, which is centrally involved at an early stage in the development of this disease in the NOD mouse. Four distinct T-cell-inducing determinants have been followed, each with its own functional pedigree, each with its own relationship to regulation. In some cases, there are further cells which can be shown to be recruited into the melee – influencing the decision between inflammatory disease versus its prevention by regulatory down-modulation. In the female NOD mouse, the regulators most often lose this battle, while in the males, they appear to win. Apparently, giving a head start to the regulators, even in the female NOD mouse, may be one of the routes to successful control. Just as a narrow inflammatory pauciclonal response can spread into a major involvement, when and if the tide turns towards regulation, there can be a variety of different types of complementary regulatory strategies that reinforce each other in firmly shutting down the inflammatory elements of disease. The potency of these potential regulators can be seen in the oft quoted idea that almost anything the experimenter does can downregulate T1D in the NOD mouse.

Acknowledgment

This work was supported in part by grants from the NIH and the JDFI.

References

1   Delovitch TL, Singh B: The nonobese diabetic mouse as a model of autoimmune diabetes: Immune dysregulation gets the NOD [published erratum appears in Immunity 1998;8:531]. Immunity 1997;7:727.

2   Kaufman DL, Clare-Salzler M, Tian J, Forsthuber T, Ting GS, Robinson P, Atkinson MA, Sercarz EE, Tobin AJ, Lehmann PV: Spontaneous loss of T-cell tolerance to glutamic acid decarboxylase in murine insulin-dependent diabetes. Nature 1993;366:69.

3   Tisch R, Yang XD, Singer SM, Liblau RS, Fugger L, McDevitt HO: Immune response to glutamic acid decarboxylase correlates with insulitis in nonobese diabetic mice. Nature 1993;366:72.

4   Daniel D, Gill RG, Schloot N, Wegmann D: Epitope specificity, cytokine production profile and diabetogenic activity of insulin-specific T cell clones isolated from NOD mice. Eur J Immunol 1995;25:1056.

5   Tian J, Lehmann PV, Kaufman DL: T cell cross-reactivity between coxsackievirus and glutamate decarboxylase is associated with a murine diabetes susceptibility allele. J Exp Med 1994; 180:1979.

6   Tian J, Clare-Salzler M, Herschenfeld A, Middleton B, Newman D, Mueller R, Arita S, Evans C, Atkinson MA, Mullen Y, Sarvetnick N, Tobin AJ, Lehmann PV, Kaufman DL: Modulating

autoimmune responses to GAD inhibits disease progression and prolongs islet graft survival in diabetes-prone mice. Nat Med 1996;2:1348.

7 Tian J, Atkinson MA, Clare-Salzler M, Herschenfeld A, Forsthuber T, Lehmann PV, Kaufman DL: Nasal administration of glutamate decarboxylase (GAD65) peptides induces Th2 responses and prevents murine insulin-dependent diabetes. J Exp Med 1996;183:1561.

8 Tian J, Lehmann PV, Kaufman DL: Determinant spreading of T helper cell 2 (Th2) responses to pancreatic islet autoantigens. J Exp Med 1997;186:2039.

9 Tian J, Kaufman DL: Attenuation of inducible Th2 immunity with autoimmune disease progression. J Immunol 1998;161:5399.

10 Tisch R, Yang XD, Liblau RS, McDevitt HO: Administering glutamic acid decarboxylase to NOD mice prevents diabetes. J Autoimmun 1994;7:845.

11 Tisch R, Liblau RS, Yang XD, Liblau P, McDevitt HO: Induction of GAD65-specific regulatory T-cells inhibits ongoing autoimmune diabetes in nonobese diabetic mice. Diabetes 1998; 47:894.

12 Tisch R, Wang B, Serreze DV: Induction of glutamic acid decarboxylase 65-specific Th2 cells and suppression of autoimmune diabetes at late stages of disease is epitope dependent. J Immunol 1999;163:1178.

13 Yoon JW, Yoon CS, Lim HW, Huang QQ, Kang Y, Pyun KH, Hirasawa K, Sherwin RS, Jun HS: Control of autoimmune diabetes in NOD mice by GAD expression or suppression in beta cells. Science 1999;284:1183.

14 Faulkner-Jones BE, Cram DS, Kun J, Harrison LC: Localization and quantitation of expression of two glutamate decarboxylase genes in pancreatic beta-cells and other peripheral tissues of mouse and rat. Endocrinology 1993;133:2962.

15 Pleau JM, Fernandez-Saravia F, Esling A, Homo-Delarche F, Dardenne M: Prevention of autoimmune diabetes in nonobese diabetic female mice by treatment with recombinant glutamic acid decarboxylase (GAD 65). Clin Immunol Immunopathol 1995;76:90.

16 Pleau JM, Esling A, Bach JF, Dardenne M: Quantitative analysis of glutamate decarboxylase (GAD 65) gene expression in NOD mouse pancreas. C R Acad Sci III 1995;318:129.

17 Pleau JM, Esling A, Van Acker C, Dardenne M: Glutamic acid decarboxylase (GAD 67) gene expression in the pancreas and brain of the nonobese diabetic mouse. Biochem Biophys Res Commun 1996;224:747.

18 Pleau JM, Esling A, Bach JF, Dardenne M: Gene expression of pancreatic glutamic acid decarboxylase in the nonobese diabetic mouse. Biochem Biophys Res Commun 1996;220:399.

19 Pleau JM, Throsby M, Esling A, Dardenne M: Ontogeny of glutamic acid decarboxylase gene expression in the mouse pancreas. Biochem Biophys Res Commun 1997;233:227.

20 Chao CC, McDevitt HO: Identification of immunogenic epitopes of GAD 65 presented by Ag7 in non-obese diabetic mice. Immunogenetics 1997;46:29.

21 Zechel MA, Elliott JF, Atkinson MA, Singh B: Characterization of novel T-cell epitopes on 65 kDa and 67 kDa glutamic acid decarboxylase relevant in autoimmune responses in NOD mice. J Autoimmun 1998;11:83.

22 Soares LR, Sercarz EE, Miller A: Vaccination of the *Leishmania major* susceptible BALB/c mouse. I. The precise selection of peptide determinant influences CD4 T cell subset expression. Int Immunol 1994;6:785.

23 Benichou G, Fedoseyeva E, Olson CA, Geysen HM, McMillan M, Sercarz EE: Disruption of the determinant hierarchy on a self-MHC peptide: Concomitant tolerance induction to the dominant determinant and priming to the cryptic self-determinant. Int Immunol 1994;6:131.

24 Maverakis E, Stevens D, Brossay L, Mendoza R, Macias L, Thai Q, Campagnoni E, Skinner D, Southwood S, Sette A, Sercarz E: Protecting the T cell repertoire from negative selection by determinant capture: Responsiveness to the amino-terminal determinant of myelin basic protein. Submitted.

25 Manca F, Kunkl A, Fenoglio D, Fowler A, Sercarz E, Celada F: Constraints in T-B cooperation related to epitope topology on *E. coli* beta-galactosidase. I. The fine specificity of T cells dictates the fine specificity of antibodies directed to conformation-dependent determinants. Eur J Immunol 1985;15:345.

26 Watts C, Lanzavecchia A: Suppressive effect of antibody on processing of T cell epitopes. J Exp Med 1993;178:1459.

27 Shivakumar S, Sercarz EE, Krzych U: The molecular context of determinants within the priming antigen establishes a hierarchy of T cell induction: T cell specificities induced by peptides of beta-galactosidase vs. the whole antigen. Eur J Immunol 1989;19:681.

28 Krzych U, Fowler AV, Sercarz EE: Repertoires of T cells directed against a large protein antigen, beta-galactosidase. II. Only certain T helper or T suppressor cells are relevant in particular regulatory interactions. J Exp Med 1985;162:311.

29 Carreno BM, Turner RV, Biddison WE, Coligan JE: Overlapping epitopes that are recognized by CD8+ HLA class I-restricted and CD4+ class II-restricted cytotoxic T lymphocytes are contained within an influenza nucleoprotein peptide. J Immunol 1992;148:894.

30 Ou D, Mitchell LA, Decarie D, Gillam S, Tingle AJ: Characterization of an overlapping CD8+ and CD4+ T-cell epitope on rubella capsid protein. Virology 1997;235:286.

31 Takahashi H, Katagiri M: DNA sequence analysis of HLA-DQB genes associated with insulin-dependent diabetes mellitus (in Japanese). Hokkaido Igaku Zasshi 1990;65:412.

32 Wong FS, Karttunen J, Dumont C, Wen L, Visintin I, Pilip IM, Shastri N, Pamer EG, Janeway CA Jr: Identification of an MHC class I-restricted autoantigen in type 1 diabetes by screening an organ-specific cDNA library. Nat Med 1999;5:1026.

33 Wegmann DR, Norbury-Glaser M, Daniel D: Insulin-specific T cells are a predominant component of islet infiltrates in pre-diabetic NOD mice. Eur J Immunol 1994;24:1853.

34 Daniel D, Wegmann DR: Protection of nonobese diabetic mice from diabetes by intranasal or subcutaneous administration of insulin peptide B-(9-23). Proc Natl Acad Sci USA 1996;93:956.

35 Schoenberger SP, Toes RE, van der Voort EI, Offringa R, Melief CJ: T-cell help for cytotoxic T lymphocytes is mediated by CD40-CD40L interactions. Nature 1998;393:480.

36 Bennett SR, Carbone FR, Karamalis F, Flavell RA, Miller JF, Heath WR: Help for cytotoxic-T-cell responses is mediated by CD40 signalling. Nature 1998;393:478.

37 Van den Elzen P, Maverakis E, Kumar V, Wilson S, Sercarz EE: A repertoire shift is a key feature in recovery from autoimmunity. Nature, submitted.

38 Quinn A, McInerney B, Reich EP, Kim O, Jensen K, Sercarz E: Regulatory and effector CD4 T cells in NOD mice recognize overlapping determinants on GAD-65 and use distinct Vβ genes. J Immunol 2001;166:2982–2991.

39 Prochazka M, Serreze DV, Frankel WN, Leiter EH: NOR/Lt mice: MHC-matched diabetes-resistant control strain for NOD mice. Diabetes 1992;41:98.

40 Quinn A, McInerney M, Sercarz E: MHC class I-restricted determinants on the GAD-65 molecule induce spontaneous CTL activity and can downregulate autoimmune diabetes. J Immunol, submitted.

41 Cetkovic-Cvrlje M, Gerling IC, Muir A, Atkinson MA, Elliot JF, Leiter EH: Retardation or acceleration of diabetes in NOD/Lt mice mediated by intrathymic administration of candidate beta-cell antigens [published erratum appears in Diabetes 1998;47:303]. Diabetes 1997;46:1975.

42 Zekzer D, Wong FS, Ayalon O, Millet I, Altieri M, Shintani S, Solimena M, Sherwin RS: GAD-reactive CD4+ Th1 cells induce diabetes in NOD/SCID mice. J Clin Invest 1998;101:68.

43 Quinn A, Sercarz EE: T cells with multiple fine specificities are used by non-obese diabetic (NOD) mice in the response to GAD(524–543). J Autoimmun 1996;9:365.

44 Zhang ZJ, Davidson L, Eisenbarth G, Weiner HL: Suppression of diabetes in nonobese diabetic mice by oral administration of porcine insulin. Proc Natl Acad Sci USA 1991;88:10252.

45 Harrison LC, Honeyman MC, Trembleau S, Gregori S, Gallazzi F, Augstein P, Brusic V, Hammer J, Adorini L: A peptide-binding motif for I-A(g7), the class II major histocompatibility complex (MHC) molecule of NOD and Biozzi AB/H mice. J Exp Med 1997;185:1013.

46 Judkowski V, Pinilla C, Schroder K, Tucker L, Sarvetnick N, Wilson D: Identification of MHC class II-restricted peptide ligands, including GAD-like sequences, that stimulate diabetogenic T cells from transgenic BDC 2.5 non-obese diabetic (NOD) mice. J Immunol 2000;166: 908–917.

47 Bhardwaj V, Kumar V, Geysen HM, Sercarz EE: Degenerate recognition of a dissimilar antigenic peptide by myelin basic protein-reactive T cells. Implications for thymic education and autoimmunity. J Immunol 1993;151:5000.

48   Hagerty DT, Allen PM: Intramolecular mimicry. Identification and analysis of two cross-reactive
     T cell epitopes within a single protein. J Immunol 1995;155:2993.
49   Wilson DB, Pinilla C, Wilson DH, Schroder K, Boggiano C, Judkowski V, Kaye J, Hemmer B,
     Martin R, Houghten RA: Immunogenicity. I. Use of peptide libraries to identify epitopes that acti-
     vate clonotypic CD4+ T cells and induce T cell responses to native peptide ligands. J Immunol
     1999;163:6424.
50   Groux H, O'Garra A, Bigler M, Rouleau M, Antonenko S, de Vries JE, Roncarolo MG: A CD4+
     T-cell subset inhibits antigen-specific T-cell responses and prevents colitis. Nature 1997;389:737.
51   Chen Y, Kuchroo VK, Inobe J, Hafler DA, Weiner HL: Regulatory T cell clones induced by oral
     tolerance: Suppression of autoimmune encephalomyelitis. Science 1994; 265:1237.
52   Kumar V, Sercarz E: Induction or protection from experimental autoimmune encephalomyelitis
     depends on the cytokine secretion profile of TCR peptide-specific regulatory CD4 T cells.
     J Immunol 1998;161:6585.
53   Eisenbarth GS: Type I diabetes mellitus. A chronic autoimmune disease. N Engl J Med 1986;314:1360.
54   Dilts SM, Lafferty KJ: Autoimmune diabetes: The involvement of benign and malignant auto-
     immunity. J Autoimmun 1999;12:229.
55   Miller BJ, Appel MC, O'Neil JJ, Wicker LS: Both the Lyt-2+ and L3T4+ T cell subsets are
     required for the transfer of diabetes in nonobese diabetic mice. J Immunol 1988;140:52.
56   Katz J, Benoist C, Mathis D: Major histocompatibility complex class I molecules are required for
     the development of insulitis in non-obese diabetic mice. Eur J Immunol 1993;23:3358.
57   Sumida T, Furukawa M, Sakamoto A, Namekawa T, Maeda T, Zijlstra M, Iwamoto I, Koike T,
     Yoshida S, Tomioka H, et al: Prevention of insulitis and diabetes in beta-2-microglobulin-deficient
     non-obese diabetic mice. Int Immunol 1994;6:1445.
58   Christianson SW, Shultz LD, Leiter EH: Adoptive transfer of diabetes into immunodeficient
     NOD-scid/scid mice. Relative contributions of CD4+ and CD8+ T-cells from diabetic versus
     prediabetic NOD.NON-Thy-1a donors. Diabetes 1993;42:44.
59   Serreze DV, Chapman HD, Varnum DS, Gerling I, Leiter EH, Shultz LD: Initiation of autoimmune
     diabetes in NOD/Lt mice is MHC class I-dependent. J Immunol 1997;158:3978.
60   Wegmann DR, Gill RG, Norbury-Glaser M, Schloot N, Daniel D: Analysis of the spontaneous
     T cell response to insulin in NOD mice. J Autoimmun 1994;7:833.
61   McInerney MF, Rath S, Janeway CA Jr: Exclusive expression of MHC class II proteins on
     CD45+ cells in pancreatic islets of NOD mice. Diabetes 1991;40:648.
62   Kay TW, Campbell IL, Oxbrow L, Harrison LC: Overexpression of class I major histocompatibility
     complex accompanies insulitis in the non-obese diabetic mouse and is prevented by anti-interferon-
     gamma antibody. Diabetologia 1991;34:779.
63   Rabinovitch A, Suarez-Pinzon WL: Cytokines and their roles in pancreatic islet beta-cell destruc-
     tion and insulin-dependent diabetes mellitus. Biochem Pharmacol 1998;55:1139.
64   Young LH, Peterson LB, Wicker LS, Persechini PM, Young JD: In vivo expression of perforin by
     CD8+ lymphocytes in autoimmune disease. Studies on spontaneous and adoptively transferred
     diabetes in nonobese diabetic mice. J Immunol 1989;143:3994.
65   Peterson JD, Haskins K: Transfer of diabetes in the NOD-scid mouse by CD4 T-cell clones.
     Differential requirement for CD8 T-cells. Diabetes 1996;45:328.
66   Romero P, Corradin G, Luescher IF, Maryanski JL: H-2Kd-restricted antigenic peptides share a
     simple binding motif. J Exp Med 1991;174:603.
67   Rammensee HG, Falk K, Rotzschke O: Peptides naturally presented by MHC class I molecules.
     Annu Rev Immunol 1993;11:213.
68   Lafaille JJ, Nagashima K, Katsuki M, Tonegawa S: High incidence of spontaneous autoimmune
     encephalomyelitis in immunodeficient anti-myelin basic protein T cell receptor transgenic mice.
     Cell 1994;78:399.
69   Katz JD, Wang B, Haskins K, Benoist C, Mathis D: Following a diabetogenic T cell from genesis
     through pathogenesis. Cell 1993;74:1089.
70   Ohashi PS, Oehen S, Buerki K, Pircher H, Ohashi CT, Odermatt B, Malissen B, Zinkernagel RM,
     Hengartner H: Ablation of 'tolerance' and induction of diabetes by virus infection in viral antigen
     transgenic mice. Cell 1991;65:305.

71 Acha-Orbea H, Mitchell DJ, Timmermann L, Wraith DC, Tausch GS, Waldor MK, Zamvil SS, McDevitt HO, Steinman L: Limited heterogeneity of T cell receptors from lymphocytes mediating autoimmune encephalomyelitis allows specific immune intervention. Cell 1988;54:263.

72 Urban JL, Kumar V, Kono DH, Gomez C, Horvath SJ, Clayton J, Ando DG, Sercarz EE, Hood L: Restricted use of T cell receptor V genes in murine autoimmune encephalomyelitis raises possibilities for antibody therapy. Cell 1988;54:577.

73 Kumar V, Sercarz EE: The involvement of T cell receptor peptide-specific regulatory CD4+ T cells in recovery from antigen-induced autoimmune disease. J Exp Med 1993;178:909.

74 Kumar V, Sercarz E: T cell regulatory circuitry: Antigen-specific and TCR-idiopeptide-specific T cell interactions in EAE. Int Rev Immunol 1993;9:287.

75 Kumar V, Stellrecht K, Sercarz E: Inactivation of T cell receptor peptide-specific CD4 regulatory T cells induces chronic experimental autoimmune encephalomyelitis (EAE). J Exp Med 1996;184:1609.

76 Kumar V, Sercarz E: Genetic vaccination: The advantages of going naked. Nat Med 1996;2:857.

77 Kumar V, Tabibiazar R, Geysen HM, Sercarz E: Immunodominant framework region 3 peptide from TCR V beta 8.2 chain controls murine experimental autoimmune encephalomyelitis. J Immunol 1995;154:1941.

78 Kumar V, Aziz F, Sercarz E, Miller A: Regulatory T cells specific for the same framework 3 region of the Vbeta8.2 chain are involved in the control of collagen II-induced arthritis and experimental autoimmune encephalomyelitis. J Exp Med 1997;185:1725.

79 Wicker LS, Miller BJ, Mullen Y: Transfer of autoimmune diabetes mellitus with splenocytes from nonobese diabetic (NOD) mice. Diabetes 1986;35:855.

80 Boitard C, Yasunami R, Dardenne M, Bach JF: T cell-mediated inhibition of the transfer of autoimmune diabetes in NOD mice. J Exp Med 1989;169:1669.

81 Hutchings PR, Cooke A: The transfer of autoimmune diabetes in NOD mice can be inhibited or accelerated by distinct cell populations present in normal splenocytes taken from young males. J Autoimmun 1990;3:175.

82 Charlton B, Bacelj A, Slattery RM, Mandel TE: Cyclophosphamide-induced diabetes in NOD/WEHI mice. Evidence for suppression in spontaneous autoimmune diabetes mellitus. Diabetes 1989;38:441.

83 Sempe P, Richard MF, Bach JF, Boitard C: Evidence of CD4+ regulatory T cells in the non-obese diabetic male mouse. Diabetologia 1994;37:337.

84 Bell RB, Lindsey JW, Sobel RA, Hodgkinson S, Steinman L: Diverse T cell receptor V beta gene usage in the central nervous system in experimental allergic encephalomyelitis. J Immunol 1993;150:4085.

85 Yang Y, Charlton B, Shimada A, Dal Canto R, Fathman CG: Monoclonal T cells identified in early NOD islet infiltrates. Immunity 1996;4:189.

86 Bacelj A, Charlton B, Mandel TE: Prevention of cyclophosphamide-induced diabetes by anti-V beta 8 T-lymphocyte-receptor monoclonal antibody therapy in NOD/Wehi mice. Diabetes 1989;38:1492.

87 McDuffie M: Diabetes in NOD mice does not require T lymphocytes expressing V beta 8 or V beta 5. Diabetes 1991;40:1555.

88 Shizuru JA, Taylor-Edwards C, Livingstone A, Fathman CG: Genetic dissection of T cell receptor V beta gene requirements for spontaneous murine diabetes. J Exp Med 1991;174:633.

89 Fry AM, Matis LA: Self-tolerance alters T-cell receptor expression in an antigen-specific MHC restricted immune response. Nature 1988;335:830.

90 Atkinson MA, Maclaren NK: The pathogenesis of insulin-dependent diabetes mellitus. N Engl J Med 1994;331:1428.

91 Tisch R, McDevitt H: Insulin-dependent diabetes mellitus. Cell 1996;85:291.

92 Weiner HL, Friedman A, Miller A, Khoury SJ, al-Sabbagh A, Santos L, Sayegh M, Nussenblatt RB, Trentham DE, Hafler DA: Oral tolerance: Immunologic mechanisms and treatment of animal and human organ-specific autoimmune diseases by oral administration of autoantigens. Annu Rev Immunol 1994;12:809.

Eli Sercarz, PhD, Chief, Division of Immune Regulation,
La Jolla Institute for Allergy, 10355 Science Center Drive, San Diego, CA 92121 (USA)
Tel. +1 858 678 4559, Fax +1 858 678 4595, E-Mail eli@liai.org

von Herrath MG (ed.) Molecular Pathology of Type 1 Diabetes mellitus.
Curr Dir Autoimmun. Basel, Karger, 2001, vol 4, pp 193–217

Cytokine Regulation of Diabetes in Experimental Models

E. Allison Green[a], Richard A. Flavell[a,b]

[a] Section of Immunobiology, Yale University School of Medicine and
[b] Howard Hughes Medical Institute, New Haven, Conn., USA

Type 1 diabetes mellitus (T1D) is an autoimmune disease characterized by the selective destruction of the insulin-producing β-cells in the islets of Langerhans by T-cell-mediated mechanisms [1]. Both genetic and environmental factors, like infection and diet, contribute to T1D [2] and addressing why these factors promote T1D has been greatly facilitated by the availability of good animal models of T1D, the BioBreeding (BB) rat and the nonobese diabetic (NOD) mouse.

Studies in NOD mice and the BB rat show that prior to β-cell destruction islets become infiltrated with numerous of immune cells; dendritic cells, macrophages, B cells and T cells [2–4]. Each of these cell types has the capacity to secrete a number of cytokines that either promote diabetes progression or prevent disease. Thus, at any one time, the islet milieu is potentially a rich source of many different cytokines. How, therefore, can we establish the relationship between a particular cytokine and the progression/prevention of T1D? Originally, strategies took the approach of injecting cytokines or antibodies that neutralized a particular cytokine and monitoring the disease process. However, with the advancement of molecular biology, two new and more powerful tools have been developed; the capacity to genetically modify mice to overexpress a particular cytokine (termed transgenic mice) [5–9], or a dominant negative form of it receptor in a particular cell or tissue [10] and the generation of mice that carry a null mutation for the desired gene (termed knockout mice) [11, 12]. More recently, transgenic technology has taken a more intricate turn, facilitating the generation of transgenic mice where specific expression of a particular gene is controlled by a transcriptional on/off switch [13–17]. This latter approach enables identification of the importance of a particular cytokine at a defined step in the progression to diabetes and overcomes the limitations of earlier transgenic models that use constitutively active promoters.

The accumulating results obtained from all these different approaches highlight the complexity of the role cytokines play in T1D. In this chapter, we will focus on some of the most relevant cytokines that have been implicated in either promotion of T1D or protection from disease in murine models and discuss the potential for cytokine-based approaches for ameliorating T1D in man.

Relationship between T Cells and the Development of Diabetes

It has been established for several years that progression to diabetes in NOD mice and BB rats requires T cells. For example, transfer of CD4+, CD8+ T-cell clones [18] or T cells from diabetic NOD mice into immunodeficient NOD SCID mice can induce diabetes [19, 20] whereas transfer of serum or B cells cannot [2]. Secondly, cyclosporin treatment of NOD mice, which specifically inhibits signaling through the IL-2 receptor (which is required for activation of T cells), prevents diabetes progression [21].

The contribution of each subset of T cells to the disease process is still debated. For example, NOD mice deficient in CD8+ T cells, due to an absence of MHC class I molecules or following depletion with anti-CD8 antibodies, do not develop insulitis nor diabetes [12, 22, 23]. In contrast, the absence of MHC class II molecules, and as a consequence, decreased numbers of peripheral CD4+ T cells, results in insulitis but not diabetes [11].

Our understanding of T-cell biology has evolved dramatically in the last several years. The same general conclusions can be drawn for the two major classes of T-cell CD4 and CD8. T cells exit the thymus as naïve precursors of the effector cells that mediate resistance to host infectious agents. CD8+ T cells differentiate into cytotoxic T cells which have as their effector molecules perforin, granzyme B and other lytic molecules. Upon recognition of the target, CD8+ T cells lyse the membrane of the target cell and inject granzyme B, triggering the induction of apoptosis. Likewise, CD4+ T cells differentiate into two classes of effector T cells known as TH1 and TH2 cells. TH1 cells secrete cytokines such as IFNγ and LTα, leading to the activation of macrophages to kill intracellular parasites, and the organization of inflammatory sites within the host. TH2 effector cells secrete interleukin (IL)-4, -5, -6, -9 -10, -13 and so on, and potentiate both the anti-parasite response and the activation of growth and differentiation of B cells [for a review of these general principles, see 24]. Recent studies show that CD4+ T cells in the mouse also include a set of so-called regulatory T cells which appear to be the equivalent of suppressor T cells which inhibit immune responses. These cells are in general CD4+ CD25+ and their mechanism of action is under intensive study [for a review, see 25].

Cytokine Modulation of/by Antigen-Presenting Cells and T1D Progression

T cells do not recognize free antigen. Instead, antigens are acquired by specialized antigen-presenting cells (APCs), degraded into peptide fragments and presented on the surface of the APCs in association with molecules from the major histocompatibility complex (MHC) [26]. APCs not only have the capacity to present antigen to T cells, but can also secrete cytokines that can modulate the T-cell response, for example, by promoting differentiation down the TH1 or TH2 pathway, or into cytotoxic T lymphocytes (CTL) [27, 28]. For these reasons, the importance of particular subgroups of APCs in T1D has been an area of active, and somewhat controversial, research.

Three major classes of APCs can present antigen to MHC class II-restricted CD4+ T cells; B cells, dendritic cells and macrophages. All three classes of APC have been implicated in either the progression to, or protection from, T1D depending on the experimental system [7, 29–31]. For example, deficiency in B cells either by depletion with antibodies, or through genetic means, protects NOD mice from both insulitis and diabetes [32–35]. Similarly, cytokines like transforming growth factor β (TGFβ) are believed to protect NOD mice from developing diabetes by favoring presentation of islet antigens by macrophages as opposed to B cells and as a result, promote TH2 responses [30]. These data, and others, suggest that B cells are important for the progression to diabetes, whereas macrophages are critical for preventing diabetes development.

Macrophages are a major source of IL-10, a potent anti-inflammatory cytokine that can downregulate production of many proinflammatory cytokines that have been implicated in β-cell damage [36, 37]. Thus it is not surprising that macrophages may have a prominent anti-inflammatory role in T1D. However, the evidence that macrophage-derived products like IL-1 and nitric oxide (NO) are strongly implicated in β-cell dysfunction and apoptosis argue that macrophages contribute to β-cell death [38]. In support of this, Alleva et al. [39] demonstrated that macrophages from NOD mice are abnormal in that they both over-produce and sustain production of the TH1-inducing cytokine IL-12 compared to macrophages from non-NOD strains of mice. Thus it remains to be determined whether macrophages have a protective or detrimental role to play in T1D.

Dendritic cells (DCs) have also come under intense focus [40, 41]. Currently, two major classes of DCs have been identified in both mice and humans based on the expression pattern of CD11b and CD8α [42–45]. For example, myeloid-derived DCs are CD11c+CD11b+CD8α− whereas lymphoid-derived DCs are CD11c+CD8α+CD11b+ [46–49]. More current data has suggested that at least in mice, these two major groups may be subdivided further based on expression of CD4 [50]. Initially, in vitro studies demonstrated that lymphoid DCs were the principal source of IL-12 and could polarize naïve T cells along the TH1 pathway, whereas myeloid DCs failed to produce IL-12 and thereby promoted differentiation of T cells into the TH2 subset [42–45, 49, 51–54]. These in vitro data were supported by in vivo studies which demonstrated antigen-pulsed lymphoid DCs generated TH1 responses following injection into mice, whereas antigen-pulsed myeloid DCs promoted TH2 responses [28, 53–55]. However, it is now evident that myeloid DCs too can induce TH1 responses, despite the fact they fail to secrete IL-12 [56]. The mechanism by which these myeloid DCs induce TH2 responses is unknown. There has been no evidence that these DCs are capable of secreting TH2-promoting cytokine IL-4. Instead it is speculated that their production of IL-10 and TGFβ prevents transcription of IL-12 and thereby inhibits development of TH1 cells [57].

In contrast to mice, in humans, myeloid-derived DCs induce TH1 responses whereas lymphoid DCs induce TH2 responses [58]. Takahashi et al. [59] demonstrated that in individuals genetically predisposed to developing diabetes, defects in activation of myeloid DCs correlated with diabetes susceptibility. This suggests that in humans TH1 responses may be beneficial in preventing diabetes progression.

The findings outlined above, highlight the importance of the types of APCs that are recruited to the islets due to the inflammatory process; not only can their antigen presentation capacity be positively or negatively regulated by the cytokines present in the inflamed islets [51, 60, 61], but the APCs themselves may contribute to the skewing of the anti-islet T-cell response from a 'protective' TH2 environment to a 'detrimental' TH1 environment, by secreting certain cytokines.

Proinflammatory Cytokines and T1D

Histological studies in isolated pancreatic sections demonstrated that the progression to diabetes in NOD mice and BB rats was linked to the downregulation of anti-inflammatory TH2 cytokines like IL-4, IL-10 and TGFβ, and the emergence of proinflammatory cytokines like IL-2, TNFα, IL-1 and IFNγ [2]. This data has led to many extensive studies determining whether this proinflammatory environment is contributing to the disease process and if so, which cytokines are the most dominant at manipulating the anti-islet T-cell response and β-cell viability. Below we shall discuss some of the most common proinflammatory cytokines that have been linked to disease development.

Interleukin-2

One of the most important cytokines for T-cell regulation is the proinflammatory cytokine IL-2 [62] which stimulates T cells after antigen activation. IL-2 can also participate in B cell, natural killer (NK) cell and monocyte stimulation. There has been a long association with IL-2 and the breakdown in tolerance to host antigens. For example, T cells rendered anergic following inappropriate stimulation through their T-cell receptor in the absence of co-stimulatory molecules, can subsequently respond to their cognate antigen if given exogenous IL-2 [63, 64]. Further, mice expressing H-2K^b molecules on their islets under control of the rat insulin promoter, rejected H-2K^b skin grafts only if IL-2 was administered at the same time as the transplant [65]. Finally, administration of reagents that specifically deplete IL-2-producing T cells [66, 67], or block IL-2 production [21] can prevent diabetes in NOD mice, whereas administration of IL-2 to BB rats promoted diabetes progression [68]. From these data it was speculated that onset of autoimmunities like diabetes may result from an inappropriate delivery of IL-2 to the T cells in the islet environment. However, transgenic expression of IL-2 on the β-cells of C57BL/6 or CDA mice, resulted in severe insulitis but not diabetes [69]. Tolerance to islet antigens could only be broken if the transgenic mice were crossed to T-cell receptor transgenic mice whose T cells were specific for islet antigen [70]. Thus it is possible that in individuals harboring the correct genetic bias towards autoimmunity (perhaps in the form of potentially autoreactive T cells) a local or systemic increase in IL-2 levels may serve as a catalyst to precipitate T1D.

An alternative role for IL-2 in T1D is not as an accelerant to disease but actually as a protector of autoimmunity. This is based on the observation that mice carrying a null mutation for IL-2, or its receptor, develop autoimmunities due to defects in Fas-mediated apoptosis of lymphocyte populations [71]. In addition to its cell growth properties, IL-2 signals are important for activation-induced cell death (AICD). Thus, the development of lymphoproliferative disorders and autoimmunity in IL-2-deficient mice is believed to be due to disruption of the AICD pathway demonstrating the importance of AICD in maintaining peripheral homeostasis. Interestingly, IL-2$^{-/-}$ and IL-2 receptor$^{-/-}$ mice do not develop putative regulatory CD4+CD25+ T cells, since the development of such regulatory T cells is dependent on IL-2 signals [72].

In the NOD mice, the gene for IL-2 has been mapped to a position on chromosome 3 (termed idd3) which has been shown to contribute to diabetes susceptibility [73–76]. For example, congenic NOD mice which carry C57BL/10 alleles as opposed to NOD alleles at this genetic locus do not develop diabetes [75]. Sequence and protein studies have determined that NOD-derived IL-2 has a polymorphic variation in its sequence as well as an altered glycosylation pattern compared to non-NOD strains of mice [76]. This has led to the suggestion that NOD-derived IL-2 may induce different signals following binding to its receptor, than the signals induced by IL-2 derived from non-NOD strains of mice. Thus, it has been speculated that alterations in NOD IL-2 may affect the signals generated following binding to its receptor where T-cell proliferation is unaffected, but T-cell death via AICD is compromised. To date, there has been no in vitro evidence to suggest this altered form of IL-2 in NOD mice is different from IL-2 from non-NOD species [77, 78]. However, the genetic association of IL-2 with the diabetes susceptibility idd3 locus and the fact that IL-2 is strongly linked to the development of autoimmunity, it is possible that the polymorphisms in the NOD IL-2 influence diabetes susceptibility.

Interleukin-1

The evidence that macrophages are one of the key cells to enter the islet in the first stages on insulitis, has prompted considerable studies into determining the importance of macrophage-derived cytokines in β-cell destruction and immune modulation. One of these cytokines, IL-1, has received considerable attention due to its pronounced toxicity for islet β-cells in vitro [38] and the evidence that administration of IL-1 following insulitis can protect NOD mice from developing diabetes [79]. IL-1 exists as a precursor that is converted to its active form following cleavage with caspase 1 and has a broad spectrum of activities, ranging from an inducer of dendritic cell maturation to a mediator of Fas-induced death. Determining which of these functions help contribute to diabetes progression has therefore been difficult. Most data pertaining to the importance of IL-1 in diabetes have been based on in vitro studies on either insulinoma cell lines or primary β-cell cultures [80–82]. Although these studies have clearly provided a link between IL-1 and β-cell death [83, 84] the mechanism by which IL-1 promotes β-cell death is still debated.

Recently, IL-18, a member of the IL-1 family which shares structural features with IL-1 and biological properties with IL-12 [85–87] has also been linked to diabetes [88] and other autoimmunities like multiple sclerosis [89]. Although IL-18 can synergize with IL-12 to promote IFNγ production [90], recent evidence has also documented that IL-18 can also promote production of potent anti-inflammatory cytokines like IL-13 [91]. Studies in NOD mice suggest that this latter property of IL-18 may help dampen the anti-islet response. For example, Rothe et al. [88] demonstrated that treatment of NOD mice with recombinant human IL-18 following insulitis development prevented the progression to diabetes. Further, this was related to a downregulation in proinflammatory cytokines levels in the pancreas. Thus, IL-18 therapy may be a potential new avenue for treatment of diabetes.

Tumor Necrosis Factor α

The first evidence that the proinflammatory cytokine, TNFα may be involved in the progression to diabetes was based on the evidence that high levels of TNFα mRNA are detectable in the islets following the initial infiltration of immune cells [92]. Originally, the source of TNFα was thought to be the islet-infiltrating CD4+ T cells but the studies of Dahlén et al. [93] have clearly documented that the first cells to infiltrate the islets, the macrophages and dendritic cells, are the major sources of intra-islet TNFα.

TNFα is a multifunctional cytokine; it can induce differentiation, promote or inhibit apoptosis, enhance or decrease T- and B-cell responses and modulate the presentation capacity of APCs to name but a few of its actions [94]. TNFα is a member of the TNF family of molecules which share structural and functional properties. There are two major receptors for TNFα; TNF receptor I (TNFRI) and TNF receptor II (TNFRII) [95]. Both receptors are ubiquitously expressed but have different functional properties.

Signaling through TNFRI can promote apoptosis of target cells – this is facilitated by the presence of a death domain which promotes apoptosis downstream of the receptor [96 98]. TNFRII does not contain a death domain, and originally it was thought that signals through TNFRII would not lead to apoptosis. However, it has been demonstrated that CD8+ T cells are susceptible to death following signals through TNFRII and thus differential signaling of this molecule on CD8+ T cells seems to play an important role in regulating CD8+ T-cell activity [99–101].

Several studies including our own have addressed the role of TNFα in diabetes progression either by the injection of the recombinant cytokine [79, 102, 103] or the generation of transgenic mice that express TNFα in their β-cells [7, 8, 29]. Such studies have demonstrated that TNFα can have conflicting roles to play depending on the timing of injection or islet-specific expression. Neonatal injection, or islet-specific expression, of TNFα accelerates diabetes onset in NOD mice [7, 103] whereas adult injection, or islet-specific expression, protects against diabetes development [8, 79, 104]. These findings are consistent with observations that neutralization of TNFα in neonatal NOD mice protects against diabetes whereas similar neutralization in adult mice accelerates the disease process.

The mechanisms by which adult administration/islet-specific expression of TNFα can protect NOD mice from diabetes are unknown, although TNFα-mediated downregulation in T-cell responses to islet antigens has been suggested [102, 105]. In contrast, our studies in TNFα-NOD mice that express TNFα in the islets from birth, have provided insights into the mechanisms by which islet-specific expression of TNFα in neonatal NOD mice can accelerate the disease process. In our transgenic model, although we did detect apoptosis of some β-cells prior to insulitis, it seems unlikely that TNFα itself is directly involved in the destruction of the β-cells through toxicity [106] since placing the transgene onto the scid background, or onto nongenetically predisposed mice like Balb/c [31] or C57BL/6, does not lead to diabetes [7, 9]. Instead our evidence suggests that TNFα creates an islet environment that promotes recruitment and activation of dendritic cells which in turn present islet antigen to infiltrating islet-reactive T cells. These events occur rapidly within the first few weeks of life as evidenced by the more rapid development of insulitis and emergence of anti-islet T-cell responses in comparison to nontransgenic littermates [7]. The T-cell population that seems critical for β-cell destruction in our model are the CD8+ T cells. This was based on the evidence that crossing TNFα-NOD mice to NOD mice deficient in CD8+ T cells due to a target disruption of the MHC class I gene, prevented diabetes progression, whereas deficiency in MHC class II molecules did not [29]. Thus, the original hypothesis suggested that the MHC class II-restricted CD4+ T cells had no role to play in the disease process in our mice. This finding was intriguing since activation of CD8+ T cells normally requires interaction of three cell types: CD4+ T cells, CD8+ T cells and APCs. In this scenario, CD4+ T cells are required to provide the necessary signals for the efficient activation of the APC which in turn can then activate the CD8+ T cell. The necessary CD4+ T-cell-derived signals for APC activation are provided by interaction of CD154 on the T cell and CD40 on the APC [107]. Interaction of CD154 with CD40 results in upregulation of MHC and co-stimulatory molecules on the APC [108] as well as induce transcription of molecules that promote APC survival [109]. Indeed, the importance of CD154−CD40 signals are clearly documented by the evidence that deficiency in either molecule results in defective humoral and T-cell-mediated responses [110–113]. Further, cross-linkage of CD40 on APCs is sufficient to induce APC activation in the absence of CD4+ T-cell help [114–116]. Since CD4+ T cells did not seem to be required for diabetes progression, we speculated that the CD154−CD40 pathway was also obsolete in our model. This indeed turned out to be the case since crossing TNFα-NOD mice to CD154-deficient mice resulted in diabetes with kinetics and penetrance similar to wild-type mice [29]. Further, the APCs in the islets of TNFα-CD154$^{-/-}$ mice had an activated phenotype and could present islet antigen to CD8+ T cells (but not CD4+ T cells). Thus, TNFα can create an intra-islet environment that overcomes the requirement for these regulatory CD154–CD40 signals thereby promoting activation of APCs and the presentation of islet antigen to infiltrating CD8+ T cells.

TNFα has also been linked to the breakdown in peripheral tolerance to host antigens in mice that are not predisposed to developing autoimmunity [16, 117, 118]. For example, Guerder et al. [117] demonstrated that TNFα can promote diabetes in mice that co-express the co-stimulatory molecule CD80 in association with TNFα, a finding that was confirmed by others [118]. In this circumstance, like the NOD studies above, TNFα seemed to promote diabetes independent of the need for CD4+ T cells [118]. Recently, we took a new look at this model, to question the temporal importance of TNFα islet-specific expression on the development of diabetes. By adapting the tetracycline-regulated gene transcription system [13, 15, 17, 119], we developed a murine model where islet-specific expression of TNFα could be mediated by a transcriptional on/off switch depending on whether tetracycline was absent or present in the drinking water [16]. By selectively expressing TNFα in the islets from birth until set times prior to the development of insulitis, or following the development of insulitis, we established that the duration of the TNF-TNFR signals is the critical parameter in the progression or prevention in the development of diabetes in this model [16]. Like others, we established that diabetes progression was CD8+ T-cell-dependent. Of particular interest was the data demonstrating that expression of TNFα from birth until 21 days of age prevented diabetes progression, whereas repression of TNFα expression on day 25 or 30 resulted in delayed or accelerated progression to diabetes (respectively). Thus, this unique transgenic model helped define a period between 21 and 25 days following initiation of TNFα expression when the autoimmune outcome of the animals is decided, and suggested that autoregulatory mechanisms which prevent anti-islet T-cell responses are generated during this 21- to 25-day period. In attempts to determine the mechanisms that may contribute to this regulatory period, we depleted CD4+ T cells during this 21- to 25-day time point. In all instances, anti-CD4-treated mice rapidly progressed to diabetes whereas mice receiving isotype-control antibody showed the expected delay in diabetes progression following [120]. Thus, contrary to previous beliefs that CD4+ T cells had no role to play in TNFα-mediated diabetes, instead they participate in regulating the anti-islet response.

It is unclear at this time whether prolonged inflammation, or prolonged TNFα expression adversely affects the generation, function or survival of regulatory T cells. If so, this may relate to change in the types of APCs in the islets should regulatory T cells require specialized APCs. This area of research will be of importance in the design of therapeutic drugs that can prevent diabetes.

Interferon γ

One of the most prototypic TH1-type cytokines is IFNγ in terms of selective expression and functional properties. Its capacity to activate macrophages, induce MHC class I and II expression on both immune cells and nonimmune cells thereby enhancing APC function and increase homing of T cells into target tissues [121–124] makes it one of the most extensively studied cytokines in many models of autoimmunity.

IFNγ is produced from CD4+ TH1 cells, but can also be secreted by CD8+ T cells, NK cells and certain subsets of dendritic cells. Although mRNA levels of IFNγ increase in the islets of NOD mice as they progress to diabetes, more definitive association between IFNγ and diabetes is provided by the evidence that administration of anti-IFNγ neutralizing antibodies to NOD mice prevents diabetes [125, 126]. In addition, overexpression of IFNγ in the islets of NOD mice results in rapid disease progression [127], and NOD mice deficient in the IFNγ receptor a chain were protected from disease [128]. These in vivo studies showing a role for IFNγ in promoting autoimmunity were supported by in vitro evidence that IFNγ treatment of insulinoma cell lines or primary β-cell cultures resulted in upregulation of MHC class I molecules, making β-cells more susceptible to CD8+ T-cell attack [122–124].

In light of all these data, it was surprising therefore, that NOD mice carrying a null mutation for IFNγ developed diabetes similar to wild-type littermates [129]. This latter result would argue against a key role of IFNγ in diabetes development in NOD mice. At present there is no explanation as to why there are discrepancies in the autoimmune outcome of mice carrying a null mutation for IFNγ and those overexpressing the cytokine in the islets. To complicate matters further, studies in murine models for multiple sclerosis, a TH1-type autoimmunity, have demonstrated that IFNγ is required for protection against disease [130]. In this instance, IFNγ is an important inducer of apoptosis of autoaggressive T cells. Thus it will be interesting to determine whether IFNγ is truly friend or foe in the fight against diabetes.

Interleukin-12

Early studies on the development of TH1 cells demonstrated the importance of the macrophage and dendritic cell-derived cytokine IL-12 for effective differentiation of CD4+ T cells into TH1 cells [131]. For example, mice deficient in IL-12 or carry a null mutation for signal transducer and activator of transcription (stat) 4, a member of the IL-12 receptor signal pathway [132], have severely impaired TH1 responses [133, 134]. These data focused attention on whether IL-12 production by islet-infiltrating macrophages and dendritic cells contributed to diabetes progression. Initially there were several studies that provided support for this hypothesis: first, mRNA transcripts for IL-12 increase as NOD mice progress to diabetes [135]; expression of TH2 cytokine IL-4 or important physiological inhibitors of IL-12 production [136] prevented diabetes in NOD mice [137] and neutralization of IL-12 in neonatal NOD mice abrogated diabetes progression [138, 139]. However, NOD mice with a targeted disruption of the IL-12 gene developed diabetes with only slightly delayed kinetics, implying that IL-12 was not absolutely required for disease progression [140].

Although these conflicting data could simply reflect the fact that IFNγ may not be important for diabetes progression, evidence that IFNγ production can occur independent of IL-12 may offer an alternative explanation as to why abrogation of IL-12 did not prevent diabetes progression [140]. Nevertheless, therapeutic strategies that modulate TH1 responses via amelioration of IL-12 production are still considered an important tool to prevent diabetes progression.

Anti-Inflammatory Cytokines and T1D

Anti-inflammatory cytokines secreted from APCs, TH2 and TH3 cells can prevent or downregulate the production of proinflammatory cytokines like TNFα, IL-12, IFNγ and inhibit production of free radicals. mRNA transcripts for anti-inflammatory cytokines like IL-4, IL-10 and TGFβ predominate in the early islet infiltrates of NOD mice [2] and decrease as mice progress to diabetes. For these reasons, there have been several studies to determine whether anti-inflammatory cytokines can protect against diabetes. Below we discuss some of the most relevant anti-inflammatory cytokines linked to diabetes protection.

Interleukin-4

IL-4, although a classical TH2 cytokine, has now been shown to be secreted from B cells and NK T cells [141]. IL-4 promotes differentiation of naïve uncommitted CD4+ T cells along the TH2 pathway and helps suppress the production of TH1 cytokines potentially through inhibition in transcription of the p40 subunit of the p40/p35 (p70) IL-12 complex [142]. Originally, defects in IL-4 production were speculated to contribute to diabetes progression when it was demonstrated that T cells from female NOD mice secrete less IL-4 than non-NOD strains of mice [143–145]. Stronger evidence on the importance of IL-4 was obtained by studies demonstrating that injection of IL-4 into neonatal NOD mice protected against diabetes development [145]. In addition, overexpression of IL-4 in the islets of NOD mice protects against diabetes progression [137]. Finally, protection from diabetes development following oral feed of insulin [146] is abrogated in IL-4 deficient or Stat 6 (a member of the IL-4 signal pathway) deficient mice [147]. Thus it was surprising that NOD mice carrying a null mutation for IL-4 had the same degree of insulitis and no evidence of enhanced disease as wild-type littermates [148]. Although this latter data does not exclude a role for IL-4 in diabetes, it is possible that other cytokines compensate for ablation of IL-4.

Recently, IL-13, a cytokine that shares many similar properties to IL-4 [149], has been shown to contribute to protection against diabetes in NOD mice [150]. This was based on the observation that prolonged treatment of NOD mice (between 5 and 16 weeks) with human IL-13 protected against disease, and short-term treatment (between 5 and 6 weeks) delayed disease progression. Furthermore, protection against diabetes was long-lived, with no evidence of diabetes 50 weeks post the last IL-13 treatment. In addition, treatment of NOD-scid mice with IL-13 prevent diabetic splenocytes from NOD mice transferring disease. In part, this potency for IL-13 to protect against diabetes is thought to relate to its strong anti-inflammatory properties. IL-13 is secreted from TH0, CD8+ T cells, CD45RA+ T cells and TH1 cells (at low levels) [149]. Similar to IL-4, IL-13 can downregulate the production of proinflammatory cytokines like TNFα, IL-1 and IFNγ [151]. Indeed, treatment of NOD mice with hIL-13 results in a decrease in the serum levels of these inflammatory cytokines and an increase in IL-4 [150]. Although IL-4 and IL-13 share the IL-4 receptor α chain and have similar functional properties, the target cell for IL-13 is believed to be APCs, not T cells since T cells do not bear IL-13 receptors [149]. For example, IL-13 can prevent IL-12 transcription [152] and enhance IL-6 transcription in macrophages [153] thereby promoting TH2 responses [154]. Similarly, IL-1β and NO production by macrophages and TNFα-mediated apoptosis are decreased following culture with IL-13 [151]. Whether these anti-inflammatory properties relate to the ability of IL-13 to suppress NFκB transcription is unknown [155]. Nevertheless, the potency of hIL-13, which has a lower affinity for the murine IL-13 receptor, to protect against diabetes in NOD mice has led to the hope treatment of humans with IL-13 may offer a new therapeutic strategy to combat diabetes.

Interleukin-10

One of the most potent anti-inflammatory cytokines is IL-10. IL-10 has been documented to decrease MHC class I and II levels of APCs, and as such decrease the priming capacity of APCs [156]. Studies in NOD mice have demonstrated that mRNA transcripts for IL-10 correlated with protection from diabetes whereas decrease in IL-10 mRNA levels correlates with disease progression. One of the most commonly documented models that demonstrates the importance IL-10 in downregulating anti-host immune responses is inflammatory bowel disease (IBD) [157]. IBD can be induced following transfer of CD4+CD45RB^{hi} T cells into scid-Balb/c mice [158, 159]. Co-transfer of CD4+CD45RB^{lo} regulatory T cells with CD4+CD45RB^{hi} T cells can prevent disease, and such protection is partly dependent on IL-10 since neither IL-10-deficient mice nor mice treated with anti-IL-10 can regulate anti-intestinal responses in this co-transfer model [157].

The potent anti-inflammatory properties of IL-10 led to the speculation that intra-islet IL-10 production may protect against diabetes. Preliminary studies in NOD mice provided support for this hypothesis. For example, injection of IL-10 into NOD mice protected from disease development [160], as did promotion of signals through the IL-10 receptor [161]. However, in transgenic NOD mice that overexpress IL-10, the islets rapidly progress to diabetes in comparison to nontransgenic littermates [6]. These data would argue that IL-10 is an important mediator for promoting diabetes in NOD mice. Pauza et al. [162] in an attempt to address this discrepancy in results generated a transgenic mouse where CD4+ T cells were induced to express IL-10 irrespective of their T-helper phenotype. In these mice, there was no evidence that IL-10 production by CD4+ T cells that infiltrated the islets resulted in acceleration to diabetes. Indeed, the incidence of diabetes in transgenic versus nontransgenic mice was similar. These findings argue against a detrimental role for IL-10 in T1D, a finding that has been further substantiated by the evidence that IL-10 deficiency does not influence diabetes progression in NOD mice [163].

Transforming Growth Factor β

One of the most prominent 'regulatory' cytokines is TGFβ. A widely distributed cytokine, TGFβ acts on almost any cell type in the body and has pleiotropic functions. In mammals, there are three isoforms of TGFβ, TGFβ$_1$, TGFβ$_2$ and TGFβ$_3$ [164]. Although there is strong homology between these three isoforms and they use the same receptor complex for signaling, TGFβ$_1$ is the most extensively studied due to its association with the lymphoid system [164]. For example, deficiency in TGF-β$_1$ results in embryonic lethality for 50% of the progeny and by 3 weeks of age, the 50% that survive experience multiple immune abnormalities [165, 166]. TGFβ in vitro has been well documented as a cytokine that can regulate cell proliferation, differentiation and activation [167, 168]. For example, the capacity of TGFβ to inhibit T- and B-cell proliferation is believed to be linked to its ability to downregulate of MHC molecules on macrophages and B cells [169] as well as decrease immunoglobulin receptor levels on B cells [170], all factors that ultimately inhibit the capacity of APCs to activate T cells.

In addition, recent evidence has demonstrated that in the absence of TGFβ, Treg cells are incapable of preventing IBD [171, 172] suggesting this cytokine is an important component of the immune regulatory network. At present it is unknown why TGFβ signals are required in this model although upregulation of cytotoxic T lymphocyte associated antigen 4 (CTLA-4) on activated T cells is a possible explanation [173].

The role for TGFβ in dampening autoimmune responses suggested that the cytokine may help protect against diabetes development. To test this, we and others generated transgenic NOD mice that overexpress human TGFβ₁ in their islets [30]. In both laboratories, TGFβ-transgenic NOD mice showed decreased incidence and kinetics for disease progression, compared to nontransgenic littermates. However, there has been some controversy as to the mechanism by which TGFβ inhibits diabetes. King et al. [30] demonstrated that in TGFβ-transgenic NOD mice, macrophage-driven presentation of islet antigens resulted in a potent protective TH2 response. In contrast, in nontransgenic mice B cells presented islet antigen to T cells resulting in TH1 responses to islets.

In our TGFβ-transgenic NOD mice, such deviation of the immune responses to a TH2 profile was not evident. Indeed both TH1 and TH2 cytokines were clearly detectable [Grewal and Flavell, unpubl. observations]. Instead an increase in apoptosis of islet-infiltrating cells was seen suggesting TGFβ overexpression in the islets may promote death of autoaggressive islet-specific T cells. In addition, we found that neonatal expression of TGFβ resulted in abnormal growth of the pancreas, with evidence of extensive fibrosis [Grewal and Flavell, unpubl. observations], whereas King et al. [30] did not document such abnormalities.

One of the major difficulties in defining the mechanisms by which TGFβ protects against diabetes is that a vast number of different cells either infiltrating the islets or residing in the islets have TGFβ receptors and as such, are potential targets for TGFβ-mediated modulation. How, therefore, can we dissect which cell is the target? One novel approach is the creation of transgenic mice that overexpress a dominant negative form of the TGFβRII on the surface of T cells [10]. Although these mice are viable, they develop immune abnormalities similar to that seen for TGFβ-deficient mice, suggesting TGFβ acts on T cells to control autoimmune responses [10]. These mice will be invaluable at determining whether TGFβ acts directly on T cells (as suggested by Grewal) or indirectly via APCs (as suggested by King et al.) to protect against diabetes in NOD mice.

Cytokine Involvement in Human Diabetes

The evidence that type 1 diabetes is a T-cell-mediated disease in humans is not as easy to determine as it is for murine models. Nevertheless, there are several reports that indicate that T cells are critically involved in the destruction of human β-cells [174]. Probably one of the strongest suggestions that diabetes in humans requires T cells is the evidence that individuals that bear HLA-DQ8 (which is homologous to NOD I-A^{g7}) molecules are more susceptible to developing diabetes that individuals that carry other MHC haplotypes [76, 175–177]. Indeed, 'humanized' C57BL/6 mice which carry HLA-DQ8 on their APCs, develop diabetes [178]. Further, similar to NOD mice, cyclosporin treatment of humans delays progression to diabetes in genetically susceptible individuals [179]. Finally, transfer of T cells from newly diagnosed diabetic patients into immunoincompetent NOD mice can induce diabetes [180, 181].

However, the question of whether the TH1 v TH2 paradigm exists in humans and contributes to disease progression is uncertain. Reports demonstrating diabetic patients have sequence polymorphisms in the genes encoding certain cytokines, like TNFα, have implicated a link between abnormal cytokine production and progression to diabetes [95]. However, stronger evidence on the importance of TH1 and TH2 cytokines in the pathology of human diabetes has been based either on the histological analysis of pancreatic sections from newly diagnosed patients or measurement of cytokine levels in the blood of diabetic and nondiabetic siblings.

In terms of the histological studies, a clear correlation TH1-type cytokine expression in the islets is associated with the development of diabetes [182–186], a finding that echoes the data from murine models. However, documentation of whether elevation of particular cytokines in the serum can serve as a marker for diabetes progression have been more controversial. For example, some reports have suggested that progression to diabetes relates to elevated serum TNFα levels, whereas others report that progression to disease is linked to a fall in serum TNFα levels [187, 188].

One important link between cytokines and diabetes in humans is not related to the modulation of the T-cell responses to islet antigens, instead relates to cytokine-mediated neurological and vascular complications. For example, microvascular disease which can lead to glomerulonephritis and diabetic retinopathy is linked to TGFβ [38, 189]. TGFβ induces synthesis of matrix proteins like collagen type IV and fibronectin and inhibits matrix degradation [190, 191]. Although these are important properties in wound healing, they can induce scarring. Studies in diabetic rats have shown breakdown in homeostasis of glucose results in increased TGFβ levels in the kidney [192, 193]. The result of this is scarring of the kidney and ultimately renal failure. Further, administration of TGFβ inhibitors like decorin can prevent glomerulonephritis and kidney scarring in animal models [194]. It is speculated that a similar mechanism for renal complications occurs in humans, since diabetics patients also have elevated glomerular TGFβ content [195].

Cytokines and Therapeutic Intervention of Diabetes

Current treatment for human diabetes is daily injections of insulin. This of course is not a cure and can have a negative impact on the quality of life for many people. In addition, for many diabetics, over a period of time, it becomes more difficult to maintain glucose homeostasis and as a consequence patients develop life-threatening vascular and neurological problems. Although the data obtained from studies in the NOD mouse and BB rat have suggested potential avenues by which cytokine modulation may ameliorate diabetes in humans, there are several obstacles which will have to be overcome: who should receive the therapy; when to give the therapy, and are there toxic consequences of cytokine-based therapies are just a few problems that will have to be addressed.

Ultimately one would like to target cytokine-based therapies to prediabetic individuals, however there is no current regime for establishing who among the population that carry MHC susceptiblity alleles, will eventually develop diabetes. Even among genetically identical twins, diabetes occurs in only 30% of siblings carrying MHC susceptibility haplotypes [76]. With the completion of sequencing the human genome, we will in time have new genetic markers that will determine most accurately individuals who will develop diabetes. At the present time, cytokine-based therapies are aimed at either newly diagnosed diabetics or diabetic patients receiving islet transplants. In the former instance, the natural history of type 1 diabetes has identified a period in humans called the 'honeymoon' phase in which the need for insulin in minimal and β-cell function improves. Ongoing clinical trials are focusing on this period with the hope that cytokine-based therapies can prolong the honeymoon phase and ultimately stop progression to complete dependence on insulin. For example, phase II clinical trials are currently underway to determine the efficacy of ingesting human recombinant IFNα (which has been successful at rejection of syngenic islet grafts in mice [196]) at prolonging the honeymoon period. The immune status of an individual will also limit the use of cytokine-based therapies. For example, results in NOD mice have led to the speculation that IL-13 treatment of humans is a potential new therapy [150]. Indeed, pathological examination of NOD mice treated with human IL-13 showed no evidence of liver or kidney damage suggesting the cytokine was not toxic at the concentrations given. However, serum IgE levels increased remarkably following IL-13 treatment and thus, this could increase the risk of patients developing allergic reactions particularly in individuals with atopic diseases.

Perhaps the most concerning data relating to the translation of cytokine-based therapies in murine models to humans in attempts to modulate autoimmunity is that seen for abrogation of TNFα signaling by injection of TNFR-Fc fusion proteins to prevent septic shock. Animal models of septic shock had clearly documented the therapeutic value of blocking the TNFα-TNFR pathway in preventing disease [197, 198]. Since injection of TNFR-Fc fusion proteins prevented arthritis in mice [199, 200] and such treatment was successfully translated to humans [201], it was speculated that TNFR-Fc treatment could prevent septic shock in man. Unfortunately, in clinical trials it was evident that this therapy either had no effect on the disease process or more unsettlingly enhanced the disease cumulating in the death of the patient [202]. Thus, translation of murine-based therapies to humans must be performed with great caution.

Nevertheless, this does not mean that cytokine-based therapies have no role to play in human diabetes. Indeed, the ability to modulate the immune response by blocking cytokine signaling [203] or genetic modification of islets to protect them from cytokine-mediated toxicity [204] are areas of active research in the hope that such treatments will prolong survival of transplanted islets into diabetic patients.

Conclusion

The study of cytokine function in mice has led to significant breakthroughs in our understanding of disease processes as well as many puzzling inconsistencies. The results of knockout mice are frequently not predicative of treatment with neutralizing antibodies to the same cytokine and overexpression of cytokines in the islets occasionally leads to conflicting results. It is likely that these aberrant results derive from the limitation of the systems. Knockout mice can develop compensatory mechanisms and overexpression of cytokines in islets can exaggerate the role of a given cytokine at that location and miss the function of a cytokine elsewhere. Further, use of antibodies can lead to non-physiologic disruption of (immune) functions at locations not directly involved in the disease. Nonetheless, substantial inroads have been made into the understanding of type 1 diabetes through the combined use of these approaches and further progress is expected.

References

1   Tisch R, McDevitt HO: Insulin-dependent diabetes mellitus. Cell 1996;85:291–297.
2   Delovitch TL, Singh B: The nonobese diabetic mouse as a model of autoimmune diabetes: immune dysregulation gets the NOD. Immunity 1997;7:727–738.

3   Jansen H, et al: Immunohistochemical characterization of monocytes-macrophages and dendritic cells involved in the initiation of insulitis and β-cell destruction in NOD mice. Diabetes 1994;43:667–675.

4   Leenen PJM, et al: Markers of mouse macrophage development detected by monoclonal antibodies. J Immunol Methods 1994;174:5–19.

5   Allison J, et al: Overexpression of beta-2-microglobulin in transgenic mouse islet beta cells results in defective insulin secretion. Proc Natl Acad Sci USA 1991;88:2070–2074.

6   Balasa B, et al: IL-10 impacts autoimmune diabetes via a CD8+ T cell pathway circumventing the requirement for CD4+ T and B lymphocytes. J Immunol 1998;161:4420–4427.

7   Green EA, Eynon EE, Flavell RA: Local expression of TNF alpha in neonatal NOD mice promotes diabetes by enhancing presentation of islet antigens. Immunity 1998;9:733–743.

8   Grewal IS, et al: Local expression of transgene encoded TNFα in islets prevents autoimmune diabetes in nonobese diabetic (NOD) mice by preventing the development of auto-reactive islet-specific T cells. J Exp Med 1996;184:1963–1974.

9   Picarella DE, et al: Transgenic tumor necrosis factor (TNF)-α production in pancreatic islets leads to insulitis, not diabetes. J Immunol 1993;150:4136–4150.

10 Gorelik L, Flavell R: Abrogation of TGFβ signaling in T cells leads to spontaneous T cell differentiation and autoimmune disease. Immunity 2000;12:171–181.

11 Mora C, et al: Pancreatic infiltration but not diabetes occurs in the relative absence of MHC class II-restricted CD4 T cells: Studies using NOD/CIITA-deficient mice. J Immunol 1999;162:4576–4588.

12 Serreze DV, et al: Major histocompatibility complex class I-deficient NOD-β2m null mice are diabetes and insulitis resistant. Diabetes 1994;43:505–509.

13 Furth P, et al: Temporal control of gene expression in transgenic mice by a tetracycline-responsive promoter. Proc Natl Acad Sci USA 1994; 91:9302–9306.

14 Gossen M, Bujard H: Tight control of gene expression in mammalian cells by tetracycline-responsive promoters. Proc Natl Acad Sci USA 1992; 89:5547–5551.

15 Gossen M, et al: Transcriptional activation by tetracyclines in mammalian cells. Science 1995; 268:1766–1769.

16 Green EA, Flavell RA: The temporal importance of TNFα expression in the development of diabetes. Immunity 2000;12:459–469.

17 Shockett P, et al: A modified tetracycline-regulated system provides autoregulatory, inducible gene expression in cultured cells and transgenic mice. Proc Natl Acad Sci USA 1995;92:6522–6526.

18 Wong FS, et al: CD8 T cell clones from young nonobese diabetic (NOD) islets can transfer rapid onset of diabetes in NOD mice in the absence of CD4 cells. J Exp Med 1996;183:67–76.

19 Christianson SW, Shultz LD, Leiter EH: Adoptive transfer of diabetes into immunodeficient NOD-scid/scid mice. Relative contributions of CD4+ and CD8+ T cells from diabetic versus prediabetic NOD.NON-Thy-1α donors. Diabetes 1993;42:44–55.

20 Miller B, et al: Both Lyt-2 and L3T4+ T cell subsets are required for the transfer of diabetes in nonobese diabetic mice. J Immunol 1988;140:52–58.

21 Wang Y, et al: Effect of cyclosporine on immunologically mediated diabetes in nonobese diabetic mice. Transplantation 1988;46(suppl):S101–S106.

22 Katz J, Benoist C, Mathis D: Major histocompatibility complex class I molecules are required for the development of insulitis in non-obese diabetic mice. Eur J Immunol 1993;23:3358–3360.

23 Wang B, et al: The role of CD8+ T cells in the initiation of insulin-dependent diabetes mellitus. Eur J Immunol 1996;26:1762–1769.

24 Janeway CA Jr, et al: Immunobiology: The Immune System in Health and Disease. London, Current Biology/Garland, 1999.

25 Shakaguchi S: Regulatory T cells: Key controllers of immunologic self-tolerance. Cell 2000;101: 455–458.

26 Abbas AK, Murphy KM, Sher A: Functional diversity of helper T lymphocytes. Nature 1996;383: 787–793.

27 Banchereau J, Steinman RM: Dendritic cells and the control of immunity. Nature 1998;392:245–252.

28 Banchereau J, et al: Immunobiology of dendritic cells. Annu Rev Immunol 2000;18:767–811.

29 Green E, et al: Neonatal tumor necrosis factor α promotes diabetes in nonobese diabetic mice by CD154-independent antigen presentation to CD8+ T cells. J Exp Med 2000;191:225–237.

30 King C, et al: TGF-β_1 alters APC preference, polarizing islet antigen responses toward a Th2 phenotype. Immunity 1998;8:601–613.

31 McSorley SJ, et al: Immunological tolerance to a pancreatic antigen as a result of local expression of TNFα by islet β cells. Immunity 1997;7:401–409.

32 Akashi T, et al: Direct evidence for the contribution of B cells to the progression of insulitis and development of diabetes in non-obese diabetic mice. Int Immunol 1997;9:1159–1164.

33 Noorchashm H, et al: B-cells are required for the initiation of insulitis and sialitis in nonobese diabetic mice. Diabetes 1997;46:941–946.

34 Serreze DV, et al: B lymphocytes are essential for the initiation of T cell-mediated autoimmune diabetes: Analysis of a new 'speed congenic' stock of NOD.Igu null mice. J Exp Med 1996;184: 2049–2053.

35 Serreze DV, et al: Subcongenic analysis of the idd13 locus in NOD/Lt mice: Evidence for several susceptibility genes including a possible diabetogenic role for β_2-microglobulin. J Immunol 1998; 160:1472–1478.

36 De Waal-Malefyt R, et al: Interleukin-10 (IL-10) inhibits cytokine synthesis by human monocytes: An autoregulatory role of IL-10 produced by monocytes. J Exp Med 1991;174:1209–1220.

37 Howard M, O'Garra A: Biological properties of interleukin-10. Immunol Today 1992;13:198–200.

38 Pankewycz O, Guan JX, Benedict JF: Cytokines as mediators of autoimmune diabetes and diabetic complications. Endocr Rev 1995;16:164–176.

39 Alleva DG, Kaser SB, Beller DI: Aberrant cytokine expression and autocrine regulation characterize macrophages from young MRL+/+ and NZB/W F1 lupus-prone mice. J Immunol 1997; 159:5610–5619.

40 Morel PA, Vasquez A, Feili-Hairi M: Immunobiology of DC in NOD mice. J Leuk Biol 1999;66: 276–280.

41 Feili-Hariri M, et al: Immunotherapy of NOD mice with bone marrow-derived dendritic cells. Diabetes 1999;48:2300–2308.

42 Winzler C, et al: Maturation stages of mouse dendritic cells in growth factor-dependent long-term cultures. J Exp Med 1997;185:317–328.

43 Inaba K, et al: Generation of large numbers of dendritic cells from mouse bone marrow cultures supplemented with granulocyte/macrophage colony-stimulating factor. J Exp Med 1992;176: 1693–1702.

44 Scheicher C, et al: Dendritic cells from mouse bone marrow: In vitro differentiation using low doses of recombinant granulocyte-macrophage colony-stimulating factor. J Immunol Methods 1992;154:253–264.

45 Ardavin C, et al: Thymic dendritic cells and T cells develop simultaneously in the thymus from a common precursor population. Nature 1993;362:761–763.

46 Wu L, et al: Mouse thymus dendritic cells: Kinetics of development and changes in surface markers during maturation. Eur J Immunol 1995;25:418–425.

47 Vremec D, Shortman K: Dendritic cell subtypes in mouse lymphoid organs. J Immunol 1997;159: 565–573.

48 Maraskovsky E, et al: Dramatic increase in the numbers of functionally mature dendritic cells in Flt3 ligand-treated mice: Multiple dendritic cell subpopulations identified. J Exp Med 1996; 184:1953–1962.

49 Pulendran B, et al: Developmental pathways of dendritic cells in vivo: Distinct function, phenotype, and localization of dendritic cell subsets in FLT3 ligand-treated mice. J Immunol 1997;159:2222–2231.

50 Vremec D, et al: CD4 and CD8 expression by dendritic cell subtypes in mouse thymus and spleen. J Immunol 2000;164:2978–2986.

51 Reis e Sousa C, et al: In vivo microbial stimulation induces rapid CD40 ligand-independent production of interleukin-12 by dendritic cells and their redistribution to T cells areas. J Exp Med 1997;186:1819–1829.

52 Reis e Sousa C, et al: Paralysis of dendritic cell IL-12 production by microbial products prevents infection-induced immunopathology. Immunity 1999;11:637–647.

53 Maldonado-Lopez R, et al: CD8α+ and CD8α− subclasses of dendritic cells direct the development of distinct T helper cells in vivo. J Exp Med 1999;189:587–592.

54 Ohteki T, et al: Interleukin-12-dependent interferon-γ production by CD8α+ lymphoid dendritic cells. J Exp Med 1999;189:1981–1986.

55 Smith AL, Groth FdS: Antigen-pulsed CD8α+ dendritic cells generate an immune response after subcutaneous injection without homing to the draining lymph node. J Exp Med 1999;189:593–598.

56 Pulendran B, et al: Distinct dendritic cell subsets differentially regulate the class of immune response in vivo. Proc Natl Acad Sci USA 1999;96:1036–1041.

57 Moser M, Murphy K: Dendritic cell regulation of TH1–TH2 development. Nat Immunol 2000;1:199–205.

58 Rissoan MC, et al: Reciprocal control of T helper cell and dendritic cell differentiation. Science 1999;283:1183–1186.

59 Takahashi K, Honeyman MC, Harrison LC: Impaired yield, phenotype, and function of monocyte-derived dendritic cells in humans at risk for insulin-dependent diabetes. J Immunol 1998;161:2629–2635.

60 Inaba K, et al: High levels of a major histocompatibility complex II-self-peptide complex on dendritic cells from the T cell areas of lymph nodes. J Exp Med 1997;186:665–672.

61 Cella M, et al: Inflammatory stimuli induce accumulation of MHC class II complexes on dendritic cells. Nature 1997;388:782–787.

62 Arai KI, et al: Cytokines: Coordinators of immune and inflammatory responses. Annu Rev Biochem 1990;59:783–836.

63 Essery G, Feldmann M, Lamb JR: Interleukin-2 can prevent and reverse antigen-induced unresponsiveness in cloned human T lymphocytes. Immunology 1988;64:413–417.

64 Schwartz RH: A cell culture model for T lymphocyte clonal anergy. Science 1990;248:1349–1356.

65 Dallman MJ, et al: Peripheral tolerance to alloantigen results from altered regulation of the interleukin-2 pathway. J Exp Med 1991;173:79–87.

66 Kelley VE, et al: Anti-interleukin-2 receptor antibody suppresses murine diabetic insulitis and lupus nephritis. J Immunol 1988;140:59–61.

67 Pacheco-Silva A, et al: Interleukin-2 receptor targeted fusion toxin (DAB486-IL-2) treatment blocks diabetogenic autoimmunity in non-obese diabetic mice. Eur J Immunol 1992;22:697–702.

68 Zielasek J, et al: Interleukin-2-dependent control of disease development in spontaneously diabetic BB rats. Immunology 1990;69:209–214.

69 Allison J, et al: Inflammation but not autoimmunity occurs in transgenic mice expressing constitutive levels of interleukin-2 in islet beta cells. Eur J Immunol 1992;22:1115–1121.

70 Heath WR, et al: Autoimmune diabetes as a consequence of locally produced interleukin-2. Nature 1992;359:547–549.

71 Horak I, et al: Interleukin-2 deficient mice: A new model to study autoimmunity and self-tolerance. Immunol Rev 1995;148:35–44.

72 Papiernik M, et al: T cell deletion induced by chronic infection with mouse mammary tumor virus spares a CD25-positive, IL-10-producing T cell population with infectious capacity. J Immunol 1997;158:4642–4653.

73 Ghosh S, et al: Polygenic control of autoimmune diabetes in nonobese diabetic mice. Nat Genet 1993;4:404–409.

74 Todd JA, et al: Genetic analysis of autoimmune type 1 diabetes mellitus in mice. Nature 1991;351:542–547.

75 Wicker LS, et al: Resistance alleles at two non-major histocompatibility complex-linked insulin-dependent loci on chromosome 3, Idd3 and Idd10, protect nonobese diabetic mice from diabetes. J Exp Med 1994;180:1705–1713.

76 Todd J: From genome to aetiology in a multifactorial disease, type 1 diabetes. Bioessays 1999;21:164–174.

77 Robb RJ, Smith KA: Heterogeneity of human T-cell growth factor(s) due to variable glycosylation. Mol Immunol 1981;18:1087–1094.

78 Matesanz F, Alcina A: Glutamine and tetrapeptide repeat variations affect the biological activity of different mouse interleukin-2 alleles. Eur J Immunol 1996;26:1675–1682.

79 Jacob CO, et al: Prevention of diabetes in nonobese diabetic mice by tumor necrosis factor (TNF): Similarities between TNF-alpha and interleukin-1. Proc Natl Acad Sci USA 1990;87:968–972.

80 Mandrup-Poulsen T, et al: Affinity-purified human interleukin-1 is cytotoxic to isolated islets of Langerhans. Diabetologia 1986;29:63–67.

81 Mandrup-Poulsen T, et al: Involvement of interleukin-1 and interleukin-1 antagonist in pancreatic beta-cell destruction in insulin-dependent diabetes mellitus. Cytokine 1993;5:185–191.

82 Sandler S, et al: Biochemical and molecular actions of interleukin-1 on pancreatic beta-cells. Autoimmunity 1991;10:241–253.

83 Spinas GA, et al: The bimodal effect of interleukin-1 on rat pancreatic beta-cells – stimulation followed by inhibition – depends upon dose, duration of exposure, and ambient glucose concentration. Acta Endocrinol (Copenh) 1988;119:307–311.

84 Hansen BS, et al: Effect of interleukin-1 on the biosynthesis of proinsulin and insulin in isolated rat pancreatic islets. Biomed Biochim Acta 1988;47:305–309.

85 Kohno K, Kurimoto M: Interleukin-18, a cytokine which resembles IL-1 structurally and IL-12 functionally but exerts its effect independently of both. Clin Immunol Immunopathol 1998; 86:11–15.

86 McInnes IB, et al: Interleukin-18: A pleiotropic participant in chronic inflammation. Immunol Today 2000;21:312–315.

87 Bazan JF, Timans JC, Kastelein RA: A newly defined interleukin-1? Nature 1996;379:591.

88 Rothe H, et al: IL-18 inhibits diabetes development in nonobese diabetic mice by counterregulation of Th1-dependent destructive insulitis. J Immunol 1999;163:1230–1236.

89 Shi FD, et al: IL-18 directs autoreactive T cells and promotes autodestruction in the central nervous system via induction of IFN-γ by NK cells. J Immunol 2000;165:3099–3104.

90 Robinson D, et al: IGIF does not drive Th1 development but synergizes with IL-12 for interferon-gamma production and activates IRAK and NF kappa B. Immunity 1997;7:571–581.

91 Hoshino T, Wiltrout RH, Young HA: IL-18 is a potent coinducer of IL-13 in NK and T cells: A new potential role for IL-18 in modulating the immune response. J Immunol 1999;162:5070–5077.

92 Held W, et al: Genes encoding tumor necrosis factor-α and granzyme A are expressed during development of autoimmune diabetes. Proc Natl Acad Sci USA 1990;87:2239–2243.

93 Dahlén E, et al: Dendritic cells and macrophages are the first and major producers of TNFα in pancreatic islets in the nonobese diabetic mouse. J Immunol 1998;160:3585–3593.

94 Fiers W: Tumor necrosis factor. FEBS 1991;285:199–212.

95 Kollias G, et al: On the role of tumor necrosis factor and receptors in models of multiorgan failure, rheumatoid arthritis, multiple sclerosis and inflammatory bowel disease. Immunol Rev 1999; 169:175–194.

96 Engelmann H, Novick D, Wallach D: Two tumor necrosis factor-binding proteins purified from human urine. Evidence for immunological cross-reactivity with cell surface tumor necrosis factor receptors. J Biol Chem 1990;265:1531–1536.

97 Tartaglia LA, et al: The two different receptors for tumor necrosis factor mediate distinct cellular responses. Proc Natl Acad Sci USA 1991;88:9292–9296.

98 Tartaglia LA, et al: A novel domain within the 55 kd TNF receptor signals cell death. Cell 1993; 74:845–853.

99 Zheng L, et al: Induction of apoptosis in mature T cells by tumor necrosis factor. Nature 1995; 377.348–351.

100 Herbein G, et al: Apoptosis of CD8+ T cells is mediated by macrophages through interaction of HIV gp120 with chemokine receptor CXCR4. Nature 1998;395:189–194.

101 Alexander-Miller MA, et al: Supraoptimal peptide-major histocompatibility complex causes a decrease in bcl-2 levels and allows tumor necrosis factor alpha receptor II-mediated apoptosis of cytotoxic T lymphocytes. J Exp Med 1998;188:1391–1399.

102 Cope A, Ettinger R, McDevitt H: The role of TNF alpha and related cytokines in the development and function of the autoreactive T-cell repertoire. Res Immunol 1997;148:307–312.

103 Yang XD, et al: Effect of tumor necrosis factor α on insulin-dependent diabetes mellitus in NOD mice. I. The early development of autoimmunity and the diabetogenic process. J Exp Med 1994;180:995–1004.

104 Yang XD, McDevitt HO: Role of TNF-α in the development of autoimmunity and the pathogenesis of insulin-dependent diabetes mellitus in NOD mice. Circ Shock 1994;43:198–201.

105 Cope AP, et al: Chronic tumor necrosis factor alters T cell responses by attenuating T cell receptor signaling. J Exp Med 1997;185:1573–1584.

106 Mandrup-Poulsen T, et al: Human tumor necrosis factor potentiates human interleukin-1-mediated rat pancreatic beta-cell cytotoxicity. J Immunol 1987;139:4077–4082.

107 Grewal IS, Flavell RA: CD40 and CD154 in cell-mediated immunity. Annu Rev Immunol 1998;16: 111–135.

108 Caux C, et al: Activation of human dendritic cells through CD40 cross-linking. J Exp Med 1994; 180:1263–1272.

109 Wang Z, et al: Induction of Bcl-x by CD40 engagement rescues slg-induced apoptosis in murine B cells. J Immunol 1995;155:3722–3725.

110 Grewal IS, et al: Requirement for CD40 ligand in co-stimulation induction, and experimental allergic encephalomyelitis. Science 1996;273:1864–1867.

111 Grewal IS, Xu J, Flavell RA: Impairment of antigen-specific T-cell priming in mice lacking CD40 ligand. Nature 1995;378:617–620.

112 Foy TM, et al: In vivo CD40–gp39 interactions are essential for thymus-dependent humoral immunity. II. Prolonged suppression of the humoral immune response by an antibody to the ligand for CD40, gp39. J Exp Med 1993;178:1567–1575.

113 Mach F, et al: Reduction of atherosclerosis in mice by inhibition of CD40 signaling. Nature 1998; 394:200–203.

114 Bennet SRM, et al: Help for cytotoxic-T-cell responses is mediated by CD40 signaling. Nature 1998;393:478–480.

115 Ridge F, Di Rosa F, Matzinger P: A conditioned dendritic cell can be a temporal bridge between a CD4+ T-helper cell and a T-killer cell. Nature 1998;393:474–478.

116 Schoenberger SP, et al: T-cell help for cytotoxic T lymphocytes is mediated by CD40-CD40L interactions. Nature 1998;393:480–483.

117 Guerder S, et al: Costimulator B7-1 confers antigen-presenting-cell function to parenchymal tissue and in conjunction with tumor necrosis factor α leads to autoimmunity in transgenic mice. Proc Natl Acad Sci USA 1994;91:5138–5142.

118 Herrera PL, et al: A CD8+ T-lymphocyte-mediated and CD4+ T-lymphocyte-independent autoimmune diabetes of early onset in transgenic mice. Diabetologia 1994;37: 1277–1279.

119 Shockett PE, Schatz DG: Diverse strategies for tetracycline-regulated inducible gene expression. Proc Natl Acad Sci USA 1996;93:5173–5176.

120 Green EA, Choi Y, Flavell RA: Pancreatic lymph node-derived CD4+CD25+ Treg cells, highly potent regulators of diabetes that require both CD154 and TRANCE signals for their generation. Submitted.

121 Pober JS: Warner-Lambert/Parke-Davis Award Lecture. Cytokine-mediated activation of vascular endothelium. Physiology and pathology. Am J Pathol 1988;133:426–433.

122 Campbell IL, et al: Interferon-gamma enhances the expression of the major histocompatibility class I antigens on mouse pancreatic beta cells. Diabetes 1985;34:1205–1209.

123 Campbell IL, et al: Interferon-gamma induces the expression of HLA-A, B, C but not HLA-DR on human pancreatic beta-cells. J Clin Endocrinol Metab 1986;62:1101–1109.

124 Hayakawa M, et al: Morphological analysis of selective destruction of pancreatic beta-cells by cytotoxic T lymphocytes in NOD mice. Diabetes 1991;40:1210–1217.

125 Kay TW, et al: Overexpression of class I major histocompatibility complex accompanies insulitis in the non-obese diabetic mouse and is prevented by anti-interferon-gamma antibody. Diabetologia 1991;34:779–785.

126 Debray-Sachs M, et al: Prevention of diabetes in NOD mice treated with antibody to murine IFN-gamma. J Autoimmun 1991;4:237–248.

127 Sarvetnick N, et al: Loss of pancreatic islet tolerance induced by beta-cell expression of interferon-gamma. Nature 1990;346:844–847.

128 Wang B, et al: Interferon-γ impacts at multiple points during the progression of autoimmune diabetes. Proc Natl Acad Sci USA 1997;94:13844–13849.

129 Vermeire K, et al: Accelerated collagen-induced arthritis in IFN-gamma receptor-deficient mice. J Immunol 1997;158:5507–5513.

130   Chu CQ, Wittmer S, Dalton DK: Failure to suppress the expansion of the activated CD4 T cell population in interferon-gamma-deficient mice leads to exacerbation of experimental autoimmune encephalomyelitis. J Exp Med 2000;192:123–128.

131   Trinchieri G, Gerosa F: Immunoregulation by interleukin-12. J Leukoc Biol 1996;59:505–511.

132   Jacobson NG, et al: Interleukin-12 signaling in T helper type 1 (Th1) cells involves tyrosine phosphorylation of signal transducer and activator of transcription (Stat)3 and Stat4. J Exp Med 1995; 181:1755–1762.

133   Kaplan MH, et al: Impaired IL-12 responses and enhanced development of Th2 cells in Stat4-deficient mice. Nature 1996;382:174–177.

134   Thierfelder WE, et al: Requirement for Stat4 in interleukin-12-mediated responses of natural killer and T cells. Nature 1996;382:171–174.

135   Rabinovitch A, Suarez-Pinzon WL, Sorensen O: Interleukin-12 mRNA expression in islets correlates with beta-cell destruction in NOD mice. J Autoimmun 1996;9:645–651.

136   Kalinski P, et al: IL-4 is a mediator of IL-12p70 induction by human Th2 cells: Reversal of polarized Th2 phenotype by dendritic cells. J Immunol 2000;165:1877–1881.

137   Mueller R, Krahl T, Sarvetnick N: Pancreatic expression of interleukin-4 abrogates insulitis and autoimmune diabetes in nonobsese diabetic (NOD) mice. J Immunol 1996;184:1093–1099.

138   Trembleau S, et al: Deviation of pancreas-infiltrating cells to Th2 by interleukin-12 antagonist administration inhibits autoimmune diabetes. Eur J Immunol 1997;27:2330–2339.

139   Trembleau S, et al: The role of IL-12 in the induction of organ-specific autoimmune diseases. Immunol Today 1995;16:383–386.

140   Trembleau S, et al: Pancreas-infiltrating Th1 cells and diabetes develop in IL-12-deficient nonobese diabetic mice. J Immunol 1999;163:2960–2968.

141   Murphy KM, et al: Signaling and transcription in T helper development. Annu Rev Immunol 2000;18:451–494.

142   Seder R, Paul W: Acquisition of lymphokine-producing phenotype by CD4+ T cells. Annu Rev Immunol 1994;12:635–673.

143   Burlinson E, Drakes M, Wood P, Differential patterns of production of granulocyte macrophage colony-stimulating factor, IL-2, IL-3 and IL-4 by cultured islets of Langerhans from non-obese diabetic and non-diabetic strains of mice. Int Immunol 1995;7:79–87.

144   Gombert JM, et al: Early quantitative and functional deficiency of NK1+-like thymocytes in the NOD mouse. Eur J Immunol 1996;26:2989–2998.

145   Rapoport M, et al: Interleukin-4 reverses T cell proliferative unresponsiveness and prevents the onset of diabetes in nonobese diabetic mice. J Exp Med 1993;178:87–99.

146   Von Herrath MG, Dyrberg T, Oldstone MB: Oral insulin treatment suppresses virus-induced antigen-specific destruction of beta cells and prevents autoimmune diabetes in transgenic mice. J Clin Invest 1996;98:1324–1331.

147   Homann D, et al: Autoreactive CD4+ T cells protect from autoimmune diabetes via bystander suppression using the IL4/Stat6 pathway. Immunity 1999;11:463–472.

148   Wang B, et al: Interleukin-4 deficiency does not exacerbate disease in NOD mice. Diabetes 1998; 47:1207–1211.

149   De Vries J: The role of IL-13 and its receptor in allergy and inflammatory responses. J Allergy Clin Immunol 1998;102:165–169.

150   Zaccone P, et al: Interleukin-13 prevents autoimmune diabetes in NOD mice. Diabetes 1999; 48:1522–1528.

151   Muchamuel T, et al: IL-13 protects mice from lipopolysaccharide-induced lethal endotoxemia: Correlation with down-modulation of TNF-alpha, IFN-gamma and IL-12 production. J Immunol 1997;158:2898–2903.

152   Doherty T, et al: Modulation of murine macrophage function by IL-13. J Immunol 1993;151: 7151–7160.

153   Di Santo E, et al: IL-13 inhibits TNF production but potentiates that of IL-6 in vivo and ex vivo mice. J Immunol 1997;1:379–382.

154   Rincon M, et al: Interleukin-6 (IL)-6 directs the differentiation of IL-4-producing CD4+ T cells. J Exp Med 1997;185:461–469.

155 Lentsch A, et al: In vivo suppression of NFκB and preservation of IκB alpha by interleukin-10 and interleukin-13. J Clin Invest 1997;100:2443–2448.

156 De Waal RWM, et al: IL-10. Curr Opin Immunol 1992;4:314.

157 Asseman C, et al: An essential role for interleukin-10 in the function of regulatory T cells that inhibit intestinal inflammation. J Exp Med 1999;7:995–1003.

158 Read S, et al: CD38+CD45RB(low) CD4+ T cells: A population of T cells with immune regulatory activities in vitro. Eur J Immunol 1998;28:3435–3447.

159 Powrie F, Mason D: OX-22high CD4+ T cells induce wasting disease with multiple organ pathology: Prevention by OX-22low subset. J Exp Med 1990;172:1701–1708.

160 Pennline KJ, Roque-Gaffney E, Monahan M: Recombinant human IL-10 prevents the onset of diabetes in the nonobese diabetic mouse. Clin Immunol Immunopathol 1994;71:169–175.

161 Zheng XX, et al: A noncytolytic IL-10/Fc fusion protein prevents diabetes, blocks autoimmunity, and promotes suppressor phenomena in NOD mice. J Immunol 1997;158:4507–4513.

162 Pauza ME, et al: T-cell production of an inducible interleukin-10 transgene provides limited protection from autoimmune diabetes. Diabetes 1999;48:1948–1953.

163 Balasa B, et al: IL-10 deficiency does not inhibit insulitis and accelerates cyclophosphamide-induced diabetes in the nonobese diabetic mouse. Cell Immunol 2000;202:97–102.

164 Massague J: The transforming growth factor-beta family. Annu Rev Cell Biol 1990;6:597–641.

165 Kulkarni A, et al: Transforming growth factor-beta-1 null mice. An animal model for inflammatory disorders. Am J Pathol 1995;146:264–275.

166 Shull M, et al: Targeted disruption of the mouse transforming growth factor-beta-1 gene results in multifocal inflammatory disease. Nature 1992;359:693–699.

167 Bright J, Kerr L, Sriram S: TGF-beta inhibits IL-2-induced tyrosine phosphorylation and activation of JAK-1 and Stat 5 in T lymphocytes. J Immunol 1997;159:175–183.

168 Kerhl J, et al: Production of transforming growth factor β by human T lymphocytes and its potential role in the regulation of T cell growth. J Exp Med 1986;163:1037–1050.

169 Cazarniecki C, et al: TGF-β modulated the expression of class II MHC antigens on human cells. J Immunol 1988;140:4217–4223.

170 Kerhl J, et al: Transforming growth factor-beta suppresses human B lymphocyte Ig production by inhibiting synthesis and the switch from membrane form to secreted form of Ig mRNA. J Immunol 1991;146:4016–4023.

171 Bridoux F, et al: Transforming growth factor beta (TGF-beta)-dependent inhibition of T helper 2 (Th2)-induced autoimmunity by self-major histocompatibility complex (MHC) class II-specific, regulatory CD4(+) T cell lines. J Exp Med 1997;185:1769–1775.

172 Powrie F, et al: A critical role for transforming growth factor-beta but not interleukin-4 in the suppression of T helper 1-mediated colitis by CD45RB(low) CD4+ T cells. J Exp Med 1996;183:2669–2674.

173 Read S, Malmstron V, Powrie F: Cytotoxic T lymphocyte-associated antigen 4 plays an essential role in the function of CD25+CD4+ regulatory cells that control intestinal inflammation. J Exp Med 2000;192:295–302.

174 Lampeter EF, et al: Transfer of insulin-dependent diabetes between HLA-identical siblings by bone marrow transplantation. Lancet 1993;341:1243–1244.

175 Thorsby E, Undlien D: The HLA associated predisposition to type 1 diabetes and other autoimmune diseases. J Pediatr Endocrinol Metab 1996;9(suppl 1):75–88.

176 Undlien DE, et al: HLA-encoded genetic predisposition in IDDM: DR4 subtypes may be associated with different degrees of protection. Diabetes 1997;46:143–149.

177 She JX: Susceptibility to type I diabetes: HLA-DQ and DR revisited. Immunol Today 1996;17:323–329.

178 Wen L, et al: In vivo evidence for the contribution of human histocompatibility leukocyte antigen (HLA)-DQ molecules to the development of diabetes. J Exp Med 2000;191:97–104.

179 Assan R, et al: Metabolic and immunological effects of cyclosporin in recently diagnosed type 1 diabetes mellitus. Lancet 1985;i:67–71.

180 Bougneres PF, et al: Limited duration of remission of insulin dependency in children with recent overt type I diabetes treated with low-dose cyclosporin. Diabetes 1990;39:1264–1272.

181 Calcinaro F, et al: Detection of cell-mediated immunity in type I diabetes mellitus. J Autoimmun 1992;5:137–147.

182 Bottazzo GF, et al: In situ characterization of autoimmune phenomena and expression of HLA molecules in the pancreas in diabetic insulitis. N Engl J Med 1985;313:353–360.

183 Hanninen A, Jaakola I, Jalkanen S: Mucosal addressin is required for the development of diabetes in nonobese diabetic mice. J Immunol 1998;160:6018–6025.

184 Foulis AK, Farquharson MA, Meager A: Immunoreactive alpha-interferon in insulin-secreting beta cells in type 1 diabetes mellitus. Lancet 1987;ii:1423–1427.

185 Foulis AK, McGill M, Farquharson MA: Insulitis in type 1 (insulin-dependent) diabetes mellitus in man – Macrophages, lymphocytes, and interferon-gamma containing cells. J Pathol 1991;165:97–103.

186 Huang X, et al: Interferon expression in the pancreases of patients with type I diabetes. Diabetes 1995;44:658–664.

187 Cavallo MG, et al: Cytokines in sera from insulin-dependent diabetic patients at diagnosis. Clin Exp Immunol 1991;86:256–259.

188 Lorini R, et al: Low serum levels of tumor necrosis factor-alpha in insulin-dependent diabetic children. Horm Res 1995;43:206–209.

189 Sharma K, Ziyadeh FN: Renal hypertrophy is associated with upregulation of TGF-beta-1 gene expression in diabetic BB rat and NOD mouse. Am J Physiol 1994;267:F1094–F2001.

190 Ebner R, et al: Determination of type I receptor specificity by the type II receptors for TGF-beta or activin. Science 1993;262:900–902.

191 Sporn MB, et al: Transforming growth factor-beta: Biological function and chemical structure. Science 1986;233:532–534.

192 Nakamura T, et al: mRNA expression of growth factors in glomeruli from diabetic rats. Diabetes 1993;42:450–456.

193 Shankland SJ, et al: Expression of transforming growth factor-beta-1 during diabetic renal hypertrophy. Kidney Int 1994;46:430–442.

194 Border WA, et al: Natural inhibitor of transforming growth factor-beta protects against scarring in experimental kidney disease. Nature 1992;360:361–364.

195 Yamamoto T, et al: Expression of transforming growth factor beta is elevated in human and experimental diabetic nephropathy. Proc Natl Acad Sci USA 1993;90:1814–1818.

196 Brod S, et al: Ingested interferon-alpha prevents allograft islet transplant rejection. Transplantation 2000;69:2162–2166.

197 Lesslauer W, et al: Recombinant soluble tumor necrosis factor receptor proteins protect mice from lipopolysaccharide-induced lethality. Eur J Immunol 1991;21:2883–2886.

198 Ashkenazi A, et al: Protection against endotoxic shock by a tumor necrosis factor receptor immunoadhesin. Proc Natl Acad Sci USA 1991;88:10535–10539.

199 Mori L, et al: Attenuation of collagen-induced arthritis in 55-kDa TNF receptor type 1 (TNFR1)-IgG1-treated and TNFR1-deficient mice. J Immunol 1996;157:3178–3182.

200 Piguet PF, et al: Evolution of collagen arthritis in mice is arrested by treatment with anti-tumour necrosis factor (TNF) antibody or a recombinant soluble TNF receptor. Immunology 1992;77:510–514.

201 Feldman M, et al: Anti-TNF-alpha therapy is useful in rheumatoid arthritis and Crohn's disease: Analysis of the mechanism of action predicts utility in other diseases. Transplant Proc 1998;30:4126–4127.

202 Fisher CJ Jr, et al: Treatment of septic shock with the tumor necrosis factor receptor: Fc fusion protein. The Soluble TNF Receptor Sepsis Study Group. N Engl J Med 1996;334:1697–1702.

203 Li XC, et al: Blocking the common gamma-chain of cytokine receptors induces T cell apoptosis and long-term islet allograft survival. J Immunol 2000;164:1193–1199.

204 Giannoukakis N, et al: Protection of human islets from the effects of interleukin-1-beta by adenoviral gene transfer of an IkappaB repressor. J Biol Chem 2000;275:36509–36513.

Richard A. Flavell, PhD, Section of Immunobiology, Yale University School of Medicine/HHMI, 310 Cedar Street, FMB 412, New Haven, CT 06520-8011 (USA)
Tel. +1 203 737 2216, Fax +1 203 737 2958, E-Mail richard.flavell@yale.edu

von Herrath MG (ed.): Molecular Pathology of Type 1 Diabetes mellitus.
Curr Dir Autoimmun. Basel, Karger, 2001, vol 4, pp 218–238

Understanding the Interaction of Genetics and Cellular Responses in Nonobese Diabetic Mice

William M. Ridgway[a], *C. Garrison Fathman*[b]

Departments of Medicine, Division of Rheumatology and Immunology,
[a] University of Pittsburgh School of Medicine, Pittsburgh, Pa. and
[b] Stanford University School of Medicine, Stanford, Calif., USA

The subject of this chapter, understanding the interaction of genetics and cellular responses in nonobese diabetic (NOD) mice, is enormous. Over 1,500 publications on NOD mice, as well as at least one book [1], convey a wealth of information on the cellular responses and biochemistry of NOD mice. In addition, over a decade of intense genetic work has identified more than 20 Idd (insulin-dependent diabetes) loci contributing (in a complex, multigenic fashion) to disease pathogenesis (see the chapter by Serreze and Leiter in this volume). It is increasingly apparent that genetic studies will provide interpretation of cellular studies. To date, however, there has been relatively little integration of the cellular and genetic bases of diabetes pathogenesis in NOD mice. Only one gene from the 20 genetic intervals (the stretches of genome containing the Idd loci) has been identified with even moderate certainty (i.e. the MHC class II molecule from Idd1, as we will discuss in this chapter). Moreover, despite hundreds of publications on NOD disease and on MHC class II in general, the exact function of MHC class II in NOD diabetes is still controversial. This example illustrates the difficulties and pitfalls of interpreting and integrating genetic and cellular data in NOD diabetes pathogenesis. Given these difficulties, we will attempt to survey some main cellular/genetic topics and identify important questions and conclusions. We view cellular responses in NOD mice as phenotypes; the question becomes which phenotypes are related to the genotypes implicated in disease pathogenesis. The NOD literature contains many instances of fascinating cellular phenotypes that subsequently have been excluded as etiologically relevant, since they map outside of disease-linked genetic intervals. The intersection of genetic and cellular approaches to NOD diabetes is an example of 'hypothesis-driven' versus hypothesis-independent approaches. Immunologically based cell culture or transfer experiments test the hypothesis that a particular cell or cellular product displays phenotypic differences in NOD versus control mouse strains, without testing whether the cell or cellular product is linked to the disease as assayed by genome wide scanning in experimental crosses. To approach an understanding of some of the conundrums raised in the NOD field, we will attempt to interpret the cellular studies according to their genetic relevance.

Hundreds of 'cures' or preventions have been reported for NOD diabetes, ranging from 'tolerization' with autoantigens to raising the temperature of the breeding colony [2, 3]. Many more agents have been reported to prevent diabetes than could possibly be implicated as direct causes genetically. Thus a subtle but important point arises: it is possible that interventions which prevent diabetes in NOD mice are unrelated to the genetically based disease pathogenesis. How is this possible? Our interpretation is that diabetes in NOD mice (and possibly in man) is mediated by a defect (or defects) in a common genetic/biochemical pathway involving central components of the immune system (fig. 1). Any intervention that can affect one or more separate steps in this pathway can prevent progression along the pathway. We can construct a hypothetical model, which integrates some of the cellular findings into this genetic structure. We are fully aware, however, that this is only a hypothetical model and that there are potentially many other models, which fit this genetic scaffold. For example, the steps in the cellular/genetic pathway leading to diabetes might consist of the following (see fig. 1): (1) A primary ('developmental') tissue abnormality occurs in NOD pancreatic development [4]. This tissue abnormality could result in the expression of an aberrant protein(s) (i.e. neoantigen(s)) which was (were) not present in the thymus during initial selection of the T-cell repertoire or might consist of the inappropriate expression of chemoattractants (chemokines) by 'altered' pancreatic tissue. (2) Due to the unique NOD MHC class II, I-A^{g7}, an abnormal repertoire of autoreactive (potentially high-affinity) CD4+ T cells populates the periphery in NOD mice [5, 6]. (3) NOD mice possess abnormal, immature hematopoietically derived antigen-presenting cells (APCs), which might present the neoantigen(s) in an unusual way to self-reactive T cells. These self-reactive T cells could cause the initial tissue damage, resulting in uptake of additional selfantigens by 'professional' APCs for presentation to yet more self-reactive CD4+ T cells [7, 8]. Alternatively, high-affinity self-reactive T cells might be initially attracted to the islet tissue by 'inappropriate' expression of chemokines. (4) Due to the unique combination of NOD non-MHC immune regulatory background genes, the high-affinity self-reactive T cells that have

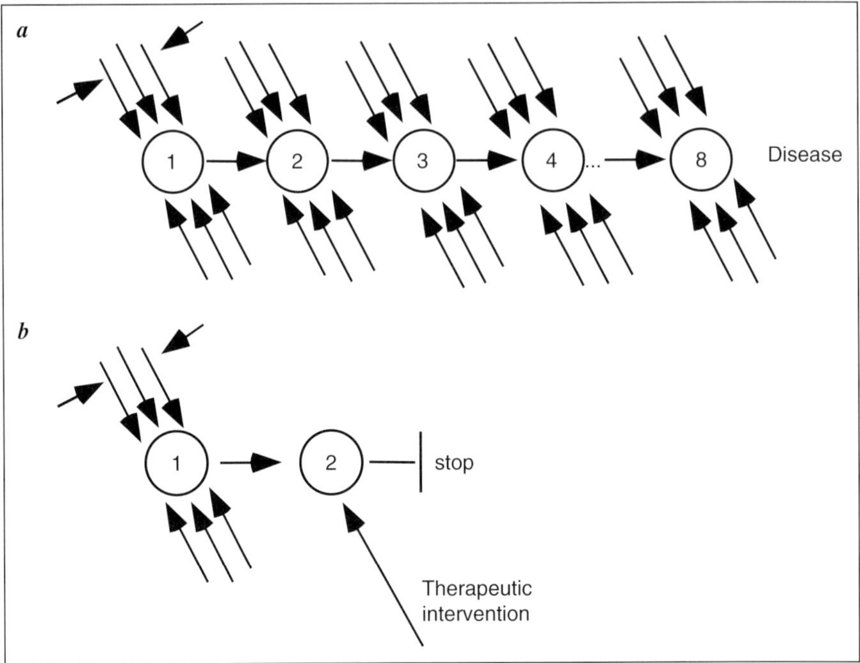

Fig. 1. A vectorial representation of a complex genetic cascade. In *a*, the 'normal' disease state, the numbers represent the phenotypes of disease-linked alleles which when co-expressed lead to disease (No. 8) (see text for one possible model of this process). The arrows impacting on the disease-associated alleles represent secondary genes which can affect the expression of the critical alleles and their phenotypes to a degree represented by the magnitude of the vector. The direction of the vector refers to whether the secondary gene modifies the disease-associated allele towards the disease phenotype (top arrows) or away from the disease phenotype (bottom arrows). 'Normally' in NOD mice the sum of the secondary gene vectors is zero, resulting in a positive direction due to the action of the disease-linked (pathological) genes (represented by the numbers in the figure) leading to spontaneous disease as indicated by the 'net' forward vector between each number. However, as in *b*, an intervention acting on a secondary gene can affect the function of a disease-related gene to an extent that stops progression towards disease and causes arrest at an intermediate phenotype (e.g. insulitis). It is easy to imagine tertiary genes (smaller arrows in *a*) affecting the function of secondary genes, etc. There may be no practical limit on the number of genes which could therefore affect a disease state without actually causing it. The phenotypes described in the text are only one set of potentially hundreds of different plausible immunological scenarios.

escaped thymic selection might react in an abnormal way upon stimulation, creating an excessive inflammatory response [9, 10]. (5) The interaction of T cells expressing the abnormal T-cell repertoire with the self-antigens presented abnormally by NOD APCs could result in a prolonged and excessive immune response, causing further tissue damage, accompanied by expression of CD4+-derived cytokines, which activate other cells in the autoimmune reaction. (6) Activated NOD CD4+ T cells activate B cells, leading to uptake and presentation of more autoantigens and resultant increase in the inflammatory response [11]. (7) The prolonged inflammatory response results in epitope spreading of the autoreactive T-cell response. (8) A chronic inflammatory phase eventually results in pancreatic β-cell destruction by late stage events, presaged by the long period of inflammation.

We emphasize that any number of such scenarios can be developed, many supported by the voluminous NOD literature, and each one needs to be tested against the existing evidence from cellular and genetic studies. A theoretical genetic rationale for multiple cellular pathways to autoimmune diabetes has also been demonstrated: the sum of predisposing genetic components needs to exceed a threshold for the disease phenotype [12]. McAleer et al. [12] demonstrated that several different combinations of predisposing genes could serve to attain this threshold, i.e. that no particular set of genes (thus no single cellular pathway) is necessary and sufficient to develop diabetes. An additional important point is that if such a 'cascade' of cellular reactions is occurring in NOD mice, then an interruption early or later in the pathway could prevent the disease from progressing without necessarily meaning that the intervention pinpoints an etiological gene in NOD disease (see fig. 1). Since any step or 'phenotype' in the pathway can either be the result of, or affected by, a multitude of genes, intervention affecting a non-linked gene can impact the function or expression of a linked gene. For example, priming NOD mice with CFA prevents diabetes [13]. CFA-induced prevention is associated with a Th2-like phenotype of T cells in the islets, as opposed to the pathogenic Th1 phenotype [14, 15]. It is possible that CFA therefore directly affects T-cell functional responses to stimulation (step 4 in fig. 1). It is also possible, however, that through a nonspecific immunostimulatory effect, CFA indirectly causes release of cytokines (such as CSF-1 [5]) which lead to maturation of the NOD APCs (step 3 in fig. 1). It is possible that if the macrophages undergo maturation, no amount of autoreactivity in the T-cell repertoire (steps 2 and 4) will result in diabetes; the 'mature' macrophages could interact with T cells in a different way and thus effect changes downstream in the disease pathway. The cascade could end at step 2 or 4 in this scenario (fig. 1b). In this review therefore we will keep a critical eye on drawing conclusions, and attempt to 'correct' immunological hypotheses with knowledge of the results of genetic linkage studies. The major conclusion we draw is that an enormous amount of genetic work remains to be performed before it will be possible to place most of the cellular studies of NOD diabetes in their proper theoretical context. We believe the major importance of studying NOD mice lies not so much in finding potential cures for human diabetes (since it seems much easier to prevent the disease in mice than in humans [2, 3]), but in identifying the genetic and cellular basis of the immune pathways resulting in autoimmune diabetes.

Autoreactive T-Cell Responses in NOD Diabetes

An enormous amount of work has implicated autoreactive T cells in NOD pathology. Early histological studies demonstrated that CD4+ T cells infiltrated the pancreatic islets [16]. Subsequently it was demonstrated that diabetic spleen cells could transfer disease to young recipients [17]. Treatments directed toward T cells, including anti-CD4 therapy, prevented disease [18, 19]. Given these facts and the association of NOD diabetes with a unique MHC class II allele (see below), it was suspected and then confirmed that islet-reactive autoimmune CD4+ T cells played an essential role in the disease processes. Important issues with respect to the role of T cells in NOD diabetes that will be discussed are the following: (1) What autoantigens do NOD T cells recognize? (2) Are any CD4+ T-cell autoantigens central to NOD pathogenesis? (3) Is there a restricted TCR Vβ response in NOD diabetes? (4) Are there 'regulatory' T-cell responses that are deficient in NOD mice?

A wide range of T-cell autoantigens has been described in NOD mice. In the last 7 years over 70 publications have investigated glutamic acid decarboxylase (GAD) as a disease causing and, used appropriately, as a disease-remitting agent. Tisch et al. [20] and Kaufman et al. [21] initially reported that NOD mice develop spontaneous T-cell reactivity to GAD65 with subsequent spread to other antigens. Tisch et al. [22] and others reported that 'tolerization' with GAD could prevent NOD diabetes. Peterson et al. [23] reported that neonatal tolerization with GAD prevented diabetes [23]. Elliot et al. [24] showed that the larger isoform GAD67 could also prevent diabetes [24]. Tisch et al. then showed that priming with GAD induced GAD-specific regulatory cells as a mechanism to inhibit diabetes [25], and that specific GAD peptides also prevented disease [26]. Tian et al. [27] showed that GAD peptides administered nasally (in a 'tolerogenic' fashion) could prevent diabetes by inducing a Th2 response. Gerling et al. [28] showed that intrathymic administration of a GAD peptide could retard diabetes, presumably through deletional effects on the TCR repertoire. There were some negative reports: Schloot et al. [29] reported that they were unable to induce GAD-reactive T-cell clones that would transfer diabetes [29] (unlike transfer of diabetic CD4+ T cells into NOD/SCID recipients, who then develop diabetes [30]). In addition, Plesner et al. [31] were unable to induce diabetes by priming with GAD65. Furthermore, it was reported that GAD65 was expressed in islets at very low levels, and was sometimes undetectable [32]. Zekzer et al. [33] answered some of these points by showing that priming with rat GAD could accelerate diabetes in NOD mice and that GAD-reactive clones derived from rat GAD priming could transfer diabetes to NOD-SCID mice. Finally, Yoon et al. [34] reported that β-cell-specific suppression of GAD (using antisense GAD expression) could prevent diabetes, and, moreover, somehow prevented GAD-specific T cells from even developing in the mice.

The weight of this evidence would seem overwhelmingly to support a central role for GAD65 and its dominant epitopes in the etiology of NOD autoimmunity. Recently, however, an experiment has cast doubt on the importance of GAD as the key diabetes autoantigen. Kash et al. [35] developed a GAD65 knockout (KO) mouse and backcrossed it onto NOD. These GAD65 KO mice developed insulitis and diabetes. Therefore, diabetes can develop in the absence of GAD65.

How is this result possible? If these findings are upheld, they are interpretable in view of the 'cascade' pathway of autoimmune diabetes suggested above. It may be that immune reactivity to GAD is an 'epiphenomenon' in the pathology of diabetes and develops after an earlier, primary event. Removing GAD from the downstream pathway would not then prevent disease. Presumably other autoreactive T-cell epitopes would still develop in these KO mice. It is possible that prevention of diabetes by treatment with GAD65 in NOD mice, then, does not indicate a truly pathogenic role for GAD65 in the disease, but interrupts the pathway via secondary immune effects to which the NOD mouse is notoriously susceptible.

Many other self-T-cell epitopes have been described in NOD mice, including response to insulin (by both CD4+ and CD8+ T cells) [36, 37]. A variety of insulin epitopes have been described; notably insulin is the only autoantigen expressed predominantly in the pancreas [38, 39].

Other potentially diabetes-related autoantigens include carboxypeptidase H [40], and HSP65 [41]. Indeed, as with GAD, HSP65 has been described both as causing and treating NOD diabetes [41]. An HSP peptide, p277, was recognized by NOD T-cell clones and could tolerize against disease development [42]. However, another group found that HSP or p277 immunization not only did not prevent diabetes, but also actually accelerated it [43].

What conclusion can be reached about NOD T-cell reactivity? One point appears undeniable: NOD mice have a large repertoire of autoreactive T cells. This point was demonstrated by Ridgway et al. [44], who showed that NOD mice possess T cells reactive to a variety of self-antigens that bore no apparent disease relevance, such as myoglobin and TCR peptides. The T-cell responses to these self-peptides could be induced in NOD mice, but not 'normal' strains, following priming with the self-peptide in adjuvant [44]. The concept of an expanded NOD autoreactive T-cell repertoire was formally demonstrated by Kanagawa et al. [45] who showed by limiting dilution analysis that NOD mice have an increased precursor frequency of autoreactive T cells. Thus T-cell responses to such apparently disease-related peptides as GAD and carboxypeptidase could be reflective of a larger defect in the NOD T-cell repertoire. We will address this issue more thoroughly in the next section.

At least ten studies have examined whether NOD TCR Vβ usage is restricted in response to islet antigens. This question has theoretical importance, since Vβ restriction at any point could suggest that response to a single antigen was driving the NOD autoimmune T-cell response. The practical point of such a determination (and a major motivation behind many of the T-cell studies in NOD mice) is that, if a single TCR structure drove autoimmune diabetes, it or the antigen it recognized could become a target for therapy. The result of such studies (reviewed by Serreze [46]) suggested against the hypothesis that a single T-cell autoantigen drove autoimmune diabetes. Nonetheless, these studies are incomplete and have not formally excluded an early disease-specific pathogenic T cell that is rapidly 'diluted' in the ensuing inflammatory response.

A final question about the NOD T-cell response is whether the NOD T-cell receptor (TCR) peripheral repertoire is unique, or whether 'normal' mice have T cells that react to the same autoantigens but in a 'regulated' way. The finding of diffuse reactivity to the autoantigens as described above could be due to a unique NOD TCR repertoire, or to lack of some peripheral regulatory processes which suppresses such reactivity in normal strains. A separate chapter in this book (by Delovitch et al.) will address the issue of NOD 'regulatory' cells. We will confine our concluding remarks to the NOD MHC and its influence on the NOD CD4+ T-cell repertoire.

The Role of MHC in NOD Autoimmune Disease and Diabetes

MHC structure and T-cell repertoire selection is a huge subject. Recent reviews have addressed aspects that we will not cover here, such as the mechanisms underlying thymocyte signaling and the developmental cascade of thymocyte selection [47–50]. We will focus on biochemical analysis of the murine (NOD) MHC class II diabetes-associated molecule, I-A^{g7}, and studies of its effect on the autoimmune T-cell repertoire. In addition, we will interpret several transgenic and KO models in light of the findings in the spontaneous disease.

Associations between particular MHC class II molecules and autoimmunity have been established in many diseases, including diabetes in both humans and NOD mice. The NOD H2^{g7} haplotype (Idd1) has the strongest effect on diabetes incidence of all the Idd loci, estimated at 40% of the genetic load [51]. Notably, Idd1 is now thought to contain, in addition to class II molecules, two additional Idd loci (one of which may be class I) [52]. Nevertheless, a large body of work makes it highly likely that the unique NOD class II MHC, I-A^{g7}, is one of the genes contributing to the diabetic linkage of the Idd1 locus. First, replacement of the H2^{g7} haplotype in NOD mice with a nondisease associated MHC class II haplotype (in NOD MHC congenic mice) markedly suppressed diabetes incidence, despite expression of a full complement of non-MHC disease associated genes [53]. Moreover, a single copy of the resistant MHC haplotype reduced diabetes incidence 40-fold in (NOD MHC congenic × NOD)F$_1$ mice [53]. Imposition of a disease-resistant MHC as a transgene also diminished diabetes incidence and implicated the I-A^{g7} molecule as a critical gene product from within the extended H2^{g7} haplotype [54–56]. NOD mice express only the I-A MHC molecule; I-E is not expressed because of a deletion in the I-E α-chain promoter. In addition, the I-A^{g7} molecule is unique in its β chain, which is characterized by a His56 and Ser57 (most mice express Pro and Asp, respectively, in these positions) of the MHC class II β-sheet, critical residues for MHC class II structure and stability [57]. Strikingly, a majority of human patients with type 1 diabetes also express class II molecules (DQ8, i.e. HLA-DQA1* 0301/DQB1*0302) characterized by a non-Asp amino acid at position 57 of the MHC class II β-sheet [58]. Nishimoto et al. [54] crossed a BL6 I-E transgenic mouse onto the NOD background; expression of I-E prevented diabetes. Lund et al. [56] generated a NOD mouse with a Pro56 I-A^{g7} transgene, as well as transgenic I-E expressing NOD mice (i.e. expressing I-E without any of the BL6 flanking genome). Both transgenic mice were protected from diabetes. Subsequently, Quartey-Papfio et al. [59] reported that a mutated I-A^{g7} expressing an Asp at position 57 reduced, but did not prevent, diabetes. Finally, Singer et al. [60] showed that no diabetes developed in a NOD transgenic mouse with an I-A^{g7} molecule mutated to Pro-Asp at p56–57. Collectively these results show that the MHC Class II molecule alone had a powerful influence on diabetes pathogenesis in NOD mice.

The question of why MHC class II molecules are associated with autoimmune diabetes is an interesting example of how difficult it can be to establish a biological mechanism despite an overwhelming amount of data. Reich et al. [61] originally concluded that transgenic I-E molecules deleted autoreactive Vβ5-expressing T cells in the thymus, since they found that diabetogenic T-cell clones from NOD mice expressed Vβ5, and that Vβ5 is deleted by I-E. Lund et al. [56] however, did not find gross evidence from FACS studies of Vβ deletion in their I-A^{g7} 56 Pro mice, and thereby concluded that deletional mechanisms were not operative. Bohme et al. [62] also showed that I-E mutated transgenes that could delete the Vβ T-cell subsets which are deleted by I-E, did not protect from diabetes, thus concluding that a deletional mechanism did not explain the protective effect of the I-E molecule. In retrospect, these studies were overly dependent upon a superantigen (e.g. mls)–based model of T-cell deletion, i.e. they examined whole Vβ subsets for evidence of deletion and, not finding it, concluded that a thymic deletional mechanism did not explain the lack of autoreactivity. If autoreactive T cells, however, can express a wide variety of TCR Vβ chains (see above section), then deleting a subset of them from the repertoire would not result in a noticeable change in peripheral autoreactivity. Nevertheless, as a result of these studies, as well as elucidation of the role of MHC class II molecules in presenting peptides to CD4+ T cells, a different interpretation of the mechanism of MHC-disease association emerged.

MHC class II molecules function by binding peptides for presentation to antigen-specific CD4 cells expressing unique TCRs. The TCRs recognize MHC/peptide complexes on the surface of APCs. There are two dominant physiological roles for MHC class II gene products: (1) selection of the T-cell repertoire in the thymus, and (2) presentation of foreign antigens in the periphery. By processing and presenting self-peptides bound to MHC molecules to developing thymocytes, thymic APCs first select the potential peripheral T-cell repertoire (positive selection) and then purge this positively selected repertoire of T cells which react with high avidity (negative selection) to self-peptide/MHC complexes [49]. Subsequently, peripheral APCs present foreign peptides on their MHC molecules to the peripheral T cells that have survived this selection process in the thymus, and initiate physiologic immune responses. So, it is inherently hard to say whether the effect of MHC class II molecules is occurring at the level of the thymus, the periphery, or both. Nevertheless, given the results in the transgenic models described above, for several years the predominant interpretation of the association between MHC and autoimmune diseases has been that disease-associated alleles are efficient binders of autoantigens presented to autoreactive T lymphocytes [63, 64]. What distinguished the disease associated alleles, this theory proposed, was that they would bind autoantigens such as GAD more strongly and therefore stimulate the GAD-reactive T cells in the periphery in an efficient manner compared to nondisease associated MHC alleles. This view was modeled after the physiology of responding vs. non-responding strains reactive to foreign peptides: responding strains had MHC molecules which could efficiently bind and present the peptide to a T-cell repertoire, while nonresponding strains did not [65].

The paradox raised by the 'good binding' theory was, if the MHC molecules were such 'good binders' of the autoantigens, why did they not efficiently delete the autoreactive cells in the thymus? The only answer to this could be that the cells were 'clonally ignorant' in the thymus, i.e. the autoantigen was not present in the thymus in sufficient quantities to mediate selection. Subsequently, however, PCR technology has allowed detection of the expression of several such autoantigens (e.g. insulin [66]) in the thymus. Furthermore, if these MHC molecules were such 'good' binders of the autoantigen, how was it possible that a 2-fold reduction in MHC concentration (an F_1 between NOD and a NOD.MHC congenic mouse) could produce a 30-fold reduction in diabetes [53]?

The interpretation that MHC molecules were associated with autoimmunity because they efficiently bound the disease causing autoantigens was first contested in an antigen-specific inducible disease model, experimental autoimmune encephalomyelitis (EAE). Wraith and co-workers [67] demonstrated that the disease-associated allele, H-2^u, bound the autoantigen (Ac1-11 of myelin basic protein) poorly. This group subsequently showed that poor binding of the autoantigen allowed Ac 1-11-specific autoreactive T cells to escape thymic negative selection [68]. These studies encouraged re-examination of the biochemistry and peptide-binding characteristics of the NOD class II molecule, I-A^{g7}, at a biochemical level.

Reich et al. [69] examined the repertoire of peptides isolated from NOD splenic I-A^{g7} molecules. The authors found that both intracellular and extracellular peptides bound to I-A^{g7}, and found that an acidic (negatively charged) residue in the carboxy-terminus of the binding peptides played a critical role in peptide binding to I-A^{g7}. Such a negatively charged amino acid is uncommon in the carboxy end of mouse class II binding peptides eluted from class II molecules, since normally ASP57 of the class II β chain would repel acidic p9 residues. They suggested that acidic residues in the carboxy-terminus could interact with an arginine left exposed in the MHC molecule due to the lack of an aspartate in position 57 [69]. Carrasco-Marin et al. [76] investigated the biochemical properties of I-A^{g7} and found it to be structurally unstable (the αβ dimer disintegrated under SDS-PAGE). Moreover, this manuscript reported that I-A^{g7} was a poor peptide binder; 12 self and foreign peptides bound poorly in their assays. Whether or not I-A^{g7} is a poor peptide binder is of critical importance in view of the paradigm of the disease-associated MHC alleles as good binders of dominant autoantigens. Reizis et al. [70] disputed the notion that I-A^{g7} was a poor peptide binder, while substantiating the finding of SDS instability. They proposed a 9-amino acid-based peptide-binding motif and agreed with Reich et al. [69] that p9 contained negatively charged amino acids. Harrison et al. [71] also did not find that I-A^{g7} was a poor peptide-binder, but their proposed peptide-binding motif for I-A^{g7} was radically different from Reizis and Reich, with a basic (lysine or arginine) p9 residue.

Several studies subsequently have examined these issues. Peterson and Sant [72] confirmed in an in vitro system that I-A^{g7} was unable to form SDS-stable dimers, but could find no defect in the ability of I-A^{g7} to bind to the invariant chain (Ii) or to DM. Thus the purported poor peptide binding of I-A^{g7}

could not be due to a structural or functional inability to interact with DM or Ii. Hausmann et al. [73] extended the biochemical characterization of I-A^{g7} by expressing it as a soluble protein in *Drosophila* cells. They examined the peptide-binding characteristics of I-A^{g7} under detergent-free conditions, and again confirmed the observation of Unanue's group that I-A^{g7} was not SDS-resistant. With respect to peptide binding, they found that some (previously identified [69]) autoantigens bound extremely well, but other, islet-associated antigens (e.g. three separate GAD65 peptides) bound poorly to I-A^{g7}. Interestingly, the murine CLIP peptide bound very poorly at endosomal pH (pH = 5) while binding well at neutral pH. CLIP bound at neutral pH, however, rapidly dissociated from I-A^{g7} at acidic pH. None of the tested peptides induced to SDS resistance in the I-A^{g7} complex [74]. Adorini and co-workers [74] confirmed elements of their original proposed peptide-binding motif by using a large, random peptide library expressed in M13 phage. They isolated 90 random phage-expressed peptides that bound to I-A^{g7}, synthesized the peptides, and performed competitive binding experiments. They concluded again that a motif containing basic (positively charged) residues at the critical P9 position fit their library of peptide sequences. In addition, less than half of the phage-derived peptides even contained an acidic residue [75]. Carrasco-Marin et al. [76] found, somewhat surprisingly, that by incorporating the TCR contacts of an I-A^{g7}-binding peptide (a self-peptide from the Eα chain) onto a polyalanine backbone, no residues were essential to binding I-A^{g7}. Only two residues were sensitive to amino acid substitutions that prevented MHC binding. They concluded that I-A^{g7} bound promiscuously, but weakly, to many peptides.

Recently an additional critical biochemical analysis of I-A^{g7} has been completed, i.e. the solution of its crystal structure [77]. Corper et al. [77] expressed I-A^{g7} in *Drosophila* cells as both 'empty' and as single peptide-MHC complexes. They used the 'empty' I-A^{g7} molecules to perform peptide-binding studies with phage libraries. In contrast to Gregori et al. [74], their phage-peptide-binding studies revealed an increase in acidic residues in the last 4 amino acids of the binding peptides. In addition, their results supported the hypothesis that I-A^{g7} was a relatively promiscuous peptide binder, since a poly-Ala peptide (an ALA 11 mer) could displace a known I-A^{g7}-binding peptide, and since I-A^{g7} bound an increased number of GAD peptides compared to I-A^{d} [77]. The crystal structure of I-A^{g7} bound to GAD 207–220 was also determined by Corper et al. [77]. A major result of this study was the characterization of the p9-binding pocket; electrostatic analysis showed the p9 pocket was positively charged. Thus the p9 position is more attractive to negatively charged residues [77]. Based on their I-A^{g7} structure, they then suggested a peptide-binding motif for I-A^{g7} which allowed a negatively charged p9 residue or Gly, Ala, or Ser at p9. These findings explain the result by Carrasco-Marin et al. [76]; according to this proposed peptide-binding motif, a poly-ALA backbone could bind reasonably well to I-A^{g7}.

The summary of the biochemical analysis of I-A^{g7} is as follows: (1) most evidence suggests that acidic residues at the p9-binding position are a critical aspect of the binding motif; (2) several groups have confirmed that I-A^{g7} is SDS unstable; (3) some groups deny that I-A^{g7} is a poor peptide binder [71, 72, 77], one group claims it is a poor peptide binder in general [70], and a third group finds it binds some peptides well (e.g. some 'foreign' peptides) and other peptides poorly (some autoantigens [74]). Independently of whether I-A^{g7} is a poor peptide binder, several groups agree it is a 'promiscuous' peptide binder, i.e. it binds more, and more diverse peptides, than other I-A molecules. Some of the discrepancies in the literature could be due to different experimental conditions, e.g. different pH or temperature conditions. Most notable in this regard is the report by Hausmann et al. [73] showing a dramatically different capacity of clip peptide binding at pH 5 vs. 7.

This detailed analysis of the biochemistry of I-A^{g7} is important insofar as it helps answer the question of the function of the class II molecule in autoimmune pathology, and why MHC class II is linked so strongly to autoimmune disease. The older theory that disease-associated MHC molecules are 'good binders' of autoimmune associated peptides is not in accord with the detailed analyses showing that I-A^{g7} is either a poor peptide binder [70] or binds critical autoantigens poorly [74]. We offer our interpretation of these results in the next section.

I-A^{g7} and Control of the Peripheral T-Cell Repertoire

We have previously reported our interpretation of the role of I-A^{g7} in NOD diabetes [8, 78, 79], but will review this theory in light of the additional studies discussed above. We examined the effects of I-A^{g7} on the functional peripheral T-cell repertoire by examining the effect of MHC type and expression levels on T-cell responses to a panel of self and foreign antigens in NOD, NOD MHC congenic, NOD MHC transgenic, and NOD MHC KO mice, as well as a variety of F$_1$ strains derived from these mice [7]. This study followed a previous one that found an abnormal number of autoreactive T-cell responses in NOD mice following priming with self-peptides in CFA [44]. The goal of the subsequent studies was to hold non-MHC NOD genes constant while varying the allotype of MHC on the APC cell surface. The surprising result was that [NOD × NOD.H2^k congenic]F$_1$ mice and NOD-I-A^k transgenic mice, which both expressed I-A^{g7} and I-A^k on an otherwise NOD background, had very different T-cell responses to I-A^{g7}-binding peptides. The F$_1$ mice were diverted to a (non-autoreactive) T-cell response pattern characteristic of the NOD.H2^k parents, while the transgenic mice had a T-cell response pattern (autoreactive) characteristic of the NOD mice [7]. The critical difference was in the quantitative ratio of I-A^{g7} to I-A^k expressed on the APCs. The F_1 mice did not lose the autoreactive T-cell response simply due to insufficient I-A^{g7} expression, since [NOD.MHC null × NOD]F_1 mice, which were hemizygous for I-A^{g7}, had the autoreactive T-cell phenotype. In addition, the [NOD × NOD.H2^k congenic]F_1 APCs could present I-A^{g7}-restricted peptides to T cells, although these F_1 mice lost I-A^{g7}-restricted T-cell responses from their repertoire. These results strongly suggested that I-A^{g7} was ineffective at negative selection of autoreactive T cells, and that imposition of a disease-resistant allele deleted the autoreactive T cells from the repertoire [7]. This suggested a defective thymic deletional mechanism for the increased number of autoreactive T cells found in the NOD periphery (see section on Autoreactive T-Cell Responses above). The Santamaria group [81] mustered additional evidence that nondisease associated MHC alleles act by deleting autoreactive T cells from their peripheral repertoire. They used an I-A^{g7}-restricted TCR transgenic, and showed that the effect of imposition of nondisease associated alleles was to decrease the number of single positive transgenic T cells [80]. This group extended these findings by providing evidence that, in the case of this particular TCR, the disease-resistant allele can neither efficiently positively select the transgenic TCR when expressed solely in the thymic cortical epithelium, nor effectively present the autoantigen in the periphery [81]. These results raise the interesting possibility that the pathogenic TCR in this case was positively and negatively selected by a different MHC [81].

Several other transgenic mouse models offer insights into the effect of the thymic-expressed MHC molecules on the composition of the peripheral T-cell repertoire. Several papers, expression of MHC class II restricted to thymic cortical epithelium ('unopposed positive selection') [82], the KO of H2-M [83–85], or the expression of a transgenic MHC linked to a single peptide [86, 87], all suggest the acquisition of a peripheral repertoire in which large numbers of T cells are autoreactive. In each model, thymic selection is incomplete, lacking deletion or negative selection of T cells retaining abnormally high reactivity towards self-peptide:self-MHC complexes. Some other transgenic models have addressed autoimmune repertoire selection. Wu et al. [88] studied double transgenic TCR-HNT/Ins-HA mice (which spontaneously develop diabetes at a 70% incidence) crossed to MHC congenic mice to produce H-2$^{d/b}$ or H-2$^{d/s}$ F_1 mice. The F_1 mice showed dramatic reduction in diabetes incidence. The F_1 mice, however, had significant insulitis, which resulted in diabetes following treatment with cyclophosphamide [88]. Thus, the large repertoire of transgenic autoreactive T cells could 'overcome' the protective MHC effect in the presence of an inciting event. Balasa et al. [89] found a similar effect in IL-10 transgenic mice: addition of an autoreactive transgenic TCR (BDC 2.5) in high numbers (increasing the precursor frequency of autoreactive T cells) overwhelmed the protective effect of both non-MHC background genes and MHC heterozygosity.

We proposed a model to explain the effect of MHC structure on thymic selection of the autoreactive T-cell repertoire, which may also explain the requirement of MHC homozygosity in the association of MHC with autoimmune diabetes [8, 78, 79]. In this model (a linear avidity model), the overall avidity (threshold) at which thymocytes are positively or negatively selected is considered a physiological constant and the effects of TCR avidity are inversely proportional to the total or effective concentration of the MHC:peptide complex. If overall MHC:T-cell avidity is proportional to the product of the concentration of MHC:peptide and the TCR affinity (i.e. (MHC:T-cell avidity) $\propto$ [MHC:peptide] $\times$ [TCR affinity]) (as suggested in several reports, [90–92]), then diminished I-A^{g7} self-peptide binding would lower the effective ligand dose of the MHC:peptide complex, necessitating a compensatory increase in TCR affinity to achieve an equivalent avidity threshold for positive (as well as negative) selection. Such a shift would result in the selection of a population of peripheral T cells with a high-affinity receptor repertoire for self; these T cells could potentially mediate autoreactivity. In other words, the instability of I-A^{g7} necessitates a correlative increase in mean population T-cell affinity for self-antigen to attain the (constant) cell signaling threshold necessary to induce thymic positive or negative selection [8, 78, 79]. The increase in mean population TCR affinity means that T cells which would normally be deleted by negative selection escape the thymus and enter the periphery as potentially autoreactive cells. This model explains the Wraith results [67, 68], as a special case where, rather than a global MHC instability (as seen in I-A^{g7}), a single peptide binds poorly to the MHC molecule and allows 'escape' from thymic negative selection of a limited repertoire of T cells with high affinity to the peptide (in that case, MBP 1-11). The findings of Hausmann et al. [73] modify the argument: not all peptides bind I-A^{g7} poorly, so a certain proportion of the NOD T-cell repertoire should react normally to foreign antigens (i.e. the whole T-cell population is not autoimmune). Indeed many reports have documented a spectrum of immune responses in the NOD mouse, including normal responses to some foreign antigens.

As discussed above, the finding of poor peptide binding by I-A^{g7} is not universally accepted. However, most reports confirm that I-A^{g7} is unstable under some conditions. Instability of the I-A^{g7} molecule would fit into the above avidity equation by reducing its half-life on the cell surface, thereby reducing the effective I-A^{g7} concentration in the thymus and necessitating higher affinity T cells to reach any given avidity threshold. It is less clear (to us) what the effect of promiscuous peptide binding would be upon the thymic selection of the peripheral repertoire.

The avidity interpretation of 'poor' peptide binding faces essentially the same criticism that the original 'good binding' paradigm did (see above), i.e., if disease-associated MHC molecules are 'good' binders of autoantigens, why do they not mediate negative selection of autoreactive T cells in the thymus? Conversely, if the disease-associated MHC molecules are 'poor' peptide binders, how can they effectively stimulate autoreactive T-cell responses in the periphery? Or, if the mechanism of defective thymic selection is due to structural instability of the class II molecule, should the same defect not apply in the periphery, preventing effective stimulation of the autoreactive T-cell repertoire? Indeed, multiple studies have suggested that even when high-affinity autoreactive T cells exist in the periphery of mice, tolerance still must be 'broken' to allow autoimmune sequelae [44, 93]. The observations by Hausmann et al. [73] suggest that inflammation (from infectious or other events that break tolerance), could alter the MHC structure compared to its thymic structure and lead to a different presentation profile of antigens in the periphery. Their data show that pH changes at the site of inflammation affect MHC structural stability, and/or change the proportion of 'empty' MHC molecules available to bind extracellular tissue antigens at sites of inflammation and or cell death. Combining these results and interpretations allows the possibility that a change in MHC structure or stability in the periphery could allow the intrinsically higher affinity TCR repertoire in NOD mice to interact with the self-antigen loaded MHC in a potentially destructive fashion, precipitating autoimmunity. Mice without disease-associated MHC alleles would not develop autoimmunity at the site of inflammation, since that MHC had not selected for a pre-existing high-affinity self-reactive (potentially autoimmune) T-cell repertoire in the thymus and therefore the available T-cell repertoire would possess an intrinsically lower affinity for self-antigens.

Conclusion

The above discussion suggests that: (1) There are many potential immunological pathways which could fit the 'genetic scaffold' in a model of NOD diabetes. (2) I-A^{g7} is probably one of the disease-associated genes within the Idd1 locus. (3) There is a superabundance of autoreactive T cells of many different specificities in NOD mice (and we do not know which autoantigen, if any, is critically important in an etiological sense). (4) Despite a large amount of knowledge of the structure and function of MHC class II molecules, the exact mechanism of the association of I-A^{g7} to the autoimmune process is still open to debate. Of the over 20 Idd loci identified by linkage to date, MHC class II is the only gene identified with even a moderate certainty. Forty years or so of study has generated thousands of publications concerning the structure and physiological function of MHC class II; 20 years of studying NOD mice has generated hundreds of papers in the NOD field concerning MHC-restricted T-cell responses and the structure:function of I-A^{g7}; yet there is still considerable controversy as to exactly how I-A^{g7} contributes to disease in NOD pathogenesis. Hence conclusions about etiological mechanism of most other cellular responses in NOD mice, or the role of other purported genetic loci, are even more likely to be preliminary and poorly understood. This should act as a cautionary tale in interpreting cellular responses in NOD mice. A variety of other cellular phenotypes considered important in NOD disease pathogenesis are not considered in this chapter, including: (1) role of specific cytokines and groups of cytokines in NOD pathogenesis; (2) role of specific APCs (e.g. macrophages and dendritic cells); (3) role of B cells; (4) role of 'regulatory' cells and cell subsets, and (5) role of target tissue polymorphisms and 'nonimmune' mechanisms. In considering these and other cellular responses, we would advise the following procedures: (1) Demonstrate that the proposed cellular response is connected to loci genetically linked to the disease pathogenesis. (2) Place less priority on reports of the form 'X prevents diabetes in NOD mice', since such reports, if not linked to a gene/genes directly implicated in the disease pathogenesis, likely have less relevance either to NOD pathogenesis or to the treatment of human disease (see fig. 1). (3) Entertain global schemes of NOD disease pathogenesis (such as presented in fig. 1) with skepticism. We are likely decades away from a convincing integrated view of the pathological genetic and cellular response cascade that results in autoimmune diabetes.

Acknowledgements

W.M. Ridgway is supported by the Juvenile Diabetes Foundation, a Pfizer Scholars Grant, and the Competitive Medical Research Fund of UPMC. Dr. Fathman is indebted to continuing support from the NIH, the JDFI and the ADA.

References

1   Leiter E, Atkinson M (eds): NOD Mice and Related Strains. Austin, Landes, 1998.
2   Rossini AA, Handler ES, Mordes JP, Greiner DL. Human autoimmune diabetes mellitus: Lessons from BB rats and NOD mice–Caveat emptor. Clin Immunol Immunopathol 1995;74:2–9.
3   Atkinson MA, Leiter EH: The NOD mouse model of type 1 diabetes: As good as it gets? Nat Med 1999;5:601–604.
4   Hoglund P, Mintern J, Waltzinger C, Heath W, Benoist C, Mathis D: Initiation of autoimmune diabetes by developmentally regulated presentation of islet cell antigens in the pancreatic lymph nodes. J Exp Med 1999;189:331–339.

5   Ridgway WM, Ito H, Fasso M, Yu C, Garrison-Fathman C: Analysis of the role of variation of major histocompatibility complex class II expression on nonobese diabetic (NOD) peripheral T cell response. J Exp Med 1998;188:2267–2275.

6   Ridgway WM, Fasso M, Fathman CG: A new look at MHC and autoimmune disease. Science 1999;284:749–751.

7   Serreze DV, Gaskins HR, Leiter EH. Defects in the differentiation and function of antigen presenting cells in NOD/Lt mice. J Immunol 1993;150:2534–2543.

8   Wong FS, Visintin I, Wen L, Flavell RA, Janeway CA Jr: CD8 T cell clones from young nonobese diabetic (NOD) islets can transfer rapid onset of diabetes in NOD mice in the absence of CD4 cells. J Exp Med 1996;183:67–76.

9   Scott B, Liblau R, Degermann S, Marconi LA, Ogata L, Caton AJ, McDevitt HO, Lo D: A role for non-MHC genetic polymorphism in susceptibility to spontaneous autoimmunity. Immunity 1994;1:73–83.

10 Fox CJ, Danska JS: Independent genetic regulation of T-cell and antigen-presenting cell participation in autoimmune islet inflammation. Diabetes 1998;47:331–338.

11 Serreze DV, Chapman HD, Varnum DS, Hanson MS, Reifsnyder PC, Richard SD, Fleming SA, Leiter EH, Shultz LD: B lymphocytes are essential for the initiation of T cell-mediated autoimmune diabetes: Analysis of a new 'speed congenic' stock of NOD.Ig mu null mice. J Exp Med 1996;184:2049–2053.

12 McAleer MA, Reifsnyder P, Palmer SM, Prochazka M, Love JM, Copeman JB, Powell EE, Rodrigues NR, Prins JB, Serreze DV et al: Crosses of NOD mice with the related NON strain. A polygenic model for IDDM. Diabetes 1995;44:1186–1195.

13 Sadelain MW, Qin HY, Lauzon J, Singh B: Prevention of type I diabetes in NOD mice by adjuvant immunotherapy. Diabetes 1990;39:583–589.

14 Shehadeh NN, LaRosa F, Lafferty KJ: Altered cytokine activity in adjuvant inhibition of autoimmune diabetes. J Autoimmun 1993;6:291–300.

15 Rabinovitch A: An update on cytokines in the pathogenesis of insulin-dependent diabetes mellitus. Diabetes Metab Rev 1998;14:129–151.

16 Miyazaki A, Hanafusa T, Yamada K, Miyagawa J, Fujino-Kurihara H, Nakajima H, Nonaka K, Tarui S: Predominance of T lymphocytes in pancreatic islets and spleen of pre-diabetic non-obese diabetic (NOD) mice: A longitudinal study. Clin Exp Immunol 1985;60:622–630.

17 Wicker LS, Miller BJ, Mullen Y: Transfer of autoimmune diabetes mellitus with splenocytes from nonobese diabetic (NOD) mice. Diabetes 1986;35:855–860.

18 Wang Y, Hao L, Gill RG, Lafferty KJ: Autoimmune diabetes in NOD mouse is L3T4 T-lymphocyte dependent. Diabetes 1987;36:535–538.

19 Koike T, Itoh Y, Ishii T, Ito I, Takabayashi K, Maruyama N, Tomioka H, Yoshida S: Preventive effect of monoclonal anti-L3T4 antibody on development of diabetes in NOD mice. Diabetes 1987;36:539–541.

20 Tisch R, Yang XD, Singer SM, Liblau RS, Fugger L, McDevitt HO: Immune response to glutamic acid decarboxylase correlates with insulitis in non-obese diabetic mice. Nature 1993;366:72–75.

21 Kaufman DL, Clare-Salzler M, Tian J, Forsthuber T, Ting GS, Robinson P, Atkinson MA, Sercarz EE, Tobin AJ, Lehmann PV: Spontaneous loss of T-cell tolerance to glutamic acid decarboxylase in murine insulin-dependent diabetes. Nature 1993;366:69–72.

22 Tisch R, Yang XD, Liblau RS, McDevitt HO: Administering glutamic acid decarboxylase to NOD mice prevents diabetes. J Autoimmun 1994;7:845–850.

23 Petersen JS, Karlsen AE, Markholst H, Worsaae A, Dyrberg T, Michelsen B: Neonatal tolerization with glutamic acid decarboxylase but not with bovine serum albumin delays the onset of diabetes in NOD mice. Diabetes 1994;43:1478–1484.

24 Elliott JF, Qin HY, Bhatti S, Smith DK, Singh RK, Dillon T, Lauzon J, Singh B: Immunization with the larger isoform of mouse glutamic acid decarboxylase (GAD67) prevents autoimmune diabetes in NOD mice. Diabetes 1994;43:1494–1499.

25 Tisch R, Liblau RS, Yang XD, Liblau P, McDevitt HO: Induction of GAD65-specific regulatory T-cells inhibits ongoing autoimmune diabetes in nonobese diabetic mice. Diabetes 1998;47: 894–899.

26 Tisch R, Wang B, Serreze DV: Induction of glutamic acid decarboxylase 65-specific Th2 cells and suppression of autoimmune diabetes at late stages of disease is epitope dependent. J Immunol 1999;163:1178–1187.

27 Tian J, Atkinson MA, Clare-Salzler M, Herschenfeld A, Forsthuber T, Lehmann PV, Kaufman DL: Nasal administration of glutamate decarboxylase (GAD65) peptides induces Th2 responses and prevents murine insulin-dependent diabetes. J Exp Med 1996;183:1561–1567.

28 Gerling IC, Atkinson MA, Leiter EH: The thymus as a site for evaluating the potency of candidate beta cell autoantigens in NOD mice. J Autoimmun 1994;7:851–858.

29 Schloot NC, Daniel D, Norbury-Glaser M, Wegmann DR: Peripheral T cell clones from NOD mice specific for GAD65 peptides: Lack of islet responsiveness or diabetogenicity. J Autoimmun 1996;9:357–363.

30 Christianson SW, Shultz LD, Leiter EH: Adoptive transfer of diabetes into immunodeficient NOD-scid/scid mice. Relative contributions of CD4+ and CD8+ T-cells from diabetic versus prediabetic NOD.NON-Thy-1a donors. Diabetes 1993;42:44–55.

31 Plesner A, Worsaae A, Dyrberg T, Gotfredsen C, Michelsen BK, Petersen JS: Immunization of diabetes-prone or non-diabetes-prone mice with GAD65 does not induce diabetes or islet cell pathology. J Autoimmun 1998;11:335–341.

32 Kim J, Richter W, Aanstoot HJ, Shi Y, Fu Q, Rajotte R, Warnock G, Baekkeskov S: Differential expression of GAD65 and GAD67 in human, rat, and mouse pancreatic islets. Diabetes 1993;42: 1799–1808.

33 Zekzer D, Wong FS, Ayalon O, Millet I, Altieri M, Shintani S, Solimena M, Sherwin RS: GAD-reactive CD4+ Th1 cells induce diabetes in NOD/SCID mice. J Clin Invest 1998;101:68–73.

34 Yoon JW, Yoon CS, Lim HW, Huang QQ, Kang Y, Pyun KH, Hirasawa K, Sherwin RS, Jun HS: Control of autoimmune diabetes in NOD mice by GAD expression or suppression in beta cells. Science 1999;284:1183–1187.

35 Kash SF, Condie BG, Baekkeskov S: Glutamate decarboxylase and GABA in pancreatic islets: Lessons from knock-out mice. Horm Metab Res 1999;31:340–344.

36 Wegmann DR, Gill RG, Norbury-Glaser M, Schloot N, Daniel D: Analysis of the spontaneous T cell response to insulin in NOD mice. J Autoimmun 1994;7:833–843.

37 Wong FS, Karttunen J, Dumont C, Wen L, Visintin I, Pilip IM, Shastri N, Pamer EG, Janeway CA Jr: Identification of an MHC class I-restricted autoantigen in type 1 diabetes by screening an organ-specific cDNA library. Nat Med 1999;5:1026–1031.

38 Daniel D, Gill RG, Schloot N, Wegmann D: Epitope specificity, cytokine production profile and diabetogenic activity of insulin-specific T cell clones isolated from NOD mice. Eur J Immunol 1995;25:1056–1062.

39 Haskins K, Wegmann D: Diabetogenic T-cell clones. Diabetes 1996;45:1299–1305.

40 Nepom GT: Glutamic acid decarboxylase and other autoantigens in IDDM. Curr Opin Immunol 1995;7:825–830.

41 Elias D, Reshef T, Birk OS, van der Zee R, Walker MD, Cohen IR: Vaccination against autoimmune mouse diabetes with a T-cell epitope of the human 65-kDa heat shock protein. Proc Natl Acad Sci USA 1991;88:3088–3091.

42 Elias D, Cohen IR: Peptide therapy for diabetes in NOD mice. Lancet 1994;343:704–706.

43 Funda DP, Hartoft-Nielsen ML, Kaas A, Buschard K: Effect of intrathymic administration of mycobacterial heat shock protein 65 and peptide p277 on the development of diabetes in NOD mice: Caution required in vaccination studies. Apmis 1998;106:1009–1016.

44 Ridgway WM, Fasso M, Lanctot A, Garvey C, Fathman CG: Breaking self-tolerance in nonobese diabetic mice. J Exp Med 1996;183:1657–1662.

45 Kanagawa O, Martin SM, Vaupel BA, Carrasco-Marin E, Unanue ER: Autoreactivity of T cells from nonobese diabetic mice: An I-Ag7- dependent reaction. Proc Natl Acad Sci USA 1998;95: 1721–1724.

46 Serreze D: The identity and ontogenic origins of autoreactive T lymphocytes in NOD mice; in Leiter E, Atkinson M (eds): NOD Mice and Related Strains. Austin, Landes, 1998, chapt 3, pp 71–101.

47 Sebzda E, Mariathasan S, Ohteki T, Jones R, Bachmann MF, Ohashi PS: Selection of the T cell repertoire. Annu Rev Immunol 1999;17:829–874.

48   Benoist C, Mathis D: Positive selection of T cells: Fastidious or promiscuous? Curr Opin Immunol 1997;9:245–249.

49   Jameson SC, Bevan MJ: T-cell selection. Curr Opin Immunol 1998;10:214–219.

50   Marrack P, Kappler J: Positive selection of thymocytes bearing alpha beta T cell receptors. Curr Opin Immunol 1997;9:250–255.

51   Leiter E: Genetics and immunogenetics of NOD mice and related strains; in Leiter E, Atkinson M (eds): NOD Mice and Related Strains. Austin, Landes, 1998, chapt 2, pp 37–71.

52   Hattori M, Yamato E, Itoh N, Senpuku H, Fujisawa T, Yoshino M, Fukuda M, Matsumoto E, Toyonaga T, Nakagawa I, Petruzzelli M, McMurray A, Weiner H, Sagai T, Moriwaki K, Shiroishi T, Maron R, Lund T: Cutting edge: Homologous recombination of the MHC class I K region defines new MHC-linked diabetogenic susceptibility gene(s) in nonobese diabetic mice. J Immunol 1999;163:1721–1724.

53   Wicker LS, Miller BJ, Coker LZ, McNally SE, Scott S, Mullen Y, Appel MC: Genetic control of diabetes and insulitis in the nonobese diabetic (NOD) mouse. J Exp Med 1987;165:1639–1654.

54   Nishimoto H, Kikutani H, Yamamura K, Kishimoto T: Prevention of autoimmune insulitis by expression of I-E molecules in NOD mice. Nature 1987;328:432–434.

55   Slattery RM, Kjer-Nielsen L, Allison J, Charlton B, Mandel TE, Miller JF: Prevention of diabetes in non-obese diabetic I-Ak transgenic mice. Nature 1990;345:724–726.

56   Lund T, O'Reilly L, Hutchings P, Kanagawa O, Simpson E, Gravely R, Chandler P, Dyson J, Picard JK, Edwards A, et al: Prevention of insulin-dependent diabetes mellitus in non-obese diabetic mice by transgenes encoding modified I-A beta-chain or normal I-E alpha-chain. Nature 1990;345:727–729.

57   Acha-Orbea H, McDevitt HO: The first external domain of the nonobese diabetic mouse class II I-A beta chain is unique. Proc Natl Acad Sci USA 1987;84:2435–2439.

58   Todd JA, Bell JI, McDevitt HO: HLA-DQ beta gene contributes to susceptibility and resistance to insulin-dependent diabetes mellitus. Nature 1987;329:599–604.

59   Quartey-Papafio R, Lund T, Chandler P, Picard J, Ozegbe P, Day S, Hutchings PR, O'Reilly L, Kioussis D, Simpson E, et al: Aspartate at position 57 of nonobese diabetic I-Ag7 beta-chain diminishes the spontaneous incidence of insulin-dependent diabetes mellitus. J Immunol 1995; 154:5567–5575.

60   Singer SM, Tisch R, Yang XD, Sytwu HK, Liblau R, McDevitt HO: Prevention of diabetes in NOD mice by a mutated I-Ab transgene. Diabetes 1998;47:1570–1577.

61   Reich EP, Sherwin RS, Kanagawa O, Janeway CA Jr: An explanation for the protective effect of the MHC class II I-E molecule in murine diabetes. Nature 1989;341:326–328.

62   Bohme J, Schuhbaur B, Kanagawa O, Benoist C, Mathis D: MHC-linked protection from diabetes dissociated from clonal deletion of T cells. Science 1990;249:293–295.

63   Vaysburd M, Lock C, McDevitt H: Prevention of insulin-dependent diabetes mellitus in nonobese diabetic mice by immunogenic but not by tolerated peptides. J Exp Med 1995;182:897–902.

64   Wucherpfennig KW, Strominger JL: Selective binding of self peptides to disease-associated major histocompatibility complex (MHC) molecules: A mechanism for MHC-linked susceptibility to human autoimmune diseases. J Exp Med 1995;181:1597–1601.

65   Benacerraf B, McDevitt HO: Histocompatibility-linked immune response genes. Science 1972; 175:273–279.

66   Hanahan D: Peripheral-antigen-expressing cells in thymic medulla: Factors in self-tolerance and autoimmunity. Curr Opin Immunol 1998;10:656–662.

67   Fairchild PJ, Wildgoose R, Atherton E, Webb S, Wraith DC: An autoantigenic T cell epitope forms unstable complexes with class II MHC: A novel route for escape from tolerance induction. Int Immunol 1993;5:1151–1158.

68   Liu GY, Fairchild PJ, Smith RM, Prowle JR, Kioussis D, Wraith DC: Low avidity recognition of self-antigen by T cells permits escape from central tolerance. Immunity 1995;3:407–415.

69   Reich EP, von Grafenstein H, Barlow A, Swenson KE, Williams K, Janeway CA Jr: Self peptides isolated from MHC glycoproteins of non-obese diabetic mice. J Immunol 1994;152:2279–2288.

70   Reizis B, Eisenstein M, Bockova J, Konen-Waisman S, Mor F, Elias D, Cohen IR: Molecular characterization of the diabetes-associated mouse MHC class II protein, I-Ag7. Int Immunol 1997;9:43–51.

71 Harrison LC, Honeyman MC, Trembleau S, Gregori S, Gallazzi F, Augstein P, Brusic V, Hammer J, Adorini L: A peptide-binding motif for I-A(g7), the class II major histocompatibility complex (MHC) molecule of NOD and Biozzi AB/H mice. J Exp Med 1997;185:1013–1021.

72 Peterson M, Sant AJ: The inability of the nonobese diabetic class II molecule to form stable peptide complexes does not reflect a failure to interact productively with DM. J Immunol 1998; 161:2961–2967.

73 Hausmann DH, Yu B, Hausmann S, Wucherpfennig KW: pH-dependent peptide binding properties of the type I diabetes-associated I-Ag7 molecule: Rapid release of CLIP at an endosomal pH. J Exp Med 1999;189:1723–1734.

74 Gregori S, Bono E, Gallazzi F, Hammer J, Harrison LC, Adorini L: The motif for peptide binding to the insulin-dependent diabetes mellitus-associated class II MHC molecule I-Ag7 validated by phage display library. Int Immunol 2000;12:493–503.

75 Carrasco-Marin E, Shimizu J, Kanagawa O, Unanue ER: The class II MHC I-Ag7 molecules from non-obese diabetic mice are poor peptide binders. J Immunol 1996;156:450–458.

76 Carrasco-Marin E, Kanagawa O, Unanue ER: The lack of consensus for I-A(g7)-peptide binding motifs: Is there a requirement for anchor amino acid side chains? Proc Natl Acad Sci USA 1999;96:8621–8626.

77 Corper AL, Stratmann T, Apostolopoulos V, Scott CA, Garcia KC, Kang AS, Wilson IA, Teyton L: A structural framework for deciphering the link between I-Ag7 and autoimmune diabetes. Science 2000;288:505–511.

78 Ridgway WM, Fathman CG: MHC structure and autoimmune T cell repertoire development. Curr Opin Immunol 1999;11:638–642.

79 Ridgway WM, Fathman CG: The association of MHC with autoimmune diseases: Understanding the pathogenesis of autoimmune diabetes. Clin Immunol Immunopathol 1998;86:3–10.

80 Schmidt D, Verdaguer J, Averill N, Santamaria P: A mechanism for the major histocompatibility complex-linked resistance to autoimmunity. J Exp Med 1997;186:1059–1075.

81 Schmidt D, Amrani A, Verdaguer J, Bou S, Santamaria P: Autoantigen-independent deletion of diabetogenic CD4+ thymocytes by protective MHC class II molecules. J Immunol 1999;162: 4627–4636.

82 Laufer TM, DeKoning J, Markowitz JS, Lo D, Glimcher LH: Unopposed positive selection and autoreactivity in mice expressing class II MHC only on thymic cortex. Nature 1996;383:81–85.

83 Fung-Leung WP, Surh CD, Liljedahl M, Pang J, Leturcq D, Peterson PA, Webb SR, Karlsson L: Antigen presentation and T cell development in H2-M-deficient mice. Science 1996;271: 1278–1281.

84 Martin WD, Hicks GG, Mendiratta SK, Leva HI, Ruley HE, Van Kaer L: H2-M mutant mice are defective in the peptide loading of class II molecules, antigen presentation, and T cell repertoire selection. Cell 1996;84:543–550.

85 Miyazaki T, Wolf P, Tourne S, Waltzinger C, Dierich A, Barois N, Ploegh H, Benoist C, Mathis D: Mice lacking H2-M complexes, enigmatic elements of the MHC class II peptide-loading pathway. Cell 1996;84:531–541.

86 Ignatowicz L, Kappler J, Marrack P: The repertoire of T cells shaped by a single MHC/peptide ligand. Cell 1996;84:521–529.

87 Fukui Y, Ishimoto T, Utsuyama M, Gyotoku T, Koga T, Nakao K, Hirokawa K, Katsuki M, Sasazuki T: Positive and negative CD4+ thymocyte selection by a single MHC class II/peptide ligand affected by its expression level in the thymus. Immunity 1997;6:401–410.

88 Wu AY, Schulman SJ, Marconi LA, Reilly CR, Scott B, Lo D: Protection against diabetes by MHC heterozygosity and reversal by cyclophosphamide. Cell Immunol 1998;184:112–120.

89 Balasa B, Lee J, Sarvetnick N: Differential impact of T cell repertoire diversity in diabetes-prone or -resistant IL-10 transgenic mice. Cell Immunol 1999;193:170–178.

90 Ashton-Rickardt PG, Tonegawa S: A differential-avidity model for T-cell selection. Immunol Today 1994;15:362–366.

91 Ashton-Rickardt PG, Bandeira A, Delaney JR, Van Kaer L, Pircher HP, Zinkernagel RM, Tonegawa S: Evidence for a differential avidity model of T cell selection in the thymus. Cell 1994; 76:651–663.

92   Kim DT, Rothbard JB, Bloom DD, Fathman CG: Quantitative analysis of T cell activation: Role of
      TCR/ligand density and TCR affinity. J Immunol 1996;156:2737–2742.
93   Ohashi PS, Oehen S, Buerki K, Pircher H, Ohashi CT, Odermatt B, Malissen B, Zinkernagel RM,
      Hengartner H: Ablation of 'tolerance' and induction of diabetes by virus infection in viral antigen
      transgenic mice. Cell 1991;65:305–317.

C. Garrison Fathman, MD, Stanford University School of Medicine,
Department of Medicine, Division of Immunology & Rheumatology,
300 Pasteur Drive, CCSR Building, Room 2225, Stanford, CA 94305-5166 (USA)
Tel. +1 650 723 7887, Fax +1 650 725 1958, E-Mail cfathman@leland.stanford.edu

von Herrath MG (ed.) Molecular Pathology of Type 1 Diabetes mellitus.
Curr Dir Autoimmun. Basel, Karger, 2001, vol 4, pp 239–251

Human T-Cell Responses to Islet Cell Antigens

Grete Sønderstrup[a], *Ivana Durinovich-Belló*[b]

[a] Department of Microbiology and Immunology, Stanford University,
Stanford, Calif., USA and
[b] Department of Internal Medicine, University of Ulm, Germany

Juvenile diabetes mellitus (type 1 diabetes, T1D) was previously considered an endocrine disease with unknown multigenic inheritance pattern. Our understanding of T1D extended rapidly after it was reported that there was a strong genetic association between susceptibility to T1D and particular HLA, human major histocompatibility (MHC), class II alleles [1, 2]. This knowledge was further advanced by the demonstration of β-islet cell-specific autoantibodies in the serum from patients with diabetes, indicating that T1D was indeed a genetic disease sharing features with a number of tissue-specific autoimmune diseases [3]. MHC class II molecules function by selecting and presenting immunogenic peptide epitopes from self and foreign protein antigens to the immune system. MHC class II molecules also determine the positive and negative selection processes shaping the CD4+ T-cell receptor (TCR) repertoire released from the thymus to the peripheral lymphoid tissues. Thus, a mechanism involving immune responses to tissue-specific self antigens would provide a sensible explanation for HLA class II-associated disease susceptibility in organ-specific autoimmune diseases in general [4, 5].

In T1D this hypothesis was further supported by the dramatically increased risk of developing T1D in apparently healthy humans carrying disease-associated HLA class II alleles if they were also positive for certain combinations of β-islet cell-specific autoantibodies [6–9]. Moreover, these islet-specific human autoantibodies included different immunoglobulin subtypes indicating that they were in fact produced with the support of distinct types of CD4+ T-helper cells, presumably specific for the same autoantigens [10]. T-helper-2-related autoantibodies of distinct IgG subclasses have been found to be pathogenic in pemphigus vulgaris and bullous pemphigoid in humans [11, 12]. Pathogenic Th2 supported autoantibodies specific for glutamic acid decarboxylase (GAD) 65 have also been found in stiff man's syndrome (SMS) in humans [13]. Yet, the autoantibodies in SMS seem to be different from the GAD65 autoantibodies seen in T1D patient, since the GAD65 antibodies in the two diseases generally are of different immunoglobulin classes or IgG isotypes [14]. In spite of the fact that islet cell-specific autoantibodies are definite markers of ongoing self-reactivity in humans, it has been difficult to link these antibodies to a distinct pathogenic function in T1D in humans. However, there has been strong evidence particularly from studies in animal models of autoimmune diabetes, that islet cell-specific T-cell responses are very important for the ongoing β-cell destruction in the prediabetic period [15, 16].

Islet-Specific CD4+ T-Cell Responses in Humans

Since the incidence of T1D for several years has been on a rise in developed countries, there is an increasing interest in devising nontoxic methods of preventing the islet cell destruction in prediabetic individuals, or protecting transplanted islet cells from destruction either by transplant rejection or recurrence of islet cell-specific autoimmune responses [17, 18]. Some of these efforts have focussed on development of approaches analogous to a protective vaccine directed at elimination of pathogenic islet cell-specific T-cell responses either during the prediabetic period, or after allogeneic islet cell transplantation. Other efforts have been centered around more generalized immunosuppressive approaches using antibodies directed against surface molecules expressed on T lymphocytes or other lymphoid cell types present in the affected β-islets [19–21]. One point these different approaches have in common has been a need for monitoring the status of the autoimmune process in the human subjects/patients under treatment, because an end point of diabetes as the only measure would not be satisfactory. Thus, there is a real need for development of reliable and informative monitoring systems of islet cell-specific autoimmunity in human subjects participating in these trials.

Pancreas biopsies have been performed in a few diabetes patients in Japan [22]. However, from a review of the literature on islet cell-specific T-cell responses in T1D there is a consensus for attempts using in vitro testing of peripheral blood mononuclear cells (BMNC) for evaluation of islet cell autoimmunity in humans.

Human β-Islet Cell Antigens

A summary of 10–20 years of studies in T1D and controls, also supported by experimental data in the nonobese diabetic (NOD) mouse strain, indicates that preproinsulin or its maturation products, proinsulin, and insulin are important candidate autoantigens with potentially pathogenic function in T1D [23–25]. Since combinations of autoantibodies against insulin, GAD65, and the islet cell tyrosine phosphatase, IA-2, have been shown to have strong prognostic value in predicting progression to overt diabetes in clinically healthy humans with disease-susceptible HLA alleles, these islet cell proteins presumably also play a role in the pathogenesis of T1D [6–9].

Other less studied islet cell proteins to consider are the 38-kD autoantigen, the p69 autoantigen, that carries the ABBOS peptide shared with the bovine casein protein, and carboxypeptidase H, that is one of the enzymes involved in insulin processing [26–28]. T cells with specific reactivity to a number of other less well characterized islet cell proteins from rodent or human insulinoma cell lines [29–33] or to the heat-shock protein 60, a non-β-cell-specific protein autoantigen, have been found in humans as well as in the NOD mouse [34, 35].

T-Cell Proliferation

Several review articles on human autoimmune T-cell responses have attempted to draw general conclusions from studies of CD4+ T-cell responses of human BMNC to islet cell antigens [36–38]. However, the experimental methodologies used in the past have often been unique to the individual study, and the interpretation of positive and negative responses in these assays has therefore been difficult. Most of the studies have included healthy human control subjects with a variety of diabetes-associated or diabetes-protective HLA alleles from nondiabetic control populations. Consequently, we must assess studies using unfractionated BMNC [39], studies using fractionated human CD4+ memory T cells [40], or studies focusing on islet cell-specific T-cell clones [41–43] in attempting to define common denominators. These reports generally studied human T-cell proliferative responses to recombinant insulin, insulin precursors or insulin sequence-derived peptides [39, 42, 44–51], GAD67 [52], GAD65 [53–57], and IA-2 [51, 58–62] derived peptides. Nevertheless, it has been a general conclusion that islet cell-specific in vitro T-cell responses against these human self proteins (a sign of previous activation of antigen-specific T cells) could be found both in humans with diabetes as well as in healthy nondiabetic control individuals.

Moreover, it has not been possible to establish any correlation between islet cell-specific T-cell proliferation and autoantibody production [46, 58, 63].

Similarly, a significant epitope-specific cytokine response, determined as secreted protein, will often not be accompanied by antigen-specific T-cell proliferation [64]. Some of these discrepancies may in fact be due to islet cell protein-specific 'anergy' as described by Dosch and co-workers [60, 65, 66] allowing for cytokine production, but not for T-cell proliferation. Studies of IL-4 production that at first seemed promising in T1D patients [67] turned out to be problematic [68]. However, solving the technical aspects of cytokine measurements, for example by using RT-PCR or Elispot at the single cell level, measuring intracellular RNA and intracellular protein respectively, is unlikely to solve the discrepancies between T-cell proliferation and cytokine responsiveness [69–72].

Mimicry and Prediction of T-Cell Epitopes

There has been an extensive focus on linking T1D to environmental triggering event such as neonatal intake of cow's milk protein [73, 74] or previous viral infection. Several so-called mimicry epitopes shared between human islet cell proteins and certain viral proteins (GAD65 and both rubella virus and coxsackie virus, or IA-2 and rotavirus) have been suggested to play a role in T1D development [75–77]. However, it is not clear what role, if any, these mimicry epitopes play in diabetes development, because such potentially cross-reactive T cells stimulated by the same or a similar peptide epitope are commonly found in normal T-cell repertoires [38].

One of the problems in studying human T-cell responses has been to produce high-quality recombinant antigen preparations in sufficient amounts. Consequently, there have been major efforts aimed at predicting the amino acid sequences of the important immunogenic T-cell epitopes of these proteins in order to use synthetic peptides instead. Many different approaches have been tried, including computer prediction using motif searches, peptide binding to soluble HLA molecules, or elution of peptides from purified HLA/peptide complexes from, for example, IA-2 protein-loaded human B-lymphoblastoid cell lines [78–81]. The last method [81] is excellent for dissection of already known immunogenic T-cell epitopes, where it offers a special opportunity to design the best naturally occurring peptides with optimal amino acid length, which may be extremely important for immunogenicity and function.

Asking a similar question, namely, what are the immunogenic T-cell epitopes of β-islet cell proteins of importance for autoimmunity and pathology in T1D, we undertook the development of an in vivo transgenic mouse model which would function as a substitute for the CD4+ compartment of a human immune system [82]. We first concentrated on the common HLA-DRB1*0401 allele, and later the HLA-DQB1*0301/DQB1*0302 (DQ8) allele, which are both together on a prevalent HLA haplotype that is highly increased in Caucasian T1D patients [83]. We have utilized these mice to determine the immunogenic T-cell epitopes of recombinant human GAD65 and preproinsulin as found after immunization in incomplete Freund's adjuvant, using in vitro culture conditions for T-helper cells [84–86]. Using these immunogenic peptide epitopes determined in HLA-DR4 or DQ8 transgenic mice, we have recently completed a series of T-cell proliferation studies in 48 newly diagnosed T1D patients and 32 healthy control subjects. This primarily pediatric patient group had a high frequency of HLA-DR4, DQ8 alleles [J. Olson and G. Sønderstrup, unpubl. data], and the patients were shown to respond in an HLA-restricted manner to the same epitopes as determined in the DR4 and DQ8 transgenic mice. We essentially confirmed what had also been shown by others, that both an individual diabetes patient and an individual healthy person with the appropriate HLA genotype would respond to a selection of these GAD65 and pre-proinsulin peptides with indistinguishable response patterns. This study also confirmed the common discrepancy between peptide-specific T-cell proliferation and cytokine responses.

The HLA transgenic mouse model offers the possibility also to study CD4+ T-cell immune responses to human islet cell proteins in the setting of HLA class II alleles which have a protective effect against development of T1D in humans [87, 88]. We have focussed on the strongly diabetes-protective DRB1*0403 allele that differs by only 3 amino acids from its close relative, the diabetes-susceptible HLA-DRB1*0405 allele, which lacks aspartic acid in position 57 of the DRB1 chain, similar to DQ8 [88–92]. During comparative studies of the TCR repertoires in these two HLA-DR4 transgenic lines, we found two rather diverse sets of immunogenic T-cell epitopes of human GAD65 [S. Parry and G. Sønderstrup, unpubl. data]. However, after crossing these two DR4 transgenic lines, the heterozygous DRB1*0405/ DRB1*0403 transgenic mice had practically excluded the T cells responding to the two major immunogenic DRB1*0405-restricted T-cell epitopes [S. Parry and G. Sønderstrup, unpubl. data]. From studies using human antigen-presenting cells (APCs) for presentation of GAD65 protein to T-cell hybridomas responding to each of these two immunodominant DRB1*0405 epitopes, we know that this is not a transgenic artifact. Since human DRB1*0405/DRB1*0403 heterozygous APCs are equal to other DRB1*0405-positive APCs in presenting synthetic peptides to these same T-cell hybridomas, we believe that this is a process operating at the intracellular level [S. Parry and G. Sønderstrup, unpubl. data]. Thus, there seems to be ample possibilities for interaction and competition between a normal person's 6–10 different HLA class II molecules. We are pursuing this subject by studying peptide/MHC binding under physical conditions mimicking either surface conditions at pH 7 or endosomal conditions at pH 5.5 +/− soluble human HLA-DM, which is a molecule that normally edits peptide loading in the endosomal compartment [F. Hall and G. Sønderstrup, submitted].

Mechanisms

What is the significance of these antigen/peptide-specific T-cell responses if they are found in both diabetes patients and healthy control subjects, and are therefore not disease-specific? Why are these T-cell responses given so much attention?

Regulatory T (Tr) Cells

It is not a new idea that autoimmune destruction of β-islet cells in T1D in humans may be based on an imbalance between autoreactive β-islet protein-specific effector T cells and a regulatory T-cell population that is capable of controlling these potentially pathogenic T cells [38, 40, 93]. There is also ample experimental evidence from NOD diabetes supporting this notion [94–100].

T-Helper (Th) Cells

The CD4+ T-cell repertoire is generally considered to be tolerant to self proteins in normal healthy individuals. This self-tolerance is attained through positive and negative selection of the T lymphocytes as they mature in the thymus. Thus, autoreactive T cells are supposed to be eliminated by means of central deletion before they leave the thymus. However, if potentially autoreactive T cells escape from the thymus, there is a second level of protection against self-reactive T cells in the periphery. Immune tolerance is an active process constantly sustained by several different mechanisms including, deletion by apoptosis, anergy induction by regulatory T cells, or Th1 to Th2 deviation. We need to accept that the state of self-tolerance is a dynamic process that repeatedly will be challenged.

Past studies in T1D have almost exclusively focussed either on the very heterogeneous peripheral blood mononuclear cell population or on the whole T-helper cell population. Our most recent study of enriched CD45RO-memory and CD45RA-naïve/regulatory T-helper cell populations, of DRB1*0401, DQB1*0302-positive individuals, revealed distinguished recognition of dominant preproinsulin epitopes by autoantibody-positive individuals with high risk for T1D in comparison to the controls or T1D patients. This epitope could be distinguished by both peptide-specific T-cell proliferation and cytokine responses [I. Durinovic-Bello et al., submitted]. At this point we do not have available a set of test systems that can provide an adequate assessment of a persons β-islet cell autoimmunity. Recently, the T Cell Workshop organized by the Immunology of Diabetes Society initiated a concerted effort to standardize testing methods for human T-cell assays including attempts to provide standardized recombinant human islet cell antigen preparations [101]. The results of this worldwide cooperative venture are not yet available, but carry promises for significant future advances in this area.

Future Goals

First will be to standardize the T-cell purification methods and culture systems and organize standardized recombinant islet cell proteins for the T-cell proliferation assays. Addition of IL-2 to the cultures may be used to circumvent potential anergy of T cells. It is unclear which cytokine assays will be the most informative, and until more knowledge is obtained, combinations of a number of different assays should be tested in parallel. These assays may include combinations of cytokine protein measurements by enzyme-linked immunostaining (ELISA) assay, reverse transcriptase-polymerase chain reaction (RT-PCR) or real-time PCR measuring the dynamic development of different cytokine mRNAs over time, enzyme-linked immunospot (Elispot) assay, or other single cell assays using labeling of intracellular cytokine after permeabilization of the T cells [71, 100, 102]. Similarly, enumeration of memory T cells or virgin T cells should be done using FACS techniques after labeling with fluorescent antibodies against CD45RB, CD25 or other surface markers of interest [100]. Lastly, it may also be helpful to determine the numbers of islet antigen-specific CD4+ T cells using MHC/peptide tetramers [103].

Conclusion

Both T1D patients and healthy nondiabetic control individuals may have proliferative T-cell responses to a variety of β-islet cell proteins as a sign of previous antigen exposure. However, these potentially autoreactive T cells are not necessarily pathogenic [38]. It is our hypothesis that the autoimmune pathology in T1D partly is due to a relative deficiency in the function of regulatory T cells of the patient. In addition, it is also part of our working hypothesis that these regulatory T cells use a different TCR repertoire and therefore recognize a different set of immunogenic peptide epitopes from the islet cell autoantigens. Regulatory T cells have a unique profile of cytokine production and make high levels of IL-10 and TGF-β, but no IL-4 or IL-2 [104]. Regulatory/suppressor

T cells also respond to a much lower dosage of specific antigen compared to T-helper cells [105]. Consequently, it is not to be expected that these T cells will be able to expand in conventional tissue culture systems that are designed for supporting T-helper cells. New culture conditions designed for survival and expansion of the regulatory T-cell subsets need to be developed in order to get a possibility to monitor these T cells.

References

1 Nerup J, Platz P, Andersen OO, Christy M, Lyngsoe J, Poulsen JE, Ryder LP, Nielsen LS, Thomsen M, Svejgaard A: HL-A antigens and diabetes mellitus. Lancet 1974;ii:864–866.
2 Todd JA, Bell JI, McDevitt HO: HLA antigens and insulin-dependent diabetes. Nature 1988; 333:710.
3 Bottazzo GF, Florin-Christensen A, Doniach D: Islet-cell antibodies in diabetes mellitus with autoimmune polyendocrine deficiencies. Lancet 1974;ii:1279–1283.
4 Todd JA, Acha-Orbea H, Bell JI, Chao N, Fronek Z, Jacob CO, McDermott M, Sinha AA, Timmerman L, Steinman L, McDevitt HO: A molecular basis for MHC class II-associated autoimmunity. Science 1988;240:1003–1009.
5 Wucherpfennig KW, Strominger JL: Selective binding of self peptides to disease-associated major histocompatibility complex (MHC) molecules: A mechanism for MHC-linked susceptibility to human autoimmune diseases. J Exp Med 1995;181:1597–1601.
6 Verge CF, Gianani R, Kawasaki E, Yu L, Pietropaolo M, Chase HP, Eisenbarth GS: Number of autoantibodies (against insulin, GAD or ICA512/IA2) rather than particular autoantibody specificities determines risk of type I diabetes. J Autoimmun 1996;9:379–383.
7 Fuchtenbusch M, Ferber K, Standl E, Ziegler AG: Prediction of type 1 diabetes postpartum in patients with gestational diabetes mellitus by combined islet cell autoantibody screening: A prospective multicenter study. Diabetes 1997;46:1459–1467.
8 Verge CF, Stenger D, Bonifacio E, Colman PG, Pilcher C, Bingley PJ, Eisenbarth GS: Combined use of autoantibodies (IA-2 autoantibody, GAD autoantibody, insulin autoantibody, cytoplasmic islet cell antibodies) in type 1 diabetes: Combinatorial Islet Autoantibody Workshop. Diabetes 1998;47:1857–1866.
9 Maclaren N, Lan M, Coutant R, Schatz D, Silverstein J, Muir A, Clare-Salzer M, She JX, Malone J, Crockett S, Schwartz S, Quattrin T, DeSilva M, Vander Vegt P, Notkins A, Krischer J: Only multiple autoantibodies to islet cells (ICA), insulin, GAD65, IA-2 and IA-2beta predict immune-mediated (type 1) diabetes in relatives. J Autoimmun 1999;12:279–287.
10 Bonifacio E, Scirpoli M, Kredel K, Fuchtenbusch M, Ziegler AG: Early autoantibody responses in prediabetes are IgG1 dominated and suggest antigen-specific regulation. J Immunol 1999;163: 525–532.
11 Bhol K, Natarajan K, Nagarwalla N, Mohimen A, Aoki V, Ahmed AR: Correlation of peptide specificity and IgG subclass with pathogenic and nonpathogenic autoantibodies in pemphigus vulgaris: A model for autoimmunity. Proc Natl Acad Sci USA 1995;92:5239–5243.
12 Budinger L, Borradori L, Yee C, Eming R, Ferencik S, Grosse-Wilde H, Merk HF, Yancey K, Hertl M: Identification and characterization of autoreactive T cell responses to bullous pemphigoid antigen 2 in patients and healthy controls. J Clin Invest 1998;102:2082–2089.
13 Hummel M, Durinovic-Bello I, Bonifacio E, Lampasona V, Endl J, Fessele S, Then Bergh F, Trenkwalder C, Standl E, Ziegler AG: Humoral and cellular immune parameters before and during immunosuppressive therapy of a patient with stiff-man syndrome and insulin-dependent diabetes mellitus. J Neurol Neurosurg Psychiatry 1998;65:204–208.
14 Lohmann T, Hawa M, Leslie RD, Lane R, Picard J, Londei M: Immune reactivity to glutamic acid decarboxylase 65 in stiff-man syndrome and type 1 diabetes mellitus. Lancet 2000;i:31–35.

15 Haskins K, McDuffie M: Acceleration of diabetes in young NOD mice with a CD4+ islet-specific T cell clone. Science 1990;249:1433–1436.

16 Hoglund P, Mintern J, Waltzinger C, Heath W, Benoist C, Mathis D: Initiation of autoimmune diabetes by developmentally regulated presentation of islet cell antigens in the pancreatic lymph nodes. J Exp Med 1999;189:331–339.

17 Braghi S, Bonifacio E, Secchi A, Di Carlo V, Pozza G, Bosi E: Modulation of humoral islet autoimmunity by pancreas allotransplantation influences allograft outcome in patients with type 1 diabetes. Diabetes 2000;49:218–224.

18 Brooks-Worrell BM, Peterson KP, Peterson CM, Palmer JP, Jovanovic L: Reactivation of type 1 diabetes in patients receiving human fetal pancreatic tissue transplants without immunosuppression. Transplantation 2000;69:1824–1829.

19 Hahn HJ, Kuttler B, Laube F, Emmrich F: Anti-CD4 therapy in recent-onset IDDM. Diabetes Metab Rev 1993;9:323–328.

20 Koulmanda M, Mandel TE: Effect of anti-CD4 and anti-ICAM MAb on survival of fetal pig pancreas grafts in NOD mice. Transplant Proc 1994;26:3466.

21 Lenschow DJ, Ho SC, Sattar H, Rhee L, Gray G, Nabavi N, Herold KC, Bluestone JA: Differential effects of anti-B7-1 and anti-B7-2 monoclonal antibody treatment on the development of diabetes in the nonobese diabetic mouse. J Exp Med 1995;181:1145–1155.

22 Parker DC, Greiner DL, Phillips NE, Appel MC, Steele AW, Durie FH, Noelle RJ, Mordes JP, Rossini AA: Survival of mouse pancreatic islet allografts in recipients treated with allogeneic small lymphocytes and antibody to CD40 ligand. Proc Natl Acad Sci USA 1995;92:9560–9564.

23 Levisetti MG, Padrid PA, Szot GL, Mittal N, Meehan SM, Wardrip CL, Gray GS, Bruce DS, Thistlethwaite JR Jr, Bluestone JA: Immunosuppressive effects of human CTLA4Ig in a non-human primate model of allogeneic pancreatic islet transplantation. J Immunol 1997;159: 5187–5191.

24 Shimada A, Imazu Y, Morinaga S, Funae O, Kasuga A, Atsumi Y: T-cell insulitis found in anti-GAD65+ diabetes with residual beta-cell function. A case report. Diabetes Care 1999;22:615–617.

25 Daniel D, Gill RG, Schloot N, Wegmann D: Epitope specificity, cytokine production profile and diabetogenic activity of insulin-specific T cell clones isolated from NOD mice. Eur J Immunol 1995;25:1056–1062.

26 Arden SD, Roep BO, Neophytou PI, Usac EF, Duinkerken G, de Vries RR, Hutton JC: Imogen 38: A novel 38-kD islet mitochondrial autoantigen recognized by T cells from a newly diagnosed type 1 diabetic patient. J Clin Invest 1996;97:551–561.

27 Gaedigk R, Karges W, Hui MF, Scherer SW, Dosch HM: Genomic organization and transcript analysis of ICAp69, a target antigen in diabetic autoimmunity. Genomics 1996;38:382–391.

28 Alcalde L, Tonacchera M, Costagliola S, Jaraquemada D, Pujol-Borrell R, Ludgate M: Cloning of candidate autoantigen carboxypeptidase H from a human islet library: Sequence identity with human brain CPH. J Autoimmun 1996;9:525–528.

29 Van Vliet E, Roep BO, Meulenbroek L, Bruining GJ, De Vries RR: Human T-cell clones with specificity for insulinoma cell antigens. Eur J Immunol 1989;19:213–216.

30 Monetini L, Cavallo MG, Barone F, Valente L, Russo M, Walker B, Thorpe R, Pozzilli P: T cell reactivity to human insulinoma cell line (CM) antigens in patients with type 1 diabetes. Autoimmunity 1999;29:171–177.

31 Kasimiotis H, Myers MA, Argentaro A, Mertin S, Fida S, Ferraro T, Olsson J, Rowley MJ, Harley VR: Sex-determining region Y-related protein SOX13 is a diabetes autoantigen expressed in pancreatic islets. Diabetes 2000;49:555–561.

32 Tree TI, O'Byrne D, Tremble JM, MacFarlane WM, Haskins K, James RF, Docherty K, Hutton JC, Banga JP: Evidence for recognition of novel islet T cell antigens by granule-specific T cell lines from new onset type 1 diabetic patients. Clin Exp Immunol 2000;121:100–105.

33 Cohen IR: The Th1/Th2 dichotomy, hsp60 autoimmunity, and type I diabetes. Clin Immunol Immunopathol 1997;84:103–106.

34 Abulafia-Lapid R, Elias D, Raz I, Keren-Zur Y, Atlan H, Cohen IR: T cell proliferative responses of type 1 diabetes patients and healthy individuals to human hsp60 and its peptides. J Autoimmun 1999;12:121–129.

35 Tisch R, Yang XD, Singer SM, Liblau RS, Fugger L, McDevitt HO: Immune response to glutamic acid decarboxylase correlates with insulitis in non-obese diabetic mice. Nature 1993;366:72–75.
36 Roep BO: T-cell responses to autoantigens in IDDM. The search for the Holy Grail. Diabetes 1996;45:1147–1156.
37 Durinovic-Bello I: Autoimmune diabetes: The role of T cells, MHC molecules and autoantigens. Autoimmunity 1998;27:159–177.
38 SL Parry, FC Hall, J Olson, T Kamradt, Sønderstrup G: Autoreactivity versus autoaggression: A different perspective on human autoantigens. Curr Opin Immunol 1998;10:633–638.
39 Naquet P, Ellis J, Tibensky D, Kenshole A, Singh B, Hodges R, Delovitch TL: T cell autoreactivity to insulin in diabetic and related non-diabetic individuals. J Immunol 1988;140:2569–2578.
40 Peterson LD, van der Keur M, de Vries RR, Roep BO: Autoreactive and immunoregulatory T-cell subsets in insulin-dependent diabetes mellitus. Diabetologia 1999;42:443–449.
41 Endl J, Otto H, Jung G, Dreisbusch B, Donie F, Stahl P, Elbracht R, Schmitz G, Meinl E, Hummel M, Ziegler AG, Wank R, Schendel DJ: Identification of naturally processed T cell epitopes from glutamic acid decarboxylase presented in the context of HLA-DR alleles by T lymphocytes of recent onset IDDM patients. J Clin Invest 1997;99:2405–2415.
42 Schloot NC, Willemen S, Duinkerken G, de Vries RR, Roep BO: Cloned T cells from a recent onset IDDM patient reactive with insulin B-chain. J Autoimmun 1998;11:169–175.
43 Tabata H, Kanai T, Yoshizumi H, Nishiyama S, Fujimoto S, Matsuda I, Yasukawa M, Matsushita S, Nishimura Y: Characterization of self-glutamic acid decarboxylase 65-reactive CD4+ T-cell clones established from Japanese patients with insulin-dependent diabetes mellitus. Hum Immunol 1998;59:549–560.
44 Scheinin T, Maenpaa J, Koskimies S, Dean BM, Bottazzo GF, Kontiainen S: Insulin responses and lymphocyte subclasses in children with newly diagnosed insulin-dependent diabetes. Clin Exp Immunol 1988;71:91–95.
45 Keller RJ: Cellular immunity to human insulin in individuals at high risk for the development of type I diabetes mellitus. J Autoimmun 1990;3:321–327.
46 Hummel M, Durinovic-Bello I, Ziegler AG: Relation between cellular and humoral immunity to islet cell antigens in type 1 diabetes. J Autoimmun 1996;9:427–430.
47 Schloot NC, Roep BO, Wegmann D, Yu L, Chase HP, Wang T, Eisenbarth GS: Altered immune response to insulin in newly diagnosed compared to insulin-treated diabetic patients and healthy control subjects. Diabetologia 1997;40:564–572.
48 Sarugeri E, Dozio N, Belloni C, Meschi F, Pastore MR, Bonifacio E: Autoimmune responses to the beta cell autoantigen, insulin, and the INS VNTR-IDDM2 locus. Clin Exp Immunol 1998;114: 370–376.
49 Dubois-LaForgue D, Carel JC, Bougneres PF, Guillet JG, Boitard C: T-cell response to proinsulin and insulin in type 1 and pretype 1 diabetes. J Clin Immunol 1999;19:127–134.
50 Semana G, Gausling R, Jackson RA, Hafler DA: T cell autoreactivity to proinsulin epitopes in diabetic patients and healthy subjects. J Autoimmun 1999;12:259–267.
51 Fuchtenbusch M, Kredel K, Bonifacio E, Schnell O, Ziegler AG: Exposure to exogenous insulin promotes IgG1 and the T-helper 2-associated IgG4 responses to insulin but not to other islet autoantigens. Diabetes 2000;49:918–925.
52 Honeyman MC, Cram DS, Harrison LC: Glutamic acid decarboxylase 67-reactive T cells: A marker of insulin-dependent diabetes. J Exp Med 1993;177:535–540.
53 Lohmann T, Leslie RD, Hawa M, Geysen M, Rodda S, Londei M: Immunodominant epitopes of glutamic acid decarboxylase 65 and 67 in insulin-dependent diabetes mellitus. Lancet 1994;343: 1607–1608.
54 Durinovic-Bello I, Hummel M, Ziegler AG: Cellular immune response to diverse islet cell antigens in IDDM. Diabetes 1996;45:795–800.
55 Lohmann T, Leslie RD, Londei M: T cell clones to epitopes of glutamic acid decarboxylase 65 raised from normal subjects and patients with insulin-dependent diabetes. J Autoimmun 1996;9:385–389.
56 Schloot NC, Roep BO, Wegmann DR, Yu L, Wang TB, Eisenbarth GS: T-cell reactivity to GAD65 peptide sequences shared with coxsackie virus protein in recent-onset IDDM, post-onset IDDM patients and control subjects. Diabetologia 1997;40:332–338.

57   Schloot NC, Batstra MC, Duinkerken G, De Vries RR, Dyrberg T, Chaudhuri A, Behan PO, Roep BO: GAD65-reactive T cells in a non-diabetic stiff-man syndrome patient. J Autoimmun 1999;12:289–296.
58   Ellis TM, Schatz DA, Ottendorfer EW, Lan MS, Wasserfall C, Salisbury PJ, She JX, Notkins AL, Maclaren NK, Atkinson MA: The relationship between humoral and cellular immunity to IA-2 in IDDM. Diabetes 1998;47:566–569.
59   Lohmann T, Halder T, Engler J, Morgenthaler NG, Khoo-Morgenthaler UY, Schroder S, Seissler J, Scherbaum WA, Kalbacher H: T cell reactivity to DR*0401- and DQ*0302-binding peptides of the putative autoantigen IA-2 in type 1 diabetes. Exp Clin Endocrinol Diabetes 1999;107:166–171.
60   Dosch H, Cheung RK, Karges W, Pietropaolo M, Becker DJ: Persistent T cell anergy in human type 1 diabetes. J Immunol 1999;163:6933–6940.
61   Hawkes CJ, Schloot NC, Marks J, Willemen SJ, Drijfhout JW, Mayer EK, Christie MR, Roep BO: T-cell lines reactive to an immunodominant epitope of the tyrosine phosphatase-like autoantigen IA-2 in type 1 diabetes. Diabetes 2000;49:356–366.
62   Schulz RM, Hawa M, Leslie RD, Sinigaglia F, Passini N, Rogge L, Picard JK, Londei M: Proliferative responses to selected peptides of IA-2 in identical twins discordant for type 1 diabetes. Diabetes Metab Res Rev 2000;16:150–156.
63   Schatz DA, Ottendorfer EW, Lan MS, Wasserfall C, Salisbury PJ, She JX, Notkins AL, Maclaren NK, Atkinson MA: The relationship between humoral and cellular immunity to IA-2 in IDDM. Diabetes 1998;47:566–569.
64   Mayer A, Rharbaoui F, Thivolet C, Orgiazzi J, Madec AM: The relationship between peripheral T cell reactivity to insulin, clinical remissions and cytokine production in type 1 (insulin-dependent) diabetes mellitus. J Clin Endocrinol Metab 1999;84:2419–2424.
65   Miyazaki I, Cheung RK, Gaedigk R, Hui MF, Van der Meulen J, Rajotte RV, Dosch HM: T cell activation and anergy to islet cell antigen in type I diabetes. J Immunol 1995;154:1461–1469.
66   Karjalainen J, Martin JM, Knip M, Ilonen J, Robinson BH, Savilahti E, Akerblom HK, Dosch HM: A bovine albumin peptide as a possible trigger of insulin-dependent diabetes mellitus. N Engl J Med 1992;327:302–307.
67   Berman MA, Sandborg CI, Wang Z, Imfeld KL, Zaldivar F Jr, Dadufalza V, Buckingham BA: Decreased IL-4 production in new onset type I insulin-dependent diabetes mellitus. J Immunol 1996;157:4690–4696.
68   Redondo MJ, Gottlieb PA, Motheral T, Mulgrew C, Rewers M, Babu S, Stephens E, Wegmann DR, Eisenbarth GS: Heterophile anti-mouse immunoglobulin antibodies may interfere with cytokine measurements in patients with HLA alleles protective for type 1A diabetes. Diabetes 1999;48:2166–2170.
69   Karlsson MG, Ludvigsson J: Determination of mRNA expression for IFN-gamma and IL-4 in lymphocytes from children with IDDM by RT-PCR technique. Diabetes Res Clin Pract 1998;40:21–30.
70   Karlsson MG, Ludvigsson J: Peptide from glutamic acid decarboxylase similar to coxsackie B virus stimulates IFN-gamma mRNA expression in Th1-like lymphocytes from children with recent-onset insulin-dependent diabetes mellitus. Acta Diabetol 1998;35:137–144.
71   Pelfrey CM, Rudick RA, Cotleur AC, Lee JC, Tary-Lehmann M, Lehmann PV: Quantification of self-recognition in multiple sclerosis by single-cell analysis of cytokine production. J Immunol 2000;165:1641–1651.
72   Karlsson MG, Lawesson SS, Ludvigsson J: Th1-like dominance in high-risk first-degree relatives of type I diabetic patients. Diabetologia 2000;43:742–749.
73   Cheung R, Karjalainen J, Vandermeulen J, Singal DP, Dosch HM: T cells from children with IDDM are sensitized to bovine serum albumin. Scand J Immunol 1994;40:623–628.
74   Sarugeri E, Dozio N, Meschi F, Pastore MR, Bonifacio E: Cellular and humoral immunity against cow's milk proteins in type 1 diabetes. J Autoimmun 1999;13:365–373.
75   Ou D, Mitchell LA, Metzger DL, Gillam S, Tingle AJ: Cross-reactive rubella virus and glutamic acid decarboxylase (65 and 67) protein determinants recognised by T cells of patients with type I diabetes mellitus. Diabetologia 2000;43:750–762.

76 Klemetti P, Hyoty H, Roivainen M, Ilonen J, Savola K, Knip M, Akerblom HK, Vaarala O: Relation between T-cell responses to glutamate decarboxylase and coxsackievirus B4 in patients with insulin-dependent diabetes mellitus. J Clin Virol 1999;14:95–105.

77 Honeyman MC, Stone NL, Harrison LC: T-cell epitopes in type 1 diabetes autoantigen tyrosine phosphatase IA-2: Potential for mimicry with rotavirus and other environmental agents. Mol Med 1998;4:231–239.

78 Harfouch-Hammoud E, Walk T, Otto H, Jung G, Bach JF, van Endert PM, Caillat-Zucman S: Identification of peptides from autoantigens GAD65 and IA-2 that bind to HLA class II molecules predisposing to or protecting from type 1 diabetes. Diabetes 1999;48:1937–1947.

79 Geluk A, van Meijgaarden KE, Schloot NC, Drijfhout JW, Ottenhoff TH, Roep BO: HLA-DR binding analysis of peptides from islet antigens in IDDM. Diabetes 1998;47:1594–1601.

80 Hiemstra HS, van Veelen PA, Geluk A, Schloot NC, de Vries RR, Ottenhoff TH, Roep BO, Drijfhout JW: Limitations of homology searching for identification of T-cell antigens with library derived mimicry epitopes. Vaccine 1999;18:204–208.

81 Peakman M, Stevens EJ, Lohmann T, Narendran P, Dromey J, Alexander A, Tomlinson AJ, Trucco M, Gorga JC, Chicz RM: Naturally processed and presented epitopes of the islet cell autoantigen IA-2 eluted from HLA-DR4. J Clin Invest 1999;104:1449–1457.

82 Sønderstrup G, Cope AP, Patel S, Congia M, Hain N, Hall FC, Parry SL, Fugger LH, Michie S, McDevitt HO: HLA class II transgenic mice: Models of the human CD4+ T-cell immune response. Immunol Rev 1999;172:335–343.

83 Buyse I, Sandkuyl LA, Zamani-Ghabanbasani M, Gu XX, Bouillon R, Bex M, Dooms L, Emonds MP, Duhamel M, Marynen P, et al: Association of particular HLA class II alleles, haplotypes and genotypes with susceptibility to IDDM in the Belgian population. Diabetologia 1994;37:808–817.

84 Patel SD, Cope AP, Congia M, Chen TT, Kim E, Fugger L, Wherrett D, Sønderstrup-McDevitt G: Identification of immunodominant T cell epitopes of human glutamic acid decarboxylase 65 by using HLA-DR(alpha1*0101,beta1*0401) transgenic mice. Proc Natl Acad Sci USA 1997;94: 8082–8087.

85 Congia M, Patel S, Cope AP, De Virgiliis S, Sønderstrup G: T cell epitopes of insulin defined in HLA-DR4 transgenic mice are derived from preproinsulin and proinsulin. Proc Natl Acad Sci USA 1998;95:3833–3838.

86 Herman AE, Tisch RM, Patel SD, Parry SL, Olson J, Noble JA, Cope AP, Cox B, Congia M, McDevitt HO: Determination of glutamic acid decarboxylase 65 peptides presented by the type I diabetes-associated HLA-DQ8 class II molecule identifies an immunogenic peptide motif. J Immunol 1999;163:6275–6282.

87 Donner H, Seidl C, Van der Auwera B, Braun J, Siegmund T, Herwig J, Weets I, Usadel KH, Badenhoop K: HLA-DRB1*04 and susceptibility to type 1 diabetes mellitus in a German/Belgian family and German case-control study. The Belgian Diabetes Registry Tissue Antigens 2000;55: 271–274.

88 Sanjeevi CB: HLA-DQ6-mediated protection in IDDM. Hum Immunol 2000;61:148–153.

89 Roep BO, Schipper R, Verduyn W, Bruining GJ, Schreuder GM, de Vries RR: HLA-DRB1*0403 is associated with dominant protection against IDDM in the general Dutch population and subjects with high-risk DQA1*0301-DQB1*0302/DQA1*0501-DQB1*0201 genotype. Tissue Antigens 1999;54:88–90.

90 Undlien DE, Friede T, Rammensee HG, Joner G, Dahl-Jorgensen K, Sovik O, Akselsen HE, Knutsen I, Ronningen KS, Thorsby E: HLA-encoded genetic predisposition in IDDM: DR4 sub-types may be associated with different degrees of protection. Diabetes 1997;46:143–149.

91 Yasunaga S, Kimura A, Hamaguchi K, Ronningen KS, Sasazuki T: Different contribution of HLA-DR and -DQ genes in susceptibility and resistance to insulin-dependent diabetes mellitus (IDDM). Tissue Antigens 1996;47:37–48.

92 Van der Auwera B, Van Waeyenberge C, Schuit F, Heimberg H, Vandewalle C, Gorus F, Flament J: DRB1*0403 protects against IDDM in Caucasians with the high-risk heterozygous DQA1*0301-DQB1*0302/DQA1*0501-DQB1*0201 genotype. Belgian Diabetes Registry. Diabetes 1995;44: 527–530.

93 Lederman MM, Ellner JJ, Rodman HM: Defective suppressor cell generation in juvenile onset diabetes. J Immunol 1981;127:2051–2055.

94 Rashba EJ, Reich EP, Janeway CA, Sherwin RS: Type 1 diabetes mellitus: An imbalance between effector and regulatory T cells? Acta Diabetol 1993;30:61–69.

95 Sempe P, Richard MF, Bach JF, Boitard C: Evidence of CD4+ regulatory T cells in the non-obese diabetic male mouse. Diabetologia 1994;37:337–343.

96 Shimada A, Rohane P, Fathman CG, Charlton B: Pathogenic and protective roles of CD45RB(low) CD4+ cells correlate with cytokine profiles in the spontaneously autoimmune diabetic mouse. Diabetes 1996;45:71–78.

97 Martins TC, Aguas AP: A role for CD45RBlow CD38+ T cells and costimulatory pathways of T-cell activation in protection of non-obese diabetic (NOD) mice from diabetes. Immunology 1999;96:600–605.

98 Bergerot I, Arreaza GA, Cameron MJ, Burdick MD, Strieter RM, Chensue SW, Chakrabarti S, Delovitch TL: Insulin B-chain reactive CD4+ regulatory T-cells induced by oral insulin treatment protect from type 1 diabetes by blocking the cytokine secretion and pancreatic infiltration of diabetogenic effector T-cells. Diabetes 1999;48:1720–1729.

99 Lepault F, Gagnerault MC: Characterization of peripheral regulatory CD4+ T cells that prevent diabetes onset in nonobese diabetic mice. J Immunol 2000;164:240–247.

100 Salomon B, Lenschow DJ, Rhee L, Ashourian N, Singh B, Sharpe A, Bluestone JA: B7/CD28 costimulation is essential for the homeostasis of the CD4+CD25+ immunoregulatory T cells that control autoimmune diabetes. Immunity 2000;12:431–440.

101 Roep BO, Atkinson MA, van Endert PM, Gottlieb PA, Wilson SB, Sachs JA: Autoreactive T cell responses in insulin-dependent (type 1) diabetes mellitus. Report of the First International Workshop for Standardization of T Cell Assays. J Autoimmun 1999;13:267–282.

102 Li X, Wang X: Application of real-time polymerase chain reaction for the quantitation of interleukin-1beta mRNA upregulation in brain ischemic tolerance. Brain Res Brain Res Protoc 2000;5:211–217.

103 Kwok WW, Liu AW, Novak EJ, Gebe JA, Ettinger RA, Nepom GT, Reymond SN, Koelle DM: HLA-DQ tetramers identify epitope-specific T cells in peripheral blood of herpes simplex virus type 2-infected individuals: Direct detection of immunodominant antigen-responsive cells. J Immunol 2000;164:4244–4249.

104 Levings MK, Roncarolo MG: T-regulatory 1 cells: A novel subset of CD4 T cells with immunoregulatory properties. J Allergy Clin Immunol 2000;106:109–112.

105 Takahashi T, Kuniyasu Y, Toda M, Sakaguchi N, Itoh M, Iwata M, Shimizu J, Sakaguchi S: Immunologic self-tolerance maintained by CD25+CD4+ naturally anergic and suppressive T cells: Induction of autoimmune disease by breaking their anergic/suppressive state. Int Immunol 1998;10:1969–1980.

Grete Sønderstrup, Department of Microbiology and Immunology,
Stanford University School of Medicine, Stanford, CA 94305 (USA)
Tel. +1 650 723 7523, Fax +1 650 723 9180, E-Mail gretes@stanford.edu

von Herrath MG (ed.): Molecular Pathology of Type 1 Diabetes mellitus.
Curr Dir Autoimmun. Basel, Karger, 2001, vol 4, pp 252–282

Autoantibodies in Human Diabetes

Massimo Pietropaolo[a], *George S. Eisenbarth*[b]

[a] Division of Immunogenetics, Diabetes Institute, Rangos Research Center,
Children's Hospital of Pittsburgh, University of Pittsburgh School of Medicine,
Pittsburgh, Pa., and
[b] Barbara Davis Center for Childhood Diabetes, University of Colorado Health Sciences
Center, Denver, Colo., USA

A shared attribute of many autoimmune diseases is a humoral response
directed against multiple target antigens [1, 2]. The immunological diagnosis of
autoimmune diseases relies mainly on the detection of autoantibodies in the
serum of patients [3]. Although their pathogenic significance is still unclear,
they have the great advantage of serving as constitutive markers for specific
autoimmune responses. They are also important tools for the molecular cloning,
identification and characterization of novel autoantigens [4, 5]. Cloned autoantigens represent an unlimited source of reagents that can be used for experimental
and diagnostic purposes. In addition to studying the immunological properties
of autoantigens, these molecules can readily be utilized to optimize fluid-phase
radioimmunoassays [6], which in turn have future diagnostic purposes [7, 8].
For some diseases, such as autoimmune diabetes, autoimmune connective tissue
diseases, systemic lupus erythematosus (SLE), autoimmune thyroid disease
(AITD), rheumatoid arthritis (RA) and other chronic systemic autoimmune diseases, recombinant proteins can be utilized to devise some of the most sensitive
and specific biochemical assays currently available for autoantibody detection
[5, 9]. In the majority of chronic autoimmune disorders for which a diagnosis has
been established, antibody laboratory testing is instrumental in decision-making
for disease management based on the identification of disease activity.

Autoantibodies are some of the most potent risk determinants for autoimmune diseases with relative risk exceeding 100 [10–12]. The quintessential
model for the application of autoantibody markers in the prediction of a selective immune-mediated tissue damage is type 1 diabetes (T1D) and this concept

can be theoretically extended to other chronic autoimmune diseases. For example, several recent studies have suggested that using a combination of humoral immunological markers gives a higher predictive value for T1D, and great sensitivity without significant loss of specificity [13–17]. In particular, an increasing number of studies have shown that multiple autoantibodies to islet autoantigens conferred a cumulative risk of developing diabetes of nearly 90% during a prospective follow-up [13, 14, 16, 18, 19]. With this degree of predictability we have developed a high-risk strategy where we can identify individuals with the highest risk for developing diseases such as T1D. These are very important prerequi-sites for designing effective intervention strategies for the prevention of T1D.

Humoral autoimmunity markers offer also the opportunity to study individuals with multiple autoimmune endocrine disorders. Some of these patients, particularly type 1 diabetics, have a high risk to develop Addison's disease as compared to the general population [20, 21]. Today, a specific assay for antibodies to 21-hydroxylase can accurately predict which patients are 'at risk' to develop Addison's disease [22]. Routine screening is recommended in 'at-risk' patients and those should be followed more closely in terms of adrenal function so that replacement glucocorticoid therapy could be initiated prior to development of acute adrenal insufficiency. The disease association and the inheritance pattern of Addison's disease justifies the detection of specific and sensitive markers, such as 21-hydroxylase antibodies [22, 23], in family members of patients with Addison's disease prior to the development of the life-threatening manifestations of adrenal insufficiency.

Theoretically, evidence of humoral autoimmunity can also be applied to cancer. Immunological abnormalities developing in association with neoplasms and paraneoplastic syndromes are termed oncogenic autoimmunity. A common charactcristic of paraneoplastic disease is its relation to ectopic protein production by tumor cells [24, 25]. The proteins produced may be peptide hormones or hormone-producing enzymes in cases of endocrine paraneoplasia, antigenic proteins, which are normally limited in their expression to immunologically privileged sites such as the CNS. Often the demonstration of autoantibodies against molecular targets localized within the neoplastic milieu may aid in the diagnosis of occult malignancy, even prior to the appearance of clinically manifested cancer in some patients [25].

The Past: Islet Cell Antibodies

In the past, physicians and researchers have relied on immunofluorescence techniques to detect autoantibodies directed against pancreatic islet cell antigens, particularly cytoplasmic islet cell antibodies (ICA), described for the first time by Bottazzo et al. [26]. ICA are predictive of type 1A (autoimmune) diabetes [27–30], but multiple international workshops have proved the marked variability between laboratories in performing these assays, despite a JDF unit standard [31, 32]. As a matter of fact, this assay is semiquantitative and remains difficult to standardize, despite improvements resulting from standardization workshops [33]. Inherent limitations of the ICA assay include the need of subjective scoring of the sections for positivity and wide variation in the results obtained with pancreatic tissues from different donors [32]. Even among the best reference laboratories in the world, ICARUS data showed that within ranges of broad agreement, some sera give widely divergent results. Such divergence appears to result not only from differences in assay sensitivity, but also from different ICA assay formats which may detect heterogeneity in the antigens recognized. In addition, ICA represents a family of autoantigens including, for subsets of patients, anti-glutamic acid decarboxylase antibodies [34–38], antibodies to a GM2-1 ganglioside [36, 39], or neither of the above. Since the discovery of ICA, multiple workshops have been held and a Proficiency Testing Service was established for laboratories measuring cytoplasmic islet cell autoantibodies [40]. The International Diabetes Society (IDS) workshops demonstrated that there existed remarkable assay variation between different laboratories and within the same laboratory. A standard for ICA was established utilizing dilutions of sera and the units assigned were termed the JDF units (named for the Juvenile Diabetes Foundation, which sponsored the meetings). With such a standard, differences in assay sensitivity could be rationalized and it became apparent that >20 JDF units of cytoplasmic islet cell autoantibodies were highly predictive of diabetes among first-degree relatives.

Prognostically significant heterogeneity of ICA has been described and first-degree relatives of patients with T1D or patients with polyendocrine autoimmunity with high titers of ICA seem to have a low risk of progression to overt diabetes [34, 37].

The Future: Autoantibodies to Molecularly Characterized Islet Antigens

During the past decade, a number of autoantibodies to islet antigens were molecularly characterized (table 1) [4, 41, 42]. As mentioned above, several recent studies have suggested that using a combination of humoral immunological markers to these islet antigens rather than a single test, gives a higher predictive value for T1D, and greater sensitivity without significant loss of specificity [13–15, 17]. These represent important prerequisites for the use of these

Table 1. Most characterized islet autoantigens associated with type 1 diabetes

	Localization	Humoral response	Cellular response
Insulin[1]	Secretory granules pancreatic β cells. Human thymus and PAE cells (peripheral antigen-expressing cells)	Insulin autoantibodies are found in virtually 100% of young children ($<$5 years of age) before the onset of T1D. Correlation with younger age and fast rate of progression to insulin requirement in first-degree relatives of T1D patients. Prophylactic subcutaneous injection of insulin, oral and intranasal administration prevent T1D in NOD mice	PBLs from humans and NOD mice react with insulin β chain
GAD65[1] and GAD67	Synaptic-like microvesicles of neuroendocrine cells. Present in testis and ovary. Human thymus and PAE cells	A subset of 64-kDa autoantibodies recognize GAD. Autoantibodies to GAD65 are present in 70–80% of prediabetic subjects or newly diagnosed diabetic patients. GAD antibodies are also detected in patients with stiff-man syndrome, and with autoimmune thyroid disease Radioimmunoassay of in vitro transcribed/translated GAD65 useful for large-scale screening	PBL responses to GAD65 in newly diagnosed diabetic patients and in NOD mice
ICA512 (IA-2)[1] and phogrin (IA-2β)	Neurosecretory granules (pancreatic β cells, CNS, pituitary, adrenal). Human thymus and PAE cells	Autoantibodies to ICA512 (IA-2) are present in ~60% of prediabetics or newly diagnosed T1D patients. Relationship between 37 and 40 kDa tryptic fragments and ICA512 (IA-2) Radioimmunoassay of in vitro transcribed/translated ICA512 (IA-2) useful for large-scale screening	PBL responses in newly diagnosed diabetic patients
Islet cell autoantigen 69 kDa (ICA69)	Neuroendocrine tissues (cytosol). Human, murine thymus and PAE cells	Autoantibodies to ICA69 can be detected in 43% of prediabetic subjects by Western blotting	Association between HLA-DR3 and PBL responsiveness in newly diagnosed diabetics. Regions of similarity with bovine serum albumin
Carboxy-peptidase H	Neurosecretory granules	Autoantibodies to carboxypeptidase H found in ~20% of prediabetics	Present

	Localization	Humoral response	Cellular response
Ganglioside GM2-1	Pancreatic islet cells	Autoantibodies to GM2-1 detected in −80% of prediabetic subjects and NOD mice	?
Imogen 38 (38 kDa)	Mitochondria; widely distributed with variable levels of expression	Presence of circulating antibodies to 38-kDa proteins. Possible presence if antibodies to imogen 38	PBLs from newly diagnosed diabetics proliferate to imogen 38
Glima 38	Amphiphilic N-Asp glycated β-cell membrane protein that is expressed in islets and neuronal cell lines	Autoantibodies to Glima 38 can be detected in 14–22.7% of newly diagnosed diabetics and prediabetics. The majority of these patients are negative for GAD65 and/or ICA512 (IA-2) autoantibodies	?
Peripherin	Neuronal cells	Autoantibody response against peripherin in NOD mice	T-cell responses against peripherin in NOD mice
Heat-shock protein (HSP60)	Ubiquitously inducible	Antibodies to HSP60 in prediabetic NOD mice	HSP60-reactive T cells can accelerate disease in prediabetic NOD mice
SOX13 (ICA12)	SRY-type HMG box transcription factor family. Expressed in the nucleus and cytoplasm of most human tissues with the highest levels in the pancreas	Autoantibodies against SOX13 are detected in ~18% of type 1 diabetics, the majority of whom are ICA-negative. Anti SOX13 Ab are more frequent in adults with T1D	?

[1]Biochemical assays readily available for large screening programs.

radioimmunoassays in screening for individuals who are at risk for developing insulin requirement as well as in the general population.

The evaluation of risk factors for the etiology of T1D is very important for understanding the natural history of this disease. The new radio-immuno-assays for detecting antibodies to GAD65, insulin and IA-2 offer the practical advantage of reproducibility and ease standardization, otherwise not possible with techniques using indirect immunofluorescence. This can form the basis for rapid analysis of the predictive value of immunological markers in prospective studies as well as in the general population. The Centers for Disease Control have evaluated filter paper collection of sera for determination of auto-antibodies with the finding that autoantibodies can readily be measured following elution from such blood spots [43]. These results imply that screening for T1D risk with whole blood samples collected onto special filter paper, may represent a feasible solution to simplify collection and reduce the costs associated with diabetes screening programs.

Currently, sequenced autoantigens with recombinant autoantibody assays include insulin [44], glutamic acid decarboxylase (GAD) [45], ICA512/IA-2 [46, 47], and I-A2β (phogrin) [48]. In addition, there are many other autoantigens in a variety of stages of characterization, including molecules with characterized sequences but without fully developed assays and proteins with known molecular masses 155, 52 and 37–38 kDa [49, 50]. Other molecules have been associated with T1D, such as anti-bovine serum albumin antibodies (anti-BSA) [51], antibodies reacting with ICA69 [52, 53] anti-insulin receptor antibodies [54], antibodies to heat-shock proteins [55–57], antibodies to jun-B,16 [58] and antibodies to a series of other autoantigens identified by screening islet libraries [4, 41].

Insulin

Insulin is a 51-amino acid disulphide-linked heterodimer that is secreted by the pancreatic β cells. Proinsulin, which is processed within the secretory granules, is the precursor to insulin. Insulin is assembled as a pre-prohormone, subsequently is processed to proinsulin and ultimately, with the cleavage of the connecting peptide (C-peptide), to mature insulin. Processing occurs within β-cell secretory granules, whereby insulin is packaged in a crystal form and equimolar concentrations of insulin and C-peptide are secreted by pancreatic β cells.

Insulin autoantibodies (IAA) are among the earliest markers to appear in young children who eventually develop T1D. This is particularly evident in infants younger than 1 year of age who begin to express these auto-antibodies shortly after birth. In addition, the micro-IAA assay readily detects insulin autoantibodies in the nonobese diabetic (NOD) mouse and the early expression of insulin autoantibodies of NOD mice correlates with the early development of T1D in the NOD mouse model [59]. Approximately 90% of NOD mice expressing insulin autoantibodies at 8 weeks of age develop diabetes by 16 weeks of age (fig.1).

To date, insulin and proinsulin are considered the only pancreatic β-cell-specific autoantigens. Of note, in both mouse and human, proinsulin, both at the mRNA and protein level, is present in the thymus. A number of independent investigations, including from our group [60], provided evidence that the

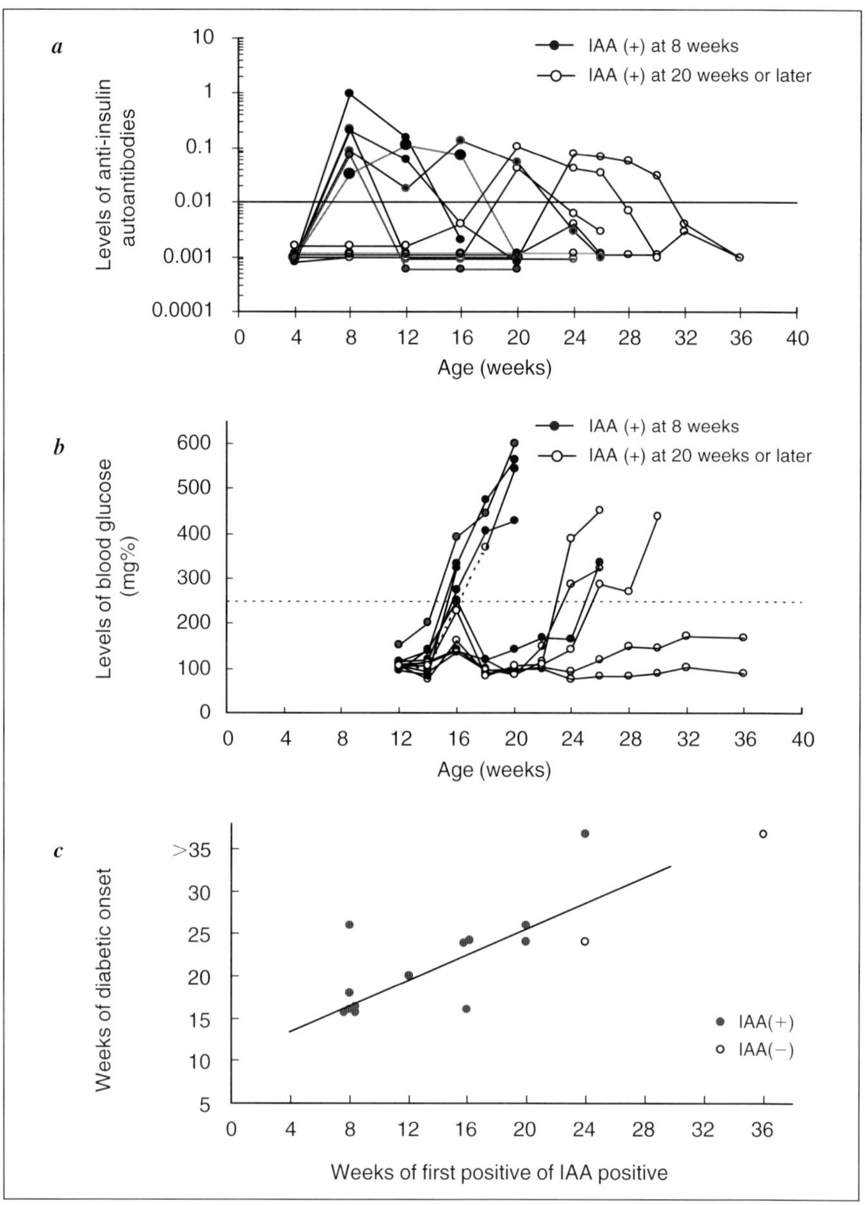

Fig. 1. Insulin autoantibodies (IAA) and blood glucose levels for NOD mice expressing IAA at 8 weeks of age and 20 weeks later. The mice were followed starting from 4 weeks of age until they developed diabetes or until 36 weeks. *a* IAA levels. *b* Blood glucose readings. *c* Age of first detection of IAA plotted out versus age of diabetes onset for all NOD mice evaluated prospectively [reprinted from 59, with permission].

expression of antigens in the thymus may result in the development of immuno-logical tolerance. It has been reported that in fetal thymus the VNTR class III alleles, normally associated with protection for T1D, are associated with 2- to 3-fold higher insulin mRNA levels [60], then thymic mRNA levels found carrying class I VNTR alleles, which is associated with susceptibility to T1D. These observations suggest that transcriptional effects of the INS VNTR may play a role in the establishment of tolerance to insulin during thymic development.

In the 1950s, Berson and Yallow developed the first radioimmunoassay utilizing sera with insulin antibodies from patients treated with bovine insulin. Bovine insulin differs from human insulin at 3 of 51 amino acid and porcine insulin differs from human insulin by 1 amino acid (at the terminus of the B chain). For most patients treated with subcutaneous insulin, the presence of anti-insulin antibodies does not interfere with insulin therapy. A subset of insulin-treated patients with extremely high levels of insulin antibodies develop insulin resis-tance. For such patients, the species of insulin used for therapy is usually substi-tuted (e.g., from bovine to human insulin) or ultimately therapy with sulfated insulin is attempted. Anti-insulin antibodies during pregnancy may facilitate transplacental passage of animal insulin and can be correlated with fetal macro-somia [61]. Insulin autoimmune syndrome, also termed Hirata syndrome [62–64], is a very rare disease associated with the DRB1*0406 allele, and patients develop extremely high levels of insulin autoantibodies. This syndrome occurs after exposure to sulfhydryl-containing medications, such as methimazole [65]. These patients usually develop symptoms related to hypoglycemia, which resolves after discontinuing of the offending drug.

In 1983, Palmer et al. [66] reported the presence of insulin autoantibodies in patients with new-onset T1D, prior to the administration of exogenous insulin. Subsequently, a large number of studies have demonstrated that anti-insulin antibodies are present for years before the development of insulin requirement [67]. A number of international workshops and sera exchanges led to the conclusion that only the antibodies detected with the radioassay formats were specific for the diagnosis and prediction of T1D [68]. In particular, the insulin autoantibodies that are associated with diabetes risk are all of extremely high affinity and, most important, of very low capacity (10^{-12} M), making their detection with plate-binding assays difficult [44]. The epitope of the insulin molecule recognized using a fluid-phase assay appears to be homo-geneous [69]. In addition, the insulin autoantibodies from relatives of diabetic patients react with a conformational epitope (do not react with either the A or B chain) and react equally well with insulin and proinsulin. To date, all insulin autoantibodies of at-risk relatives react with des B23 to B30 insulin lacking the terminal 8 amino acids of the B chain. Des B23 to B30 insulin in turn fails to bind to the insulin receptor. Figure 2 illustrates the three-dimensional structure

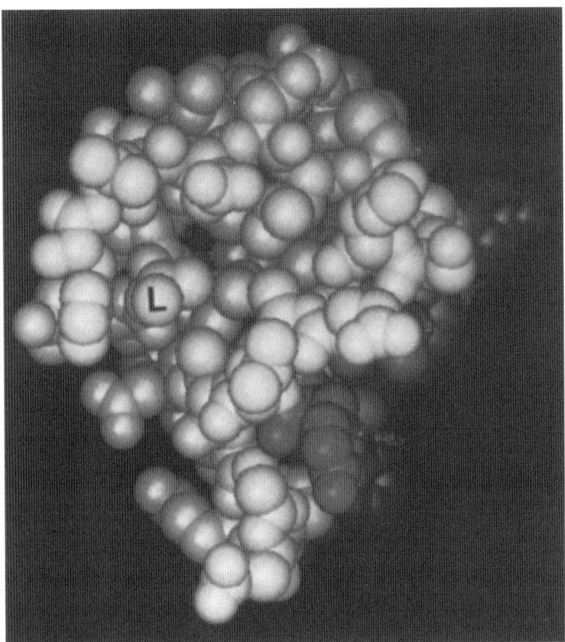

Fig. 2. Three-dimensional structure of insulin. Pocket to the left of molecule is centered at A13 leucine. The receptor binding face at right of molecule is highlighted.

of insulin with the A13 leucine in center of 'pocket' and receptor binding region highlighted. The A13 leucine is in the base of a pocket, and if this leucine is replaced by a tryptophan that extends out of the pocket, insulin autoantibody reactivity is abrogated. The reason for this marked specificity of insulin autoantibodies reacting with the face opposite the insulin-binding domain is unknown. It is, however, very likely that the antibody response to insulin is 'antigen driven', giving rise to high-affinity autoantibodies. Insulin may react with B cells while bound to insulin receptors, either of neighboring immunocytes or perhaps following its transport on insulin receptors through the endothelium. The ability to produce insulin analogues that are recognized by insulin autoantibodies but not by the insulin receptor suggests that one can selectively target B cells producing such antibodies. Whether elimination of such B cells which might be extremely efficient at presenting insulin to T cells, would influence β-cell destruction is unknown. Levels of insulin autoantibodies, similar to GAD auto-antibodies, appear to be regulated at different levels over long periods of time in prediabetic first-degree relatives. The levels of antibodies correlate inversely with the age at which T1D develops. Hence, levels >2,000 nU/ml are

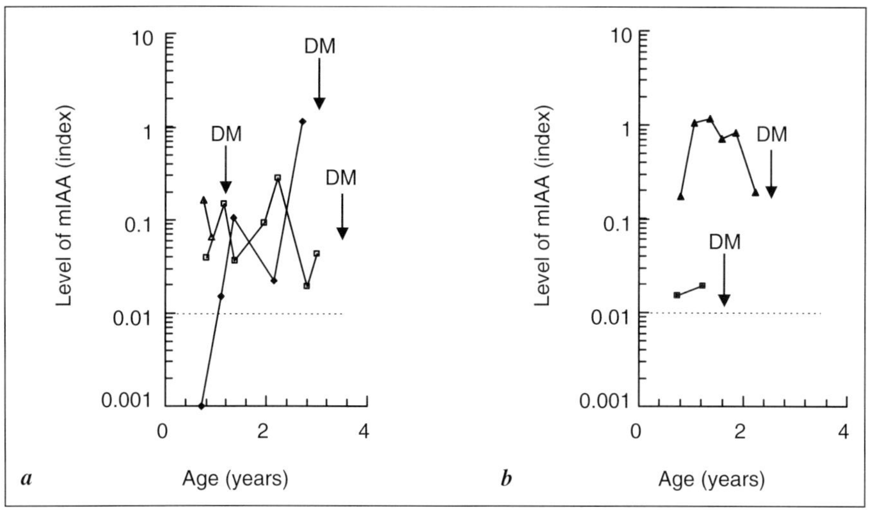

Fig. 3. Insulin autoantibodies (IAA) plotted versus age in participants of the DAISY study. The children were followed to the development of insulin requirement. *a* Relatives of patients with T1D. *b* Individuals from the general population without a diabetic relative that were HLA typed at birth and followed to T1D onset [reprinted from 59, with permission].

almost exclusively found in patients who progress to T1D prior to age 5, and less than half of individuals developing T1D after age 15 have detectable levels of anti-insulin autoantibodies. In particular, investigators from the DAISY study [59, 70] reported that 5 children were found to have persistent IAA before 1 year of age, and 4 of them went on to develop the clinical onset of T1D (all before 3.5 years of age) (fig. 3). In contrast, of 929 children not expressing persistent IAA before age 1, only 1 developed insulin requirement. Although these observations emphasize the utility of IAA particular in younger populations and justify the need to design trials in such a young group, the success of these intervention strategies depends on the safety and effectiveness of therapeutical regimens.

High levels of such antibodies may be genetically influenced. An association was reported between high levels of IAA and HLA-DR4 [67], and with other haplotypes, such as HLA-DQA1 alleles of lineage 2, DQA1*0102,0201,0301,0401 [71]. The presence of high-titer insulin autoantibodies appears to be associated with these two alleles in a dominant manner. In particular, individuals developing type 1A diabetes who are homozygous for DQA1*0501/DQB1*0201 (DR3 homozygotes) either are often negative for insulin autoantibodies or have very low levels.

Glutamic Acid Decarboxylase

GAD is the enzyme that catalyzes the synthesis of the inhibitory neurotransmitter γ-aminobutyric acid (GABA). There are at least three genes and two isoforms of GAD in humans. The 65-kDa isoform, GAD65, which is the auto-antigen that can be identified serologically [45, 72], and the 67-kDa isoform or GAD67 [73]. Much work has been devoted into characterizing the antibody response to GAD65, including the development of sensitive and specific screen-ing assays suitable for large-scale testing. The two isoforms of GAD, GAD65 and GAD67, appear to share an identical exon-intron organization and are 76% identical in amino-acid sequences with only 22% identity in the N-terminal region and over 90% identity in the C-terminal regions [74].

In 1990, the groups of DeCamilli and Baekkeskov [45] reported that auto-antibodies against a 64 kDa reacted at low titer with GAD and that a subset of the anti-GAD autoantibodies of new-onset diabetics were similar to anti-GAD antibodies present in stiff-man syndrome (SMS). Previously, Baekkeskov and co-workers [75, 76] reported that serum from type 1 diabetics can immunoprecipitate a 64-kDa molecule using metabolically labeled human islets. The two isoforms of GAD, termed GAD65 and GAD67, are also expressed in testes, ovary, and neurons [77, 78].

Antibodies to GAD65 can be detected in up to 80% of the patients with T1D [14]. These patients have 11–18% antibodies to GAD67 [79]. The auto-antibody binding to GAD67 in type 1 diabetic patients seems to share epitopes with GAD65 [79] although it was reported that patients with Graves' disease carry GAD67 antibodies in the absence of GAD65 antibodies [80]. GAD65/GAD67 chimeric proteins were widely utilized in most of the laboratories to define the epitope regions of GAD65. Monoclonal antibodies derived from individuals with new-onset diabetes have been particularly useful in mapping GAD epitopes [81–83]. Two major epitopes targeted by human diabetic sera have been identified [83]: one of them corresponds the epitope in the middle part of GAD65 between amino acids 240–435 (termed T1D-E1) and the other one is localized in the C-terminal region of GAD65 between amino acids 451–570 (termed T1D-E2). Recently, Schwartz et al. [84] used homologue-scanning mutagenesis to identify GAD65-specific amino acid residues which form autoreactive antibody epitopes in this molecule. Detailed mapping of 13 conformational epitopes, recognized the by human monoclonal antibodies derived from patients, revealed a remarkable spectrum of human autoreactivity to GAD65, targeting almost the entire molecule. These data confirm previously published findings [88] that type 1 diabetic patients recognize highly conformation-dependent epitopes. It will be of importance to investigate whether progression to T1D coexists with a reduction of recognized epitopes, and whether in subjects with broad epitope recognition a GAD65 autoantibody response is associated with nonprogressive chronic autoimmunity. These results may have implications for the application of restricted GAD65 autoantibody epitope specificity and the prediction of the disease.

GAD autoantibodies are also present in patients with SMS, a rare neurological disorder characterized by spasms of muscular rigidity. In both T1D and SMS, GAD antibodies are preferentially directed against the M_r 65,000 isoform. However, they differ in several respects. Their levels are 10- to 100-fold higher in SMS sera than in T1D sera. Secondly, GAD65 antibodies, unlike the majority of those from type 1 diabetic patients, usually react with GAD65 on Western blotting. Thirdly, GAD65 autoantibodies present in SMS patients react mainly with two epitopes at the N-terminus and C-terminus of GAD65, while GAD65 autoantibodies detected in T1D recognize epitopes in the middle and C-terminus of the protein. Immunoblotting analyses using deletion mutants of GAD65 suggest that the C-terminus epitope, recognized by specific antibodies from patients with SMS, is conformational [78]. From a clinical perspective, it is of interest that antibodies against the T1D-related autoantigen ICA512 (IA-2), unlike GAD65 antibodies, are exclusively detected in SMS patients who have coexisting T1D, raising the concept that they may be used as predictive markers in SMS-affected individuals who have a significant risk of developing T1D [25].

ICA512 (IA-2) and IA-2β (Phogrin)

ICA512 (IA-2) and its homologue IA-2β (phogrin) are both neuroendocrine molecules that belong to the family of protein tyrosine phosphatase (PTP)-like proteins. ICA512 (IA-2) seems to be the only single ICA marker that is associated with substantial risk for T1D as compared to the other conventional markers. As an example, compared to GAD65 autoantibodies, the ICA512 (IA-2) antibody radioimmunoassay has a low prevalence in nondiabetic first-degree relatives, a very high specificity, and a sensitivity of approximately 60% in newly diagnosed diabetic patients. This marker is commonly used in combination with both IAA and GAD65 biochemical assays to enhance both sensitivity and positive predictive value.

ICA512 (IA-2) was initially cloned by screening a human islet cDNA expression library using a pool of sera from T1D patients [46, 85]. The same molecule, termed IA-2, was independently cloned using a subtraction library (subtracting glucagonoma from insulinoma cDNAs) [47]. Subsequent analysis of ICA512 clones showed that the published discrepancies of the deduced cDNA sequence between IA-2 and ICA512 were due to technical reasons. Therefore, with regard to the coding region, ICA512 and IA-2 appeared to be identical in sequence. The deduced ICA512 (IA-2) cDNA sequence predicts a

979-amino acid protein with a single transmembrane region encompassing residues 577–601 and with significant homology to receptor-type PTP (RT-PTPase).

An insulin granule membrane PTP homologue, termed phogrin, was subsequently cloned [48, 86, 87] by screening a rat insulinoma cDNA library with polyclonal antisera to highly enriched insulinoma secretary granule membranes. The deduced polypeptide sequence of 1,004 amino acids encoded a protein of 111,876 Da) a of a moderately acidic character (predicted pI 5.66), with a single-functioned NH_2-linked glycosylation site. A consensus sequence for the catalytic site of a PTP appears at amino acids 931–942 close to the COOH-terminus of the molecule. EMBL (European Molecular Biology Laboratory) and GenBank database searches revealed homology with a large number of different PTPs over a stretch of ~260 residues commencing 100 amino acids after the transmembrane domain and extending to the COOH-terminus. Among the 250 positively scoring PTP homologues, phogrin showed the highest homology with ICA512 (IA-2). In vitro translation of mRNA transcribed from the full-length cDNA generated a protein of 128 kDa, which is compatible with the deduced protein sequence given its acidic character.

Double immunofluorescence studies have been performed using anti-glucagon and anti-somatostatin and specific rabbit antisera generated against either the cytoplasmic (amino acid residues 89–59) or the luminal (amino acid residues 92–18) domain of the human ICA512 (IA-2), which indicate that the molecule is detectable in β, α and δ cells [88]. Other neuroendocrine tissues expressing ICA512 (IA-2) included pituitary and adrenal medulla. The highest levels of ICA512 (IA-2) are found in the infundibular tract and posterior pituitary. Subcellular fractionation of insulinoma tissue showed that the 60- and 64-kDa products of the breakdown of phogrin had a very similar distribution of insulin and carboxypeptidase H, and it was concluded that phogrin was predominantly localized in the secretory granules [48, 86, 88].

It has been reported that antibodies to the 37,000- and 40,000-M_r proteolytic fragment were found in up to 80% of recent onset type 1 diabetic patients [89], and similar frequencies were reported in twins who developed the overt disease [67–75]. Several reports [90–92] suggest that trypsinization of the recombinant intracellular domain of IA-2 generated fragments identical in size to the 40-kDa insulinoma trypsinized fragment. Sera from recent-onset T1D patients reacting to the 40,000-M_r insulinoma fragment gave strong reactivity with the ICA512 (IA-2). In particular, a strong association (r = 0.85, p < 0.001) was detected between the two assays that measured antibodies to the 40,000-M_r insulinoma fragment and the IA-2 antigen respectively [90]. It is likely that the 37,000-M_r fragment derives from a different protein, one of which is related to ICA512 (IA-2) [90–92]. Additional studies performed with anti-phogrin antibodies suggest that the 37,000-M_r tryptic fragment recognized by sera from diabetic patients corresponds to the phogrin molecule [48, 86, 91].

Glycolipid Autoantigens

The initial studies suggesting that antibodies directed against islet glycolipids may be a component of islet autoantibodies came from the observation that monoclonal antibodies reacting with neuronal gangliosides can also react with pancreatic islets. Monoclonal antibodies A2B5 [93], 3G5 [94], R2D6 [95], and tetanus toxin all react with complex gangliosides and all react with all pancreatic islet cells. Unlike anti-islet monoclonal antibodies reacting with islet proteins (e.g., HISL19 [96], 4F2 [97]), the above anti-ganglioside antibodies are not species-specific. Gangliosides are characterized by a ceramide group bound to a complex polysaccharide with at least one sialic acid group, gangliosides are amphiphilic [36]. Buschard and co-workers have reported in abstract form that sulfatides may be a relevant autoantigen as they detect anti-sulfatide antibodies (utilizing immunochemical detection of the staining of sulfatides following thin-layer chromatography) in the majority of patients with recent-onset diabetes. Finally, Dotta et al. [98] have analyzed human islets and islets from NOD mice for gangliosides. They have isolated from human pancreatic tissue and sequenced a molecule named GM2-1, which contains a terminal sialic acid residue. In addition, they have developed a TLC-based assay for detecting autoantibodies reacting with GM2-1 and find the majority of prediabetics express antibodies reacting with this molecule.

Islet Cell Autoantigen 69 kDa (ICA69)

ICA69 is a neuroendocrine molecule of unknown function [52]. Data from at least five independent groups suggest that ICA69 may represent a target molecule of T1D-related islet autoimmunity for the following reasons: (1) Detection of autoantibodies in both first-degree relatives of T1D patients who were followed to overt diabetes in a prospective follow-up [52] and in newly diagnosed diabetic patients [99–101]. (2) Demonstration of autoreactive T cells in newly diagnosed diabetic children [102, 103]. (3) The finding that over 80% of patients with recent-onset T1D had either ICA69 autoreactive T cells or autoantibodies against this molecule [102]. (4) Tolerogenic effect in NOD mice conferred by neonatal injection of the murine peptide of ICA69 termed Tep69 [104], but the mechanism of this protection was not yet ascertained [105]. (5) Tep69-specific T cells play a driving role in the acceleration of T-cell-mediated islet cell destruction in NOD mice [106]. The latter studies are promising, and are based on the generation of a NOD mouse T-cell hybridoma that recognizes the murine peptide Tep69. To test the functional significance of the Tep69-specific T cells in NOD mice, adoptive transfer models of diabetes development were utilized. Intra-venous administration of Tep69 peptide into adoptively transferred recipients dramatically accelerated the development of diabetes in NOD mice.

ICA69 was identified by screening of a human islet expression λgt11 library with ICA-positive sera from relatives of T1D patients who developed the overt disease. This protein, which on SDS gel migrates at 69 kDa, has been sequenced and found to have in molecular mass of 54,600. The aberrant gel migra-tion at 69 kDa on a SDS-polyacrylamide gel (SDS-PAGE) can be explained by the presence of several highly charged region throughout the whole molecule, as ICA69 is not glycosylated. ICA69 is present, at least at the levels of mRNA, in brain, lung, kidney and heart, with high levels in islet and other neuroendocrine tissues. ICA69 is identical in sequence to the cow's milk-related protein p69 described by Dosch and co-workers [53]. It has been reported that diabetic indi-viduals have antibodies to BSA and there are a number of epidemiological stud-ies indicating that the neonatal ingestion of cow's milk increases the development of T1D. ICA69 and BSA share two short regions (5 amino acids) of identity and in the region of the ABBOS albumin peptide (potential T-cell epitope), four of nine amino acids are identical.

In the initial assay format, antibodies to ICA69 were detected by Western blot utilizing recombinant human ICA69. The Western blot assay format relative to radioassays for antibodies react with insulin, GAD65, and ICA512 is inade-quate, and approximately 5% of normal controls react with the recombinant molecule. ICA69 autoantibodies seem to be diabetes-related, although were also detected in patients with RA by Western blotting [99].

There is evidence that T1D may develop as a result of a failure of central and peripheral tolerance to islet cell autoantigens. We have recently demon-strated by immunostaining of thymic sections and immunoprecipitation of thymic extracts that ICA69 is expressed in human thymus. By RT-PCR [M. Pietropaolo, unpubl. observations], we have shown that dendritic cells express ICA69 in the thymus. This completes our previous observation of expression of ICA69 transcripts in both fetal and postnatal human thymus [107]. The latter findings open a number of questions as to whether normal constituents of pan-creatic β cells expressed in the thymus, such as ICA69, insulin, or both, may contribute to resistance to an organ-specific autoimmune disease such as T1D because they are expressed at levels sufficient to induce central tolerance against these molecules.

Carboxypeptidase H

Carboxypeptidase H is an autoantigen identified by screening a human islet expression library using sera from prediabetics with high ICA titer. In the secretory granules of pancreatic islets, carboxypeptidase H3 cleaves the carboxy-terminal amino acids during the processing of proinsulin to insulin [108]. This molecule is not islet-specific, being present also in bovine adrenal, pituitary and kidney [109]. In β cells, carboxypeptidase H is localized in the insulin secretory granule, where it exists both in a membrane bound (52 kDa) and soluble form (50 kDa). Carboxypeptidase H is probably the most abundant islet protein after proinsulin–insulin. In the original assay format, reactivity with the recombinant carboxypeptidase H expressed in *Escherichia coli* plaques was found in one out of four serum samples from prediabetics.

Partially Characterized Autoantigens

52 kDa

A 52-kDa antigen distinct from carboxypeptidase has been identified by Western blotting of human islet extracts using serum from diabetic patients. Interestingly, this antigen that appears to be islet cell-specific shares an antigenic determinant with the protein PC2, a component of the rubella virus capsid [110]. The rubella virus is as yet the only virus that has been clearly associated with the development of diabetes and congenital rubella infection is associated with diabetes. The 52-kDa antigen appears to be expressed within β-cell secretory granules [111] and with future characterization may contribute to a better definition of anti-islet autoimmunity.

37–38 kDa

Several investigators have reported antibodies to 38-kDa molecules in patients with T1D and their first-degree relatives. In the original description of antibodies precipitating the 64-kDa molecule form islets, it was shown that some sera also precipitated a 38-kDa molecule. Honeyman et al. [58] identified a 38-kDa autoantigen as the nuclear transcription protein jun-B by screening an islet expression library with sera from diabetic patients. The relevance of this antigen to autoantibodies of prediabetics is unclear. An additional 38-kDa molecule has been described by Pak et al. [112]. Following their observation that some new-onset diabetic patients have cytomegalovirus infection, they immunized BALB/c mice with cytomegalovirus and obtained islet-reactive monoclonal antibodies that by Western blot recognize a 38-kDa human islet protein. Roep et al. [49] have identified a 38-kDa islet granule autoantigen and peripheral blood lymphocytes (PBL) from diabetic patients proliferate against this molecule. The relationship between this antigen recognized by T cells and autoantibodies to 38-kDa molecules is unknown.

In both newly diagnosed diabetic patients and prediabetics, a prevalence of 14–22.7% for antibodies to a 38-kDa neuroendocrine membrane protein, termed

Glima 38, was recently reported [113, 114]. Interestingly, most of these patients were negative for GAD65 and/or ICA512 (IA-2) [113]. Glima 38 is an amphiphilic N-Asp glycated β-cell membrane protein that is expressed in islets and neuronal cell lines. The tertiary structure of the protein moiety itself rather than the carbohydrate structure seems to be crucial for antibody binding.

155 kDa

In 1990, McEvoy and co-workers [115] used a novel method to detect diabetes-associated autoantigens. They developed a panel of mouse monoclonal antibodies (mAbs) reacting with the rat insulinoma cell line RIN5F and displaced the binding of the mAbs to the tumor cells with diabetic sera. The binding of one of these mAbs, termed 1A2, was selectively displaced by sera from diabetic children. The 1A2 monoclonal binds specifically to the insulinoma cell line RIN5F but not the other rat or human tissues. Subsequently, the authors reported that monoclonal 1A2 recognizes by Western blot a 150-kDa membrane-bound protein DAP1 (diabetes-associated protein 1) [115]. Using an assay based on displacement of the radiolabeled monoclonal binding to rat insulinoma cells, 159 of 169 sera of children with T1D versus 10 of 351 age-matched control sera were positive for autoantibodies to DAP1. However, antibodies to this antigen can be detected in a large percentage of nondiabetic first-degree relatives (>60%).

Other Autoantigens

Elias et al. [116] suggested a role for T-cell clones reacting to the heat-shock protein 65 kDa (HSP65) in the pathogenesis of diabetes in the NOD mouse. In particular, these authors identified a peptide within the sequence of the human HSP that was recognized by T-cell clones that could transfer a transient form of diabetes and hyperglycemia. The administration of irradiation-attenuated HSP-specific T cells, or of the HSP peptide, presented diabetes in NOD mouse. Despite initial reports [117], HSP65 is apparently not an autoantigen of humoral autoimmunity in man [118].

In 1990, Johnson et al. [119] suggested that an islet-specific glucose transporter is an autoantigen in type 1A diabetes. The evidence supporting this conclusion derived from the ability of IgG fractions from diabetic patients to inhibit glucose uptake from rat islet cells. Sera from 26 of 27 patients with T1D inhibited the rat islet glucose uptake versus 0 of 5 sera from patients with type 2 diabetes. Finally, it has been recently reported that there is an increased frequency (~18%) of autoantibody reactivity to ICA12, which is the putative transcription factor SOX13, in sera from type 1 diabetic subjects compared with type 2 diabetics and healthy controls [120]. However, it is worth noting that antibodies to SOX13 can occur independently of these other conventional autoantibody markers, and so may prove a useful addition to a panel of biochemical markers used to diagnose and predict T1D.

Humoral Autoimmunity in Type 2 Diabetes

Type 2 diabetes affects in the United States an estimated 15 million individuals and costs over USD 90 billion/year [121–123]. By the time the disease is diagnosed by conventional markers, a significant number of individuals already suffer from a wide variety of complications related to diabetes. Several lines of evidence indicate that type 2 (noninsulin-dependent) diabetes mellitus is a heterogeneous disease that results from a combination of abnormalities in both insulin secretion and insulin action [124]. The causes of decreased insulin secretion in type 2 diabetes are still not completely understood, but in a subgroup of type 2 diabetic patients they may be related to an autoimmune destruction of the pancreatic β cells. Islet cell autoimmunity, which is characteristic of type 1A diabetes [125], appears to be present in up to 10–33% of subjects diagnosed clinically with type 2 diabetes [123, 126–129]. In the UK Prospective Diabetes Study (UKPDS) [130], it was reported that patients diagnosed with in type 2 diabetes, the presence of GAD and ICA was a predictor of subsequent insulin requirement compared to patients carrying both autoantibodies. These results are strikingly similar to a number of prospective studies carried out in childhood diabetes. If islet cell autoimmunity is truly present in 10–33% of subjects clinically diagnosed with type 2 diabetes [123, 128], up to 4 million Americans might have an unidentified autoimmune form of type 2 diabetes, a prevalence greater than that of recent-onset childhood diabetes. Therefore, the identification of a subgroup of individuals at risk of developing type 2 diabetes using autoantibody markers is of public health interest, not only for the identification of unclassified forms of diabetes, but also because immunomodulatory therapeutic strategies could potentially be instituted early enough in a large number of patients diagnosed as having type 2 diabetes in order to prevent complications related with hyperglycemia and, possibly, the time of onset of insulin requirement.

There has been gathering evidence that in type 2 diabetes there is a cytokine-associated acute-phase reaction, which is considered part of the innate immune response [131]. We examined the prevalence of GAD65 and IA-2 autoantibodies in elderly type 2 diabetic patients, participants of the Cardiovascular Health Study [129, 132]. Among 196 type 2 diabetic patients evaluated, GAD65 and IA-2 autoantibodies exceeded the limit of normal in 10.2% (20 out of 196) and 2.6% (5 out of 196), respectively. Autoantibodies to GAD65 and/or IA-2 exceeded the cut-off point for these two assays in 12.2% (24 of 196; p = 0.001) of the diabetic patients, a prevalence which is significantly higher than that of the nondiabetic control group 1.1% (1 out of 94). In addition, race-specific analysis revealed that autoantibodies to GAD65 and/or IA-2 were present in 13.2% (n = 18/136) of White diabetic participants and 10.2% (6/59) of Black diabetic participants. None of the Black nondiabetic controls exceeded the normal range for either autoantibody. Type 2 diabetic patients who were positive for GAD65 and/or IA-2 autoantibodies had significantly higher median 2-hour blood glucose levels after an oral glucose tolerance test at baseline as compared to GAD65 and/or IA-2 autoantibody-negative type 2 diabetic patients (302.5 vs. 220.0 mg/dl, respectively, p = 0.030). In type 2 diabetic patients with evidence of humoral islet cell autoimmunity as compared to the autoantibody-negative patients, there was a statistically significant increased frequency of oral hypoglycemic treatment over time (OHGA) (p = 0.004). Moreover, this rise in the percentage from baseline to year 7 of OHGA treatment, analyzed by McNemar test, was statistically significant in the autoantibody-positive group (p = 0.031), but not in the autoantibody-negative group (p = 0.200).

A statistically significant increase in fibrinogen (fig. 4) and C-reactive protein levels (fig. 5) was found in high-titer GAD65 and/or IA-2 autoantibody type 2 diabetic patients as compared to low-titer (p = 0.004, p = 0.03, respectively), autoantibody-negative diabetic patients (p = 0.006, p = 0.03, respectively) and nondiabetic subjects (p = 0.0006, p = 0.001, respectively). Moreover, type 2 diabetic patients with evidence of islet cell autoimmunity had reduced serum albumin levels as compared to autoantibody-negative patients (mean 3.91 ± 0.32 Ab+; 4.05 ± 0.26 Ab−; p = 0.033).

Based on our observations [129], we propose that islet cell autoimmunity in the elderly may be part of the pathogenic process involved in the generation of glucose intolerance (fig. 6). In fact, we found a pronounced activation of the acute-phase response that seems to be associated with islet cell autoimmunity [131, 133]. These results may in part explain the defect in insulin secretion as well as insulin resistance seen in type 2 diabetes.

As a model for the pathogenesis of type 2 diabetes (fig. 6), it has been proposed that type 2 diabetes is an acute-phase disease in which there are increased serum levels of cytokines that are secreted from many cells such as macrophages, adipose tissue, endothelium, etc., which may be genetically preprogrammed [131, 134–143]. Cytokines such as IL-β can act on the pancreatic β cell contributing to impaired insulin secretion. TNF-α might contribute to the development of insulin resistance by inhibiting the tyrosine kinase activity of the insulin receptor [131]. Likewise, the soluble fraction of TNF receptor 2 (sTNFR2) appears to be involved in the pathogenesis of insulin resistance [144].

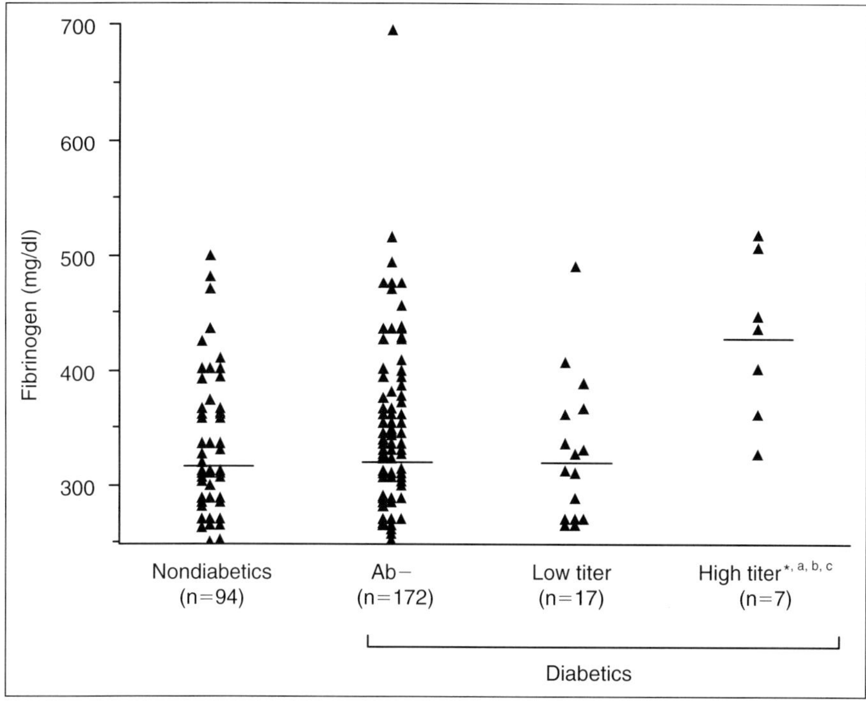

Fig. 4. Fibrinogen levels in relation to GAD65 and/or IA-2 autoantibody titer. High antibody titer was defined as 3-fold above the threshold for the GAD65 and IA-2 autoantibody assays. The median of fibrinogen values is represented by a horizontal bar. *Nonparametric analysis of variance (Kruskal-Wallis test) was used to test the difference in fibrinogen across the four groups ($p < 0.05$). A statistically significant increase in fibrinogen levels was found in high-titer GAD65 and/or IA-2 autoantibody type 2 diabetic patients (range 328–518 mg/dl) compared to: [a]low titer ($p = 0.004$; range 235–490 mg/dl); [b]autoantibody-negative diabetic patients ($p = 0.006$; range 185–695 mg/dl), and [c]nondiabetic subjects ($p = 0.0006$; range 212–499 mg/dl) [reprinted from 129, with permission].

A number of observations have indicated that TNF-α might play a key role in the pathogenesis of autoimmune diabetes [145, 146]. Of importance, McDevitt and co-workers [145] demonstrated that TNF-α-treated NOD mice exhibited T cell as well as autoantibody responses to islet autoantigens including GAD65. TNF-α may also be involved in the development of insulin resistance by inhibiting the tyrosine kinase activity of the insulin receptor [131]. An insulin secretory defect, which resembles that of type 2 diabetes, is also seen in transgenic mice expressing elevated serum TNF-α levels [147]. Other recent observations suggest that TNF-α may activate intra-islet resident macrophages, resulting in the release of IL-1β, which generates inducible nitric oxide

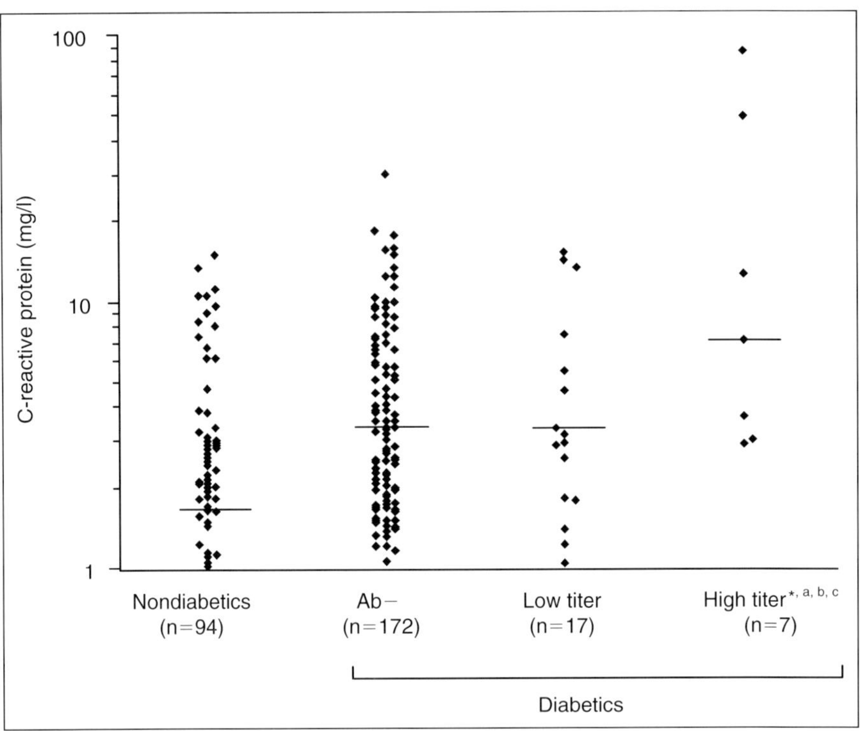

Fig. 5. C-reactive protein (CRP) levels in relation to GAD65 and/or IA-2 autoantibody titer. High antibody titer was defined as 3-fold above the threshold for the GAD65 and IA-2 autoantibody assays. The median of CRP values is represented by a horizontal bar. *Nonparametric analysis of variance (Kruskal-Wallis test) was used to test the difference in CRP across the four groups (p < 0.05). A statistically significant increase in CRP levels was found in high-titer GAD65 and/or IA-2 autoantibody type 2 diabetic patients (range 2.93–86.36 mg/l) compared to: [a]low titer (p = 0.03; range 0.48–15.46 mg/l); [b]autoantibody-negative diabetic patients (p = 0.03; range 0.18–62.79 mg/l), and [c]nondiabetic subjects (p = 0.001; range 0.17–15.09 mg/l) [reprinted from 129, with permission].

synthase (iNOS) expression and the overproduction of nitric oxide (NO) in β cells [148]. The free radical, NO, then down-regulates cellular metabolism primarily by interfering with mitochondrial function that leads to β-cell insulin resistance [148]. In addition, IL-1β may promote an impairment of insulin secretion in pancreatic β cells [149]. One attractive hypothesis is that β-cell insulin resistance may be a major determinant in creating the insulin secretory defect, which is present in type 2 diabetes [150]. In type 2 diabetic patients these β-cell abnormalities may be the result of elevated serum levels of TNF-α.

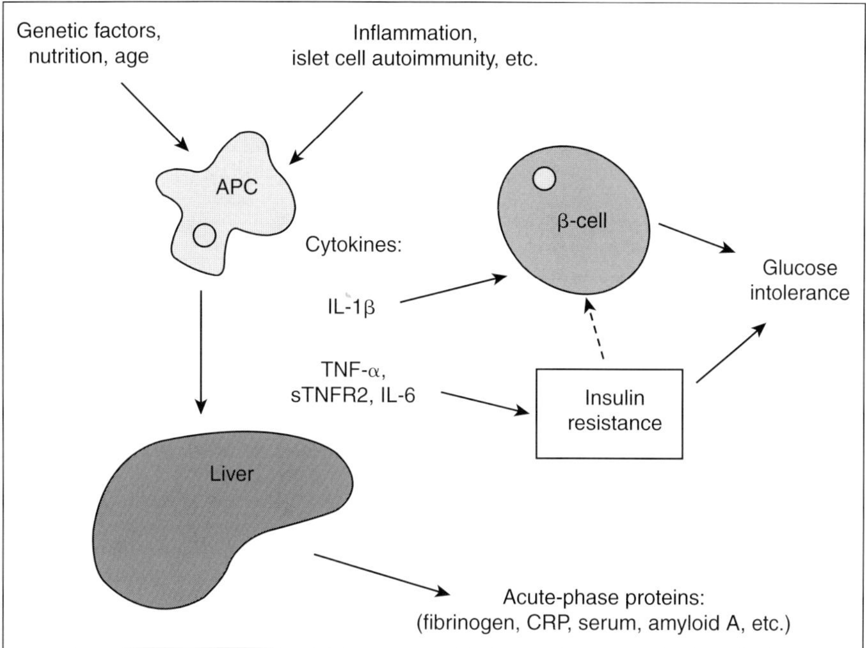

Fig. 6. A model for the role of islet cell autoimmunity, inflammation and cytokines in the pathogenesis of type 2 diabetes. Since a pronounced activation of the acute-phase response was found to be associated with islet cell autoimmunity [137], this [133] may in part explain the defect in insulin secretion as well as insulin resistance in a subset of type 2 diabetic patients.

Concluding Remarks

Despite uncertainty about the pathogenic role of islet autoantigens, the ability to biochemically detect a series of autoantibodies to islet antigens greatly facilitates the prediction of T1D. Utilizing only three radioimmunoassays for detecting autoantibodies to insulin, GAD65 and ICA512 (IA-2), 70–80% of either newly diagnosed diabetics or prediabetics (first-degree relatives who developed the overt disease during follow-up) express two or more autoantibodies. An important concept is that the occurrence of autoantibodies assorts independently in prediabetic individuals as well as in new-onset diabetics, and this justifies the utilization of a panel of biochemical markers rather than one marker alone (e.g. GAD65 autoantibodies) to diagnose and predict T1D.

However, recent observations from the Pittsburgh and Denver groups [151, 152] indicate that HLA-DQ/DR typing can identify a seronegative diabetes phenotype associated with the development of insulin requirement and the presence of

HLA-DQ/DR high-risk alleles. For instance, the fact that two HLA-DQ high-risk haplotypes can be found in a subset of autoantibody-negative prediabetics [151] might be of importance because a subset of these relatives, although negative for the most widely used conventional antibody markers, might have an autoimmune form of the disease with as yet unidentified immunological abnormalities.

The utilization of combined analysis of specific humoral autoantibody markers represents a unique tool for risk assessment for T1D in ongoing intervention trials proposed to prevent the disease process. This strategy should also facilitate the stage for the utilization of these markers in the screening for T1D in the general population. To this end, we also believe that the biochemical markers of humoral autoimmunity offer the potential to establish the proportion of type 2 diabetes cases that have islet cell autoimmunity and therefore aid in distinguishing an autoimmune form of diabetes from the classical type 2 diabetes and aid in a more appropriate classification of autoimmune diabetes.

Acknowledgements

This work was supported by NIH DK32083 (G.S.E.), NIH R01 DK53456 and an American Diabetes Association Career Development Award (M.P.).

References

1   Theofilopoulos AN: The basis of autoimmunity. I. Mechanisms of aberrant self-recognition. Immunol Today 1995;16:90–98.
2   Bellgrau D, Eisenbarth GS: Immunobiology of autoimmunity; in Eisenbarth GS (ed): Molecular Mechanisms of Endocrine and Organ-Specific Autoimmunity. Georgetown Tex, Landes, 1999, pp 1–18.
3   Rose NR: Immunologic diagnosis of autoimmuity; in Leffel MS, Donnenberg AD, Rose NR (eds): Handbook of Immunology. New York, CRC Press, 1997, pp 111–123.
4   Pietropaolo M, Eisenbarth GS: Molecular targets of the autoimmunity of type I diabetes; in Draznin B, LeRoith D (eds): Molecular Biology of Diabetes. Totowa NJ, Humana Press, 1994, vol 1, pp 1–33.
5   Galperin C, Coppel RL, Gershwin ME: Use of recombinant proteins in the diagnosis of autoimmune connective tissue diseases. Int Arch Allergy Immunol 1996;111:337–347.
6   Grubin CE, Daniels T, Toivola B, Landin-Olsson M, Hagopian WA, Li L, Karlsen AE, Boel E, Michelsen B, Lernmark A: A novel radiobinding assay to determine diagnostic accuracy of isoform-specific glutamic acid decarboxylase antibodies in childhood IDDM. Diabetologia 1994;37:344–350.
7   Gianani R, Rabin DU, Verge CF, Yu L, Babu SR, Pietropaolo M, Eisenbarth GS: ICA512 autoantibody radioassay. Diabetes 1995;44:1340–1344.
8   Beever K, Bradbury J, Phillips D, McLachlan SM, Pegg SM, Goral C, Overbeck W, Feifel G, Rees Smith B: Highly sensitive assays of autoantibodies to thyroglobulin and thyroid peroxidase. Clin Chem 1989;35:1949.

9 Osterland CK: Laboratory diagnosis and monitoring in chronic systemic autoimmune diseases. Clin Chem 1994;40:2146–2153.

10 Dorman JS, McCarthy BJ, O'Leary AL, Koehler AN: Risk factors for insulin-dependent diabetes: Diabetes in America. National Institute of Diabetes and Digestive and Kidney Diseases, 1995, pp 165–177.

11 LaPorte RE, Matsushima M, Chang YF: Prevalence and incidence of insulin-dependent diabetes: Diabetes in America. 1995; in Diabetes in America, ed 2. National Institute of Diabetes and Digestive and Kidney Diseases, NIH Publ No 95–1468, pp 37–46.

12 Palosuo T, Virtamo J, Taylor RR, Aho K, Puurunen M, Vaarala O: High antibody levels to prothrombin imply a risk of deep venous thrombosis and pulmonary embolism in middle-aged men. A nested case-control study. Thromb Haemost 1997;78:1178–1182.

13 Bingley PJ, Christie MR, Bonifacio E, Bonfanti R, Shattock M, Fonte MT, Bottazzo GF, Gale EAM: Combined analysis of autoantibodies improves prediction of IDDM in islet cell antibody-positive relatives. Diabetes 1994;43:1304–1310.

14 Verge CF, Gianani R, Kawasaki E, Yu L, Pietropaolo M, Jackson RA, Chase PH, Eisenbarth GS: Prediction of type I diabetes mellitus in first-degree relatives using a combination of insulin, glutamic acid decarboxylase and ICA512bdc/IA-2 autoantibodies. Diabetes 1996;45:926–933.

15 Pietropaolo M, Peakman M, Pietropaolo SL, Zanone MM, Foley TP, Becker DJ, Trucco M: Combined analysis of GAD65 and ICA512(IA-2) autoantibodies in organ and non-organ specific autoimmune diseases confers high specificity for insulin-dependent diabetes mellitus. J Autoimmun 1998;11:1–10.

16 Ziegler AG, Hummel M, Schenker M, Bonifacio E: Autoantibody appearance and risk for development of childhood diabetes in offspring parents with type 1 diabetes. The 2-year analysis of the German BABYDIAB study. Diabetes 1999;48:460–468.

17 Wiest-Ladenburger U, Hartmann R, Hartmann U, Berling K, Bohm BO, Richter W: Combined analysis and single-step detection of GAD65 and IA2 autoantibodies can replace the histochemical islet cell antibody test. Diabetes 1997;46:565–571.

18 Kulmala P, Savola K, Petersen JS, Vähäsalo P, Karjalainen J, Dyrberg T, Knip M, and the Childhood Diabetes in Finland (DiMe) Study Group: Prediction of insulin-dependent diabetes mellitus in siblings of children with diabetes. J Clin Invest 1998;101:327–336.

19 Maclaren NK, Lan M, Coutant R, Schatz D, Silverstein J, Muir A, Clare-Salzler M, She JX, Malone J, Crockett S, Schwartz S, Quattrin T, DeSilva M, Vander Vegt P, Notkins AL, Krischer J: Only multiple autoantibodies to islet cells (ICA), insulin, GAD65, IA-2 and IA-2β predict immune-mediated (type 1) diabetes in relatives. J Autoimmun 1999;12:279–287.

20 Eisenbarth GS, Verge CF: The immunoendocrinopathy syndromes; in Wilson J, Foster DW, Kronenberg HM, Larsen PR (eds): Williams Textbook of Endocrinology. Philadelphia, Saunders, 1998, pp 1651–1662.

21 Betterle C, Volpato M, Greggio AN, Presotto F: Type 2 polyglandular autoimmune disease (Schmidt's syndrome). J Pediatr Endocrinol Metab 1996;9(suppl 1):113–123.

22 Falorni A, Nikoshokov A, Laureti S, Grenback E, Hulting A, Casucci G, Santeusanio F, Brunetti P, Luthman H, Lernmark Å: High diagnostic accuracy for idiopathic Addison's disease with a sensitive radiobinding assay for autoantibodies against recombinant human 21-hydroxylase. J Clin Endocrinol Metab 1995;80:2752–2755.

23 Colls J, Betterle C, Volpato M, Prentice L, Smith BR, Furmaniak J: Immunoprecipitation assay for autoantibodies to steroid 21-hydroxylase in autoimmune adrenal diseases. Clin Chem 1995; 41:375–380.

24 Darnell RB: Onconeural antigens and the paraneoplastic neurologic disorders: At the intersection of cancer, immunity, and the brain. Proc Natl Acad Sci USA 1996;93:4529–4536.

25 Friday RP, Pietropaolo M: Oncogenic autoimmunity; in Eisenbarth GS (ed): Molecular Mechanisms of Endocrine and Organ-Specific Autoimmunity. Georgetown Tex, Landes, 1998, pp 63–84.

26 Bottazzo GF, Florin-Christensen A, Doniach D: Islet-cell antibodies in diabetes mellitus with autoimmune polyendocrine deficiencies. Lancet 1974;ii:1279–1282.

27 Bottazzo GF, Genovese S, Bosi E, Dean BM, Christie MR, Bonifacio E: Novel considerations on the antibody/autoantigen system in type I (insulin-dependent) diabetes mellitus. Ann Med 1991; 23:453–461.

Autoantibodies in Human Diabetes

28 Srikanta S, Rabizadeh A, Omar MAK, Eisenbarth GS: Assay for islet cell antibodies: Protein A-monoclonal antibody method. Diabetes 1985;34:300–305.

29 Riley WJ, Maclaren NK, Krischer J, Spillar RP, Silverstein JH, Schatz DA, Schwartz S, Malone J, Shah S, Vadheim C, Rotter JL: A prospective study of the development of diabetes in relatives of patients with insulin-dependent diabetes. N Engl J Med 1990;323:1167–1172.

30 Bonifacio E, Bingley PJ, Shattock M, Dean BM, Dunger D, Gale EAM, Bottazzo GF: Quantification of islet-cell antibodies and prediction of insulin-dependent diabetes. Lancet 1990;335:147–149.

31 Bonifacio E, Lernmark A, Dawkins RL: Serum exchange and use of dilutions have improved precision of measurement of islet cell antibodies. J Immunol Methods 1988;106:83–88.

32 Landin-Olsson M: Precision of the islet-cell antibody assay depends on the pancreas. J Clin Lab Anal 1990;4:289–294.

33 Bonifacio E, Boitard C, Gleichman H, Shattock MA, Molenaar JL, Bottazzo GF: Assessment of precision, concordance, specificity, and sensitivity of islet cell antibody measurements in 41 assays. Diabetologia 1990;33:731–736.

34 Gianani R, Pugliese A, Bonner-Weir S, Shiffrin AJ, Soeldner JS, Erlich H, Awdeh ZL, Alper CA, Jackson RA, Eisenbarth GS: Prognostically significant heterogeneity of cytoplasmic islet cell antibodies in relatives of patients with type I diabetes. Diabetes 1992;41:347–353.

35 Dotta F, Previti M, Tiberti C, Lenti L, Pugliese G, DiMario U: Identification of the GM2-1 islet ganglioside: Similarities with a major neuronal autoantigen. Diabetes 1992;41:365A.

36 Colman PG, Nayak RC, Campbell IL, Eisenbarth GS: Binding of cytoplasmic islet cell antibodies is blocked by human pancreatic glycolipid extracts. Diabetes 1988;37:645–652.

37 Genovese S, Bonifacio E, McNally JM, Dean BM, Wagner R, Bosi E, Gale EAM, Bottazzo GF: Distinct cytoplasmic islet cell antibodies with different risks for type I (insulin-dependent) diabetes mellitus. Diabetologia 1992;35:385–388.

38 Kaufman DL, Erlander MG, Clare-Salzler M, Atkinson MA, Maclaren NK, Tobin AJ: Autoimmunity to two forms of glutamate decarboxylase in insulin-dependent diabetes mellitus. J Clin Invest 1992;89:283–292.

39 Dotta F, Previti M, Lenti L, Dionisi S, Casetta B, D'Erme M, Eisenbarth GS, DiMario U: GM2-1 pancreatic islet ganglioside: Identification and characterization of a novel islet-specific molecule. Diabetologia 1995;38:1117–1121.

40 Greenbaum CJ, Palmer JP, Nagataki S, Yamaguchi Y, Molenaar JL, Van Beers WAM, Maclaren NK, Lernmark A, and Participating Laboratories: Improved specificity of ICA assays in the Fourth International Immunology of Diabetes Exhange Workshop. Diabetes 1992;41: 1570–1574.

41 Atkinson MA, Maclaren NK: Islet cell autoantigens in insulin-dependent diabetes. J Clin Invest 1993;92:1608–1616.

42 Tisch R, McDevitt H: Insulin-dependent diabetes mellitus. Cell 1996;85:291–297.

43 Henderson LO, Atkinson M, Bingley PJ, Bonifacio E, Bosi E, Ellis TE, Eisenbarth GS, Hagopian WA, Hannon WH, Jackson R, Lernmark A, Lou D, Mei J, Orban T, Smith SJ, Vogt RJ, Wasserfall C, Williams AJK, Yu L, Study Group-DBSTAR Group: Dried blood-spots as a matrix for type 1-diabetes autoantibody detection. Diabetes 2000;49(suppl 1):A36.

44 Greenbaum CJ, Palmer JP, Kuglin B, Kolb H, and Participating Laboratories: Insulin autoantibodies measured by radioimmunoassay methodology are more related to insulin-dependent diabetes mellitus than those measured by enzyme-linked immunosorbent assay: Results of the Fourth International Workshop on the Standardization of Insulin Autoantibody Measurement. J Clin Endocrinol Metab 1992;74:1040–1044.

45 Baekkeskov S, Aanstoot H, Christgau S, Reetz A, Solimena MS, Cascalho M, Folli F, Richter-Olsen H, DeCamilli P: Identification of the 64K autoantigen in insulin dependent diabetes as the GABA-synthesizing enzyme glutamic acid decarboxylase. Nature 1990;347:151–156.

46 Rabin DU, Pleasic S, Palmer-Crocker R, Shapiro JA: Cloning and expression of IDDM-specific human autoantigens. Diabetes 1992;41:183–186.

47 Lan MS, Lu J, Goto Y, Notkins AL: Molecular cloning and identification of a receptor-type protein tyrosine phosphatase, IA-2, from human insulinoma. DNA Cell Biol 1994;13: 505–514.

48 Wasmeier C, Hutton JC: Molecular cloning of phogrin, a protein tyrosyne phosphatase homologue localized to insulin secretory granule membranes. J Biol Chem 1996;271:18161–18170.

49 Roep BO, Kallan AA, Hazenbos WLW, Bruining GJ, Bailyes EM, Arden S, Hutton JC, DeVries RRP: T-cell reactivity to a 38 kD insulin-secretory-granule protein in patients with recent-onset type 1 diabetes. Lancet 1991;337:1439–1441.

50 Christie MR, Hollands JA, Browm TJ, Michelsen BK, Delovitch TL: Detection of pancreatic islet 64,000 M_r autoantibodies in insulin-dependent diabetes distinct from glutamate decarboxylase. J Clin Invest 1993;92:240–248.

51 Karjalainen J, Martin JM, Knip M, Ilonen J, Robinson BH, Savilahti E, Akerblom HK, Dosch HM: A bovine albumin peptide as a possible trigger of insulin-dependent diabetes mellitus. N Engl J Med 1992;327:302–307.

52 Pietropaolo M, Castano L, Babu S, Buelow R, Kuo YLS, Martin S, Martin A, Powers AC, Prochazka M, Naggert J, Leiter EH, Eisenbarth GS: Islet cell autoantigen 69 kDa (ICA69): Molecular cloning and characterization of a novel diabetes associated autoantigen. J Clin Invest 1993;92:359–371.

53 Miyazaki I, Gaedigk R, Hui MF, Cheung RK, Morkowski J, Rajotte RV, Dosch HM: Cloning of human and rat p69, a candidate autoimmune target in type I diabetes. Biochem Biophys Acta 1994;1227:101–104.

54 Elias D, Cohen IR, Shechter Y, Spirer Z, Golander A: Antibodies to insulin receptor followed by anti-idiotype. Diabetes 1987;36:348–354.

55 Birk OS, Douek DC, Elias D, Dewchand H, Gur SL, Walker MD, Van der Zee R, Cohen IR, Altmann DM: A role of Hsp60 in autoimmune diabetes: Analysis in a transgenic model. Proc Natl Acad Sci USA 1996;93:1032–1037.

56 Atkinson MA, Maclaren NK, Scharp DW: No evidence for HSP65 in insulin dependent diabetes. Lancet 1990;337:263–264.

57 Georgopoulos C, McFarland H: Heat shock proteins in multiple sclerosis and other autoimmune diseases. Immunol Today 1993;14:373–375.

58 Honeyman MC, Cram DS, Harrison LC: Transcription factor jun-B is target of autoreactive T-cells in IDDM. Diabetes 1993;42:626–630.

59 Yu L, Robles DT, Abiru N, Rewers M, Kelemen K, Eisenbarth GS: Early expression of anti-insulin autoantibodies of humans and the NOD mouse: Evidence for early determination of subsequent diabetes. Proc Natl Acad Sci USA 2000;97:1701–1706.

60 Pugliese A, Zeller M, Fernandez A Jr, Zalcberg LJ, Barlett RJ, Ricordi C, Pietropaolo M, Eisenbarth GS, Benett ST, Patel DD: The insulin gene is transcribed in the human thymus and transcription levels correlate with allelic variation at the *INS* VNTR-*IDDM2* susceptibility locus for type 1 diabetes. Nature Genet 1997;15:293–297.

61 Menon RK, Cohen RM, Sperling MA, Gutfield WS, Mimouni F, Khoury JC: Transplacental passage of insulin in pregnant women with insulin dependent diabetes mellitus. N Engl J Med 1990;323:309–315.

62 Hirata Y, Ishizu H, Ouchi N, Motumura S, Abe M, Hara Y, Wakasugi H, Takahashi I, Sakano H, Tanaka M, Kawano H, Kanesaki T: Insulin autoimmunity in a case with spontaneous hypoglycaemia. Jpn J Diabetes 1970;13:312–319.

63 Uchigata Y, Tokunaga K, Takayama-Hasumi S, Miyamoto M, Juji T, Kuwata S, Eguchi Y, Omori Y, Hirata Y: Strong association of insulin autoimmune syndrome with HLA-DR4. Lancet 1992; 339:393–394.

64 Uchigata Y, Hirata Y, Omori Y, Iwamoto Y, Tokunaga K: Worldwide differences in the incidence of insulin autoimmune syndrome (Hirata disease) with respect to the evolution of HLA-DR4 alleles. Hum Immunol 2000;61:154–157.

65 Hirata Y: Methimazole and insulin autoimmune syndrome with hypoglycemia. Lancet 1983;99: 182–184.

66 Palmer JP, Asplin CM, Clemons P, Lyen K, Tatpati O, Raghu PK, Paquette TL: Insulin antibodies in insulin-dependent diabetics before insulin treatment. Science 1983;222:1337–1339.

67 Ziegler AG, Ziegler R, Vardi P, Jackson RA, Soeldner JS, Eisenbarth GS: Life table analysis of progression to diabetes of anti-insulin autoantibody-positive relatives of individuals with type I diabetes. Diabetes 1989;38:1320–1325.

68 Verge CF, Stenger D, Bonifacio E, Colman PG, Pilcher C, Bingley PJ, Eisenbarth GS: Combined use of autoantibodies (IA-2 autoantibody, GAD autoantibody, insulin autoantibody, cytoplasmic

islet cell antibodies) in type 1 diabetes: Combinatorial Islet Autoantibody Workshop. Diabetes 1998;47:1857–1866.

69 Castano L, Ziegler AG, Ziegler R, Shoelson S, Eisenbarth GS: Characterization of insulin auto-antibodies in relatives of patients with type 1 diabetes. Diabetes 1993;42:1202–1209.

70 Rewers M, Norris JM, Eisenbarth GS, Erlich HA, Beaty B, Klingensmith G, Yu L, Bugawan TL, Hamman M, Groshek M, McDuffie RS Jr: Beta-cell autoantibodies in infants and toddlers without IDDM relatives: Diabetes autoimmunity study in the young (DAISY). J Autoimmun 1996; 9:405–410.

71 Pugliese A, Bugawan T, Moromisato R, Awdeh ZL, Alper CA, Jackson RA, Erlich HA, Eisenbarth GS: Two subsets of HLA-DQA1 alleles mark phenotypic variation in levels of insulin autoanti-bodies in first-degree relatives at risk for insulin-dependent diabetes. J Clin Invest 1993;93: 2447–2452.

72 Lernmark A: Glutamic acid decarboxylase – Gene to antigen to disease. J Intern Med 1996; 240:259–277.

73 Bu DF, Erlander MG, Hitz BC, Tillakaratne NJK, Kaufman DL, Wagner-McPherson CB, Evans GA, Tobin AJ: Two human glutamate decarboxylases, 65-kDa GAD and 67-kDa GAD, are each encoded by a single gene. Proc Natl Acad Sci USA 1992;89:2115–2119.

74 Bu DF, Tobin AJ: The exon-intron organization of the genes (GAD1 and GAD2) encoding two human glutamate decarboxylases (GAD67 and GAD65) suggests that they derive from a common ancestral GAD. Genomics 1994;21:222–228.

75 Baekkeskov S, Landin M, Kristensen JK, Srikanta S, Bruining GJ, Mandrup-Poulsen T, Beaufort C, Soeldner JS, Eisenbarth GS, Lindgren F, Sundquist G, Lernmark A: Antibodies to a 64,000 M, human islet cell antigen precede the clinical onset of insulin dependent diabetes. J Clin Invest 1987;79:926–934.

76 Christie MR, Landin-Olsson M, Sundkvist G, Dahlquist G, Lernmark A, Baekkeskov S: Antibodies to a M_r-64,000 islet cell protein in Swedish children with newly diagnosed Type I (insulin-dependent) diabetes. Diabetologia 1988;31:597–602.

77 Erdo SL, Wolff JR: Gamma-aminobutyric acid outside the mammalian brain. J Neurochem 1990; 54:363–372.

78 Solimena M, Butler MH, De Camilli P: GAD, diabetes, and stiff man syndrome: Some progress and more questions. J Endocrinol Invest 1994;17:509–520.

79 Hagopian WA, Michelsen B, Karlsen AE, Larsen F, Moody A, Grubin CE, Rowe R, Petersen J, McEvoy R, Lernmark A: Autoantibodies in IDDM primarily recognize the 65,000-M_r rather than the 67,000-M_r isoform of glutamic acid decarboxylase. Diabetes 1993;42:631–636.

80 Hallengren B, Falorni A, Landin-Olsson M, Lernmark A, Papadopoulos KI, Sundkvist G: Islet cell and glutamic acid decarboxylase antibodies in hyperthyroid patients: At diagnosis and following treatment. J Intern Med 1996;239:63–68.

81 Richter W, Endl J, Eiermann TH, Brandt M, Kientsch-Engel R, Thivolet C, Jungfer H, Scherbaum WA: Human monoclonal islet cell antibodies from a patient with insulin-dependent diabetes mellitus reveal glutamate decarboxylase as the target antigen. Proc Natl Acad Sci USA 1992;89: 8467–8471.

82 Richter W, Shi Y, Baekkeskov S: Autoreactive epitopes in glutamic acid decarboxylase defined by diabetes-associated human monoclonal antibodies. Proc Natl Acad Sci USA 1992;90:2832–2836.

83 Daw K, Powers AC: Two distinct glutamic acid decarboxylase autoantibody specificities in IDDM target different epitopes. Diabetes 1995;44:216–220.

84 Schwartz HL, Chandonia JM, Kash SF, Kanaani J, Tunnell E, Domingo A, Cohen FE, Banga JP, Madec AM, Richter W, Baekkeskov S: High-resolution autoreactive epitope mapping and struc-tural modeling of the 65 kDa form of human glutamic acid decarboxylase. J Mol Biol 1999; 287:983–999.

85 Rabin DU, Pleasic SM, Shapiro JA, Yoo-Warren H, Oles J, Hicks JM, Goldstein DE, Rae PMM: Islet cell antigen 512 is a diabetes-specific islet autoantigen related to protein tyrosine phos-phatases. J Immunol 1994;152:3183–3188.

86 Notkins AL, Lan MS, Leslie RD: IA-2 and IA-2 beta: The immune response in IDDM. Diab Metab Rev 1998;14:85–93.

87 Pietropaolo M, Hutton JC, Eisenbarth GS: Protein-tyrosine phosphatase-like proteins: Link with insulin-dependent diabetes mellitus. Diabetes Care 1997;20:208–214.

88 Solimena M, Dirkx R Jr, Hermel JM, Pleasic-Williams S, Shapiro JA, Caron L, Rabin DU: ICA512, an autoantigen of type I diabetes, is an intrinsic membrane protein of neurosecretory granules. EMBO J 1996;15:2102–2114.

89 Christie MR, Tun RYM, Lo SSS, Cassidy D, Brown TJ, Hollands J, Shattock M, Bottazzo GF, Leslie DG: Antibodies to GAD and tryptic fragments of islet 64K antigen as distinct markers for development of IDDM: Studies with identical twins. Diabetes 1992;41:782–787.

90 Payton MA, Hawkes CJ, Christie MR: Relationship of the 37,000- and 40,000-M_r tryptic fragments of islet antigens in insulin-dependent diabetes to the protein tyrosine phosphatase-like molecule IA-2 (ICA512). J Clin Invest 1995;96:1506–1511.

91 Hawkes CJ, Wasmeier C, Christie MR, Hutton JC: Identification of the 37 kD antigen in IDDM as the tyrosine phosphatase-like protein (phogrin) related with IA-2. Diabetes 1996;45: 1187–1192.

92 Bonifacio E, Lampasona V, Genovese S, Ferrari M, Bosi E: Identification of protein tyrosine phosphatase-like IA2 (islet cell antigen 512) as the insulin-dependent diabetes-related 37/40K autoantigen and a target of islet-cell antibodies. J Immunol 1995;155:5419–5426.

93 Kundu SK, Pleatman MA, Redwine WA, Boyd AE, Marcus DM: Binding of monoclonal antibody A2B5 to gangliosides. Biochem Biophys Res Commun 1983;116:836–842.

94 Nayak RC, Attawia NA, Cahill CJ, King GL, Ohashi O, Moromisato R: Expression of a monoclonal antibody (3G5) defined ganglioside antigen in renal cortex. Kidney Int 1992;41:1638–1645.

95 Alejandro R, Shienvold FL, Hajek SAV, Pierce M, Paul R, Mintz DH: A ganglioside antigen on the rat pancreatic B cell surface identified by monoclonal antibody R2D6. J Clin Invest 1984;74: 25–38.

96 Srikanta S, Krisch K, Eienbarth GS: Islet cell protein defined by monoclonal islet cell antibody HISL-19. Diabetes 1986;35:300–305.

97 Srikanta S, Telen M, Posillico JT, Dolinar R, Krisch K, Haynes BF, Eisenbarth GS: Monoclonal antibodies to a human islet cell surface glycoprotein: 4F2 and LC7-2. Endocrinology 1987;120: 2240–2244.

98 Dotta F, Colman PG, Nayak RC, Lombardi D, Scharp DW, Andreani D, Pontieri GM, DiMario U, Lenti L, Eisenbarth GS: Ganglioside expression in human pancreatic islets. Diabetes 1989; 38:1478–1483.

99 Martin S, Kardorf J, Schulte B, Lampeter EF, Gries FA, Melchers I, Wagner R, Bertrams J, Roep BO, Pfutzner A, Pietropaolo M, Kolb H: Autoantibodies to the islet antigen ICA69 occur in IDDM and in rheumatoid arthritis. Diabetologia 1995;38:351–355.

100 Ronningen KS, Atrazhev A, Luo L, Luo C, Smith DK, Korbutt G, Rajotte RV, Elliot JF: Anti-BSA antibodies do not cross react with the 69 kDa islet cell autoantigen ICA69. J Autoimmun 1998; 11:223–231.

101 Kerokoski P, Ilonen J, Gaedigk R, Dosch HM, Knip M, Hakala M, Hinkkanen A: Production of the islet cell antigen ICA69 (p69) with baculovirus expression system: Analysis with a solid-phase time-resolved fluorescence method of sera from patients with IDDM and rheumatoid arthritis. Autoimmunity 1999;29:281–289.

102 Roep BO, Duinkerken G, Schreuder GMTh, Kolb H, De Vries RRP, Martin S: HLA-associated inverse correlation between T cell and antibody responsiveness to islet autoantigen in recent-onset insulin dependent diabetes mellitus. Eur J Immunol 1996;26:1285–1289.

103 Roep BO: T-cell responses to autoantigens in IDDM. The search for the holy grail. Diabetes 1996;45:1147–1156.

104 Karges W, Hammond-McKibben D, Cheung RK, Visconti M, Shibuya N, Kemp D, Dosch HM: Immunological aspects of nutritional diabetes prevention in NOD mice. A pilot study for the cow's milk-based IDDM prevention. Diabetes 1997;46:557–564.

105 Karges W, Hammond-McKibben D, Gaedigk R, Shibuya N, Cheung R, Dosch HM: Loss of self-tolerance to ICA69 in nonobese diabetic mice. Diabetes 1997;46:1548–1556.

106 Gunaratnam L, Wiener S, Hammond-McKibben D, Cheung R, Dosch HM: Precipitation of IDDM by the ICA69 T cell epitope peptide, Tep69. Diabetes 1998;47(suppl 1):A34.

107 Friday RP, Jaquin-Gerstl A, Mathie T, Rudert WA, Trucco M, Pietropaolo M: Potential role for thymic expression of the IDDM-related autoantigen ICA69 in central immune tolerance. Am J Hum Genet 1998;63(suppl 4):A181.
108 Castano L, Russo E, Zhou L, Lipes MA, Eisenbarth GS: Identification and cloning of a granule autoantigen (carboxypeptidase H) associated with type I diabetes. J Clin Endocrinol Metab 1991; 73:1197–1201.
109 Davidson HW, Hutton JC: The insulin secretory granule carboxtypeptidase H. Biochem J 1993; 245:575–582.
110 Karounos DG, Wolinsky JS, Thomas JW: Monoclonal antibody to rubella virus capsid protein recognizes a β-cell antigen. J Immunol 1993;150:3080–3085.
111 Karounos DG, Simmerman L, Hickman SL, Jacob RJ: Identification of the p52-rubella related autoantigen as an insulin secretory granule protein. Diabetes 1993;42(suppl 1):221A.
112 Pak CY, Cha CY, Rajotte RV, McArthur RG, Yoon JW: Human pancreatic islet cell specific 38 kD autoantigen identified by cytomegalovirus-induced monoclonal islet cell autoantibody. Diabetologia 1990;33:569–572.
113 Aanstoot HJ, Kang SM, Kim J, Lindsay LA, Roll U, Knip M, Atkinson A, Mose-Larsen P, Fey S, Ludvigsson J, Landin M, Bruining J, Maclaren N, Akerblom HK, Baekkeskov S: Identification and characterization of Glima 38, a glycosylated islet cell membrane antigen, which together with GAD65 and IA2 marks the early phases of autoimmune response in type I diabetes. J Clin Invest 1996;97:2772–2783.
114 Roll U, Turck CW, Gitelman SE, Rosenthal SM, Nolte MS, Masharami U, Ziegler AG, Baekkeskov S: Peptide mapping and characterization of glycation patterns of the glima 38 antigen recognized by autoantibodies in type I diabetic patients. Diabetologia 2000;43:598–608.
115 Pan YX, Thomas NM, McEvoy RC: Purification and partial characterization of 160 kD and 155 kD beta cell membrane antigens detected by MAb 1A2. Autoimmunity 1993;15(suppl):77.
116 Elias D, Reshef T, Birk OS, Zee R, Walker MD, Cohen IR: Vaccination against autoimmune mouse diabetes with a T-cell epitope of the human 65-kDa heat shock protein. Proc Natl Acad Sci USA 1991;88:3088–3091.
117 Jones DB, Hunter NR, Duff GW: Heat-shock protein 65 as a beta cell antigen of insulin-dependent diabetes. Lancet 1990;336:583–585.
118 Atkinson MA, Holmes LA, Sharp DW, Lacy PE, Maclaren NK: No evidence for serological autoimmunity toward islet cell heat shock proteins in insulin-dependent diabetes. J Clin Invest 1991;7:721–724.
119 Johnson TH, Crider BP, McCorkle K, Alford M, Unger RH: Inhibition of glucose transport into rat islet cells by immunoglobulins from patients with new-onset insulin-dependent diabetes mellitus. N Engl J Med 1990;322:653–659.
120 Kasimiotis H, Myers MA, Argentaro A, Mertin S, Fida S, Ferraro T, Olsson J, Rowley MJ, Harley VR: Sex-determining region Y-related protein SOX13 is a diabetes autoantigen expressed in pancreatic islets. Diabetes 2000;49:555–561.
121 Harris MI, Flegal KM, Cowie CC, Eberhardt MS, Goldstein DE, Little RR, Wiedmeyer HM, Byrd-Holt DD: Prevalence of diabetes, impaired fasting glucose, and impaired glucose tolerance in US adults. The Third National Health and Nutrition Examination Survey, 1988–1994. Diabetes Care 1998;21:518–525.
122 Ray NF, Wills S, Thamer M, and Medical Technology and Practice Patterns Institute; in American Diabetes Association: Direct and Indirect Costs of Diabetes in the United States in 1992. Alexandria Va 1993.
123 Rewers M, Hamman RF: Risk factors for non-insulin dependent diabetes; in National Institute of Diabetes and Digestive and Kidney Diseases: Diabetes in America. National Institutes of Health, 1995, pp 179–220.
124 Kahn CR: Insulin action, diabetogenes and the cause of type II diabetes. Banting Lecture. Diabetes 1994;43:1066–1084.
125 The Expert Committee on the Diagnosis and Classification of Diabetes mellitus: Report of the Expert Committee on the Diagnosis and Classification of Diabetes mellitus. Diabetes Care 1997; 20:1183–1197.

126 Tuomi T, Groop LC, Zimmet PZ, Rowley MJ, Knowles WJ, Mackay IR: Antibodies to glutamic acid decarboxylase reveal latent autoimmune diabetes in adults with a non-insulin-dependent onset of diabetes. Diabetes 1993;42:359–362.

127 Tuomilehto J, Zimmet P, Mackay IR, Koskela P, Vidgren G, Toivanen L, Tuomilehto-Wolf E, Kohtamaki K, Stengard J, Rowley MJ: Antibodies to glutamic acid decarboxylase as predictors of insulin-dependent diabetes mellitus before clinical onset of disease. Lancet 1994;343:1383–1385.

128 Zimmet PZ, Tuomi T, Mackay IR, Rowley MJ, Knowles W, Cohen M, Lang A: Latent autoimmune diabetes mellitus in adults (LADA): The role of antibodies to glutamic acid decarboxylase in diagnosis and prediction of insulin dependency. Diabet Med 1994;11:299–303.

129 Pietropaolo M, Barinas-Mitchell E, Pietropaolo SL, Kuller LH, Trucco M: Evidence of islet cell autoimmunity in elderly patients with type 2 diabetes mellitus. Diabetes 2000;49:32–38.

130 Turner R, Stratton I, Horton V, Manley S, Zimmet P, Mackay IR, Shattock M, Bottazzo GF, Holman R, for UK Prospective Diabetes Study (UKPDS) Group: UKPDS 25: Autoantibodies to islet-cell cytoplasm and glutamic acid decarboxylase for prediction of insulin requirement in type 2 diabetes. Lancet 1997;350:1288–1293.

131 Pickup JC, Crook MA: Is type II diabetes mellitus a disease of the innate immune system? Diabetologia 1998;41:1241–1248.

132 Fried LP, Borhani NO, Enright P, Furberg CD, Gardin JM, Kronmal RA, Kuller LH, Manolio TA, Mittelmark MB, et al: The cardiovascular health study: Design and rationale. Ann Epidemiol 1991;1:263–276.

133 Pickup JC, Mattock MB, Chusney GD, Burt D: NIDDM as a disease of the innate immune system: Association of acute phase reactants and interleukin-6 with metabolic syndrome X. Diabetologia 1997;40:1286–1292.

134 Fearon DT, Locksley RM: The instructive role of innate immunity in the acquired immune response. Science 1996;272:50–54.

135 Medzhitov R, Janeway CA: Innate immunity: Impact on the adaptive immune response. Curr Opin Immunol 1997;9:4–9.

136 Ganda OP, Arkin CF: Hyperfibrinogenemia. An important risk factor for vascular complications in diabetes. Diabetes Care 1992;15:1245–1250.

137 Kuller LH, Eichner JE, Orchard TJ, Grandits GA, McCallum L, Tracy RP, for the Multiple Risk Factor Intervention Trial Research Group: The relation between serum albumin levels and risk of coronary heart disease in the Multiple Risk Factor Intervention Trial. Am J Epidemiol 1991; 134:1266–1277.

138 Bruno G, Cavallo-Perin P, Bargero G, Borra M, D'Errico N, Pagano G: Association of fibrinogen with glycemic control and albumin excretion rate in patients with non-insulin-dependent diabetes mellitus. Ann Intern Med 1996;125:653–657.

139 Kuller LH, Tracy RP, Shaten J, Meilahn EN, for the Multiple Risk Factor Intervention Trial Research Group: Relation of C-reactive protein and coronary heart disease in the MRFIT nested case-control study. Am J Epidemiol 1996;144:537–547.

140 Corti MC, Salive ME, Guralnik JM: Serum albumin and physical function as predictors of coronary heart disease mortality and incidence in older persons. J Clin Epidemiol 1996;49:519–526.

141 Luoma PV, Näyhä S, Sikkilä K, Hassi J: High serum alpha-tocopherol, albumin, selenium and cholesterol, and low mortality from coronary heart disease in northern Finland. J Intern Med 1994;237:49–54.

142 Macy E, Hayes T, Tracy RP: Variability in the measurement of C-reactive protein in healthy subjects: Implications for reference interval and epidemiological applications. Clin Chem 1997;43:52–58.

143 Mantovani A, Bussolino F, Introna M: Cytokine regulation of endothelial cell function: From molecular level to the bedside. Immunol Today 1997;18:231–240.

144 Fernàndez-Real JM, Broch M, Casamitjana R, Gutirrez C, Vendrell J, Richart C: Plasma levels of the soluble fraction of tumor necrosis factor receptor 2 and insulin resistance. Diabetes 1998;47:1757–1762.

145 Yang XD, Tisch R, Singer SM, Cao ZA, Liblau R, Schreiber RD, McDevitt HO: Effect of tumor necrosis factor α on insulin-dependent diabetes mellitus in NOD mice. I. The early development of autoimmunity and the diabetogenic process. J Exp Med 1994;180:995–1004.

146 Green EA, Flavell RA: Tumor necrosis factor-α and the progression of diabetes in nonobese mice. Immunol Rev 1999;169:11–22.
147 Augustine KA, Rossi RM, Van G, Housman J, Stark K, Danilenko D, Varnum B, Medlock E: Noninsulin-dependent diabetes mellitus occurs in mice ectopically expressing the human *axl* tyrosine kinase receptor. J Cell Physiol 1999;181:443–447.
148 Kwon G, Xu G, Marshall CA, McDaniel ML: Tumor necrosis factor-α-induced pancreatic β-cell insulin resistance is mediated by nitric oxide and prevented by 15-deoxy-$\Delta^{12,14}$-prostaglandin J_2 and aminoguanidine. J Biol Chem 1999;274:18702–18708.
149 Arnush M, Heitmeier MR, Scarim AL, Marino MH, Manning PT, Corbett JA: IL-1 produced and released endogenously within human islets inhibits β cell function. J Clin Invest 1998;102:516–526.
150 Kulkarni RN, Winnay JN, Postic C, Magnuson MA, Kahn CR: Tissue-specific knockout of the insulin receptor in pancreatic β cells creates an insulin secretory defect similar to that in type 2 diabetes. Cell 1999;96:329–339.
151 Pietropaolo M, Becker DJ, Dorman JS, LaPorte RE, Mazumdar S, Rudert WA, Trucco M: Are GAD65 and IA-2 autoantibodies sufficient to predict type 1 diabetes? Diabetes 1999;48 (suppl 1):A45.
152 Cornell CN, Stankiewicz W, Asher D, Babu S, Yu L, Bao F, Yu J, Nelson J, Bugawan T, Korman E, Eisenbarth GS, Rewers MJ: Absence of islet autoantibodies at diabetes onset does not rule out type 1A diabetes. Diabetes 2000;49(suppl 1):A68.

Dr. George S. Eisenbarth, Executive Director, Barbara Davis Center for Childhood Diabetes, University of Colorado Health Sciences Center,
Box B140, 4200 East 9th Avenue, Denver, CO 80262 (USA)
Tel. +1 303 315 4891, Fax +1 303 315 4892, E-Mail george.eisenbarth@uchsc.edu

von Herrath MG (ed): Molecular Pathology of Type 1 Diabetes mellitus.
Curr Dir Autoimmun. Basel, Karger, 2001, vol 4, pp 283–307

Transplantation Tolerance

Jason L. Gaglia, David M. Harlan

National Institute of Diabetes and Digestive and Kidney Diseases,
Navy Transplantation and Autoimmunity Branch, Naval Medical Research Center,
Bethesda, Md., USA

The Challenge of Organ Transplantation

There is a currently available cure for type 1 diabetes mellitus, transplantation with either whole pancreas or isolated pancreatic islets. Major advances in clinical organ transplantation over the past 50 years have changed its status from that of an experimental procedure to a widely accepted treatment modality. In the last 10 years alone, the number of solid organ transplants in the USA has nearly doubled with over 23,000 Americans receiving organ transplants in 1998 [1]. This increase has been driven primarily by improvements in immunosuppressive therapies to prevent acute rejection. It is now unusual to lose a primary renal allograft to rejection within the first 6 months after transplant [2]. Improvements in early graft survival however have not been without cost. Significant morbidity and side effects result from the use of immunosuppressive agents. Depending upon the specific agent used, these side effects can include infection, malignancy, osteoporosis, hyperlipidemia, diabetes mellitus, cataracts and renal dysfunction [3–5]. Given the comorbidities associated with type 1 diabetes mellitus, these potential side effects are particularly devastating for persons who undergo transplantation as a result of diabetic end organ failure. Additionally, transplant recipients often must follow a complex daily drug-dosing schedule to find a balance between adequate immunosuppression and complication risks. Even when current immunosuppressive therapy is associated with relatively little additional morbidity, the therapy remains costly, and usually must be taken for life.

Unfortunately, the great strides achieved in improving early graft survival have not translated into similar marked improvements in long-term graft survival. While recent long-term survival statistics for organ allografts have shown significant improvements [6], chronic rejection remains a difficult problem [7]. Indeed, most of the gains have been secondary to a decrease in acute rejection while chronic rejection rates remain a major problem; most pancreatic allografts fail within 10 years of transplantation. Recent UNOS statistics indicate a 1-year-lone pancreatic graft survival rate of 77% while the 5-year survival rate is a mere 35% (the pancreatic graft survival rates for combined kidney pancreas transplants are somewhat better at 89 and 65% respectively) [1]. These statistics emphasize the need for improved ways to prevent immune mediated chronic allograft rejection. Considering the shortcomings of immunosuppressive approaches, there has been growing interest in inducing what is somewhat flippantly referred to as 'graft tolerance'. The ideal outcome would be to prevent graft rejection and thus permit indefinite graft survival without the need for maintenance immunosuppression. The theoretical benefits of this approach are obvious. The requirement for continued immunosuppressive regimens and associated side effects would be eliminated. Instead, the graft would function as a native organ without a decline in function secondary to acute or chronic rejection.

What Is Tolerance?

Transplant tolerance is broadly defined in operational terms. It is the specific functional persistence of a transplanted tissue or organ without the development of a detrimental immune response or the need for continued therapeutic intervention. Unlike immunosuppression, organ survival secondary to tolerance does not come at the expense of protective immunity. To be clinically relevant, tolerance must also be durable. That is not to say that transplant tolerance must be forever immutable or unbreakable. Indeed, self-tolerance is itself fragile as reflected by the fact that individuals may develop any number of autoimmune diseases. Instead, evidence suggests strongly that both transplant tolerance and self-tolerance are dynamic processes that must be reinforced. In more general terms, tolerance is a physiologic process that prevents autoimmune illnesses and transplant tolerance is an attempt to utilize these processes to prevent the otherwise destructive immune response to donor tissue.

Tolerance itself is a learned phenomenon. More than 50 years ago, Owen [8] observed that dizygotic bovine twins can on occasion share a common placental circulation leading to 'a mixture of two distinct types of lymphocytes'. Shortly thereafter, Medawar's group [9, 10] showed that as with monozygotic twins, these twins were tolerant to skin grafts from each other while dizygotic twins which did not share a common placenta were not. Later,

Medawar's group [11] was able to experimentally induce tolerance by injection of tissue into embryos or neonatal mice. These mice then became tolerant to subsequent skin grafts from the same donor strain as the injection source. Shortly thereafter, Burnet [12] introduced the clonal selection theory which postulated that clones of lymphocytes capable of recognizing self-antigens are selectively deleted during development. This deletion of autoreactive clones proceeds efficiently in utero, as the developing immune system is programmed to accept all its own tissues. The extent to which this clonal deletion process continues after birth is less clear. However, new antigens appear throughout life commensurate with physiological changes, such as puberty and pregnancy. It makes teleological sense that the immune system must possess some mechanism to induce tolerance toward these newly arising self-antigens so as to prevent the induction of autoimmune disease. By embracing this physiology of learned tolerance it should be possible for the immune system to become tolerized to transplanted organs.

Mechanisms of Tolerance

Self-tolerance is initiated during T-cell development in the thymus (so-called 'central' tolerance) and is maintained in the periphery. Experimental data supports the existence of several redundant mechanisms of tolerance, these include deletion, anergy, ignorance, indifference and suppression or regulation. These terms can be simply defined. Tolerance through *deletion* means that the lymphocytes capable of recognizing that target have been eliminated. *Anergy*, on the other hand, allows for the peaceful coexistence of an antigen and lymphocytes that recognize that antigen, but means that those lymphocytes have actively 'decided' to leave that antigen alone. Tolerance through *ignorance* means that lymphocytes cannot react against an antigen if the two never meet and *indifference* is a subset of ignorance. As lymphocytes recognize antigens presented by major histocompatibility complex (MHC) proteins, if MHC expression is very low, a cell could be invisible to a cell-specific lymphocyte even if the two did meet. Finally, tolerance can be mediated through *suppression*. That is, a lymphocyte that would normally destroy an antigen could be prevented from carrying out that destruction by a third cell called a suppressor or regulatory cell. Interestingly, anergic cells may themselves act as a regulatory population by competing for antigen on the antigen-presenting cell (APC) surface, consumption of locally produced cytokines, or direct modulation of the APC [13–15].

The cellular immune system is typically divided into two subsets, the innate and adaptive immune systems. The innate system consists of phagocytic cells such as polymorphonuclear cells, monocytes, macrophages and dendritic cells that use a variety of mechanisms to recognize, engulf and digest pathogens. The adaptive immune system consists of T and B lymphocytes that possess antigen specificity. The specificity required for tolerance can only be mediated through these antigen receptors, namely the T-cell receptor (TCR) and antibody. No other known molecules possess sufficient diversity for this task. Thus antigen receptors allow the immune system to recognize its world. Clearly, agents that block the antigen receptors in form or function would lead to a temporary lack of recognition (immunosuppression) but cannot lead to tolerance. At some point, there must be antigen recognition before a specific tolerant state can develop. For this reason, tolerance induction may be mutually exclusive from immunosuppression. That is, current immunosuppressive agents typically work by attempting to 'hide' the organ from the immune system and to thereby interrupt specific recognition and thus the overall immune response, but in a nonspecific fashion.

The immune system is itself defensive in nature and must be provoked to attack. The alternative, a reactive immune system, would lead to frequent inappropriate immune activation and autoimmunity. Similarly, even when an immune attack is warranted, it must be limited. Anything more than a transient attack would be deleterious and lead to a gradual increase in the number of lymphocytes over time. Therefore, a mechanism to limit or terminate a response once antigen has been cleared must exist. Otherwise, and especially if there is any risk that some of the response could be maintained against cross-reacting environmental or self-antigens, the protective immune response could turn destructive. A clinical example of this is infection with the chronic nonlytic viral infection hepatitis B. When immunity downregulates, the patient lives as a chronic carrier. When the immune system does *not* downregulate, much of the liver is destroyed and the patient may die of fulminant hepatic failure. Another illustrative clinical example is that of infectious mononucleosis caused by the Epstein-Barr virus (EBV). During the acute infection, a very high proportion of an individual's T cells are engaged in fighting that infection. Once the infection is controlled, the vast majority of those anti-EBV T cells die because they are no longer needed [16]. The immune system thus must be able to rise to a proper defense, then retreat or we would all become one giant lymph node!

There is growing evidence that immune responses to both self and foreign antigens are subject to regulatory T-cell responses. A classic example of such immune regulation is how T cells can regulate each other. Depending upon a number of microenvironmental factors such as cytokine milieu, antigen dose, and costimulatory signals, upon activation naive T cells can differentiate into T-helper type 1 (Th1) or T-helper type 2 (Th2) effector cells. The Th1 cells promote a proinflammatory response by secreting a number of cytokines including interleukin-2 (IL-2) and interferon-γ (IFN-γ). This arm of the effector mechanism has been associated with a vigorous immune response including various autoimmune diseases and transplant rejection. On the other hand, the Th2 cells provide cognate help to B cells and secrete IL-4, IL-5 and IL-10. These two types of T cells are able to 'cross-regulate' each other, such that Th2 can dampen Th1 responses, and vice versa.

The Role of T Lymphocytes in Antigraft Immune Responses and Tolerance

Although many cell types can participate in the immune response, only T cells are absolutely required for rejection. More than 20 years ago, experiments by Hall et al. [17] demonstrated that T cells orchestrate both the induction and effector functions of allograft rejection. His group showed that previously irradiated rodents were incapable of allograft rejection without the infusion of T cells and that other immune cells could not reconstitute this activity. An adult human has approximately 10^{10}–10^{12} T lymphocytes, each of which can recognize one target based upon its TCR. It is estimated that a normal immunocompetent person has a T-cell repertoire capable of recognizing approximately 10^8–10^{10} different antigens [18]. However, T lymphocytes do not recognize whole antigens but peptides presented by MHCs. Class I MHCs, present on nearly all cells, primarily present fragments of endogenous proteins made within the cell while class II MHCs, found only on specialized APCs, present fragments of phagocytosed exogenous proteins. A limited amount of cross-presentation between these two pathways is known to occur. Approximately 0.1–1% of an individual's T cells are cross-reactive with an unmatched donor's MHC complex proteins and can potentially directly recognize transplanted tissue as non-self. It is believed that the allogeneic MHC molecule resembles a self-MHC molecule plus foreign peptide at the three-dimensional level [19]. In addition, transplanted tissue can also be indirectly recognized when the host's innate immune system encounters damaged tissue or cells from within the transplant (especially after surgical trauma or ischemic injury) leading to phagocytosis of the damaged transplanted cells. Proteins then digested from these phagocytosed cells can be presented via class II molecules to the host T cells.

The Two-Signal Model of T-Cell Activation

For an appropriate response to occur, the immune system must address both the specificity of the antigen and the context in which the antigen is encountered. This specificity is mediated through antigen receptors such as TCRs and antibodies and is often referred to as signal 1. Similarly, the context is provided by a second signal that is referred to as signal 2. This second signal is required for full T-cell activation. There are two important corollaries to these concepts. First, delivery of signal 1 without signal 2 will prevent T-cell proliferation by rendering the T cell anergic or causing it to undergo apoptosis. Even when restimulated with competent APCs and antigen, these T cells will not proliferate or produce IL-2 [20]. Second, delivery of signal 2 without signal 1 is a neutral event for the T cell, having no effect. This two-signal model was initially formulated by Bretscher and Cohn [21] and later generalized by Lafferty and Cunningham [22] in the early 1970s and is the basis for many of the techniques designed to prevent allograft rejection. The majority of clinical transplant interventions have focused on signal 1 by preventing TCR expression or signaling. As previously discussed, blockade of signal 1 is unfortunate since induction of specific tolerance requires functional receptors for recognition.

From the two-signal model it should be apparent that the T cell by itself cannot be the initiator of an immune response. Instead it is the context provided by the delivery of the second signal through interactions with APCs that control the T-cell response. Since the description of the first T-cell costimulatory receptor, CD28, by June et al. [23], modern immunology has better defined these costimulatory requirements. Although it is clear that there are many potential costimulatory molecules, under normal circumstances two T-cell-based molecules are believed to dominate these responses: CD28 and CTLA4 (CD152) [24, 25]. In general, simultaneous ligation of CD28 and TCR leads to T-cell activation. Once activated, T cells also express CTLA4 which plays a role in downregulating immune responses. Two ligands for CD28/CTLA4 have thus far been identified, B7-1 (CD80) and B7-2 (CD86) [26, 27]. Both B7 molecules can bind to either CD28 or CTLA4. The balance between CD28 and CTLA4 costimulation can provide direction to an adaptive immune response. Mice lacking the negative regulatory element of CTLA4 eventually die from a gradual accumulation of T cells and massive lymphoproliferation [28, 29]. The B7s are generally not found at high levels on resting parenchymal cells but rather are expressed on activated cells including dendritic cells, T cells, B cells, macrophages and endothelium. Resting B cells express no detectable B7-1 and very low levels of B7-2, while both are upregulated following activation [30]. This pattern of expression should reinforce quiescence of any autoreactive T cells that have escaped central tolerization while strengthening the response to antigens in areas of APC activation. However, the temporal modulation of B7/CD28/CTLA4 molecules alone probably does not provide enough context for an appropriate immune response. Instead, other regulators of the adaptive immune response such as the recently identified ICOS and PD-1 receptors for

B7 family members or costimulation via CD40L (CD154) may also play a role [31–33].

The Immune Context of Transplantation

Organ transplantation is not a benign process for the organ being transplanted. There is a great potential for ischemic and vascular damage to the organ. Further, after any allograft is implanted, the organ must be reperfused by the recipient's blood. Upon reperfusion a whole cascade of receptor : ligand interactions are triggered leading to downstream activation of the local inflammatory response. One such interaction is that between CD40 and CD40L (CD154). This interaction has already been noted to provide additional context to the immune system and may play an integral role in activating the immune system in a setting of endothelial damage. From a teleological standpoint it would make sense for the immune system to be brought to a heightened state when the probability of infection is high; one such situation is trauma. When endothelium is traumatized, the basement membrane is often exposed causing platelet activation. The integrin receptors on the platelets bind the RGD motifs of the underlying structural proteins that make up the extracellular matrix thus causing platelet activation [34]. Upon activation, platelets express a number of inflammatory mediators including CD154 [35]. This CD154 is itself a potent activator of endothelial cells and APCs causing an upregulation of MHC class I and II expression in addition to costimulators such as the CD80, CD86 and ICAM-1 (CD54) [36, 37]. This platelet-endothelial cell interaction also increases APC survival by upregulating antiapoptotic genes like bcl2 and conferring resistance to fas-mediated cell death. Given the ischemic injury and endothelial damage inherent in tissue transplantation, the transplant system is already biased towards an aggressive immune response. For the remainder of this chapter, we will discuss current experimental approaches to turn this reactive recognition into a nonaggressive one.

Given the critical role that T lymphocytes play in allograft rejection, current efforts towards the induction of tolerance have focused on these cells. As previously noted, there are three fundamental processes which may lead to T-cell tolerance: deletion, anergy and regulation or suppression. These processes are not mutually exclusive and multiple mechanisms may be responsible for the establishment or maintenance of tolerance. Although one mechanism may predominate to induce tolerance, another may be required for its maintenance. As such, the therapies described in the remainder of this chapter should not be viewed in isolation but instead in terms of their underlying mechanisms and how they may influence the dynamic state called tolerance.

Central Tolerance: Intrathymic Deletion and Chimerism

During maturation, T-cell precursors enter the subcapsular region of the thymus and migrate from the cortex towards the medulla. As the T-cell precursors or thymocytes mature and traverse the thymus they go through a process of both positive and negative selection. Thymocytes unable to bind self-MHC are ignored and die by apoptosis while thymocytes able to bind self-MHC are positively selected. Although positive selection restricts the TCR repertoire to those T cells capable of interacting with self-MHC, this process leaves some T cells with a very strong affinity for MHC and a potential for autoreactivity. Thus, during a negative selection process these potentially autoreactive T cells are eliminated. Importantly, both bone marrow-derived dendritic cells as well as thymic stromal cells mediate negative selection. In the end, only 2–5% of the original T-cell precursors ultimately leave the thymus as mature T cells after undergoing these selection processes [38]. Despite the rigorous nature of selection, autoreactive cells can escape thymic deletion and peripheral methods of tolerance are required to prevent autoimmunity.

An understanding of this potent form of self-tolerance has led investigators to attempt to harness central tolerance as a method of inducing transplant tolerance. Theoretically, one could first deplete peripheral T lymphocytes already in the circulation, then one could directly inject antigen into the thymus to negatively select new T lymphocytes from developing with any undesired specificity. Thus the periphery would be reconstituted with a T-cell repertoire deficient in cells capable of responding to the transplanted tissue. Indeed, Posselt et al. [39] demonstrated that injection of pancreatic islets directly into the thymus after treatment with antilymphocyte serum led to donor-specific unresponsiveness in a rodent model. Similar results can be obtained in a non-human primate model by intrathymic injection of hematopoietic stem cells [40]. However, given the need for intrathymic injection, the temporal limitations of cadaveric transplantation, and that the thymus is known to involute dramatically after puberty, this approach has not proven clinically practical.

Instead, investigators have attempted to use the ability of hematopoietic cells to home to the thymus and become thymic dendritic cells to overcome these problems. The aforementioned studies of Owen and Medawar provide the basis for using bone marrow transplantation (BMT) to induce donor-specific tolerance in adult recipients through chimerism. The most promising studies in the field have attempted to establish a durable multilineage chimerism known as macrochimerism. There are two types of macrochimerism: full chimerism, wherein donor cells replace the entire hematopoietic system of the recipient, and mixed chimerism, wherein the hematopoietic populations of both the recipient and donor coexist in the recipient. There is also a third type of chimerism, microchimerism, that may occur spontaneously after solid organ transplant and will be discussed later. Mixed chimerism offers several advantages over full chimerism in that it requires less toxic regimens to prepare the host for the BMT and results in both greater immunocompetence and reduced susceptibility to graft-versus-host disease (GVHD).

Mixed macrochimerism attempts to recapitulate natural central tolerance. First, since in an adult recipient mature T cells with anti-donor reactivity already exist in the periphery, the endogenous T-cell population must be eliminated or inactivated to prevent acute rejection of the organ or infused donor bone marrow. Next, donor bone marrow is infused and donor cells are encouraged to engraft in the recipient's bone marrow compartment. For sufficient levels of engraftment to occur, some form of recipient pretreatment such as irradiation or myelosuppression is usually required. Once the stem cells have engrafted, they coexist with the recipient's stem cells throughout the lifetime of the recipient, giving rise to cells of all hematopoietic lineages and from both donor and recipient. Similarly, hematopoietic stem cells seed the thymus giving rise to dendritic cells that mediate clonal deletion. The antigens expressed on cells of hematopoietic origin within the thymus help mediate negative selection leading to elimination of both host- and donor-reactive T cells. Consequently, a state of central deletional tolerance is achieved for the duration of the chimerism whereby the T-cell repertoire in the mixed chimera is tolerant towards both donor and host. As with the case of natural central tolerance, peripheral methods of tolerance reinforcement must also play a role in mixed chimerism since presumably there is a population of cells with donor or host reactivity that naturally escape thymic deletion. As noted, this general T-cell depletion also decreases allospecifc cells, thus lessening the potential number of effectors attempting to eliminate the graft and possibly aiding in the development of tolerance. This is of particular importance for patients with autoimmune diseases such as type 1 diabetes mellitus. Other strategies for T-cell depletion and their proposed mechanisms of action will be subsequently discussed in more detail.

Although mixed-chimerism protocols appear reasonably well tolerated in experimental models, the need for recipient pretreatment make it untenable for cadaveric transplantation. Chimeric tolerance has also proven difficult to translate to higher animals, and successful tolerance induction through mixed chimerism in non-human primates has required splenectomy to prevent alloantibody production [41]. Future therapies may utilize costimulatory blockade or high doses of peripheral blood stem cells to replace the pretreatment steps required for initial engraftment and chimerism [42–45]. These strategies could more easily establish chimerism at the time of surgery. The same costimulatory blockade may also help establish B-cell tolerance otherwise lacking in non-human primates without additional interventions. Costimulatory blockade will be discussed in detail in a later section.

Mechanisms of Peripheral Tolerance

A massive deletion of T cells occurs during thymic maturation. Nevertheless, it is clear that some autoreactive T cells escape into the periphery. Fortunately, most people do not develop autoimmune disease because there are also strong mechanisms of peripheral tolerance to control these potentially hazardous T cells.

Depletion of Donor APCs from the Graft
The bulk of APC:lymphocyte interactions occur in the lymph nodes. Approximately 20 years ago investigators showed that if one depletes APCs from a graft so that APCs cannot migrate to the regional nodes, antigraft immunity is prevented and peripheral tolerance is established. This tolerance can subsequently be broken with the re-introduction of donor APCs [46–48]. The majority of these experiments were performed under conditions with minimal reperfusion injury thereby eliminating most host APC recruitment. Not surprisingly, a well-vascularized reperfused organ like a whole pancreas or kidney is more likely to recruit host APCs, leading to indirect presentation. This probably explains, at least in part, why depleting APCs from vascularized organs has been less successful than hoped at prolonging allograft survival. However, such an approach (donor APC depletion) may still prove useful as evidenced by the Edmonton group's recently reported success with pancreatic islet transplantation in humans (a cellular graft) [49].

Microchimerism
Migration of graft-derived APCs has been documented in murine models by Larsen et al. [50] and more recently by Starzl et al. [51–53] in human transplant recipients. Starzl's observations led to a resurgence in microchimerism research since he reported evidence for microchimerism in transplant recipients with long-surviving renal or hepatic allografts [51–53]. This chimerism occurred spontaneously in a percentage of transplant recipients and is thought to result from the migration of persistent donor hematopoietic cells from the transplanted organ. These cells can frequently be found in secondary lymphoid organs, but the level of chimerism in these cases is significantly lower than that seen in macrochimeras. One potential technique for induction of this microchimerism is the simultaneous infusion of donor bone marrow with the transplant. In a recent trial by Starzl's group, adjuvant infusion of donor bone marrow correlated with a higher steroid withdrawal rate, a decreased acute cellular rejection rate, and fewer episodes of chronic rejection in pancreas transplant recipients [54]. However, the concept is itself controversial since others have not been able to demonstrate a causal role for this microchimerism in the

induction of graft acceptance or tolerance [55–57]. Thus, while APC migration may occur in almost all cases following transplantation, tolerance is probably more a product of the specific cellular interactions that follow than the mere presence of donor cells in the secondary lymphoid tissues alone. In any case, current immunosuppressive regimens certainly effect and may well hamper interactions that would support tolerance induction.

Depleting Strategies of Tolerance Induction

Calne [58] and Jonker et al. [59] proposed the concept that a 'window of opportunity for immunological engagement (WOFIE)' may exist immediately posttransplant. The notion is that if one could prevent rejection in the immediate posttransplant period while the graft is 'healing in', and yet allow immune system engagement with alloantigen, tolerance may ensue. Clearly however, immunosuppression using conventional agents that block TCR signaling discourages effective engagement with alloantigen and results only rarely in tolerance. Calne [60] has continued along this theoretical line to speculate that by reducing the number of lymphocytes during the peritransplant period, one could remove a sufficient amount of T-cell 'help' to predispose the immune system to tolerogenic encounters with the graft.

Immunotoxin
Treatment with T-cell immunotoxin (IT) is proving to be a promising method for reliably inducing long-term graft survival in non-human primates. Neville et al. [61] have developed an IT by coupling a mutant diphtheria toxin to an anti-CD3-specific Mab. The reagent, FN18-CRM9, was designed specifically for use with non-human primates and does not bind human CD3. Unlike other anti-CD3 regimens, T-cell death and not CD3 internalization occurs following IT administration. The agent is quite efficient as it depletes $>99\%$ of all CD3-positive cells from the peripheral and central lymphoid organs within 48 h. T-cell repopulation takes 2 weeks to 2 months, depending upon the age of the animal. When the IT is given 7 days prior to transplant to deplete T cells below detectable levels at the time of surgery, renal allografts are usually accepted without further immunosuppression. The investigators have also questioned whether injecting donor lymphocytes into the recipient's thymus could enhance IT's tolerogenic effect but the data was inconclusive, as IT was so effective alone. This tolerance is also specific. The investigators have demonstrated that third-party skin grafts are rejected, but not donor skin grafts [62].

When IT is given immediately posttransplant, acute rejection is avoided but most animals go on to develop an accelerated chronic rejection [63]. However, improved results are seen when DR/CD3-depleted bone marrow is given peri-transplant, with or without lymphoid irradiation before the BMT, and some animals develop donor-specific tolerance [64]. Chimerism has been noted to follow renal allotransplantation with IT therapy whether or not there is accompanying BMT. However, this chimerism has not correlated with rejection-free survival.

The exact mechanisms underlying IT's efficacy remain unclear and are probably more complex than simple T-cell depletion. Although IT is relatively well tolerated, it can cause a vascular leak syndrome with accompanying cytokine release. The discordant results observed between the animals given IT 1 week before transplant versus those treated immediately peritransplant suggest that IT-induced cytokine release in the presence of the newly transplanted organ may antagonize tolerance induction. Interestingly, deoxyspergualin and methylprednisolone given together in combination with IT completely prevented the peritransplant IT-induced vascular leak syndrome and proinflammatory cytokine release, and also resulted in rejection-free graft survival in 75% of the recipients [65]. Subsequent mechanistic studies found that the DSG-treated animals had a significant decrease in mature lymph node dendritic cells, and displayed a Th2-like cytokine profile [66]. Thus the ability of the IT plus DSG combination to induce stable tolerance may rely upon (alone or in combination) a reduced lymph node T-cell mass, mature dendritic cell depletion, and/or a change in the cytokine milieu.

Campath 1H

Campath 1H is a depleting humanized anti-CD52 monoclonal antibody currently being tested in small clinical trials. It is a powerful lytic agent for both T and B lymphocytes but does not affect bone marrow stem cells [67]. This antibody can reverse rejection episodes [68] and has recently proven successful as an induction phase agent followed by monotherapy to prevent renal allograft rejection [60]. To be more specific, patients were given induction therapy consisting of two 20-mg doses of Campath 1H in the perioperative period (methylprednisolone was given 30 min before the first dose to minimize cytokine release) and then were maintained on a half-dose monotherapy with cyclosporine indefinitely. This 'prope tolerance' has proven very effective with only 6 episodes of steroid-responsive rejection in 31 renal allograft recipients [60].

Anti-CD3 Monoclonal Antibodies

OKT3 is a murine IgG2a monoclonal antibody (mAb) directed against human CD3. The agent has shown greater efficacy, when compared with conventional steroid therapy, in the treatment of acute rejection [69, 70]. OKT3 exerts its immunosuppressive effects by inducing rapid and profound peripheral T-cell depletion, within minutes to hours after initial dosage. During treatment with OKT3, T cells devoid of CD3 but expressing other T cells surface molecules

(CD2, CD4, CD8) reappear in the peripheral circulation as well as the graft [71]. These cells quickly recover CD3 expression after treatment cessation or after in vitro incubation sans OKT3 [72]. OKT3 is mitogenic in vitro. When the agent is initially administered in vivo it frequently causes a cytokine release syndrome. This is due to the fact that mAb binding to the T cell triggers the release of cytokines, including TNF-α, IL-2 and IFN-γ. As noted with anti-CD3 immunotoxin, such a cytokine release may be detrimental to the development of tolerance. In general, anti-CD3 therapy suffers from many of the same deficiencies as other signal 1 directed therapies. Although anti-CD3 is an effective immunosuppressant, it does not have a lasting tolerizing effect as graft function was only slightly prolonged in non-human primate allograft models employing skin [73] and liver [74] grafts. In these studies, rejection correlated with peripheral T-cell repopulation.

The activation-related adverse effects associated with these mitogenic anti-CD3 therapies led to interest in developing less toxic, nonmitogenic anti-CD3 mAb therapies. In murine models, these nonmitogenic antibodies cause a proximal signaling defect after delivering a partial TCR signal [75]. Studies suggest that this partial TCR signal induces Th1 cell clonal anergy while selectively promoting the development of T cells with a Th2 phenotype [75, 76]. Such activity may help generate an environment favorable for the development of tolerance. Two humanized anti-CD3 antibodies have been designed and generated to be nonmitogenic: an aglycosyl form of CAMPATH 3 [77] and HuM291 [78]. Both of these humanized anti-CD3 antibodies have been engineered for low Fc binding efficacy and thus low mitogenic activity. HuM291 has been shown to mediate complete and reversible circulating T-cell depletion in chimpanzees [79], while the aglycosyl form of CAMPATH 3 has been shown to be nondepleting in humans. Rather, aglycosyl CAMPATH 3 results in the downmodulation of CD3 expression without a decrease in the CD4+ or CD8+ expression in lymphocyte populations [77]. Clinical experience with these reagents has been limited, but in a small scale clinical trial, the aglycosyl CAMPATH 3 was considered efficacious for the treatment of acute renal rejection [77]. Although mitogenic anti-CD3 has not proven successful in the induction of transplant tolerance, these nonmitogenic forms may change the peritransplant cytokine milieu and directly effect different Th subtypes. These agents therefore warrant further investigation, either alone or in combination with other agents such as DSG.

Nondepleting Strategies for Immunomodulation

Waldmann et al. [80] have proposed a 'civil service model' of tolerance in which anergized T cells play the dominant role. In this model, an isolated T cell can become activated and respond to antigens only through collaboration with other helpful T cells. The model further proposes that the same isolated T cell encountering antigen but in the absence of help, will follow a default pathway to 'switch off' or become tolerant. In the absence of help these T cells do not die, but are functionally maintained in an 'impotent' (anergized) state competing for antigen at sites of presentation and reinforcing the tolerant state. Thus, tolerance is maintained by increasing meaningless activity to block positive action, analogous to one popular view of government civil service [81].

Anti-CD4/8 Monoclonal Antibodies

Several groups have investigated the use of antibodies targeted against the TCR accessory molecules CD4 or CD8. In general, the anti-CD4 antibodies have promoted prolongation of allograft survival. Anti-CD8 antibodies have shown variable efficacy in preventing the initiation of rejection [82–84], but have shown efficacy in treating rejection and extending survival in a non-human primate kidney transplant model [85]. Interestingly, the immunosuppressive effect in most studies with anti-CD4 has not been dependent on CD4 T-cell clearance. Instead, antibodies that merely coat the CD4+ T cells without causing their severe depletion have been equally efficacious at prolonging allograft survival [86].

The ability of nondepleting anti-CD4 antibodies to induce tolerance in naive or primed animals has been demonstrated in several transplant and autoimmunity models [87–91]. In lower animal models this tolerance is maintained by a population of CD4+ T cells that have developed the capacity to specifically disable other naive lymphocytes [87]. It has been suggested that this form of dominant or 'infectious' tolerance may develop in the presence of the antibody by selectively promoting antigen-specific regulatory T cells at the expense of Th1 CD4 cells [reviewed in 81, 92]. The term 'infectious tolerance' was first coined by Richard Gershon [93] and later expanded by Cobbold and Waldman. The term describes the phenomena of suppression or tolerance and, importantly, the conferral of this ability onto new T-cell cohorts (i.e. tolerance and suppression are *infectious*) [92]. The mechanisms by which these regulatory T cells are generated or prevent the emergence of new effector cells remains to be elucidated. One of the striking features of such tolerance is the ability to arrest destruction by primed effectors and even act through linked suppression, eventually generating tolerance to third-party antigens [94].

Recently, Cooke's group [95] demonstrated a direct action in vivo of nondepleting anti-CD4 on activated effector cells as well. In a nonobese diabetic-*scid* (NOD-*scid*) model they were able to halt the destruction of pancreatic islets caused by the adoptive transfer of a diabetogenic CD4+ Th1 clone. Using this model, they noted an immediate decrease in proinflammatory cytokine production, cessation of β-cell destruction, and disappearance of infiltrating cells from the pancreas after administration of anti-CD4. Based upon this data it is likely that nondepleting anti-CD4 therapy has both direct and indirect mechanisms of preventing anti-graft effector function.

The findings in non-human primates have not been as dramatic. Early studies did not see a profound prolongation in graft survival, perhaps because the murine anti-CD4 antibodies were rapidly cleared [82]. Subsequent studies utilizing humanized antibodies however have only prolonged graft survival to around 45 days, i.e. they have not reproduced the tolerant state observed in rodent models [86, 96]. Consistent with the rodent studies, depleting and nondepleting anti-CD4 antibodies were equally efficacious in the non-human primate model system [86]. In animals receiving depleting antibodies, rejection was noted soon after the graft was infiltrated by CD4+ cells, even if circulating CD4 cell numbers remained depressed.

Anti-IL-2 Receptor

Two anti-IL-2 receptor α chain (CD25) antibodies have been recently approved to be prophylactic against acute rejection in renal transplantation: basiliximab (Simulect) [97] and daclizumab (Zenapax) [98]. Neither of these agents have demonstrated an ability to treat acute rejection. Due to CD25's short cytoplasmic tail, these antibodies do not trigger cellular activation or a cytokine release syndrome upon binding. Rather these antibodies block the IL-2 receptor and inhibit T-cell proliferation and differentiation. As with anti-CD4 therapies, these anti-CD25-based therapies have demonstrated reliable immunosuppression without evidence that they induce tolerance. The agents warrant mentioning however as specific inhibition of cytokine action can profoundly affect the immune system and therefore may find use similar to that previously described for DSG [65, 66].

Costimulatory Pathway Modulation

Strategies that modulate costimulatory pathway signaling are showing promise in establishing a prolonged rejection-free state. In the two-signal model of T-cell activation, the costimulatory signal is essential for naive T-cell activation. Under this paradigm, if a T cell were to encounter antigen in the presence of an agent interfering with costimulatory signaling (referred to as 'signal 1 without signal 2'), the responding T cell is rendered anergic. Interest in utilizing such an approach greatly increased after Lenschow et al. [99] reported a landmark study demonstrating that administering an agent designed to block costimulatory ligand specifically prevented xenograft rejection in mice. The investigators studied mice with chemically induced diabetes into which they transplanted under one kidney capsule human pancreatic islets. Not surprisingly, the mice were only temporarily cured of diabetes as the disease recurred when the human islets were rejected. However, when Lenschow et al. administered the B7-blocking agent CTLA4-Ig to the mice for the 2 weeks immediately following the islet transplantation, the mice never rejected the islets. CTLA4-Ig is a fusion protein that contains the ligand-binding domain of CTLA4 (CD152) and therefore effectively binds to both CD80 and CD86, preventing the B7s from interacting with both CD28 and CD152. The transplanted human islets were shown to be responsible for maintaining the mouse blood sugar within the normal range by surgically removing the islet-bearing kidney, a procedure that led to prompt diabetes recurrence. Most dramatic, Lenschow et al. observed that after removal of the islet-bearing kidney, islets injected under the contralateral kidney capsule could rescue the mouse from diabetes only if the islets came from the original donor. Islets from a different human donor were once again promptly rejected. Thus the animals appear to have been specifically tolerized to tissue from the original donor. More recent reports using rodent models have suggested that costimulatory blockade with CTLA4-Ig can even abrogate the vasculopathy characteristic of chronic rejection [100–102].

As has too often been the case however, efforts to translate these impressive results from the rodent models to non-human primates have been disappointing. When Bluestone's group [103] attempted islet transplantation in non-human primate experiments using CTLA4-Ig to prevent rejection, prolonged islet function was observed in only 2 out of 5 rhesus monkeys and that function was maintained for only 2–3 months total. Similarly, in a non-human primate renal transplant model, Kirk et al. [104] were able to delay acute rejection with CTLA4-Ig but could not induce a tolerant phenotype with CTLA4-Ig alone.

In the mid-1990s, reports first appeared to suggest that CD40:CD154 modulating agents could similarly induce long-term graft survival. Parker et al. [105] found that donor-specific small lymphocytes and anti-CD154 given before and for 7 weeks after allogeneic pancreatic islet transplantation could permanently prevent graft rejection and relapse of diabetes. Rossini et al. [106] proposed that the protective mechanism was similar to that induced by CTLA4-Ig. They proposed that anti-CD154 prevented APCs from upregulating B7 expression, a response that normally occurs after the interaction of the recipient's T cell's CD154 with the transfused donor small B cell's CD40. This allowed the peptide-MHC complex present on the donor small lymphocytes to interact with the host's anti-donor specific T cells without B7/CD28 costimulation, thus rendering the host T cells tolerant. Unlike other costimulatory strategies, therapy directed at this receptor ligand pair does not just target T-cell activation directly. Instead, preventing upregulation of other costimulators interrupts the context for an immune reaction.

Reasoning that anti-CD154 prevented B7 upregulation and that CTLA4-Ig bound to B7 thus further inhibited B7/CD28 signaling, Larsen et al. [107] tested the effect of both agents together in a mouse full-thickness skin allograft model. Even in the relatively permissive rodent, skin transplantation is a rigorous test for therapies designed to prevent rejection as no simple, nontoxic regimen had ever been reported to effectively prevent rejection. While neither anti-CD154 nor CTLA4-Ig alone prevented rejection, a short course of both agents together induced long-term allograft survival. Interestingly, this graft-sparing effect was abrogated when cyclosporine A was added to the treatment regimen. Recalling that the two-signal model requires a TCR-mediated signal in the absence of the costimulatory signal to induce tolerance, and that cyclosporine interferes with TCR-mediated signal transduction, one would expect use of such an agent to interfere with tolerance induction as was seen. In this setting, modulating costimulatory pathway signaling appeared to help maintain a benign-appearing context, while allowing the immune system to recognize alloantigen so that a tolerant state was induced.

With the success of this dual therapy in the murine full-thickness skin allograft model, a similar regimen was tested in a rhesus monkey renal allograft model. Rhesus monkeys treated with both CTLA4-Ig and anti-CD154 for 28 days posttransplant maintained normal graft function for at least 6 months. Surprisingly, anti-CD154 alone was as effective at preventing acute rejection as the combination therapy [104]. Subsequent experiments with anti-CD154 monotherapy have demonstrated the agent to be very effective at promoting long-term renal allograft function [108]. Interestingly, long-term survivors lost their mixed lymphocyte reactivity in a donor-specific manner, but still formed donor-specific antibody. Moreover, allograft biopsies showed an active T-cell infiltrate within the graft, yet those cells did not invade the renal blood vessels, tubules or glomeruli [108]. Anti-CD154 has induced similar graft survival in a rhesus monkey islet allograft model [109].

Early clinical trials utilizing a humanized anti-CD154 manufactured by Biogen were halted after a number of thromboembolic events were noted. Given the presence of CD154 on platelets and endothelia, it is plausible that the anti-CD154 antibody could propagate thrombosis in an epitope-specific fashion. However, trials utilizing another anti-CD154 antibody produced by IDEC Pharmaceuticals have proceeded without note of increased thromboembolic events [110]. Although the underlying cause of these thromboembolic events is not clear, many of the agents discussed in this chapter powerfully influence the immune system and may have collateral effects. These potential effects must be carefully scrutinized.

Adhesion molecules may also play a role in costimulation. One of the best studied is leukocyte function-associated antigen-1 (LFA-1 composed of

CD11a/CD18), a β-2 integrin expressed on most leukocytes. LFA-1 is involved in a variety of adhesion-dependent lymphocyte functions including T-cell-APC interactions and lymphocyte adhesion to activated endothelium. In a murine cardiac allograft model, short-term administration of anti-LFA-1 and anti-ICAM-1 (intracellular adhesion molecule-1, a ligand for LFA-1) antibodies led to a tolerant state [111]. Presumably, such treatment interferes with T-cell and APC interactions required for transplant rejection. However, recent evidence suggests that many classic adhesion molecules such as ICAM-1 may also act as costimulators themselves [reviewed in 112].

Unfortunately, antibodies against ICAM-1 are also known to disrupt cell-to-cell adhesion and thus inhibit MHC-peptide-TCR interactions. Thus, these antibodies would also be expected to interfere with signal 1 and, if current thinking is accurate, to thereby prevent both specific recognition and tolerance [113, 114]. In view of this potential liability, an antisense intercellular adhesion molecule-1 oligonucleotide may prove more efficacious since selective deletion of ICAM-1 alone does not render APCs incompetent [115]. In murine models, this approach has been shown to prolong graft survival in a dose-dependent fashion that is abrogated by concurrent use of cyclosporin A [116, 117]. ISIS 2302 [118] is a phosphorothioate oligodeoxynucleotide designed to hybridize to the 3′ untranslated region of human ICAM-1 mRNA. The formation of a double-stranded RNA complex provokes the action of RNase H which degrades the complex into nucleotides and thereby selectively prevents additional ICAM-1 expression following activation. Such an approach, when used in conjunction with B7 blockade, may provide the requisite environment for tolerance induction as a recent in vitro study showed that APC ICAM-1 was critical for T-cell activation and costimulation in the absence of CD28 signaling [115]. This approach is particularly appealing as recent experimental evidence suggests that B7 blockade may inhibit CD4- but not CD8-mediated rejection [119, 120], while ICAM-mediated costimulation may preferentially affect CD8 T cells [121, 122].

Learning from Experience and Moving Ahead to the Future

The therapeutic induction of tolerance has proven difficult. Several therapies shown to be effective in rodent models have not had the same efficacy when tested in non-human primates or man. Nevertheless, several groups have described individuals who have discontinued all immunosuppressive agents after transplant and who have maintained graft function [123–125]. Using a human-to-mouse 'trans vivo' delayed-type hypersensitivity (DTH) assay, VanBuskirk et al. [126] have demonstrated that these 'tolerant' patients harbor immune cells with a regulatory phenotype. This trans vivo assay system is performed by injecting both peripheral blood mononuclear cells (PBMCs) and challenge antigen into the footpad or ear of a naive mouse. VanBuskirk et al. [127] have reported that the PBMCs from tolerant patients display a lack of donor-reactive DTH responses in this model system. Remarkably, these patients also demonstrate a lack of antigen-linked DTH responses. Such unresponsiveness was not seen in patients who had rejected their transplants or were undergoing acute or chronic rejection. Thus, these individuals with long-term allograft survival without immunosuppression appeared unique in that they displayed evidence for active immune regulation with linked-antigen suppression. If clinicians could reliably induce such a dominant immune regulation state, one would not need to guarantee tolerance to each antigenic peptide from the outset. Rather, in such patients with an established dominant regulatory mechanism, alloantigen would serve to colocalize competent cells in the vicinity of regulatory cells. Thus, these competent cells would be influenced by the regulatory population and over time, there would be an epitope-dependent spreading of tolerance. The end result would be that all T cells would eventually interact with antigen in the lymphoid organs draining the tissues and full tolerance would ensue.

As we begin to experimentally achieve long-term graft survival, one fundamental concept is becoming abundantly clear. Immune tolerance is an active process that is best achieved through directed immune function rather than indiscriminate immune suppression. While traditional immunosuppressive agents focused on preventing antigen recognition (proliferation), the new schemes attempt to induce tolerance through a number of mechanisms designed to mimic central or peripheral tolerance. Future clinicians may be able to recapitulate self-tolerance, by reprogramming the immune system through (alone or in combination) depleting alloreactive cells, modulating the cellular environment to avoid an aggressive immune responses, and/or costimulatory pathway modulation. The future of allotransplantation may well be based upon harnessing natural regulatory or active suppressive mechanisms to provide a durable tolerogenic state.

References

1   Anonymous: 1999 Annual Report of the US Scientific Registry for Transplant Recipients and the Organ Procurement and Transplantation Network: Transplant Data: 1989–1998. Rockville, Md Richmond, Va, US Department of Health and Human Services, Health Resources and Services Administration, Office of Special Programs, Division of Transplantation; United Network for Organ Sharing, 1999.
2   Cecka JM: The UNOS Scientific Renal Transplant Registry. Clin Transpl 1998;1–16.

3   Jindal RM, Sidner RA, Milgrom ML: Post-transplant diabetes mellitus. The role of immuno-
    suppression. Drug Saf 1997;16:242–257.
4   Lucey MR: Neurological and psychiatric complications associated with immunosuppression in
    liver transplantation. Liver Transpl Surg 1995;1:39–44.
5   Shaw LM, Kaplan B, Kaufman D: Toxic effects of immunosuppressive drugs: Mechanisms and
    strategies for controlling them. Clin Chem 1996;42:1316–1321.
6   Hariharan S, Johnson CP, Bresnahan BA, Taranto SE, McIntosh MJ, Stablein D: Improved graft sur-
    vival after renal transplantation in the United States, 1988–1996. N Engl J Med 2000;342:605–612.
7   Nagano H, Tilney NL: Chronic allograft failure: The clinical problem. Am J Med Sci 1997;
    313:305–309.
8   Owen RD: Immunogenetic consequences of vascular anastomoses between bovine twins. Science
    1945;102:400–401.
9   Anderson D, Billingham RE, Lampkin GH, Medawar PB: The use of skin grafting to distinguish
    between monozygotic and dizygotic twins in cattle. Heredity 1951;5:379.
10 Billingham RE, Lampkin GH, Medawar PB, Williams HLI: Tolerance to homografts, twin diag-
    nosis, and the freemartin condition in cattle. Heredity 1952;6:201.
11 Billingham RE, Brent L, Medawar PB: Actively acquiring tolerance. Nature 1953;172:603.
12 Burnet FM: A modification of Jerne's theory of antibody production using the concept of clonal
    selection. Aust J Sci 1957;20:67–69.
13 Gunther J, Haas W, Von Boehmer H: Suppression of T cell responses through competition for
    T cell growth factor (interleukin-2). Eur J Immunol 1982;12:247–249.
14 Lombardi G, Sidhu S, Batchelor R, Lechler R: Anergic T cells as suppressor cells in vitro. Science
    1994;264:1587–1589.
15 Taams LS, van Rensen AJ, Poelen MC, van Els CA, Besseling AC, Wagenaar JP, et al: Anergic
    T cells actively suppress T cell responses via the antigen-presenting cell. Eur J Immunol 1998;
    28:2902–2912.
16 Cohen JI: Epstein-Barr virus infection. N Engl J Med 2000;343:481–492.
17 Hall BM, Dorsch S, Roser B: The cellular basis of allograft rejection in vivo. I. The cellular
    requirements for first-set rejection of heart grafts. J Exp Med 1978;148:878–889.
18 Arstila TP, Casrouge A, Baron V, Even J, Kanellopoulos J, Kourilsky P: A direct estimate of the
    human alpha beta T cell receptor diversity. Science 1999;286:958–961.
19 Lechler RI, Lombardi G, Batchelor JR, Reinsmoen N, Bach FH: The molecular basis of alloreac-
    tivity. Immunol Today 1990;11:83–88.
20 Quill H, Schwartz RH: Stimulation of normal inducer T cell clones with antigen presented by puri-
    fied Ia molecules in planar lipid membranes: Specific induction of a long-lived state of prolifera-
    tive nonresponsiveness. J Immunol 1987;138:3704–3712.
21 Bretscher P, Cohn M: A theory of self-nonself discrimination. Science 1970;169:1042–1049.
22 Lafferty KJ, Cunningham AJ: A new analysis of allogeneic interactions. Aust J Exp Biol Med Sci
    1975;53:27–42.
23 June CH, Ledbetter JA, Gillespie MM, Lindsten T, Thompson CB: T-cell proliferation involving
    the CD28 pathway is associated with cyclosporine-resistant interleukin-2 gene expression. Mol
    Cell Biol 1987;7:4472–4481.
24 Greenfield EA, Nguyen KA, Kuchroo VK: CD28/B7 costimulation: A review. Crit Rev Immunol
    1998;18:389–418.
25 Lu P, Wang YL, Linsley PS: Regulation of self-tolerance by CD80/CD86 interactions. Curr Opin
    Immunol 1997;9:858–862.
26 Freeman GJ, Freedman AS, Segil JM, Lee G, Whitman JF, Nadler LM: B7, a new member of the
    Ig superfamily with unique expression on activated and neoplastic B cells. J Immunol 1989;
    143:2714–2722.
27 Freeman GJ, Gribben JG, Boussiotis VA, Ng JW, Restivo VAJ, Lombard LA, et al: Cloning of B7-2:
    A CTLA-4 counter-receptor that costimulates human T cell proliferation. Science 1993;262:909–911.
28 Tivol EA, Borriello F, Schweitzer AN, Lynch WP, Bluestone JA, Sharpe AH: Loss of CTLA-4
    leads to massive lymphoproliferation and fatal multiorgan tissue destruction, revealing a critical
    negative regulatory role of CTLA-4. Immunity 1995;3:541–547.

29 Waterhouse P, Penninger JM, Timms E, Wakeham A, Shahinian A, Lee KP, et al: Lymphoproliferative disorders with early lethality in mice deficient in CTLA-4. Science 1995; 270:985–988.

30 Hathcock KS, Laszlo G, Pucillo C, Linsley P, Hodes RJ: Comparative analysis of B7-1 and B7-2 costimulatory ligands: Expression and function. J Exp Med 1994;180:631–640.

31 Hutloff A, Dittrich AM, Beier KC, Eljaschewitsch B, Kraft R, Anagnostopoulos I, et al: ICOS is an inducible T-cell co-stimulator structurally and functionally related to CD28. Nature 1999; 397:263–266.

32 Freeman GJ, Long AJ, Iwai Y, Bourque K, Chernova T, Nishimura H, et al: Engagement of the PD-1 immunoinhibitory receptor by a novel B7 family member leads to negative regulation of lymphocyte activation. J Exp Med 2000;192:1027–1034.

33 Blair PJ, Riley JL, Harlan DM, Abe R, Tadaki DK, Hoffmann SC, et al: CD40 ligand (CD154) triggers a short-term CD4(+) T cell activation response that results in secretion of immunomodulatory cytokines and apoptosis. J Exp Med 2000;191:651–660.

34 Barnes MJ, Knight CG, Farndale RW: The collagen-platelet interaction. Curr Opin Hematol 1998; 5:314–320.

35 Henn V, Slupsky JR, Grafe M, Anagnostopoulos I, Forster R, Muller-Berghaus G, et al: CD40 ligand on activated platelets triggers an inflammatory reaction of endothelial cells. Nature 1998; 391:591–594.

36 Caux C, Massacrier C, Vanbervliet B, Dubois B, van Kooten C, Durand I, et al: Activation of human dendritic cells through CD40 cross-linking. J Exp Med 1994;180:1263–1272.

37 Kennedy MK, Mohler KM, Shanebeck KD, Baum PR, Picha KS, Otten-Evans CA, et al: Induction of B cell costimulatory function by recombinant murine CD40 ligand. Eur J Immunol 1994;24: 116–123.

38 Kappler JW, Roehm N, Marrack P: T cell tolerance by clonal elimination in the thymus. Cell 1987; 49:273–280.

39 Posselt AM, Barker CF, Tomaszewski JE, Markmann JF, Choti MA, Naji A: Induction of donor-specific unresponsiveness by intrathymic islet transplantation. Science 1990;249:1293–1295.

40 Allen MD, Weyhrich J, Gaur L, Akimoto H, Hall J, Dalesandro J, et al: Prolonged allogeneic and xenogeneic microchimerism in unmatched primates without immunosuppression by intrathymic implantation of CD34+ donor marrow cells. Cell Immunol 1997;181:127–138.

41 Kawai T, Poncelet A, Sachs DH, Mauiyyedi S, Boskovic S, Wee SL, et al: Long-term outcome and alloantibody production in a non-myeloablative regimen for induction of renal allograft tolerance. Transplantation 1999;68:1767–1775.

42 Fuchimoto Y, Huang CA, Yamada K, Shimizu A, Kitamura H, Colvin RB, et al: Mixed chimerism and tolerance without whole body irradiation in a large animal model. J Clin Invest 2000; 105:1779–1789.

43 Wekerle T, Sayegh MH, Hill J, Zhao Y, Chandraker A, Swenson KG, et al: Extrathymic T cell deletion and allogeneic stem cell engraftment induced with costimulatory blockade is followed by central T cell tolerance. J Exp Med 1998;187:2037–2044.

44 Wekerle T, Kurtz J, Ito H, Ronquillo JV, Dong V, Zhao G, et al: Allogeneic bone marrow transplantation with co-stimulatory blockade induces macrochimerism and tolerance without cytoreductive host treatment. Nat Med 2000;6:464–469.

45 Wekerle T, Sayegh MH, Ito H, Hill J, Chandraker A, Pearson DA, et al: Anti-CD154 or CTLA4Ig obviates the need for thymic irradiation in a non-myeloablative conditioning regimen for the induction of mixed hematopoietic chimerism and tolerance. Transplantation 1999;68: 1348–1355.

46 Faustman D, Hauptfeld V, Lacy P, Davie J: Prolongation of murine islet allograft survival by pretreatment of islets with antibody directed to Ia determinants. Proc Natl Acad Sci USA 1981; 78:5156–5159.

47 Welsh KI, Batchelor JR, Maynard A, Burgos H: Failure of long surviving, passively enhanced kidney allografts to provoke T-dependent alloimmunity. II. Retransplantation of (AS × AUG)F$_1$ kidneys from AS primary recipients into (AS × WF)F$_1$ secondary hosts. J Exp Med 1979; 150:465–470.

48 Lechler RI, Batchelor JR: Restoration of immunogenicity to passenger cell-depleted kidney allografts by the addition of donor strain dendritic cells. J Exp Med 1982;155:31–41.

49 Shapiro AM, Lakey JR, Ryan EA, Korbutt GS, Toth E, Warnock GL, et al: Islet transplantation in seven patients with type 1 diabetes mellitus using a glucocorticoid-free immunosuppressive regimen. N Engl J Med 2000;343:230–238.

50 Larsen CP, Morris PJ, Austyn JM: Migration of dendritic leukocytes from cardiac allografts into host spleens. A novel pathway for initiation of rejection. J Exp Med 1990;171:307–314.

51 Starzl TE, Demetris AJ, Trucco M, Ramos H, Zeevi A, Rudert WA, et al: Systemic chimerism in human female recipients of male livers. Lancet 1992;340:876–877.

52 Starzl TE, Demetris AJ, Murase N, Ildstad S, Ricordi C, Trucco M: Cell migration, chimerism, and graft acceptance. Lancet 1992;339:1579–1582.

53 Starzl TE, Demetris AJ, Trucco M, Zeevi A, Ramos H, Terasaki P, et al: Chimerism and donor-specific nonreactivity 27 to 29 years after kidney allotransplantation. Transplantation 1993;55:1272–1277.

54 Corry RJ, Chakrabarti PK, Shapiro R, Rao AS, Dvorchik I, Jordan ML, et al: Simultaneous administration of adjuvant donor bone marrow in pancreas transplant recipients. Ann Surg 1999;230:372–379.

55 Suberbielle C, Caillat-Zucman S, Legendre C, Bodemer C, Noel LH, Kreis H, et al: Peripheral microchimerism in long-term cadaveric-kidney allograft recipients. Lancet 1994;343:1468–1469.

56 Hisanaga M, Hundrieser J, Boker K, Uthoff K, Raddatz G, Wahlers T, et al: Development, stability, and clinical correlations of allogeneic microchimerism after solid organ transplantation. Transplantation 1996;61:40–45.

57 Sivasai KS, Alevy YG, Duffy BF, Brennan DC, Singer GG, Shenoy S, et al: Peripheral blood microchimerism in human liver and renal transplant recipients: Rejection despite donor-specific chimerism [published erratum appears in Transplantation 1997;64:1636]. Transplantation 1997;64:427–432.

58 Calne SR: Progress toward tolerance and xenografting. Transplant Proc 1997;29:16–18.

59 Jonker M, Slingerland W, Ossevoort M, Kuhn E, Neville D, Friend P, et al: Induction of kidney graft acceptance by creating a window of opportunity for immunologic engagement (WOFIE) in rhesus monkeys. Transplant Proc 1998;30:2441–2443.

60 Calne R, Moffatt SD, Friend PJ, Jamieson NV, Bradley JA, Hale G, et al: Campath IH allows low-dose cyclosporine monotherapy in 31 cadaveric renal allograft recipients. Transplantation 1999;68:1613–1616.

61 Neville DMJ, Scharff J, Hu HZ, Rigaut K, Shiloach J, Slingerland W, et al: A new reagent for the induction of T-cell depletion, anti-CD3-CRM9. J Immunother Emphasis Tumor Immunol 1996;19:85–92.

62 Knechtle SJ, Vargo D, Fechner J, Zhai Y, Wang J, Hanaway MJ, et al: FN18-CRM9 immunotoxin promotes tolerance in primate renal allografts. Transplantation 1997;63:1–6.

63 Armstrong N, Buckley P, Oberley T, Fechner JJ, Dong Y, Hong X, et al: Analysis of primate renal allografts after T-cell depletion with anti-CD3-CRM9. Transplantation 1998;66:5–13.

64 Thomas JM, Neville DM, Contreras JL, Eckhoff DE, Meng G, Lobashevsky AL, et al: Preclinical studies of allograft tolerance in rhesus monkeys: A novel anti-CD3-immunotoxin given peritransplant with donor bone marrow induces operational tolerance to kidney allografts. Transplantation 1997;64:124–135.

65 Contreras JL, Wang PX, Eckhoff DE, Lobashevsky AL, Asiedu C, Frenette L, et al: Peritransplant tolerance induction with anti-CD3-immunotoxin: A matter of proinflammatory cytokine control. Transplantation 1998;65:1159–1169.

66 Thomas JM, Contreras JL, Jiang XL, Eckhoff DE, Wang PX, Hubbard WJ, et al: Peritransplant tolerance induction in macaques: Early events reflecting the unique synergy between immunotoxin and deoxyspergualin. Transplantation 1999;68:1660–1673.

67 Hale G, Dyer MJ, Clark MR, Phillips JM, Marcus R, Riechmann L, et al: Remission induction in non-Hodgkin lymphoma with reshaped human monoclonal antibody CAMPATH-1H. Lancet 1988;ii:1394–1399.

68   Friend PJ, Rebello P, Oliveira D, Manna V, Cobbold SP, Hale G, et al: Successful treatment of renal allograft rejection with a humanized antilymphocyte monoclonal antibody. Transplant Proc 1995;27:869–870.

69   Anonymous: A randomized clinical trial of OKT3 monoclonal antibody for acute rejection of cadaveric renal transplants. Ortho Multicenter Transplant Study Group. N Engl J Med 1985;313: 337–342.

70   Cosimi AB, Burton RC, Colvin RB, Goldstein G, Delmonico FL, LaQuaglia MP, et al: Treatment of acute renal allograft rejection with OKT3 monoclonal antibody. Transplantation 1981;32: 535–539.

71   Caillat-Zucman S, Blumenfeld N, Legendre C, Noel LH, Bach JF, Kreis H, et al: The OKT3 immunosuppressive effect. In situ antigenic modulation of human graft-infiltrating T cells. Transplantation 1990;49:156–160.

72   Chatenoud L, Baudrihaye MF, Kreis H, Goldstein G, Schindler J, Bach JF: Human in vivo antigenic modulation induced by the anti-T cell OKT3 monoclonal antibody. Eur J Immunol 1982; 12:979–982.

73   Nooij FJ, Jonker M: The effect of skin allograft survival of a monoclonal antibody specific for a polymorphic CD3-like cell surface molecule in rhesus monkeys. Eur J Immunol 1987;17: 1089–1093.

74   Steinhoff G, Jonker M, Gubernatis G, Wonigeit K, Lauchart W, Bornscheuer A, et al: The course of untreated acute rejection and effect of repeated anti-CD3 monoclonal antibody treatment in rhesus monkey liver transplantation. Transplantation 1990;49:669–674.

75   Smith JA, Tso JY, Clark MR, Cole MS, Bluestone JA: Nonmitogenic anti-CD3 monoclonal antibodies deliver a partial T cell receptor signal and induce clonal anergy. J Exp Med 1997;185: 1413–1422.

76   Smith JA, Tang Q, Bluestone JA: Partial TCR signals delivered by FcR-nonbinding anti-CD3 monoclonal antibodies differentially regulate individual Th subsets. J Immunol 1998;160:4841–4849.

77   Friend PJ, Hale G, Chatenoud L, Rebello P, Bradley J, Thiru S, et al: Phase I study of an engineered aglycosylated humanized CD3 antibody in renal transplant rejection. Transplantation 1999; 68:1632–1637.

78   Cole MS, Stellrecht KE, Shi JD, Homola M, Hsu DH, Anasetti C, et al: HuM291, a humanized anti-CD3 antibody, is immunosuppressive to T cells while exhibiting reduced mitogenicity in vitro. Transplantation 1999;68:563–571.

79   Hsu DH, Shi JD, Homola M, Rowell TJ, Moran J, Levitt D, et al: A humanized anti-CD3 antibody, HuM291, with low mitogenic activity, mediates complete and reversible T-cell depletion in chimpanzees. Transplantation 1999;68:545–554.

80   Waldmann H, Qin S, Cobbold S: Monoclonal antibodies as agents to reinduce tolerance in autoimmunity. J Autoimmun 1992;5(suppl A):93–102.

81   Waldmann H, Cobbold S: How do monoclonal antibodies induce tolerance? A role for infectious tolerance? Annu Rev Immunol 1998;16:619–644.

82   Jonker M, Goldstein G, Balner H: Effects of in vivo administration of monoclonal antibodies specific for human T cell subpopulations on the immune system in a rhesus monkey model. Transplantation 1983;35:521–526.

83   Conti DJ, Cosimi AB: Effect of monoclonal antibodies on primate allograft rejection. Crit Rev Immunol 1990;10:113–130.

84   Jonker M, Nooij FJ, van Suylichem P, Neuhaus P, Goldstein G: The influence of OKT8F treatment on allograft survival in rhesus monkeys. Transplantation 1986;41:431–435.

85   Wright JKJ, Barrett LV, Delmonico FL, Fuller TC, Cosimi AB: Preclinical evaluation of immunosuppression selective for T cells recognizing class I histocompatibility antigens. Transplant Proc 1987;19:1106–1109.

86   Mourad GJ, Preffer FI, Wee SL, Powelson JA, Kawai T, Delmonico FL, et al: Humanized IgG1 and IgG4 anti-CD4 monoclonal antibodies: Effects on lymphocytes in the blood, lymph nodes, and renal allografts in cynomolgus monkeys. Transplantation 1998;65:632–641.

87   Qin S, Cobbold SP, Pope H, Elliott J, Kioussis D, Davies J, et al: 'Infectious' transplantation tolerance. Science 1993;259:974–977.

88   Cobbold SP, Martin G, Waldmann H: The induction of skin graft tolerance in major histocompatibility complex-mismatched or primed recipients: Primed T cells can be tolerized in the periphery with anti-CD4 and anti-CD8 antibodies. Eur J Immunol 1990;20:2747–2755.

89   Marshall SE, Cobbold SP, Davies JD, Martin GM, Phillips JM, Waldmann H: Tolerance and suppression in a primed immune system. Transplantation 1996;62:1614–1621.

90   Chu CQ, Londei M: Induction of Th2 cytokines and control of collagen-induced arthritis by nondepleting anti-CD4 Abs. J Immunol 1996;157:2685–2689.

91   Hutchings PR, Cooke A, Dawe K, Waldmann H, Roitt IM: Active suppression induced by anti-CD4. Eur J Immunol 1993;23:965–968.

92   Cobbold S, Waldmann H: Infectious tolerance. Curr Opin Immunol 1998;10:518–524.

93   Gershon RK, Kondo K: Infectious immunological tolerance. Immunology 1971;21:903–914.

94   Chen ZK, Cobbold SP, Waldmann H, Metcalfe S: Amplification of natural regulatory immune mechanisms for transplantation tolerance. Transplantation 1996;62:1200–1206.

95   Phillips JM, Harach SZ, Parish NM, Fehervari Z, Haskins K, Cooke A: Nondepleting anti-CD4 has an immediate action on diabetogenic effector cells, halting their destruction of pancreatic beta cells. J Immunol 2000;165:1949–1955.

96   Powelson JA, Knowles RW, Delmonico FL, Kawai T, Mourad G, Preffer FK, et al: CDR-grafted OKT4A monoclonal antibody in cynomolgus renal allograft recipients. Transplantation 1994; 57:788–793.

97   Nashan B, Moore R, Amlot P, Schmidt AG, Abeywickrama K, Soulillou JP: Randomised trial of basiliximab versus placebo for control of acute cellular rejection in renal allograft recipients. CHIB 201 International Study Group [published erratum appears in Lancet 1997;350:1484]. Lancet 1997;350:1193–1198.

98   Vincenti F, Nashan B, Light S: Daclizumab: Outcome of phase III trials and mechanism of action. Double therapy and the triple therapy study groups. Transplant Proc 1998;30:2155–2158.

99   Lenschow DJ, Zeng Y, Thistlethwaite JR, Montag A, Brady W, Gibson MG, et al: Long-term survival of xenogeneic pancreatic islet grafts induced by CTLA4lg. Science 1992;257: 789–792.

100 Azuma H, Chandraker A, Nadeau K, Hancock WW, Carpenter CB, Tilney NL, et al: Blockade of T-cell costimulation prevents development of experimental chronic renal allograft rejection. Proc Natl Acad Sci USA 1996;93:12439–12444.

101 Chandraker A, Azuma H, Nadeau K, Carpenter CB, Tilney NL, Hancock WW, et al: Late blockade of T cell costimulation interrupts progression of experimental chronic allograft rejection. J Clin Invest 1998;101:2309–2318.

102 Glysing-Jensen T, Raisanen-Sokolowski A, Sayegh MH, Russell ME: Chronic blockade of CD28-B7-mediated T-cell costimulation by CTLA4Ig reduces intimal thickening in MHC class I and II incompatible mouse heart allografts. Transplantation 1997;64:1641–1645.

103 Levisetti MG, Padrid PA, Szot GL, Mittal N, Meehan SM, Wardrip CL, et al: Immunosuppressive effects of human CTLA4Ig in a non-human primate model of allogeneic pancreatic islet transplantation. J Immunol 1997;159:5187–5191.

104 Kirk AD, Harlan DM, Armstrong NN, Davis TA, Dong Y, Gray GS, et al: CTLA4-Ig and anti-CD40 ligand prevent renal allograft rejection in primates. Proc Natl Acad Sci USA 1997; 94:8789–8794.

105 Parker DC, Greiner DL, Phillips NE, Appel MC, Steele AW, Durie FH, et al: Survival of mouse pancreatic islet allografts in recipients treated with allogeneic small lymphocytes and antibody to CD40 ligand. Proc Natl Acad Sci USA 1995;92:9560–9564.

106 Rossini AA, Parker DC, Phillips NE, Durie FH, Noelle RJ, Mordes JP, et al: Induction of immunological tolerance to islet allografts. Cell Transplant 1996;5:49–52.

107 Larsen CP, Elwood ET, Alexander DZ, Ritchie SC, Hendrix R, Tucker-Burden C, et al: Long-term acceptance of skin and cardiac allografts after blocking CD40 and CD28 pathways. Nature 1996; 381:434–438.

108 Kirk AD, Burkly LC, Batty DS, Baumgartner RE, Berning JD, Buchanan K, et al: Treatment with humanized monoclonal antibody against CD154 prevents acute renal allograft rejection in non-human primates. Nat Med 1999;5:686–693.

109 Kenyon NS, Chatzipetrou M, Masetti M, Ranuncoli A, Oliveira M, Wagner JL, et al: Long-term survival and function of intrahepatic islet allografts in rhesus monkeys treated with humanized anti-CD154. Proc Natl Acad Sci USA 1999;96:8132–8137.
110 Kirk AD, Harlan DM: Thromboembolic complications after treatment with monoclonal antibody against CD40 ligand. Nat Med 2000;6:114.
111 Isobe M, Yagita H, Okumura K, Ihara A: Specific acceptance of cardiac allograft after treatment with antibodies to ICAM-1 and LFA-1. Science 1992;255:1125–1127.
112 Watts TH, DeBenedette MA: T cell co-stimulatory molecules other than CD28. Curr Opin Immunol 1999;11:286–293.
113 Gribben JG, Guinan EC, Boussiotis VA, Ke XY, Linsley L, Sieff C, et al: Complete blockade of B7 family-mediated costimulation is necessary to induce human alloantigen-specific anergy: A method to ameliorate graft-versus-host disease and extend the donor pool. Blood 1996;87:4887–4893.
114 Calhoun RF, Oppat WF, Duffy B, Mohanakumar T: Intercellular adhesion molecule-1/leukocyte function associated antigen-1 blockade inhibits alloantigen specific human T cell effector functions without inducing anergy. Transplantation 1999;68:1144–1152.
115 Gaglia JG, Greenfield EA, Mattoo A, Sharpe AH, Freeman GJ, Kuchroo VK: ICAM-1 is critical for activation of CD28-deficient T cells. J Immunol 2000;165:6091–6098.
116 Stepkowski SM, Tu Y, Condon TP, Bennett CF: Blocking of heart allograft rejection by intercellular adhesion molecule–1 antisense oligonucleotides alone or in combination with other immunosuppressive modalities [published erratum appears in J Immunol 1995;154:1521]. J Immunol 1994;153:5336–5346.
117 Stepkowski SM, Wang ME, Condon TP, Cheng-Flournoy S, Stecker K, Graham M, et al: Protection against allograft rejection with intercellular adhesion molecule-1 antisense oligodeoxynucleotides. Transplantation 1998;66:699–707.
118 Glover JM, Leeds JM, Mant TG, Amin D, Kisner DL, Zuckerman JE, et al: Phase I safety and pharmacokinetic profile of an intercellular adhesion molecule-1 antisense oligodeoxynucleotide (ISIS 2302). J Pharmacol Exp Ther 1997;282:1173–1180.
119 Newell KA, He G, Guo Z, Kim O, Szot GL, Rulifson I, et al: Cutting edge: Blockade of the CD28/B7 costimulatory pathway inhibits intestinal allograft rejection mediated by CD4+ but not CD8+ T cells. J Immunol 1999;163:2358–2362.
120 Honey K, Cobbold SP, Waldmann H: CD40 ligand blockade induces CD4+ T cell tolerance and linked suppression. J Immunol 1999;163:4805–4810.
121 Chen T, Goldstein JS, O'Boyle K, Whitman MC, Brunswick M, Kozlowski S: ICAM-1 costimulation has differential effects on the activation of CD4+ and CD8+ T cells. Eur J Immunol 1999;29:809–814.
122 Deeths MJ, Mescher MF: ICAM-1 and B7-1 provide similar but distinct costimulation for CD8+ T cells, while CD4+ T cells are poorly costimulated by ICAM-1. Eur J Immunol 1999;29:45–53.
123 Uehling DT, Hussey JL, Weinstein AB, Wank R, Bach FH: Cessation of immunosuppression after renal transplantation. Surgery 1976;79:278–282.
124 Zoller KM, Cho SI, Cohen JJ, Harrington JT: Cessation of immunosuppressive therapy after successful transplantation: A national survey. Kidney Int 1980;18:110–114.
125 Burlingham WJ, Grailer AP, Fechner JHJ, Kusaka S, Trucco M, Kocova M, et al: Microchimerism linked to cytotoxic T lymphocyte functional unresponsiveness (clonal anergy) in a tolerant renal transplant recipient. Transplantation 1995;59:1147–1155.
126 Carrodeguas L, Orosz CG, Waldman WJ, Sedmak DD, Adams PW, VanBuskirk AM: Trans vivo analysis of human delayed-type hypersensitivity reactivity. Hum Immunol 1999;60:640–651.
127 VanBuskirk AM, Burlingham WJ, Jankowska-Gan E, Chin T, Kusaka S, Geissler F, et al: Human allograft acceptance is associated with immune regulation. J Clin Invest 2000;106:145–155.

David M. Harlan, MD, Associate Professor of Medicine, Uniformed Services University of the Health Sciences, National Institute of Diabetes and Digestive and Kidney Diseases, Navy Transplantation and Autoimmunity Branch, NIDDK, Bethesda, MD 20892 (USA)
Tel. +1 301 295 7654, Fax +1 301 295 6484, E-Mail davidmh@intra.niddk.nih.gov

von Herrath MG (ed.): Molecular Pathology of Type 1 Diabetes mellitus.
Curr Dir Autoimmun. Basel, Karger, 2001, vol 4, pp 308–332

Role of Regulatory T Cells in the Pathogenesis of Autoimmune Diabetes

Guillermo A. Arreaza, Shayan Sharif, Mark J. Cameron,
Wei Chen, Terry L. Delovitch

Autoimmunity/Diabetes Group, The John P. Robarts Research Institute, and
Departments of Microbiology and Immunology and Medicine, University of
Western Ontario, London, Ont., Canada

The immune system can be considered as an intricate set of cell–cell inter-actions initiated by exposure to antigen and regulated by multiple positive and negative signals derived from lymphocytes, antigen-presenting cells (APCs), and stroma cells located in primary and secondary lymphoid tissues. Thus, tolerance and immunity are not discrete entities but rather the poles of a continuum of immune responsiveness, the former relevant to self-antigens and the latter to foreign antigens. Currently, self-tolerance is conceived as a syndrome with multiple mechanisms ranging from deletion and anergy to suppression and immune deviation or split tolerance [1]. Infectious or transferable tolerance mediated by T cells with suppressive activity was originally demonstrated for foreign antigens more than 20 years ago [2–4]. Suppression is not a property of a specific T-cell subset and can be mediated under different conditions by CD4+, CD8+, and double negative CD4−CD8− T-cell subpopulations. Consequently, the mechanisms respon-sible for the suppression of a particular immune response may differ. Accumulating evidence indicates that T-cell-dependent active suppression or 'dominant immunological tolerance' plays an active role in regulating T- and B-cell responses to self-antigens [5, 6]. Specifically, tolerance may regulate those responses associ-ated with an organ-specific autoimmune disease, e.g. type 1 diabetes (T1D), by immunoregulation via the cytokine network, anergy and immune deviation [7].

Here we describe how a breakdown of tolerance may be manifested by disruption of the balance between pathogenic effector T cells (T_{eff}) and regulatory

T (T_{reg}) (or suppressor T (T_s)) cells. This breakdown can lead to a state of immune dysregulation, which may be a principal causal factor for the development of an autoimmune disease such as T1D. Further, we present several intervention protocols that restore immune regulation in experimental animal models of T1D and evaluate their therapeutic potential.

Role of T_{reg} and T_s Cells in T1D

Considerable evidence supports an active role for T_{reg} and/or T_s cells in the prevention of spontaneous murine and rat models of T1D [7–9]. First, cyclophosphamide (CY) induces acute T1D in young nonobese diabetic (NOD) mice, which is mediated both by the selective deletion of a subpopulation of T_{reg} cells and resistance to CY-induced apoptosis of T_{eff} cells [10, 11]. Second, sublethal irradiation [12] or thymectomy and CD4+ T-cell depletion [13] are required for the transfer of T1D by diabetogenic T cells. In NOD females, thymectomy at weaning prevents the generation of protective cells and thus accelerates the onset of T1D [14]. Third, CD4+ T cells from nondiabetic young NOD mice prevent the transfer of T1D by splenocytes from diabetic NOD mice [15–17]. Fourth, distinct subpopulations of thymocytes in young healthy mice, such as CD4+CD62L+ [18] and TCRαβ+ DN thymocytes [19], function as T_{reg} cells by inhibiting the adoptive transfer of T1D. The latter thymocyte population includes natural killer T (NKT) cells and peripheral NKT cells that may regulate susceptibility to T1D [20], as will be discussed below. Several therapies prevent T1D in NOD mice during the early phase of development of the immune system, since they appear to promote the development of T_{reg} cells [7–9].

A role for thymus-derived CD4+CD25+ T cells as T_{reg} cells has also been the subject of many recent studies in various models of autoimmune disease [reviewed in 7]. These CD4+CD25+ T_{reg} cells represent a minor (~10%) subset of CD4+ T cells that coexpress CD25 (IL-2R α-chain) and suppress the activity of CD4+CD25− T cells [21, 22]. Deletion of these T_{reg} cells elicits an autoimmune syndrome evident after either neonatal thymectomy (3 days of age) or ionizing radiation in adult mice [21, 22]. The absence of CD4+CD25+ T_{reg} cells promotes the spontaneous development of several autoimmune diseases, including gastritis and thyroiditis [21–24]. Although these T_{reg} cells are hyporesponsive (anergic) to stimulation by TCR-derived signals, they inhibit polyclonal T-cell activation in vitro and adoptively suppress the responses of CD4+CD25− T-cells [7, 23]. Suppression requires cell contact, is cytokine-independent, and CD4+CD25+ T_s cells completely inhibit IL-2 gene transcription and IL-2 production by responsive T cells in an antigen-nonspecific manner [25]. CD4+CD25+ T_s cells are absent in IL-2-deficient mice, and this deficiency may contribute to the development of spontaneous autoimmunity in mice congenitally defective in the IL-2 and IL-2R genes [26]. These T_s cells do not express the NK1.1 surface antigen and appear to be distinct from NKT cells [7]. Importantly, a defect in the development and/or maintenance of CD4+ CD25+ T_s cells may play a relevant role in the pathogenesis of T1D in NOD mice [27].

CD4+CD62L^{high} T_{reg} cells in the spleen of NOD mice may also regulate susceptibility to T1D [28]. Upon activation, these T cells secrete low levels of IL-2 and IFN-γ, but not IL-4 or IL-10. Moreover, these T cells transfer protection from T1D in a cytokine-independent manner similar to that of CD4+ CD25+ T_{reg} cells by homing to the pancreas and reducing the severity of insulitis by blocking the migration of diabetogenic T_{eff} cells to the islets. CD4+ CD62L^{high} T_{reg} cells are detectable after weaning and persist until the onset of T1D, supporting the idea that destructive insulitis in NOD mice is a late and acute event. CD4+CD62L+ thymocytes also inhibit induction of T1D, a property not shared by CD4+CD62L− thymocytes [5, 6]. While CD4+CD62L+ T cells differ from other widely studied subsets of T_{reg} cells in NOD mice, including T-helper 2 (Th2), Th3, Tr1 and NKT cells, CD4+CD62L+ T cells share certain characteristics with CD4+CD25+ T cells. Both T_{reg} cell subsets lack antigen specificity and thymocytes with the same phenotype are more potent suppressors than peripheral CD4+ T cells [5–7]. It is possible that CD4+CD62L+ T_{reg} cells are precursors of CD4+CD25+ T_{reg} cells in peripheral lymphoid tissues [27], since the latter T cells are derived from the thymus and most of them express high levels of CD62L [23–25, 27]. Furthermore, T_{reg} cells that prevent T1D in rats subsequent to adult thymectomy and γ-irradiation are TCRαCD4+CD45RC^{low} RT6+ [29]. The corresponding T_{reg} cells in mice are CD4+CD45RB^{low} [5, 6].

Additional support for the ability of T_{reg} cells to modulate the development of T1D is provided by studies conducted using T-cell clones and TCR transgenic (Tg) mice. An NOD CD4+ autoreactive T_{reg} cell clone isolated from a population of pancreatic islet infiltrating T cells and characterized by a Th3-like profile (secretes mainly TGF-β) can prevent spontaneous and recurrent (after islet transplant) autoimmune T1D [30]. In NOD Tg mice that express the TCR found on islet autoreactive BDC2.5 T cells, the onset of T1D is delayed on the NOD.*BDC2.5* immune-sufficient genetic background whereas the onset of insulitis and T1D is markedly accelerated on the NOD.*Scid* immunodeficient background [31]. This result indicates that 'T1D-protective' T cells that express an endogenous TCR in NOD.*BDC2.5* mice actively control the activity of CD4+ T_{eff} cells, and that these T_s cells are absent in NOD.*Scid* mice. Thus, the onset of T1D may be regulated by several subsets of phenotypically and functionally different T_{reg} cells. It remains to be determined whether any of these T_{reg} subsets are related to each other developmentally via a common precursor T cell.

Immune Stimulation and the Emergence of T$_{reg}$ Cells

The development of pathogenic T cells destined to mediate autoimmune disease may be modified by immune stimulation at an early age in autoimmune-prone individuals [8, 9]. This stimulation may allow for the preferential secretion of certain cytokines/chemokines and other growth factors that influence the generation of T cells that modulate autoimmunity [8, 9, 32]. The observation that there is a north/south gradient that correlates with a higher incidence of autoimmune disease in humans in northern countries and that the cleaner the environment the higher the incidence of T1D in NOD mouse colonies [33] supports this hypothesis. Interestingly, the early immunization of NOD mice and BB rats with complete Freund's adjuvant (CFA) or bacillus Calmette-Guérin (BCG) is very effective in inducing long-term protection from T1D [32]. This induced protection is associated with a Th2/Th3 profile, and the adjuvant treatment has to be administered before the appearance of intra-islet insulitis [32]. However, although early human clinical trials showed promising results, the most recent trials have not reproduced the results obtained in the animal model. It has been argued that a repeated BCG immunization or a different route of administration may be necessary to obtain a more satisfactory outcome [34]. In this regard, note that vaccination with a measles virus seems to induce a predominant Th2 immune response that mediates protection from T1D in humans [35].

We and others have found that, beginning at 3–5 weeks of age, NOD thymocytes and peripheral T cells display a proliferative hyporesponsiveness (anergy) in response to TCR stimulation, which is mediated by low IL-2 and IL-4 secretion [36, 37]. Interestingly, anergy and decreased IL-4 production by TCR-stimulated human T cells from patients with new onset T1D has also been demonstrated [38, 39]. Furthermore, early provision of IL-4 and or CD28 co-stimulation may prevent T1D in the NOD mouse by the induction of not only a predominant Th2 splenic and intra-islet environment but also other regulatory T$_{reg}$ cells [40, 41].

Coincident with appearance of insulitis at about 4–6 weeks of age, a defective syngeneic mixed lymphocyte reaction (SMLR) response is detectable in NOD mice [42, 43]. This age-related defect, which resides in an SMLR responder spleen and mesenteric lymph node-derived CD4+ T-cell population [42], is characterized by reduced IL-2 production by these T cells [42, 44] and correlates closely with increased progression to T1D [45]. Furthermore, immune stimulation in vivo with IL-2 and/or Poly[I:C] restores the SMLR response, enhances the suppressive capacity and prevents diabetes onset in NOD mice [44]. In a SMLR response, T cells proliferate in response to self-MHC class II and this induces the activation of regulatory T cells. Thus, a deficiency in, rather than an absence of, regulatory CD4+ T cells is manifested during an SMLR response in NOD mice older than 6 weeks of age. A similar peripheral immunoregulatory defect occurs in T1D patients and other autoimmune diseases [46].

The onset of deficient regulatory CD4+ T-cell function at an early age may disrupt the T-cell balance required to maintain self-tolerance, and thereby augment early T-cell autoreactivity to islet cell autoantigens [47, 48]. Such an imbalance may elicit a loss of immunoregulation of pathogenic islet-reactive T cells and trigger the development of destructive insulitis and T1D. As mentioned above, young nondiabetic NOD mice possess normal levels of functional regulatory CD4+ T cells until about 4–6 weeks of age, and the number and function of these cells then rapidly decline as the first islet antigen autoreactive T cells are detected. Defects in cytokine-induced differentiation and maturation of APCs from the bone marrow of NOD mice may lead to inefficient presentation of self-antigens and impair a possible tolerogenic capacity of these cells [42–44]. Thus, early age immune stimulation may contribute to overcome these defects and protect NOD mice from T1D.

T$_s$ Cells and Human T1D

Only a paucity of evidence currently exists in support of a role for T$_s$ cells in human T1D. However, a potential role for an antigen-specific defect in T$_s$-cell function in the pathogenesis of human autoimmune thyroid disease has emerged [49, 50]. Furthermore, studies of T-cell activation to insulin [51] and glutamic acid decarboxylase 65 (GAD65) [52, 53] point to a specific defect in the response of CD8+ T cells from T1D patients to these islet autoantigens. Other studies of the distribution of T-cell subsets suggest a possible deficient suppressor T-cell function in humans [54–58]. A possible role for NKT cells as T$_s$ cells in the context of human T1D has also been reported [58, 59].

Immune Regulation of T1D by Th2 Cells. Role of Cytokines/Chemokines

CD4+ Th1 cells secrete interferon-γ (IFN-γ) and tumor necrosis factor-α (TNF-α), whereas CD4+ Th2 cells secrete IL-4, IL-5, IL-6, IL-10 and IL-13 [60–65]. Th1 cells support cell-mediated immunity and as a consequence promote inflammation, cytotoxicity and delayed-type hypersensitivity. Th2 cells support humoral immunity and serve to downregulate the inflammatory actions of Th1 cells and other effector cells such as macrophages. Therefore, Th2 cells are candidate T$_{reg}$ cells [60–65].

The main factors affecting the development of T-helper subsets are the nature of antigenic stimulation and cytokines/chemokines encountered in the stimulatory milieu. IFN-γ production via IL-12 stimulation inhibits the differentiation and functions of Th2 cells and can lead to a dominant Th1 response, whereas IL-4 plays a dominant role in the differentiation of naive T cells to Th2 cells [60–65]. The pathophysiological significance of the functional division of T-helper cells into subsets has been clearly demonstrated in animal models of organ-specific experimental autoimmunity, including T1D [9, 60–65].

Chemokines are a superfamily of low-molecular-weight (8–14 kDa) chemotactic cytokines that regulate leukocyte migration through interactions with seven-transmembrane, rhodopsin-like G protein-coupled receptors [66]. These molecules may be divided into several families, among which the two families of CC and CXC chemokines are more predominant. CC chemokine targets include monocytes, T cells, dendritic cells and NK cells [66, 67]. Recently, it has become clear that in addition to their role in the recruitment of leukocytes to sites of inflammation, chemokines play a fundamental role in mediating innate and adaptive immune responses by their ability to recruit, activate, and costimulate cells of the immune system [67, 68].

We have demonstrated that IL-4 prevents insulitis and T1D when administered in vivo to prediabetic NOD mice [36, 40]. Protection from T1D is mediated by the modulation of homing of lymphocytes to pancreatic islets during the inductive phase of T1D and the potentiation of regulatory Th2 cell function in the thymus, spleen and pancreatic islets. This result has been recently confirmed in NOD mice that express IL-4 transgenically in pancreatic β cells [69]. We also demonstrated that T1D onset after transfer of T cells from IL-4-treated NOD mice is delayed in NOD.*Scid* recipients compared with the transfer of T cells from control treated mice. This finding is compatible with a previous report that Th1 cells, but not Th2 cells, can transfer T1D to neonatal NOD recipients [62]. This suggests that systemically administered IL-4 activates regulatory Th2 cells, which remain functional and prevent or reduce T1D over a prolonged period. Moreover, we observed that T cells from IL-4-treated mice suppressed the transfer of T1D by diabetogenic effector T cells from diabetic NOD mice upon cotransfer to NOD.*Scid* recipients. This observation further supports the notion that functionally active regulatory T cells are generated upon treatment with IL-4 [65]. The persistent effect of an apparent Th2 class shift noted is also consistent with reports demonstrating that the presence of specific cytokines at the initiation of an immune response can lead to the generation of both effector and long-lived memory T-cell populations that produce restricted patterns of cytokines [70, 71]. Therefore, immune deviation via cytokine immunotherapy that results in polarization towards regulatory Th2 cells may be most effective in preventing destructive autoimmunity when initiated at an early age. Indeed, this is evident with our IL-4 administration program in NOD mice, including IL-4 gene therapy.

In addition to the inhibition of lymphocyte migration to the pancreas, IL-4 treatment also prevents the infiltration of other notable target sites of inflammation in NOD mice, viz. the salivary glands and thyroid [40]. Our previous genetic mapping studies revealed that the control of T-cell hyporesponsiveness in NOD mice is linked to a central region on chromosome 11, which includes the *Idd4* diabetogenic locus and the CC chemokine gene family [72]. Together with the well-established role of chemokines in inflammation and the effects of IL-4 [73] and CD28 signaling [74] on chemokine expression, these findings raise the possibility that different CC chemokines may be associated with T-cell differentiation as well as T1D susceptibility in NOD mice. Indeed, we have identified candidate CC chemokines that may contribute to mediate the establishment of insulitis [75]. Macrophage inflammatory protein-1α (MIP-1α) and MCP-1 play an early effector role in the establishment of insulitis in NOD mice. Interestingly, intrapancreatic MCP-1 may contribute to early islet infiltration by attracting lymphocytes, the outcome of which depends on the presence or subsequent expression, of other chemokines. The relative ratio of MIP-1α:MIP-1β is important during the initial stages of islet mononuclear cell infiltration in determining the nature of insulitis progression in NOD mice. More evidence regarding the role of MIP-1α in diabetogenesis was provided by monitoring the spontaneous incidence of T1D in NOD.*MIP-1α*$^{-/-}$ mice. The incidence of T1D was significantly reduced and delayed in NOD.*MIP-1α*$^{-/-}$ mice compared to NOD.*MIP-1α*$^{+/+}$ indicating that MIP-1α is an important effector chemokine in the pathogenesis of T1D [75].

The effector role of MIP-1α we have noted in the pathogenesis of T1D in NOD mice is consistent with its association with Th1 responses and its role in other models of organ-specific autoimmune diseases [76–80]. Recently it was demonstrated that its expression is increased in active brain lesions from patients with multiple sclerosis [80]. Indeed, we found that a close correlation exists between Th2 cytokine responses and a high MIP-1β plus MCP-1:MIP-1α chemokine ratio elicited by IL-4 treatment in the pancreas of NOD mice protected from T1D [75]. In accordance with this finding, CCR5 a receptor that is preferentially expressed on Th1 cells [80–83] was the only CC chemokine receptor that was downregulated in the pancreas upon IL-4 treatment, reflecting the immune deviation described above [36, 40, 65]. These data are in support of many reports that have linked the expression of certain chemokines and chemokine receptors to a Th1/Th2 paradigm [76–83]. The interrelationship of cytokines, chemokines and chemokine receptors in modulating and stabilizing effector Th1 cell versus regulatory Th2 cell function may determine the extent of leukocyte migration to, and the nature of inflammation at, different autoimmune target sites (fig. 1). Temporal modulation of CC chemokines, and their

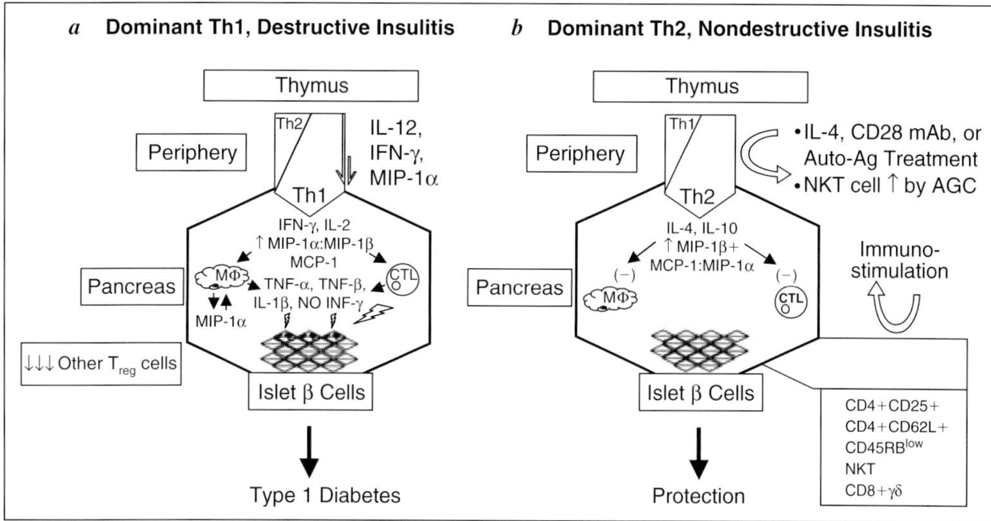

Fig. 1. Immune regulation of T1D in NOD mice. *a* A Th1/Th2 imbalance develops in the thymus and periphery of female diabetes-prone NOD mice. This leads to a progressive and age-dependent impaired function of regulatory Th2 cells. Other regulatory T cells may also improperly develop. The imbalance is further influenced by the presence of Th1- and Th2-derived cytokines and chemokines. Autoreactive T cells become activated and mediate islet β-cell destruction by participating in the recruitment of activated macrophages (Mφ) and cytotoxic T lymphocytes (CTL). A high MIP-1α:MIP-1β ratio in the pancreas may aid in lymphocyte recruitment and establishment of destructive insulitis. *b* IL-4 replacement therapy and CD28 costimulation establishes a Th2-dominant environment in central, peripheral, and pancreatic islet locations. Th2-associated cytokines and a high MIP-1β + MCP-1:MIP-1α chemokine ratio in the pancreas may mediate a nondestructive milieu in islets. The immune stimulation of other regulatory T cells (i.e., NK T cells, CD4+CD25+) may also help stabilize this islet-protective environment limiting the extent of β cell damage during an inflammatory response.

interactions with CC chemokine receptors, may influence the activity of regulatory cells and represent a novel therapeutic approach in the prevention of the development of insulitis and spontaneous T1D [75].

Role of Costimulation in Immune Deviation and Generation of Regulatory Th2 Cells in the Prevention of T1D

Optimal T-cell activation requires signaling through the TCR and CD28 costimulatory receptor [84, 85]. The costimulatory pathway of T-cell activation involves the interaction of CD28 with its ligands B7-1 and B7-2 on an APC, with B7-2 considered as the primary ligand for CD28 [85–86]. When costimulation is blocked by either CTLA4-Ig, anti-B7-1 or anti-B7-2 mAbs or by the deletion of costimulatory molecules by homologous recombination, differential effects on the incidence of various autoimmune diseases (e.g. T1D) and development of Th1 and Th2 cells are observed [27, 87–92]. The generation of Th2 cells is more dependent upon the CD28-B7 pathway than the priming of Th1 cells, and it is now clear that the development of Th subsets in vivo may be influenced by limiting CD28-B7 costimulation [27, 90–92].

These conclusions were confirmed using the CTLA-4Ig transgenic and CD28, B7-1/B7-2-deficient mouse models [27, 90–92] in which autoimmunity and particularly exacerbation of T1D were observed [27, 91–92]. Analyses of human Th2 cell development have yielded results similar to those observed in the mouse [93–95]. Interestingly, CD28/B7-2 interactions are essential for the costimulation of an IL-4-dependent CD4+ T-cell response, and IL-4 increases B7-1 and B7-2 surface expression on certain professional APCs (e.g. Langerhans' cells) and B cells [96–99]. IL-4 receptor sensitivity is also increased in T cells after CD28 accelerated IL-4-mediated Th2 differentiation [100]. Thus, failure to activate NOD thymocytes and peripheral T cells sufficiently may be due to functional and/or differentiation defects in NOD APCs, which optimally activate islet β-cell autoreactive CD4+ T_{eff} but not T_{reg} cells [42, 101]. Functional defects that compromise antigen presentation by NOD APCs, such as deficient CD28 costimulation, may lower their ability to stimulate regulatory Th2 cells without compromising their ability to stimulate autoreactive effector Th1 cells. Furthermore, CTLA-4 seems to play a fundamental role as a negative regulator of the immune response. Recent studies on murine and human T cells suggest that CTLA-4 blockade of CD28-dependent costimulation determines whether a preferential Th1 or Th2 response is elicited. It appears that while CD28 promotes Th2 differentiation, CTLA-4 limits Th2 differentiation [102, 103].

CD28 mAb Treatment Restores Proliferative Responsiveness and Prevents T1D by an IL-4-Dependent Mechanism

We have shown that anti-CD28 mAb-mediated costimulation completely restores the proliferative responsiveness of NOD thymocytes and peripheral T cells by augmenting their levels of secretion of IL-2 and IL-4 [41]. The stimulated increase in IL-4 secretion is predominant in this response. This result raised the possibility that treatment of NOD mice with a CD28 agonist that stimulates marked IL-4 secretion and is more stable than IL-4 in vivo may be more efficacious than IL-4 in protecting NOD mice from T1D.

Therefore, NOD mice were treated with the 37.51 anti-CD28 mAb [41] or control hamster IgG to determine whether CD28 costimulation protects them from T1D. We demonstrated that anti-CD28 treatment effectively prevents destructive insulitis and protects completely from T1D in NOD mice, provided that treatment is administered neonatally at a sufficiently early age (2–4 weeks) and prior to the onset of insulitis [41]. Similar treatment of mice after the onset of insulitis (5–7 weeks of age) did not protect from T1D. Our studies of the cytokine secretion profiles of stimulated peripheral (spleen) and islet infiltrating T cells and the ex vivo detection of the expression of intrapancreatic cytokines of anti-CD28-treated (2- to 4-week-old) mice displayed a significant upregulation of IL-4 production. The levels of IFN-γ expression remained unchanged. Thus, anti-CD28 mediated costimulation and protection from T1D may be mediated by the preferential induction of a Th2-like response in NOD mice. Indeed, we found that anti-CD28 mAb treatment in vivo leads to an increased production of IgG1 (which reflects increased IL-4 production by T cells and its in vivo activity) rather than IgG2a anti-GAD67 autoantibodies in the sera of treated animals [41]. Furthermore, coadministration of anti-IL-4 with anti-CD28 to 2- to 4-week-old NOD mice blocked the anti-CD28-induced protection from T1D. Thus, the protective effect of anti-CD28 treatment is likely mediated by the expansion and survival of IL-4 producing Th2 cells. However, the induction of T_{reg} cells such as CD4+CD25+ T cells is a possibility given the recent reports that indicate the dependence of the generation and homeostasis of these cells on an efficient CD28/B7 costimulation [27].

Our data reveal that T1D-prone NOD mice may be relatively deficient in their ability to generate a Th2-like mediated immune response, and therefore generate a predominant Th1-like response. When a sufficient threshold level of costimulation is provided, the capacity to generate a Th2-like response is recovered and the consequent IL-4 produced mediates protection from destructive insulitis and T1D. This notion is consistent with a 'strength of signal' hypothesis previously proposed [84, 85], which states that the intensity of T-cell signaling not only determines the initiation of a response but also affects the balance of function between Th1 and Th2 cell subsets.

In support of this hypothesis, our results demonstrate that activation of the CD28/B7 pathway before the onset of insulitis in NOD mice stabilizes a protective Th2 cell-enriched environment in pancreatic islets. This early CD28 activation may also induce the activity of other T_{reg} cells (e.g. CD4+CD25+, NKT) at the same or different site(s). Modification of the microenvironment and the quality of the infiltration appears to protect against destructive insulitis and progression to T1D. The dose, duration, route and, perhaps most importantly, the immunological milieu into which the anti-CD28 mAb is introduced, may affect the outcome of the response potentiated by anti-CD28 mAb treatment. Our data indicate that NOD T_{reg} cells may not receive a sufficient CD28 costimulation signal during the inductive phase of development of T1D, and that this CD28 signal is requisite for the activation of IL-4 producing NOD Th2 cells and/or other T_{reg} cells [41].

In conclusion, augmenting costimulation at an early age can completely prevent the spontaneous development of T1D by inducing the activity of T_{reg} cells. Nonetheless, it should be noted that in other antigen-induced autoimmune diseases CD28 costimulation can exacerbate the onset of disease [104]. Thus, the chronic evolution of the inflammatory process in NOD mice and the possibility of intervention at different stages may determine the outcome of the treatment when costimulation is modulated either by anti-CD28 mAb or otherwise.

Role of NKT Cells as T_{reg} Cells in T1D

NKT cells comprise a unique subset of T cells, whose development is differentially regulated from mainstream T cells [105]. These cells preferentially use an invariant TCRα chain consisting of the Vα14-Jα281 gene segments and also have a bias in TCRβ gene usage [105, 106]. The majority of NKT cells are CD4+ or CD4−CD8−, but they all express various NK cell-related surface markers, including NK1.1 [106]. Cells with similar phenotypic characteristics have also been described in humans, which preferentially use the invariant Vα24-JαQ TCR gene segments [107]. NKT cells are detected in primary and secondary lymphoid organs, and their development may be thymus dependent or independent [108].

NKT cells show restriction and autoreactivity to the CD1 surface glycoprotein, whose structure resembles that of a major histocompatibility complex (MHC) class I molecule. Recently, a glycolipid, α-galactosylceramide (α-GalCer), has been identified as a specific ligand for both mouse and human NKT cells [109]. Presentation of α-GalCer by CD1 stimulates a vigorous response by NKT cells marked by their proliferation, expression of activation molecules, increase in cytotoxic activity and secretion of various cytokines [110]. The response requires cell-cell contact between NKT cells and α-GalCer loaded dendritic cells and is dependent on CD40-CD40L interaction [110]. Secretion of Th1- or Th2-type cytokines may be differentially regulated by the dose of α-GalCer administered in vivo. While a single dose of α-GalCer stimulates a Th1 response, a multiple dose regime promotes a Th2 response [111]. Although NKT cells are primarily triggered by α-GalCer, bystander cells including NK, CD8+, CD4+ and B cells may also become nonspecifically activated as a consequence of NKT cell activity [112].

Regulatory Activity
NKT cells possess T_{reg} cell activity [113], and exert this activity by promoting a Th1 or a Th2 response rapidly after primary stimulation [114, 115].

Interestingly, several in vivo studies have attributed immunoregulatory functions to NKT cells in the amelioration of allogeneic graft-versus-host disease, induction of anterior chamber-associated immune deviation and induction of tolerance in recipients of hepatic allografts [116–118].

NKT Cells and Autoimmunity

A potential role for NKT cells in the regulation of autoimmunity was initially recognized in SJL mice that are susceptible to experimental allergic encephalomyelitis (EAE), a Th1-mediated autoimmune disease. SJL mice are deficient in their Th2-mediated responses, and their NKT cells are significantly reduced in number and function in various tissues [119]. A functional deficit in NKT cells, especially in the production of IL-4 and other regulatory cytokines, in autoimmune disease-prone mouse strains can lead to a dominant pathogenic autoreactive Th1 response. NKT cells have also been implicated in the regulation of susceptibility to autoimmune disease in humans. Patients with systemic sclerosis [120], rheumatoid arthritis [121] and multiple sclerosis [122] have a reduced frequency of peripheral blood NKT (Vα24-JαQ+) cells.

NKT Cells and T1D

The strongest evidence in support of a role for NKT cells in the regulation of autoimmunity stems from studies of T1D in humans [58, 59, 123], NOD mice [124, 125] and BB rats [126]. Considerable evidence suggests that NKT cells are deficient in number and function in NOD mice and T1D patients [58, 59, 123–125]. In NOD mice between 3 and 5 weeks of age, the proportion of NKT-like thymocytes is reduced and TCR cross-linking induced IL-4 production is significantly impaired. This deficiency in NKT cell number and function is restored to normal by 12–15 weeks of age [124, 125]. We have also shown that NKT cells in the thymus and spleen of NOD mice are deficient in IL-4 production in response to in vitro stimulation with α-GalCer, but they secrete normal amounts of IFN-γ [127]. This deficiency in IL-4 production may be corrected by exposure to IL-7 [125], and transfer of NKT-like thymocytes into young NOD prevents T1D in an IL-4- and IL-10-dependent manner [19]. In addition, Vα14-Jα281 TCR transgenic NOD mice are protected from T1D [20], and protection is associated with the number of NKT cells and the level of IL-4 secretion by NKT cells induced by TCR stimulation. In contrast, it has been reported that TCR- or IL-12-stimulated NOD NKT cells are severely deficient in IFN-γ but not in IL-4 production [128]. In the latter studies, purified NOD NKT cells stimulated by IL-7 not only did not prevent T1D but rather accelerated the onset of T1D. Further experimentation is required to resolve these apparent discrepancies. Nonetheless, it is noteworthy that these quantitative and qualitative deficiencies first appear at an age that represents an important checkpoint in the progression to T1D, i.e. immediately prior to the development of insulitis in 3- to 4-week-old mice [8, 9].

A role for NKT cells in pathogenesis of human T1D has been recently reported [58, 59, 123]. In studies of a group of twin/triplet sets discordant for T1D, the diabetic sibs had a lower frequency of CD4−CD8− NKT cells in their peripheral blood in comparison to their nondiabetic nonprogressor sibs. More importantly, CD4−CD8−Vα14JαQ+ NKT cell clones from these discordant sets differed appreciably in their production of IL-4 and IFN-γ. Whereas all clones from diabetic siblings only produced IFN-γ, 76 of 79 clones derived from at-risk nonprogressor siblings simultaneously produced IL-4 and IFN-γ [58]. A follow-up study of these clones pointed to distinct gene expression patterns following TCR cross-linking [59]. A recent study comparing first-degree relatives of islet cell antibody-positive patients with T1D and normal controls also suggests a primary defect of NKT cells in human T1D that may reflect defective regulatory T-cell functions [123].

It is conceivable that correction of quantitative and qualitative deficiencies of NKT cells in individuals at risk for T1D may prevent occurrence or recurrence of disease. Thus, given that NKT cells respond to α-GalCer, the activation of NKT cells by α-GalCer may prevent the onset and/or recurrence of T1D. Our results indicate that treatment with α-GalCer results in protection against the onset of CY-induced T1D [127]. Furthermore, administration of α-GalCer prior to or after the establishment of insulitis (4 and 10 weeks of age, respectively) can significantly delay the onset and increase protection against the occurrence of spontaneous T1D. The protection conferred by α-GalCer is associated with a polarized Th2 environment both in the spleen and in the pancreas [127].

We have also explored whether α-GalCer can modulate the onset of a recurrent autoimmune destructive response in overtly diabetic NOD recipients of syngeneic pancreatic islet transplants. In this model, grafts generally fail owing to a recurrent autoimmune response. Prolongation of pancreatic duodenal graft survival in nonrecurrent spontaneous diabetic BB rats is associated with the proliferation of donor-derived NKT cells in hepatic and splenic tissues and higher serum levels of IL-4 [129, 130]. Moreover, NKT cells are required for acceptance of xenogeneic islet grafts in mice [131]. We have obtained evidence that stimulation of NKT cells by α-GalCer prolongs islet graft survival by downmodulation of the recurrent autoimmune response. Our data demonstrate that multiple doses of α-GalCer administered pre- and posttransplantation in recipients prolongs islet graft survival in newly diabetic NOD mice from the typical 1 week to as long as 16 weeks [127]. Overall, these findings raise the possibility that α-GalCer treatment may be used therapeutically to prevent the onset and recurrence of human T1D.

Induction of T$_{reg}$ Cells with Islet β-Cell Autoantigens

As outlined above, the induction of tolerance to self-antigens currently represents a fundamental strategy to prevent or treat autoimmune disease. While modulation of central tolerance in the thymus is not easily achieved in humans, reconstitution of peripheral tolerance in peripheral lymphoid tissues appears to be more effective for the prevention and treatment of autoimmune diseases [132]. Thus, induction of tolerance by administration of autoantigens by different routes of delivery, such as the mucosal (oral, intranasal) and systemic routes, has been studied extensively [133, 134].

Oral tolerance has been used as an intervention strategy to treat organ-specific autoimmune disease in mice and humans [134]. However, the various mechanisms and pathways of induction of tolerance are incompletely understood. Among the many identified autoantigens in T1D, (pro)insulin and GAD65 appear to be the most important autoantigens involved in the development of T1D [8, 9, 47, 48, 135]. Oral administration of low doses of insulin can protect NOD mice from both spontaneous and CY-accelerated T1D in NOD mice, with this protection being mediated by peripheral CD4+ T$_s$ cells [134, 136–138]. These CD4+ T$_s$ cells can also adoptively transfer protection from T1D [137, 138]. Administration of GAD65 and its peptides [139, 140] yields similar results. Follow-up studies revealed that these autoantigen-induced T$_{reg}$ cells secrete several anti-inflammatory cytokines, including IL-4, IL-10 and TGF-β [137, 138, 141–144] in the pancreatic lymph nodes (PLN) and pancreas of protected mice. These observations indicate that immune deviation from a Th1 to Th2 phenotype occurs at the target tissue site of inflammation (i.e. pancreatic islets).

We have demonstrated that oral insulin-induced CD4+ T$_{reg}$ cells are reactive to an insulin B chain epitope(s) in female NOD mice. These T$_{reg}$ cells compete with diabetogenic CD4+ T$_{eff}$ cells for the recognition of insulin when transferred into NOD.*Scid* recipient mice [143]. As a result, these T$_{eff}$ cells lose their ability to produce IL-2 and IFN-γ and are unable to expand and migrate to pancreatic islets. TGF-β$_1$ does not seem to contribute to the suppression of T$_{eff}$ cells, as T1D-protective splenic T$_{reg}$ cells secrete relatively little TGF-β$_1$. Transfer of protection from T1D by insulin B-chain autoreactive CD4+ T$_{reg}$ cells results in a significant increase in the intrapancreatic expression of MIP-1β, which correlates closely with a nondestructive insulitis in NOD mice. Thus, oral insulin administration overcomes a deficiency in T$_{reg}$ cells and protects against T1D. This protection appears to be mediated by a functional inactivation and reduced migration of diabetogenic effector T cells to the pancreas [143].

Additional studies show that oral feeding of insulin conjugated to the cholera toxin B subunit, a potent mucosal adjuvant, reduces about 5,000-fold the amount of antigen necessary for delaying the onset of T1D in NOD mice [144]. These results suggest that this intervention may effectively induce antigen-specific CD4+ T_{reg} cells in the vicinity of the mucosal barrier and in draining PLN. The appearance of such T_{reg} cells in PLN presumably impairs the activity of islet autoreactive T_{eff} cells en route to the pancreatic islets [143].

Proinsulin has recently received considerable attention as an islet autoanti-gen, since T-cell reactivity to both proinsulin and GAD65 peptides is detectable in individuals at risk for T1D [145, 146]. Furthermore, transgenic expression of mouse proinsulin by MHC class II bearing cells prevents T1D in NOD mice [147]. Our recent findings show that splenic CD4+ T cells from naive NOD mice are not tolerant to proinsulin despite the expression of proinsulin in their thymus [148]. Although both oral administration and parenteral immunization of NOD female (6-week-old) mice with proinsulin do not induce autoreactive T_{reg} cells that prevent the onset of T1D, multiple parenteral vaccination with proinsulin neonatally beginning at 18 days delays the onset and reduces the incidence of T1D [148]. These results suggest that autoreactive T_{reg} cells to proinsulin may be inducible in NOD mice and highlight the relevancy of proin-sulin to the pathogenesis of T1D [147, 148].

The induction of T_{reg} cells that confer protection from T1D has also been achieved in other experimental animal models of T1D by either oral tolerance or insulin DNA immunization [149–150; also see Homan et al., this volume]. For example, in a model of T1D induced in lymphocytic choriomeningitis virus (LCMV) Tg mice, the oral feeding of insulin B chain beginning either 1 week prior to or 10 days after LCMV infection prevents T1D in >50% of these Tg mice [149]. Usually >95% of such LCMV-infected Tg mice, but not uninfected control mice, become diabetic by 2 months after challenge with LCMV [149]. These data demonstrated that insulin-specific regulatory cells are induced in Peyer's patches and act in an antigen nonspecific way at the site of inflamma-tion, i.e. pancreatic islets, by the expression of the IL-4, IL-10 and TGF-β anti-inflammatory cytokines. Although the cell type directly responsible for this regulation has not been identified, a role for a bystander suppression effect mediated by T_{reg} cells was elegantly shown [149].

More recent investigations have substantiated this evidence for bystander suppression, and have demonstrated that autoreactive CD4+ T_{reg} cells protect from T1D by suppressing the activity of autoreactive islet-destructive CD8+ T cells of different antigenic specificity in PLN [151; also see Homan et al., this volume]. This suppression requires the presence of IL-4 and/or the IL-4 signal-ing pathway. Furthermore, this investigation suggests that the activity of T_{reg} cells may be mediated not by a direct effect on T_{eff} cells but rather by the inac-tivation of APCs and the subsequent lack of expansion of LCMV-specific T_{eff} cells [151]. The dependency of antigen-induced tolerance on the presence of

IL-4 has also been shown using specific hGAD65 epitopes to prevent insulitis and T1D in NOD mice [152]. Immunization with relevant GAD65 peptides prevents T1D in normal NOD mice but does not prevent T1D in NOD mice deficient in IL-4 expression [152]. These findings illustrate that GAD65-specific peptide immunotherapy effectively suppresses progression to overt T1D and requires the production of IL-4.

Intranasal and aerosolized autoantigen delivery to induce systemic tolerance is an effective intervention to reduce the incidence of T1D in NOD mice. Intranasal administration of proinsulin peptide 24–36, which binds to the NOD mouse I-A^{g7} MHC class II molecule, induces CD4+ T_{reg} cells [133]. The intranasal administration of GAD65 T-cell epitope peptides also generates an active regulatory response mediated by type 2 CD4+ T_{reg} cells [153]. Unlike oral insulin that induces CD4+ T_{reg} cells, intranasal or aerosolized insulin administration induces TCRγδ+ CD8+ T_{reg} cells [133]. Small numbers ($1–2 \times 10^5$) of these CD8+ T_{reg} cells efficiently block the adoptive transfer of T1D. Phenotypically, these TCRγδ+ CD8+ T_{reg} cells express CD8αα and are intraepithelial lymphocytes (IELs) [133]. Note that the activation of these IELs requires stimulation by a conformationally intact but not biologically active insulin. The mechanisms responsible for the induction of γδ+ CD8+ T cells by aerosolized or intranasal insulin are not completely understood. These γδ+ CD8+ T cells can localize to PLN and produce IL-10, which may account for the antidiabetic and bystander suppressive effects of intranasal and aerosolized insulin [133]. Interestingly, a recent report suggests that TCRγδ+ CD8+ T cells induced by exogenous insulin may function as T_{reg} cells in human T1D [154].

In summary, several islet β-cell autoantigens may induce tolerance mediated by regulatory T cells in experimental animal models and afford protection from disease. Some of these autoantigens are currently being tested in ongoing clinical trials [155].

Concluding Remarks

We have summarized the experimental evidence that identify a crucial role of T_{reg} cells in the maintenance of peripheral tolerance in normal individuals, and have also outlined how immune dysregulation leading to dysfunction of T_{reg} cells may lead to autoimmune T1D. Several subpopulations of non-antigen-specific T_{reg} cells, including CD4+CD25+, CD4+CD62L+ and NKT cells, may function in a bystander suppression fashion by downregulation of an antigen-specific immune response [6, 7]. In this context, the cytokines and chemokine profiles of the milieu in secondary lymphoid organs and sites of inflammation and their influence in the development and maintenance of T_{reg} cells must be emphasized [8, 9, 75]. Although certain of the T_{reg} cells do not function by secretion of particular soluble factors, soluble factors may nevertheless determine their development and functional capacity [1, 6–9]. Furthermore, the interplay between T cells, APCs and the stroma in peripheral tissues determines the outcome of an immune response and its regulation [1, 6–9]. In addition to T cells, B cells, macrophages, dendritic cells and even target cells that function as APCs or otherwise produce particular cytokines/chemokines and provide costimulation may be critical for the induction/modulation of T_{reg} cells [27, 32, 37, 41–44, 101, 156, 157]. Defects in cytokine-induced differentiation and maturation of APCs may lead to inefficient presentation of self-antigens and impair a possible tolerogenic capacity of these cells [37, 41–44, 101]. Alternatively, such non-T cells may function themselves as regulatory cells and the mechanisms by which this may occur deserve further consideration.

Our studies, as well as those of others, demonstrate that autoantigen, cytokine and costimulation-based immunotherapies may correct a Th1/Th2 imbalance via the reprogramming of an organ-specific immune response [8, 9]. These interventions may also induce the development and activity of the above-described T_{reg} cells as well as other antigen-specific autoreactive T_{reg} or T_s cells (fig. 1) [6–9, 40, 41, 65, 127, 136–153]. Treatments that involve a combination of an autoantigen(s), cytokine(s) and adjuvant-like molecule(s) that guide an antigen-specific effector immune response toward one that is anti-inflammatory may provide minimally invasive therapies to prevent and/or reverse T1D in humans.

References

1 Nossal GJ: Cellular mechanisms of immunologic tolerance. Annu Rev Immunol 1983;1:33–62.
2 Gershon RK, Kondo K: Cell interactions in the induction of tolerance: The role of thymic lymphocytes. Immunology 1970;18:723–737.
3 Gershon RK, Kondo K: Infectious immunological tolerance. Immunology 1971;21:903–914.
4 McCullagh P: The abrogation of immunological tolerance of sheep erythrocytes by means of subcellular extracts. Aust J Exp Biol Med Sci 1973;51:783–791.
5 Saoudi A, Seddon B, Heath V, Fowell D, Mason D: The physiological role of regulatory T cells in the prevention of autoimmunity: The function of the thymus in the generation of the regulatory T cell subset. Immunol Rev 1996;149:195–216.
6 Mason D, Powrie F: Control of immune pathology by regulatory T cells. Curr Opin Immunol 1998;10:649–655.
7 Shevach EM: Regulatory T cells in autoimmunity. Annu Rev Immunol 2000;18:423–449.
8 Bach JF: Insulin-dependent diabetes mellitus as a beta-cell targeted disease of immunoregulation. J Autoimmun 1995;8:439–463.
9 Delovitch TL, Singh B: The nonobese diabetic mouse as a model of autoimmune diabetes: Immune dysregulation gets the NOD. Immunity 1997;7:727–738.
10 Charlton B, Bacelj A, Slattery RM, Mandel TE: Cyclophosphamide-induced diabetes in NOD/WEHI mice. Evidence for suppression in spontaneous autoimmune diabetes mellitus. Diabetes 1989;38:441–447.

11   Colucci F, Bergman ML, Penha-Goncalves C, Cilio CM, Holmberg D: Apoptosis resistance of nonobese diabetic peripheral lymphocytes linked to the Idd5 diabetes susceptibility region. Proc Natl Acad Sci USA 1997;94:8670–8674.

12   Wicker LS, Miller BJ, Mullen Y: Transfer of autoimmune diabetes mellitus with splenocytes from nonobese diabetic (NOD) mice. Diabetes 1986;35:855–860.

13   Sempe P, Richard MF, Bach JF, Boitard C: Evidence of CD4+ regulatory T cells in the non-obese diabetic male mouse. Diabetologia 1994;37:337–343.

14   Dardenne M, Lepault F, Bendelac A, Bach JF: Acceleration of the onset of diabetes in NOD mice by thymectomy at weaning. Eur J Immunol 1989;19:889–895.

15   Shimada A, Rohane P, Fathman CG, Charlton B: Pathogenic and protective roles of CD45RB(low) CD4+ cells correlate with cytokine profiles in the spontaneously autoimmune diabetic mouse. Diabetes 1996;45:71–78.

16   Boitard C, Yasunami R, Dardenne M, Bach JF: T cell-mediated inhibition of the transfer of autoimmune diabetes in NOD mice. J Exp Med 1989;169:1669–1680.

17   Hutchings PR, Cooke A: The transfer of autoimmune diabetes in NOD mice can be inhibited or accelerated by distinct cell populations present in normal splenocytes taken from young males. J Autoimmun 1990;3:175–185.

18   Herbelin A, Gombert JM, Lepault F, Bach JF, Chatenoud L: Mature mainstream TCR alpha beta+CD4+ thymocytes expressing L-selectin mediate 'active tolerance' in the nonobese diabetic mouse. J Immunol 1998;161:2620–2628.

19   Hammond KJL, Poulton LD, Palmisano LJ, Silveira PA, Godfrey DI, Baxter AG: alpha/beta-T cell receptor (TCR)+CD4-CD8- (NKT) thymocytes prevent insulin-dependent diabetes mellitus in nonobese diabetic (NOD)/Lt mice by the influence of interleukin (IL)-4 and/or IL-10. J Exp Med 1998;187:1047–1056.

20   Lehuen A, Lantz O, Beaudoin L, Laloux V, Carnaud C, Bendelac A, Bach JF, Monteiro RC: Overexpression of natural killer T cells protects Valpha14-Jalpha281 transgenic nonobese diabetic mice against diabetes. J Exp Med 1998;188:1831–1839.

21   Asano M, Toda M, Sakaguchi N, Sakaguchi S: Autoimmune disease as a consequence of developmental abnormality of a T cell subpopulation. J Exp Med 1996;184:387–396.

22   Takahashi T, Kuniyasu Y, Toda M, Sakaguchi N, Itoh M, Iwata M, Shimizu J, Sakaguchi S: Immunologic self-tolerance maintained by CD25+CD4+ naturally anergic and suppressive T cells: Induction of autoimmune disease by breaking their anergic/suppressive state. Int Immunol 1998;10:1969–1980.

23   Itoh M, Takahashi T, Sakaguchi N, Kuniyasu Y, Shimizu J, Otsuka F, Sakaguchi S: Thymus and autoimmunity: Production of CD25+CD4+ naturally anergic and suppressive T cells as a key function of the thymus in maintaining immunologic self-tolerance. J Immunol 1999;162:5317–5326.

24   Suri-Payer E, Amar AZ, Thornton AM, Shevach EM: CD4+CD25+ T cells inhibit both the induction and effector function of autoreactive T cells and represent a unique lineage of immunoregulatory cells. J Immunol 1998;160:1212–1218.

25   Thornton AM, Shevach EM: Suppressor effector function of CD4+CD25+ immunoregulatory T cells is antigen nonspecific. J Immunol 2000;164:183–190.

26   Hunig T, Schimpl A: Systemic autoimmune disease as a consequence of defective lymphocyte death. Curr Opin Immunol 1997;9:826–830.

27   Salomon B, Lenschow DJ, Rhee L, Ashourian N, Singh B, Sharpe A, Bluestone JA: B7/CD28 costimulation is essential for the homeostasis of the CD4+CD25+ immunoregulatory T cells that control autoimmune diabetes. Immunity 2000;12:431–440.

28   Lepault F, Gagnerault MC: Characterization of peripheral regulatory CD4+ T cells that prevent diabetes onset in nonobese diabetic mice. J Immunol 2000;164:240–247.

29   Fowell D, Mason D: Evidence that the T cell repertoire of normal rats contains cells with the potential to cause diabetes. Characterization of the CD4+ T cell subset that inhibits this autoimmune potential. J Exp Med 1993;177:627–636.

30   Han HS, Jun HS, Utsugi T, Yoon JW: A new type of CD4+ suppressor T cell completely prevents spontaneous autoimmune diabetes and recurrent diabetes in syngeneic islet-transplanted NOD mice. J Autoimmun 1996;9:331–339.

31 Kurrer MO, Pakala SV, Hanson HL, Katz JD: Beta cell apoptosis in T cell-mediated autoimmune diabetes. Proc Natl Acad Sci USA 1997;94:213–218.

32 Singh B: Stimulation of the developing immune system can prevent autoimmunity. J Autoimmun 2000;14:15–22.

33 Pozzilli P, Signore A, Williams AJ, Beales PE: NOD mouse colonies around the world – Recent facts and figures. Immunol Today 1993;14:193–196.

34 Allen HF, Klingensmith GJ, Jensen P, Simoes E, Hayward A, Chase HP: Effect of bacillus Calmette-Guérin vaccination on new-onset type 1 diabetes. A randomized clinical study. Diabetes Care 1999;22:1703–1707.

35 Blom L, Nystrom L, Dahlquist G: The Swedish childhood diabetes study. Vaccinations and infections as risk determinants for diabetes in childhood. Diabetologia 1991;34:176–181.

36 Rapoport MJ, Jaramillo A, Zipris D, Lazarus AH, Serreze DV, Leiter EH, Cyopick P, Danska JS, Delovitch TL: Interleukin-4 reverses T cell proliferative unresponsiveness and prevents the onset of diabetes in nonobese diabetic mice. J Exp Med 1993;178:87–99.

37 Dahlen E, Hedlund G, Dawe K: Low CD86 expression in the nonobese diabetic mouse results in the impairment of both T cell activation and CTLA-4 up-regulation. J Immunol 2000;164:2444–2456.

38 Berman MA, Sandborg CI, Wang Z, Imfeld KL, Zaldivar FJ, Dadufalza V, Buckingham BA: Decreased IL-4 production in new onset type I insulin-dependent diabetes mellitus. J Immunol 1996;157:4690–4696.

39 Dosch H, Cheung RK, Karges W, Pietropaolo M, Becker DJ: Persistent T cell anergy in human type 1 diabetes. J Immunol 1999;163:6933–6940.

40 Cameron MJ, Arreaza GA, Zucker P, Chensue SW, Strieter RM, Chakrabarti S, Delovitch TL: IL-4 prevents insulitis and insulin-dependent diabetes mellitus in nonobese diabetic mice by potentiation of regulatory T helper-2 cell function. J Immunol 1997;159:4686–4692.

41 Arreaza GA, Cameron MJ, Jaramillo A, Gill BM, Hardy D, Laupland KB, Rapoport MJ, Zucker P, Chakrabarti S, Chensue SW, Qin HY, Singh B, Delovitch TL: Neonatal activation of CD28 signaling overcomes T cell anergy and prevents autoimmune diabetes by an IL-4-dependent mechanism. J Clin Invest 1997;100:2243–2253.

42 Serreze DV, Leiter EH: Defective activation of T suppressor cell function in nonobese diabetic mice. Potential relation to cytokine deficiencies. J Immunol 1988;140:3801–3807.

43 Bergerot I, Arreaza G, Cameron M, Chou H, Delovitch TL: Role of T-cell anergy and suppression in susceptibility to IDDM. Res Immunol 1997;148:348–358.

44 Serreze DV, Hamaguchi K, Leiter EH: Immunostimulation circumvents diabetes in NOD/Lt mice. J Autoimmun 1989;2:759–776.

45 Baxter AG, Adams MA, Mandel TE: Comparison of high- and low-diabetes-incidence NOD mouse strains. Diabetes 1989;38:1296–1300.

46 Bowman MA, Leiter EH, Atkinson MA: Prevention of diabetes in the NOD mouse: Implications for therapeutic intervention in human disease. Immunol Today 1994;15:115–120.

47 Tisch R, Yang XD, Singer SM, Liblau RS, Fugger L, McDevitt HO: Immune response to glutamic acid decarboxylase correlates with insulitis in non-obese diabetic mice. Nature 1993;366:72–75.

48 Kaufman DL, Clare-Salzler M, Tian J, Forsthuber T, Ting GS, Robinson P, Atkinson MA, Sercarz EE, Tobin AJ, Lehmann PV: Spontaneous loss of T-cell tolerance to glutamic acid decarboxylase in murine insulin-dependent diabetes. Nature 1993;366:69–72.

49 Yoshikawa N, Morita T, Resetkova E, Arreaza G, Carayon P, Volpé R: Reduced activation of T lymphocytes by specific antigens in autoimmune thyroid disease. J Endocrinol Invest 1992;15:609–617.

50 Volpe R: Suppressor T cell function is important in the pathogenesis of autoimmune thyroid disease. Thyroid 1994;3:345–352.

51 Naquet P, Ellis J, Tibensky D, Kenshole A, Singh B, Hodges R, Delovitch TL: T cell autoreactivity to insulin in diabetic and related non-diabetic individuals. J Immunol 1988;140:2569–2578.

52 Mukuta T, Yoshikawa N, Arreaza G, Resetkova E, Leushner J, Song YH, Akasu F, Onaya T, Volpe R: Activation of T lymphocyte subsets by synthetic TSH receptor peptides and recombinant glutamate decarboxylase in autoimmune thyroid disease and insulin-dependent diabetes. J Clin Endocrinol Metab 1995;80:1264–1272.

53 Togun RA, Resetkova E, Kawai K, Enomoto T, Volpe R: Activation of CD8+ T lymphocytes in insulin-dependent diabetes mellitus. Clin Immunol Immunopathol 1997;82:243–249.

54 Bizzarro A, De Bellis A, Amoresano PV, Auletta M, Gattoni A, Altucci P, Iacono G: CD11B lymphocytes in type 1 diabetes mellitus of recent onset. Horm Metab Res Suppl 1992;26:121–122.

55 Wang SH, Chen G, Zhang K: Effects of thymosin and insulin on suppressor T cell in type 1 diabetes. Diabetes Res 1992;19:21–29.

56 Ilonen J, Surcel HM, Kaar ML: Abnormalities within CD4 and CD8 T lymphocytes subsets in type 1 (insulin-dependent) diabetes. Clin Exp Immunol 1991;85:278–281.

57 Johnston C, Alviggi L, Millward BA, Leslie RD, Pyke DA, Vergani D: Alterations in T-lymphocyte subpopulations in type I diabetes. Exploration of genetic influence in identical twins. Diabetes 1988;37:1484–1488.

58 Wilson SB, Kent SC, Patton KT, Orban T, Jackson RA, Exley M, Porcelli S, Schatz DA, Atkinson MA, Balk SP, Strominger JL, Hafler DA: Extreme Th1 bias of invariant Vα24JαQ T cells in type 1 diabetes. Nature 1998;391:177–181.

59 Wilson SB, Kent SC, Horton HF, Hill AA, Bollyky PL, Hafler DA, Strominger JL, Byrne MC: Multiple differences in gene expression in regulatory Vα24JαQ T cells from identical twins discordant for type I diabetes. Proc Natl Acad Sci USA 2000;97:7411–7416.

60 Seder RA, Paul WE: Acquisition of lymphokine-producing phenotype by CD4+ T cells. Annu Rev Immunol 1994;12:635–673.

61 Romagnani S: The Th1/Th2 paradigm. Immunol Today 1997;18:263–266.

62 Katz JD, Benoist C, Mathis D: T helper cell subsets in insulin-dependent diabetes. Science 1995;268:1185–1188.

63 Liblau RS, Singer SM, McDevitt HO: Th1 and Th2 CD4+ T cells in the pathogenesis of organ-specific autoimmune diseases. Immunol Today 1995;16:34–38.

64 Nicholson LB, Kuchroo VK: Manipulation of the Th1/Th2 balance in autoimmune disease. Curr Opin Immunol 1996;8:837–842.

65 Arreaza GA, Cameron MJ, Delovitch TL: Interleukin-4: Potential immunoregulatory agent in therapy of insulin-dependent diabetes mellitus. Clin Immunother 1996;6:251–260.

66 Zlotnik A, Yoshie O: Chemokines: A new classification system and their role in immunity. Immunity 2000;12:121–127.

67 Ward SG, Bacon K, Westwick J: Chemokines and T lymphocytes: More than an attraction. Immunity 1998;9:1–11.

68 Xia Y, Pauza ME, Feng L, Lo D: RelB regulation of chemokine expression modulates local inflammation. Am J Pathol 1997;151:375–387.

69 Gallichan WS, Balasa B, Davies JD, Sarvetnick N: Pancreatic IL-4 expression results in islet-reactive Th2 cells that inhibit diabetogenic lymphocytes in the nonobese diabetic mouse. J Immunol 1999;163:1696–1703.

70 Swain SL: Generation and in vivo persistence of polarized Th1 and Th2 memory cells. Immunity 1994;1:543–552.

71 Yoshimoto K, Swain SL, Bradley LM: Enhanced development of Th2-like primary CD4 effectors in response to sustained exposure to limited rIL-4 in vivo. J Immunol 1996;156:3267–3274.

72 Gill BM, Jaramillo A, Ma L, Laupland KB, Delovitch TL: Genetic linkage of thymic T-cell proliferative unresponsiveness to mouse chromosome 11 in NOD mice. A possible role for chemokine genes. Diabetes 1995;44:614–619.

73 Standiford TJ, Kunkel SL, Liebler JM, Burdick MD, Gilbert AR, Strieter RM: Gene expression of macrophage inflammatory protein-1-alpha from human blood monocytes and alveolar macrophages is inhibited by interleukin-4. Am J Respir Cell Mol Biol 1993;9:192–198.

74 Herold KC, Lu J, Rulifson I, Vezys V, Taub D, Grusby MJ, Bluestone JA: Regulation of C-C chemokine production by murine T cells by CD28/B7 costimulation. J Immunol 1997;159:4150–4153.

75 Cameron MJ, Arreaza GA, Grattan M, Meagher C, Sharif S, Burdick MD, Strieter RM, Cook DN, Delovitch TL: Differential expression of CC chemokines and the CCR5 receptor in the pancreas is associated with progression to type 1 diabetes. J Immunol 2000;165:1102–1110.

76 Karpus WJ, Lukacs NW, McRae BL, Strieter RM, Kunkel SL, Miller SD: An important role for the chemokine macrophage inflammatory protein-1-alpha in the pathogenesis of the T cell-mediated autoimmune disease, experimental autoimmune encephalomyelitis. J Immunol 1995; 155:5003–5010.

77 Kunkel SL: Th1- and Th2-type cytokines regulate chemokine expression. Biol Signals 1996;5: 197–202.

78 Karpus WJ, Kennedy KJ, Kunkel SL, Lukacs NW: Monocyte chemotactic protein 1 regulates oral tolerance induction by inhibition of T helper cell 1-related cytokines. J Exp Med 1998;187:733–741.

79 Karpus WJ, Lukacs NW, Kennedy KJ, Smith WS, Hurst SD, Barrett TA: Differential CC chemokine-induced enhancement of T helper cell cytokine production. J Immunol 1997;158: 4129–4136.

80 Balashov KE, Rottman JB, Weiner HL, Hancock WW: CCR5(+) and CXCR3(+) T cells are increased in multiple sclerosis and their ligands MIP-1alpha and IP-10 are expressed in demyelinating brain lesions. Proc Natl Acad Sci USA 1999;96:6873–6878.

81 Loetscher P, Uguccioni M, Bordoli L, Baggiolini M, Moser B, Chizzolini C, Dayer JM: CCR5 is characteristic of Th1 lymphocytes. Nature 1998;391:344–345.

82 Sallusto F, Lenig D, Mackay CR, Lanzavecchia A: Flexible programs of chemokine receptor expression on human polarized T helper 1 and 2 lymphocytes. J Exp Med 1998;187:875–883.

83 Bonecchi R, Bianchi G, Bordignon PP, D'Ambrosio D, Lang R, Borsatti A, Sozzani S, Allavena P, Gray PA, Mantovani A, Sinigaglia F: Differential expression of chemokine receptors and chemotactic responsiveness of type 1 T helper cells (Th1s) and Th2s. J Exp Med 1998;187:129–134.

84 Thompson CB: Distinct roles for the costimulatory ligands B7-1 and B7-2 in T helper cell differentiation? Cell 1995;81:979–982.

85 Lenschow DJ, Walunas TL, Bluestone JA: CD28/B7 system of T cell costimulation. Ann Rev Immunol 1996;14:233–258.

86 Kuchroo VK, Das MP, Brown JA, Ranger AM, Zamvil SS, Sobel RA, Weiner HL, Nabavi N, Glimcher LH: B7-1 and B7-2 costimulatory molecules activate differentially the Th1/Th2 developmental pathways: Application to autoimmune disease therapy. Cell 1995;80:707–718.

87 Lenschow DJ, Ho SC, Sattar H, Rhee L, Gray G, Nabavi N, Herold KC, Bluestone JA: Differential effects of anti-B7-1 and anti-B7-2 monoclonal antibody treatment on the development of diabetes in the nonobese diabetic mouse. J Exp Med 1995;181:1145–1155.

88 Corry DB, Reiner SL, Linsley PS, Locksley RM: Differential effects of blockade of CD28-B7 on the development of Th1 or Th2 effector cells in experimental leishmaniasis. J Immunol 1994;153: 4142–4148.

89 Rulifson IC, Sperling AI, Fields PE, Fitch FW, Bluestone JA: CD28 costimulation promotes the production of Th2 cytokines. J Immunol 1997;158:658–665.

90 Ronchese F, Hausmann B, Hubele S, Lane P: Mice transgenic for a soluble form of murine CTLA-4 show enhanced expansion of antigen-specific CD4+ T cells and defective antibody production in vivo. J Exp Med 1994;179:809–817.

91 Bachmaier K, Pummerer C, Shahinian A, Ionescu J, Neu N, Mak TW, Penninger JM: Induction of autoimmunity in the absence of CD28 costimulation. J Immunol 1996;157:1752–1757.

92 Lenschow DJ, Herold KC, Rhee L, Patel B, Koons A, Qin HY, Fuchs E, Singh B, Thompson CB, Bluestone JA: CD28/B7 regulation of Th1 and Th2 subsets in the development of autoimmune diabetes. Immunity 1996;5:285–293.

93 King CL, Stupi RJ, Craighead N, June CH, Thyphronitis G: CD28 activation promotes Th2 subset differentiation by human CD4+ cells. Eur J Immunol 1995;25:587–595.

94 Kalinski P, Hilkens CM, Wierenga EA, van der Pouw-Kraan TC, van Lier RA, Bos JD, Kapsenberg ML, Snijdewint FG: Functional maturation of human naive T helper cells in the absence of accessory cells. Generation of IL-4-producing T helper cells does not require exogenous IL-4. J Immunol 1995;154:3753–3760.

95 Webb LM, Feldmann M: Critical role of CD28/B7 costimulation in the development of human Th2 cytokine-producing cells. Blood 1995;86:3479–3486.

96 Freeman GJ, Boussiotis VA, Anumanthan A, Bernstein GM, Ke XY, Rennert PD, Gray GS, Gribben JG, Nadler LM: B7-1 and B7-2 do not deliver identical costimulatory signals, since

B7-2 but not B7-1 preferentially costimulates the initial production of IL-4. Immunity 1995; 2:523–532.

97   Ranger AM, Das MP, Kuchroo VK, Glimcher LH: B7-2 (CD86) is essential for the development of IL-4-producing T cells. Int Immunol 1996;8:1549–1560.

98   Kawamura T, Furue M: Comparative analysis of B7-1 and B7-2 expression in Langerhans cells: Differential regulation by T helper type 1 and T helper type 2 cytokines. Eur J Immunol 1995; 25:1913–1917.

99   Stack RM, Lenschow DJ, Gray GS, Bluestone JA, Fitch FW: IL-4 treatment of small splenic B cells induces costimulatory molecules B7-1 and B7-2. J Immunol 1994;152:5723–5733.

100 Kubo M, Yamashita M, Abe R, Tada T, Okumura K, Ransom JT, Nakayama T: CD28 costimulation accelerates IL-4 receptor sensitivity and IL-4-mediated Th2 differentiation. J Immunol 1999;163:2432–2442.

101 Serreze DV, Gaskins HR, Leiter EH: Defects in the differentiation and function of antigen presenting cells in NOD/Lt mice. J Immunol 1993;150:2534–2543.

102 Oosterwegel MA, Mandelbrot DA, Boyd SD, Lorsbach RB, Jarrett DY, Abbas AK, Sharpe AH: The role of CTLA-4 in regulating Th2 differentiation. J Immunol 1999;163:2634–2639.

103 Anderson DE, Bieganowska KD, Bar-Or A, Oliveira EM, Carreno B, Collins M, Hafler DA: Paradoxical inhibition of T-cell function in response to CTLA-4 blockade: Heterogeneity within the human T-cell population. Nat Med 2000;6:211–214.

104 Perrin PJ, June CH, Maldonado JH, Ratts RB, Racke MK: Blockade of CD28 during in vitro activation of encephalitogenic T cells or after disease onset ameliorates experimental autoimmune encephalomyelitis. J Immunol 1999;163:1704–1710.

105 Taniguchi M, Koseki H, Tokuhisa T, Masuda K, Sato H, Kondo E, Kawano T, Cui J, Perkes A, Koyasu S, Makino Y: Essential requirement of an invariant V alpha 14 T cell antigen receptor expression in the development of natural killer T cells. Proc Natl Acad Sci USA 1996;93: 11025–11028.

106 Bendelac A, Rivera MN, Park SH, Roark JH: Mouse CD1-specific NK1 T cells: development, specificity, and function. Annu Rev Immunol 1997;15:535–562.

107 Porcelli S, Yockey CE, Brenner MB, Balk SP: Analysis of T cell antigen receptor (TCR) expression by human peripheral blood CD4-8- alpha/beta T cells demonstrates preferential use of several V beta genes and an invariant TCR alpha chain. J Exp Med 1993;178:1–16.

108 Legendre V, Boyer C, Guerder S, Arnold B, Hammerling G, Schmitt-Verhulst AM: Selection of phenotypically distinct NK1.1+ T cells upon antigen expression in the thymus or in the liver. Eur J Immunol 1999;29:2330–2343.

109 Kawano T, Cui J, Koezuka Y, Toura I, Kaneko Y, Motoki K, Ueno H, Nakagawa R, Sato H, Kondo E, Koseki H, Taniguchi M: CD1d-restricted and TCR-mediated activation of Valpha14 NKT cells by glycosylceramides. Science 1997;278:1626–1629.

110 Kitamura H, Iwakabe K, Yahata T, Nishimura S, Ohta A, Ohmi Y, Sato M, Takeda K, Okumura K, Van Kaer L, Kawano T, Taniguchi M, Nishimura T: The natural killer T (NKT) cell ligand alpha-galactosylceramide demonstrates its immunopotentiating effect by inducing interleukin (IL)-12 production by dendritic cells and IL-12 receptor expression on NKT cells. J Exp Med 1999;189: 1121–1128.

111 Burdin N, Brossay L, Kronenberg M: Immunization with alpha-galactosylceramide polarizes CD1-reactive NK T cells towards Th2 cytokine synthesis. Eur J Immunol 1999;29:2014–2025.

112 Kitamura H, Ohta A, Sekimoto M, Sato M, Iwakabe K, Nakui M, Yahata T, Meng H, Koda T, Nishimura S, Kawano T, Taniguchi M, Nishimura T: Alpha-galactosylceramide induces early B-cell activation through IL-4 production by NKT cells. Cell Immunol 2000;199:37–42.

113 Sumida T, Takei I, Taniguchi M: Activation of acceptor-suppressor hybridoma with antigen-specific suppressor T cell factor of two-chain type: Requirement of the antigen- and the I-J-restricting specificity. J Immunol 1984;133:1131–1136.

114 Yoshimoto T, Paul WE: CD4+, NK1.1+ T cells promptly produce interleukin-4 in response to in vivo challenge with anti-CD3. J Exp Med 1994;179:1285–1295.

115 Kawamura T, Takeda K, Mendiratta SK, Kawamura H, Van Kaer L, Yagita H, Abo T, Okumura K: Critical role of NK1+ T cells in IL-12-induced immune responses in vivo. J Immunol 1998; 160:16–19.

116 Zeng D, Lewis D, Dejbakhsh-Jones S, Lan F, Garcia-Ojeda M, Sibley R, Strober S: Bone marrow NK1.1(−) and NK1.1(+) T cells reciprocally regulate acute graft versus host disease. J Exp Med 1999;189:1073–1081.

117 Sonoda KH, Exley M, Snapper S, Balk SP, Stein-Streilein J: CD1-reactive natural killer T cells are required for development of systemic tolerance through an immune-privileged site. J Exp Med 1999;190:1215–1226.

118 Ohkawa A, Ito T, Yumiba T, Maeda A, Tori M, Sawai T, Kiyomoto T, Akamaru Y, Matsuda H: Immunological characteristics of intragraft NKR-P1+ TCR alpha beta+ T (NKT) cells in rat hepatic allografts. Transplant Proc 1999;31:2699–2700.

119 Yoshimoto T, Bendelac A, Hu-Li J, Paul WE: Defective IgE production by SJL mice is linked to the absence of CD4+, NK1.1+ T cells that promptly produce interleukin 4. Proc Natl Acad Sci USA 1995;92:11931–11934.

120 Sumida T, Sakamoto A, Murata H, Makino Y, Takahashi H, Yoshida S, Nishioka K, Iwamoto I, Taniguchi M: Selective reduction of T cells bearing invariant V alpha 24J alpha Q antigen receptor in patients with systemic sclerosis. J Exp Med 1995;182:1163–1168.

121 Yanagihara Y, Shiozawa K, Takai M, Kyogoku M, Shiozawa S: Natural killer (NK) T cells are significantly decreased in the peripheral blood of patients with rheumatoid arthritis (RA). Clin Exp Immunol 1999;118:131–136.

122 Illes Z, Kondo T, Newcombe J, Oka N, Tabira T, Yamamura T: Differential expression of NK T cell V alpha 24J alpha Q invariant TCR chain in the lesions of multiple sclerosis and chronic inflammatory demyelinating polyneuropathy. J Immunol 2000;164:4375–4381.

123 Kukreja A, Sung Z, Chen Q, Vargas A, Chalew S, Gomez R, Maclaren N: Altered regulation of T cells associated with NK-T cell deficiency in immune mediated (type 1) diabetes. Diabetes 2000;49(suppl 1):A36, 148 OR.

124 Gombert JM, Herbelin A, Tancrede-Bohin E, Dy M, Carnaud C, Bach JF: Early quantitative and functional deficiency of NK1+ -like thymocytes in the NOD mouse. Eur J Immunol 1996;26: 2989–2998.

125 Gombert JM, Tancrede-Bohin E, Hameg A, Leite-de-Moraes MC, Vicari A, Bach JF, Herbelin A: IL-7 reverses NK1+ T cell-defective IL-4 production in the non-obese diabetic mouse. Int Immunol 1996;8:1751–1758.

126 Iwakoshi NN, Greiner DL, Rossini AA, Mordes JP: Diabetes prone BB rats are severely deficient in natural killer T cells. Autoimmunity 1999;31:1–14.

127 Sharif S, Arreaza GA, Zucker P, Delovitch TL: Activation of NKT cells by alpha-galactosylce-ramide treatment prevents the incidence and recurrence of diabetes in the NOD mouse model. Diabetes Res Clin Practice 2000;50:S165, P654.

128 Falcone M, Yeung B, Tucker L, Rodriguez E, Sarvetnick N: A defect in interleukin-12-induced activation and interferon-gamma secretion of peripheral natural killer T cells in nonobese diabetic mice suggests new pathogenic mechanisms for insulin-dependent diabetes mellitus. J Exp Med 1999;190:963–972.

129 Tori M, Ito T, Yumiba T, Okawa A, Maeda A, Sawai T, Kiyomoto T, Akamaru Y, Miyasaka M, Kiyono H, Matsuda H, Nozawa M, Shirakura R: Proliferation of donor-derived NKR-P1+TCR alpha beta + (NKT) cells in the nonrecurrent spontaneous diabetic BB rats transplanted with pancreaticoduodenal grafts of Wistar-Furth donors. Transplant Proc 1999;31:2741–2742.

130 Tori M, Ito T, Yumiba T, Ohkawa A, Maeda A, Sawai T, Kiyomoto T, Akamaru Y, Miyasaka M, Kiyono H, Matsuda H, Nozawa M, Shirakura R: IL-4 production in IDDM-nonrecurrent pancreas-transplanted BB rats with donor-derived NKR-P1+TCR alpha beta + (NKT) cells, but not in IDDM-recurrent BB rats. Transplant Proc 1999;31:1940–1941.

131 Ikehara Y, Yasunami Y, Kodama S, Maki T, Nakano M, Nakayama T, Taniguchi M, Ikeda S: CD4(+) Valpha14 natural killer T cells are essential for acceptance of rat islet xenografts in mice. J Clin Invest 2000;105:1761–1767.

132 Salojin KV, Zhang J, Madrenas J, Delovitch TL: T-cell anergy and altered T-cell receptor signaling: Effects on autoimmune disease. Immunol Today 1998;19:468–473.

133 Hanninen A, Harrison LC: Gamma delta T cells as mediators of mucosal tolerance: The autoimmune diabetes model. Immunol Rev 2000;173:109–119.

134 Weiner HL, Friedman A, Miller A, Khoury SJ, al-Sabbagh A, Santos L, Sayegh M, Nussenblatt RB, Trentham DE, Hafler DA: Oral tolerance: Immunologic mechanisms and treatment of animal and human organ-specific autoimmune diseases by oral administration of autoantigens. Annu Rev Immunol 1994;12:809–837.

135 Yoon JW, Yoon CS, Lim HW, Huang QQ, Kang Y, Pyun KH, Hirasawa K, Sherwin RS, Jun HS: Control of autoimmune diabetes in NOD mice by GAD expression or suppression in beta cells. Science 1999;284:1183–1187.

136 Zhang ZJ, Davidson L, Eisenbarth G, Weiner HL: Suppression of diabetes in nonobese diabetic mice by oral administration of porcine insulin. Proc Natl Acad Sci USA 1991;88:10252–10256.

137 Bergerot I, Fabien N, Maguer V, Thivolet C: Oral administration of human insulin to NOD mice generates CD4+ T cells that suppress adoptive transfer of diabetes. J Autoimmun 1994;7:655–663.

138 Ploix C, Bergerot I, Fabien N, Perche S, Moulin V, Thivolet C: Protection against autoimmune diabetes with oral insulin is associated with the presence of IL-4 type 2 T-cells in the pancreas and pancreatic lymph nodes. Diabetes 1998;47:39–44.

139 Ma SW, Zhao DL, Yin ZQ, Mukherjee R, Singh B, Qin HY, Stiller CR, Jevnikar AM: Transgenic plants expressing autoantigens fed to mice to induce oral immune tolerance. Nat Med 1997; 3:793–796.

140 Maron R, Melican NS, Weiner HL: Regulatory Th2-type T cell lines against insulin and GAD peptides derived from orally- and nasally-treated NOD mice suppress diabetes. J Autoimmun 1999; 12:251–258.

141 Hancock WW, Polanski M, Zhang J, Blogg N, Weiner HL: Suppression of insulitis in non-obese diabetic (NOD) mice by oral insulin administration is associated with selective expression of interleukin-4 and -10, transforming growth factor-beta, and prostaglandin-E. Am J Pathol 1995;147: 1193–1199.

142 Polanski M, Melican NS, Zhang J, Weiner HL: Oral administration of the immunodominant B-chain of insulin reduces diabetes in a co-transfer model of diabetes in the NOD mouse and is associated with a switch from Th1 to Th2 cytokines. J Autoimmun 1997;10:339–346.

143 Bergerot I, Arreaza GA, Cameron MJ, Burdick MD, Strieter RM, Chensue SW, Chakrabarti S, Delovitch TL: Insulin B-chain reactive CD4+ regulatory T-cells induced by oral insulin treatment protect from type 1 diabetes by blocking the cytokine secretion and pancreatic infiltration of diabetogenic effector T-cells. Diabetes 1999;48:1720–1729.

144 Ploix C, Bergerot I, Durand A, Czerkinsky C, Holmgren J, Thivolet C: Oral administration of cholera toxin B-insulin conjugates protects NOD mice from autoimmune diabetes by inducing CD4+ regulatory T-cells. Diabetes 1999;48:2150–2156.

145 Rudy G, Stone N, Harrison LC, Colman PG, McNair P, Brusic V, French MB, Honeyman MC, Tait B, Lew AM: Similar peptides from two beta cell autoantigens, proinsulin and glutamic acid decarboxylase, stimulate T cells of individuals at risk for insulin-dependent diabetes. Mol Med 1995; 1:625–633.

146 Dubois-LaForgue D, Carel JC, Bougneres PF, Guillet JG, Boitard C: T-cell response to proinsulin and insulin in type 1 and pretype 1 diabetes. J Clin Immunol 1999;19:127–134.

147 French MB, Allison J, Cram DS, Thomas HE, Dempsey CM, Silva A, Georgiou HM, Kay TW, Harrison LC, Lew AM: Transgenic expression of mouse proinsulin II prevents diabetes in nonobese diabetic mice. Diabetes 1997;46:34–39.

148 Chen W, Bergerot I, Elliott JF, Delovitch TL: Lack of neonatal tolerance to proinsulin elicits the pathogenesis IDDM in NOD mice; in Mechanisms of Immunologic Tolerance and Its Breakdown. Keystone Symposia, Steamboat Springs, Colo, 2000, p 92.

149 Von Herrath MG, Dyrberg T, Oldstone MB: Oral insulin treatment suppresses virus-induced antigen-specific destruction of beta cells and prevents autoimmune diabetes in transgenic mice. J Clin Invest 1996;98:1324–1331.

150 Coon B, An LL, Whitton JL, von Herrath MG: DNA immunization to prevent autoimmune diabetes. J Clin Invest 1999;104:189–194.

151 Homann D, Holz A, Bot A, Coon B, Wolfe T, Petersen J, Dyrberg TP, Grusby MJ, von Herrath MG: Autoreactive CD4+ T cells protect from autoimmune diabetes via bystander suppression using the IL-4/Stat6 pathway. Immunity 1999;11:463–472.

152 Tisch R, Wang B, Serreze DV: Induction of glutamic acid decarboxylase 65-specific Th2 cells and suppression of autoimmune diabetes at late stages of disease is epitope dependent. J Immunol 1999;163:1178–1187.
153 Tian J, Atkinson MA, Clare-Salzler M, Herschenfeld A, Forsthuber T, Lehmann PV, Kaufman DL: Nasal administration of glutamate decarboxylase (GAD65) peptides induces Th2 responses and prevents murine insulin-dependent diabetes. J Exp Med 1996;183:1561–1567.
154 Kretowski A, Mysliwiec J, Kinalska I: Abnormal distribution of gamma delta T lymphocytes in Graves' disease and insulin-dependent diabetes type 1. Arch Immunol Ther Exp (Warsz) 2000; 48:39–42.
155 Julius MC, Schatz DA, Silverstein JH: The prevention of type I diabetes mellitus. Pediatr Ann 1999;28:585–588.
156 Chakraborty NG, Li L, Sporn JR, Kurtzman SH, Ergin MT, Mukherji B: Emergence of regulatory CD4+ T-cell response to repetitive stimulation with antigen-presenting cells in vitro: Implications in designing antigen-presenting cell-based tumor vaccines. J Immunol 1999;162:5576–5583.
157 Clare-Salzler M, Brooks J, Chai A, Van Herle K, Anderson C: Prevention of diabetes in nonobese diabetic mice by dendritic cell transfer. J Clin Invest 1992;90:741–748.

Dr. Terry L. Delovitch, Director, Autoimmunity/Diabetes Group,
The John P. Robarts Research Institute, 1400 Western Road, London, Ont. N6G 2V4 (Canada)
Tel. +1 519 663 3972, Fax +1 519 663 3847, E-Mail del@rri.on.ca

von Herrath MG (ed): Molecular Pathology of Type 1 Diabetes mellitus.
Curr Dir Autoimmun. Basel, Karger, 2001, vol 4, pp 333–350

Immunotherapy of Type 1 Diabetes mellitus

Lucienne Chatenoud

INSERM U25 – Hôpital Necker, Paris, France

Type 1 diabetes mellitus (T1D) is a T-cell-mediated autoimmune disease [1, 2]. It is thus logical to modulate the immune system to stop the progression of the disease as it is widely done for other autoimmune diseases. T1D presents, however, with three major differences as compared to other autoimmune diseases:

(1) The clinical onset of the disease occurs late in the natural history of the pathologic process. This unquestionably represents a handicap for immunotherapy since late intervention, once the autoimmune response has reached its maximum intensity, is obviously more difficult than in the early stages of the autoimmune response.

(2) The therapeutic window is of short duration. A large (although not clearly known) proportion of β cells have already been destroyed when diabetes emerges clinically and this destruction then progresses rapidly leading to destruction of all insulin secreting cells within a few months rendering immunointervention useless. This is a major difference with most other autoimmune diseases where the pathologic process is chronic thus allowing to successfully intervene over several years even if early intervention is also preferable.

(3) There is an alternative substitutive therapy, insulin, which is sufficiently efficacious and safe not to accept any treatment exposing to severe toxicity or hazard, notably overimmunosuppression (infections, tumors).

Although, as we shall discuss, satisfactory strategies have been proposed to address these issues, it remains though that any form of immunotherapy applied to T1D should always aim at: (a) attempting to intervene as early as possible; (b) achieving a rapid blockade of the autoimmune process to stop the destruction of remnant β cells; and (c) presenting with a quasi absolute safety and especially to avoid chronic treatment with immunosuppressive drugs.

Our knowledge on immunotherapeutic strategies applicable to T1D has gained a lot from preclinical studies performed in spontaneous animal models of the disease, especially the nonobese diabetic (NOD) mouse. Much criticism has been raised, however, on this model because of the very large number of maneuvers shown to be efficacious. As we shall discuss below, this criticism only applies to interventions administered to NOD mice at an early stage of the disease which clearly represents a different situation from that encountered in the clinics.

Nevertheless, a fair number of immunotherapeutic tools have been shown in NOD mice to be highly effective in treating T1D at late stages and even at reversing established disease. Some of these data have been considered sufficiently convincing to promote their testing in the context of clinical trials.

Objectives of Immunotherapy of T1D

Insulin does not represent a satisfactory treatment of T1D since it does not afford the total prevention of degenerative complications. Of course the rate and rapidity of appearance of these complications is significantly delayed upon administration of adequate insulin therapy, but one has to admit that in a significant proportion of patients it is difficult to achieve a perfect metabolic control. Administration of fixed doses of insulin cannot mimic the endogenous insulin response in response to glucose stimuli. One major advantage of immunotherapy is to preserve the endogenous insulin production which allows a better metabolic control.

Insulin treatment is also associated with more 'acute' complications, notably hypoglycemia, which again are very much reduced when endogenous insulin production is preserved even in an incomplete manner [3].

Lastly, insulin treatment implies a daily and compulsory control of metabolic control (glycemia) and usually the administration multiple injections, two sets of constraints that should be reduced by an effective immunotherapy.

One may wonder which are the best parameters for evaluating the benefit afforded by immunotherapy of T1D. Insulin dosage has been used as an endpoint in some trials but it may be criticized because of the (relative) subjectivity of insulin dosage selection, the more so since metabolic control cannot be perfectly standardized. Better metabolic control is certainly the final end-point for reasons discussed above. Indeed, better metabolic control was observed in patients successfully treated with cyclosporin [3]. However, it cannot be used as a primary end-point in double-blind controlled clinical trials since optimal metabolic control should be aimed at in all patient groups (drug or placebo). Finally, C-peptide production is the best end-point even if it may appear less 'appealing' in the first instance. It is in fact a good correlate of the main parameter which immunotherapy aims at protecting, e.g. β-cell function. It has the drawback though not to take into account some clinically relevant factors such as insulin resistance. Insulin resistance explained that recently diagnosed diabetics treated with cyclosporin relapsed in terms of insulin requirement despite the fact a good C-peptide production was maintained [3].

Stages of T1D Progression in Relation to Immunointervention

Schematically, T1D natural history can be divided into four phases [1, 2]:

(1) *Noninvasive insulitis:* Islet-reactive autoimmunity is already present both at B- and T-cell levels but without harm to β cells. (2) *Clinically latent invasive insulitis:* The pathological process is sufficiently aggressive to induce β-cell lesion, either functional inhibition or β-cell destruction but sufficient β-cell function persists to maintain normal metabolic control. When the process progresses though, abnormal metabolic control may be observed after glucose loading (intravenous glucose infusion) or during pregnancy. (3) *Recently diagnosed clinical diabetes with persisting β cells:* This phase corresponds to the major therapeutic window discussed above. (4) *Long-term chronic diabetes:* Most β cells have been destroyed (with absence of remnant C-peptide production).

Immunotherapy of human diabetes can be envisioned at stages 2 or 3. Treatment of prediabetic patients (phase 2) presents with the major advantage of intervening at a stage when sufficient β cells are still present to let aim at a durable disease remission. It presents, however, three major pitfalls: (1) the selection of prediabetics is laborious (requiring the screening of a very large number of subjects); (2) it only applies so far to members of families having already presented a case of diabetes (thus excluding approximately 85% of new T1D cases) and (3) the risk of diabetes acceleration is hardly acceptable since it would precipitate the disease in subjects in whom the risk of developing the disease was not necessarily imminent or even absolute.

Treatment at phase 3 (recent-onset T1D) has to be rapidly efficacious. Data obtained with cyclosporin have shown that it is still time to preserve long-term endogenous insulin secretion in a large proportion of patients [3, 4], particularly if they are adequately selected in relation to disease duration. In fact, one has even observed partial recovery of C-peptide production due to clear-ance of insulitis including islet-reactive T cells that inhibit β-cell function [5] (in addition to preserving islets from glucotoxicity because of the insulin treatment that is usually instituted not long before immunotherapy) [3, 4].

Treatment at phase 1 (noninvasive insulitis) is practically not feasible at present but the situation could evolve since systematic screening of DR3/4 or DR4/4 neonates has been started in some countries [6]. This is a particularly attractive opportunity since it corresponds to the time of application of the immunotherapy shown to be efficacious in the NOD mouse.

One should also mention here the case of latent autoimmune diabetes of the adult (LADA) [7, 8] which represents a slowly progressing form of T1D where insulin dependency is preceded by a long phase of non-insulin-dependent diabetes. One may assume that immunotherapy should be efficient in such patients although the enrollment of subjects at risk of progressing to T1D in a reasonably short period of time may prove to be difficult (posing similar problems as those raised for prediabetics).

Tools of Immunotherapy

Immunosuppressive Drugs

A multitude of chemical and biological agents have been used in experimental models and in the clinic to suppress immune responses in a non-antigen-specific way. Chemicals showing a preferential effect on T cells are logical candidate drugs in T-cell-mediated autoimmune disease such as T1D. Such chemicals include cyclosporin A, FK-506, rapamycin, mycophenolic acid and azathioprine that are extensively used in organ transplantation. One might also mention amethopterin (methotrexate), which is now widely used in rheumatoid arthritis and psoriasis at low dosages.

Monoclonal Anti-T-Cell Antibodies

Therapeutic monoclonal antibodies recognizing functionally relevant T-cell surface receptors were introduced in clinical practice in the early 1980s [9, 10]. At that time they were essentially of rodent origin (mouse or rat) and were produced by hybridomas. As compared to polyclonal antilymphocyte serum, they represented homogeneous antibody preparations with well-characterized biochemical (isotype) and fine specificity characteristics.

One major advantage which rapidly appeared from the use of anti-T-cell monoclonals in the experimental setting is that some of them clearly expressed tolerogenic properties (e.g. the capacity to mediate long-term antigen-specific effects upon short-term treatment) which were in no way shared by the majority of the chemical immunosuppressants mentioned above. Importantly, these tolerogenic properties could be demonstrated not only in rodent but also in non-human primate models which for the purpose of clinical application are considered more meaningful [11–18].

The antibody specificities which showed the most relevant effects in these settings were those directed at the receptors involved in antigen recognition, that are CD3, CD4, class II histocompatibility molecules and those involved in the transduction of co-stimulatory signals namely, CD40Ligand.

First-generation therapeutic monoclonal antibodies were produced by mouse or rat hybridomas. One major pitfall linked to the in vivo use of these xenogeneic antibodies was the sensitization. Anti-monoclonal globulins expressed two major specificities: anti-isotypic and anti-idiotypic [19, 20]. Anti-idiotypic antibodies expressed a potent neutralizing capacity due to their ability to compete with the antigen for the monoclonal antibody binding [19]. At variance, anti-isotypic antibodies are essentially non-neutralizing [21]. Studies performed on immunized patients and monkey sera using purified anti-idiotypic antibodies showed that the response is oligoclonal [22]. The neutralizing potential of the humoral response to monoclonal antibodies is thus more dependent on its specificity than on the overall amount of antibodies produced, since only a few specific clones are recruited. This may explain why serum sickness was a rare event in patients immunized against murine monoclonals.

The advent of 'humanized' monoclonal antibodies produced by molecular engineering of rodent molecules allowed to circumvent this problem. Two sorts of humanized antibodies have been derived that are chimeric or complementarity determining region (CDR)-grafted monoclonal antibodies [23, 24]. Chimeric antibodies that are composed of the intact variable regions (Fab) from the parental rodent antibody coupled to a human immunoglobulin constant portion. Recombinant CDR-grafted antibodies express the parental hypervariable regions, that carry the antigen specificity, within human heavy and light chain immunoglobulin frameworks [23]. In the case of both the chimeric and the CDR-grafted antibodies, one may select for the desired human Fc fragment that will impact on the final antibody effector capacities (complement fixation, opsonization and antibody-dependent cell cytotoxicity (ADCC). We shall see further that in the particular case of CD3 antibodies the humanization is an excellent way to also get rid of their mitogenic potential.

Various humanized, chimeric or reshaped, monoclonal antibodies have been administered in the transplantation and autoimmunity setting. The data confirmed that the sensitization rate is significantly reduced as compared to what was seen with rodent monoclonals [25–28]. A humoral anti-idiotypic response is still observed with humanized antibodies in particular situations namely, autoimmune patients in whom the monoclonal is administered alone (in the absence of associated immunosuppressants), and after more than 2–3 repeated antibody courses [26, 29]. At variance, when single antibody courses are applied associated to other immunosuppressants, as it is the rule in transplantation, no sensitization is reported [25, 27, 28].

Other Biological Agents

Immunoregulatory cytokines or their analogues have been used in a number of preclinical models of autoimmune diseases, such as IL-4 [30], IL-12 [31] or IL-15 analogues. CTLA4-Ig [32] or LFA3-Ig have also been successfully used in some models and are presently tested in the clinic in various autoimmune diseases.

Preclinical Data in NOD Mice

Interest and Limitations of the NOD Mouse Model

The NOD mouse develops a form of autoimmune insulin-dependent diabetes that closely resembles human T1D. It was thus logical to use this model to screen immunotherapy strategies that would be ultimately applicable to diabetic patients. This has been extensively done (table 1) with spectacular success. The list of the main agents used shown to prevent or delay diabetes onset in NOD mice (table 1) is impressive but also worrisome since one may reasonably wonder if the demonstrated efficacy of so many and diverse agents including some 'mild' therapies such as streptococcal extracts is of any value for an extrapolation to the clinic.

There are mainly two ways to explain the unexpected easiness with which diabetes is prevented in NOD mice. One might argue that the development of T1D in NOD mice is exceptionally sensitive to external agents due to a unique feature of the model wherever the uniqueness may rely. Alternatively, and perhaps more simply, we would like to propose that this is due to the very early time at which most of the therapeutic maneuvers tested have been applied in the natural history of the disease, most usually before 12 weeks of age. In fact, looking for treatments shown to be efficacious after this age, one can only mention heat-shock protein peptide p277 [33], polyclonal antilymphocyte serum [34] and CD3 antibodies [35, 36], which in a way makes sense. The contrast between the large number of agents that are active early in the disease progression and the very limited number of agents still active late probably suggests that in early stages, the autoimmune process is not fully committed and still very plastic. Later on, in the context of a larger pool of highly committed T-cell effectors, only agents directly targeting either the effector pathways or the immune mechanisms promoting their regulation may still be active.

At a more mechanistic level, one has to distinguish therapeutic approaches based on non-antigen-specific immunomodulation that exposes to the risk of overimmunosuppression (e.g. chemical immunosuppressants) and agents inducing tolerance to β-cell autoantigen with the perspective of downregulating the islet-specific autoimmune response without altering whatsoever the host immune reactivity to infectious agents.

Table 1. Immunotheraphy of T1D

	NOD mouse	BB rat	Human T1D
Nonspecific chemical immunosuppression			
Cyclosporin A	+[78]	+[79]	+[3–4]
Azathioprine	+[80]		+[81]
FK-506	+[82]		
Rapamycin	+[84]		
Deoxyspergualine	+	+[85]	
Monoclonal antibodies			
Anti-CD3	+[35–36]		
Anti-CD4	+[51, 86]		
Anti-CD8	+[87]		
Anti-CD28	+[88]		
Anti-CD40L	+[112]		
Anti-IFNγ	+[89]	+[90]	
Anti-MHC class II	+[91]	+[92]	
Antilymphocyte serum	+[34]	+[94]	
Fusion proteins			
CTLA4-Ig	+[93]		
β-Cell Autoantigens			
Insulin			
Native molecule	+		+
Subcutaneous	+[37, 95]		+
Oral	+[39]		+[69, 96]
Intranasal	+[40]		
B chain (i.m. + adjuvant)	+[97]	+[98]	
Peptide (intranasal)	+		
APL	+		+
Glutamic acid decarboxylase			
Native molecule	+[99]		
Peptide	+[100]		+
Heat-shock protein 60 p277	+[101]		
Cytokines or related compounds			
IL-4	+		
Systemic	+		
Gene therapy	+		
IL-10	+		
Systemic	+[102]		
Gene therapy	+[103]		
IL-12 antagonists	+[31]		
α-Galactosylceramide			

	NOD mouse	BB rat	Human T1D
Miscellaneous			
Nicotinamide	+[104]		+[105]
Vaccine therapy	+[106, 107]	+[108]	+
CFA/BCG			
Q fever vaccine	+[109]		
Streptococcal extract	+[110]	+[111]	

We shall not come back to non-antigen-specific immunosuppression but rather elaborate on clinically applicable tolerance induction regimens which so far essentially include the administration of β-cell autoantigens and CD3 monoclonal antibodies.

β-Cell Autoantigen-Induced Tolerance

Administration of selected major β-cell autoantigens such as insulin and glutamic acid decarboxylase (GAD) to young NOD mice delays very significantly the appearance of diabetes [37–43]. Insulin or GAD may be administered intravenously [37, 41], subcutaneously [38], orally [39] or nasally [40, 42, 43] with similar results. The tolerogen has to be administered before 10–12 weeks of age to induce diabetes prevention. The same protection is obtained after parenteral administration of heat-shock protein 60 (hsp60) or of one of its constitutive peptides p277 [33, 44]. Hsp60 is not specific to β cells but its expression is enhanced in inflamed islets and one may assume that there is thus an overexpression of hsp60 in β cells following the first wave of T-cell infiltration. Importantly, diabetes prevention with p277 can be observed until 17–18 weeks of age which is later than reported for insulin or GAD.

Autoantigen-induced tolerance is not restricted to the antigen used for the induction of tolerance since it extends to other β-cell antigens (bystander suppression) [41, 43]. It is associated with Th2 polarization as assessed by the shift in autoantibody isotype towards a Th2 cell-dependent isotype (IgG1) and a Th2 cytokine preferential production in response to the tolerogen [45, 46].

CD3 Antibody-Induced Tolerance

As previously mentioned, the experimental and clinical experience in using polyclonal and some monoclonal anti-T-cell antibodies predominantly in transplantation, but also in autoimmunity, has demonstrated their ability to promote an acquired and durable state of antigen-specific unresponsiveness namely, immune tolerance [11–18, 35, 47–51]. For quite a long time the idea prevailed that only few antibody specificities were able to afford robust tolerance and in particular great attention was devoted to CD4 antibodies.

Experimental evidence has been accumulated to support the conclusion that antibodies to CD3 are also very good candidates to promote, upon limited treatment, specific unresponsiveness to both alloantigens and autoantigens. The initial experiments were performed, by the group of B. Hall, and demonstrated that permanent engraftment of histoincompatible vascularized heart grafts were obtained in CD3 antibody treated rats [11]. The presence in the recipients of a state of immune tolerance to the alloantigens was confirmed through the survival of second donor skin grafts while third-party grafts were normally rejected [11]. More recent data from the same group demonstrated that Th2 immune deviation of alloreactive effectors could be demonstrated in the CD3-treated tolerant hosts [52].

The strategy was extended to non-human primates using a CD3 immunotoxin. Rhesus monkeys treated with the CD3 diphtheria toxin conjugate associated either to donor bone marrow cells or to 15-deoxyspergualin exhibited long-term survival of fully mismatched renal allografts [17, 53]. Interestingly, in 3 monkeys presenting with spontaneous insulin-dependent diabetes, xenogeneic pancreatic islets were transplanted under the cover of two injections of CD3 immunotoxin (day 0 and on day +1). Cyclosporin and steroids were also administered but only for 4 days (days 0–4) and no further immunosuppression was given. The three recipients remained euglycemic at 410, 255, and 100 days post-transplant [18].

Our laboratory was the first to demonstrate some years ago that CD3 antibodies could reverse established diabetes and restore self-tolerance in adult NOD mice [35, 36]. The initial observations were that NOD females, over 15 weeks of age, presenting with established diabetes (e.g. glycosuria and glycemia $\geq$4 g/l) showed disease remission upon a short (5-days) treatment with low doses (5–20 µg) of a CD3 monoclonal antibody (145 2C11, hamster IgG) [54]. Stable remission, a return to permanent normoglycemia in the absence of insulin treatment, was observed in 60–80% of treated mice as opposed to 0% remission observed in control animals treated with hamster immunoglobulins [35, 36].

In most animals the remission was durable and not related to generalized immunosuppression. In fact, CD3 antibody-protected NOD mice responded normally to foreign antigens as shown by their ability to reject mismatched skin grafts. The elicited unresponsiveness was specific for β-cell-associated antigens since CD3 antibody-treated mice did not destroy syngeneic islet grafts unlike control NOD females [35].

In order to ensure the metabolic reconstitution, the treatment had to be started while a significant β-cell mass was still present, namely within 7 days from the detection of the first signs of overt diabetes. In case the beginning of the treatment was protracted, the transplantation of syngeneic islets was an indispensable step to achieve normoglycemia.

Concerning the immune mechanisms mediating the effect, our data argue against a massive deletion of autoreactive cells in protected animals. They favor instead a partial removal of intra-islet autoreactive effectors and the restoration of T-cell-mediated immunoregulation or dominant tolerance closely resembling that described in prediabetic 6- to 8-week-old NOD mice [1, 2, 55–59].

Within the first few days of treatment there is a complete clearance of the insulitis that probably explains the rapid reversal of hyperglycemia. Subsequently, in 3–4 weeks the islet infiltration reappears but, at variance to what usually observed in overtly diabetic mice that show destructive/invasive insulitits, it remains confined to the periphery of the islets, i.e., peripheral insulitis [35, 36].

CD3 antibody-treated animals also harbor immunoregulatory T cells as shown by the recurrence of disease upon cyclophosphamide treatment. Moreover, in co-transfer experiments we could show the presence in the spleen of protected mice of CD3+CD62L+ cells that very effectively inhibit the transfer of diabetes by diabetogenic T lymphocytes. The active nature of this immune tolerance is also suggested by the fact that when cyclosporin is administered concomitantly to CD3, the tolerogenic effect of the antibody is completely abrogated [36].

CD3 antibodies are potent T-cell mitogens which explains that when injected in vivo they promote a massive though transient cytokine release which, in the patients, causes the acute 'flu-like' syndrome, lasting for 48–72 h after the first injection and including high fever, chills, headache, vomiting and diarrhea (for further details see below) [60–66].

Importantly, we could confirm that in NOD mice nonmitogenic F(ab')$_2$ fragments of 145 2C11 were as effective as the whole mitogenic antibody in promoting permanent remission of overt diabetes [36].

Clinical Trials

Two sets of clinical trials have been started with the aim of inducing tolerance to β-cell antigens following the strategies described above.

β-Cell Antigens

The strategy is very attractive due to the very limited toxicity expected from the autoantigen preparation. Moreover, in most studies the autoantigen is used alone, in the absence of a combined immunosuppressive treatment, which avoids the risk of overimmunosuppression.

Some problems must however be considered that differ depending on the population studied (prediabetic subjects or recently diagnosed T1D patients). In the case of prediabetes, one major problem could be the risk of disease acceleration if the antigen therapy promotes sensitization rather than tolerance [67]. In the case of recently diagnosed patients the problem is mainly that of lack of effectiveness in the context of a rapidly progressing and advanced disease.

Various clinical trials have been performed or are in progress using different candidate autoantigens. Insulin is presently administered intranasally to prediabetic subjects [68]. Based on the results of encouraging pilot trials [69, 70], parenteral or oral insulin is also being used on a fairly large scale here again in prediabetic individuals (DPT1). Oral insulin has also been used in recently diagnosed T1D patients with little or no effect. Trials are also ongoing using whole recombinant GAD65 and the p277 peptide of hsp60.

CD3 Antibody

The CD3 approach is very attractive due to dual mode of action namely, the rapid clearing of insulitis upon beginning of treatment and the stimulation of T-cell-mediated regulatory circuits. The treatment works best in advanced disease and even in established T1D. The treatment is effective upon short administration of the antibody which reduces the risk of overimmunosuppression.

Aside to the risk of sensitization we discussed in detail above, another important side effect linked to the in vivo use of rodent CD3 antibodies was that of the acute 'flu-like' syndrome linked to their mitogenicity which triggered a massive release of several cytokines including TNF, IFNγ, IL-2, IL-3, IL-6, IL-10 and GM-CSF [60–66]. This mitogenic capacity that tightly correlates with the capacity of the Fc antibody portion to interact with Fc receptors on monocyte/macrophages [71, 72]. This is why in the case of CD3 antibodies the humanization provided the possibility to derive antibodies devoid of both their immunogenic and mitogenic potential. At present two of them are being tested in recent onset T1D. One has been characterized by the group of J. Bluestone and is derived from OKT3. This humanized γOKT3–5 antibody has been mutated in the Fc region and expresses a 100-fold decrease in its affinity for human Fc receptors as compared to the parental monoclonal. The γOKT3–5 antibody is not mitogenic either in vitro or in vivo [73, 74].

The other one, CAMPATH3, has been characterized by the group of H. Waldmann and is derived from the rat YTH 12.5 [75, 76]. CAMPATH3 is a human IgG1 expressing a constant region which lacks the CH2 domain glycosylation site. Therefore the antibody is aglycosylated. Here again in vitro and in vivo studies confirmed the lack of mitogenicity of CAMPATH3 [75, 77].

Conclusions

Immunotherapy of T1D is the major clinical application of the considerable knowledge acquired on the immunopathology of the disease in the context of increasingly efficient and safe immunointervention. One may now hope that the three intrinsic difficulties associated with immunotherapy of T1D relatively to that of other autoimmune diseases will be circumvented. If successful, the treatment of recently diagnosed diabetics by CD3 antibodies will open new perspectives. In fact, one may hope that once CD3 antibody efficacy is demonstrated, recently diagnosed diabetics will be submitted earlier to treatment which will become a real medical emergency. The case of prediabetes will also have to be reconsidered once safe treatments with long-term efficacy are available. More generally, the tolerance-inducing regimens will provide the total safety required for diabetics who are not ready to substitute toxic or hazardous therapies to insulin. One may reasonably aim at disease eradication by inducing tolerance in prediabetics or even in subjects at risk before they start to develop the diabetogenic process, using a strategy in the spirit of vaccinations against infectious agents.

References

1 Bach JF: Insulin-dependent diabetes mellitus as an autoimmune disease. Endocr Rev 1994;15: 516–542.
2 Delovitch TL, Singh B: The nonobese diabetic mouse as a model of autoimmune diabetes: Immune dysregulation gets the NOD. Immunity 1997;7:727–738.
3 Feutren G, Papoz L, Assan R, Vialettes B, Karsenty G, Vexiau P, Du Rostu H, Rodier M, Sirmai J, Lallemand A, Bach JF: Cyclosporin increases the rate and length of remissions in insulin-dependent diabetes of recent onset. Results of a multicentre double-blind trial. Lancet 1986;ii: 119–124.
4 The Canadian-European Randomized Control Trial Group: Cyclosporin-induced remission of IDDM after early intervention. Association of 1 year of cyclosporin treatment with enhanced insulin secretion. Diabetes 1988;37:1574–1582.
5 Strandell E, Eizirik DL, Sandler S: Reversal of beta-cell suppression in vitro in pancreatic islets isolated from nonobese diabetic mice during the phase preceding insulin-dependent diabetes mellitus. J Clin Invest 1990;85:1944–1950.
6 Rewers M, Norris JM, Eisenbarth GS, Erlich HA, Beaty B, Klingensmith G, Hoffman M, Yu L, Bugawan TL, Blair A, Hamman RF, Groshek M, McDuffie JR: Beta-cell autoantibodies in infants and toddlers without IDDM relatives: Diabetes autoimmunity study in the young (DAISY). J Autoimmun 1996;9:405–410.
7 Schranz DB, Bekris L, Landin-Olsson M, Torn C, Nilang A, Toll A, Sjostrom J, Gronlund H, Lernmark A: Newly diagnosed latent autoimmune diabetes in adults (LADA) is associated with low level glutamate decarboxylase (GAD65) and IA-2 autoantibodies. Diabetes Incidence Study in Sweden (DISS). Hormone Metab Res 2000;32:133–138.
8 Zimmet PZ, Tuomi T, MacKay IR, Rowley MJ, Knowles W, Cohen M, Lang DA: Latent autoimmune diabetes mellitus in adults (LADA): The role of antibodies to glutamic acid decarboxylase in diagnosis and prediction of insulin dependency. Diabet Med 1994;11:299–303.

9 Cosimi AB, Burton RC, Colvin RB, Goldstein G, Delmonico FL, Laquaglia MP, Tolkoff-Rubin N, Rubin RH, Herrin JT, Russell PS: Treatment of acute renal allograft rejection with OKT3 monoclonal antibody. Transplantation 1981;32:535–539.

10 Cosimi AB, Colvin RB, Burton RC, Rubin RH, Goldstein G, Kung PC, Hansen WP, Delmonico FL, Russell PS: Use of monoclonal antibodies to T-cell subsets for immunologic monitoring and treatment in recipients of renal allografts. N Engl J Med 1981;305:308–314.

11 Nicolls MR, Aversa GG, Pearce NW, Spinelli A, Berger MF, Gurley KE, Hall BM: Induction of long-term specific tolerance to allografts in rats by therapy with an anti-CD3-like monoclonal antibody. Transplantation 1993;55:459–468.

12 Qin S, Cobbold SP, Pope H, Elliott J, Kioussis D, Davies J, Waldmann H: 'Infectious' transplantation tolerance. Science 1993;259:974–977.

13 Cobbold SP, Adams E, Marshall SE, Davies JD, Waldmann H: Mechanisms of peripheral tolerance and suppression induced by monoclonal antibodies to CD4 and CD8. Immunol Rev 1996;149:5–33.

14 Larsen CP, Elwood ET, Alexander DZ, Ritchie SC, Hendrix R, Tuckerburden C, Cho HR, Aruffo A, Hollenbaugh D, Linsley PS, Winn KJ, Pearson TC: Long-term acceptance of skin and cardiac allografts after blocking CD40 and CD28 pathways. Nature 1996;381:434–438.

15 Kirk AD, Burkly LC, Batty DS, Baumgartner RE, Berning JD, Buchanan K, Fechner JR Jr, Germond RL, Kampen RL, Patterson NB, Swanson SJ, Tadaki DK, Tenhoor CN, White L, Knechtle SJ, Harlan DM: Treatment with humanized monoclonal antibody against CD154 prevents acute renal allograft rejection in nonhuman primates. Nat Med 1999;5:686–693.

16 Thomas JM, Carver FM, Cunningham PR, Olson LC, Thomas FT: Kidney allograft tolerance in primates without chronic immunosuppression – the role of veto cells. Transplantation 1991;51:198–207.

17 Hamawy MM, Knechtle SJ: Strategies for tolerance induction in nonhuman primates. Curr Opin Immunol 1998;10:513–517.

18 Thomas FT, Ricordi C, Contreras JL, Hubbard WJ, Jiang XL, Eckhoff DE, Cartner S, Bilbao G, Neville DM JR, Thomas JM: Reversal of naturally occurring diabetes in primates by unmodified islet xenografts without chronic immunosuppression. Transplantation 1999;67:846–854.

19 Chatenoud L, Baudrihaye MF, Chkoff N, Kreis H, Goldstein G, Bach JF: Restriction of the human in vivo immune response against the mouse monoclonal antibody OKT3. J Immunol 1986;137:830–838.

20 Benjamin RJ, Cobbold SP, Clark MR, Waldmann H: Tolerance to rat monoclonal antibodies. Implications for serotherapy. J Exp Med 1986;163:1539–1552.

21 Baudrihaye MF, Chatenoud L, Kreis H, Goldstein G, Bach JF: Unusually restricted anti-isotype human immune response to OKT3 monoclonal antibody. Eur J Immunol 1984;14:686–691.

22 Chatenoud L, Jonker M, Villemain F, Goldstein G, Bach JF: The human immune response to the OKT3 monoclonal antibody is oligoclonal. Science 1986;232:1406–1408.

23 Riechmann L, Clark M, Waldmann H, Winter G: Reshaping human antibodies for therapy. Nature 1988;332:323–327.

24 Morrison SL, Johnson MJ, Herzenberg LA, Oi VT: Chimeric human antibody molecules: Mouse antigen-binding domains with human constant region domains. Proc Natl Acad Sci USA 1984;81:6851–6855.

25 Lazarovits AI, Rochon J, Banks L, Hollomby DJ, Muirhead N, Jevnikar AM, White MJ, Amlot PL, Beauregard-Zollinger L, Stiller CR: Human mouse chimeric CD7 monoclonal antibody (SDZCHH380) for the prophylaxis of kidney transplant rejection. J Immunol 1993;150:5163–5174.

26 Elliott MJ, Maini RN, Feldmann M, Long-Fox A, Charles P, Bijl H, Woody JN: Repeated therapy with monoclonal antibody to tumour necrosis factor alpha (cA2) in patients with rheumatoid arthritis. Lancet 1994;344:1125–1127.

27 Vincenti F, Kirkman R, Light S, Bumgardner G, Pescovitz M, Halloran P, Neylan J, Wilkinson A, Ekberg H, Gaston R, Backman L, Burdick J: Interleukin-2-receptor blockade with daclizumab to prevent acute rejection in renal transplantation. Daclizumab Triple Therapy Study Group. N Engl J Med 1998;338:161–165.

28   Nashan B, Moore R, Amlot P, Schmidt AG, Abeywickrama K, Soulillou JP: Randomised trial of basiliximab versus placebo for control of acute cellular rejection in renal allograft recipients. CHIB 201 International Study Group [published erratum appears in Lancet 1997;350:1484]. Lancet 1997;350:1193–1198.

29   Isaacs JD, Watts RA, Hazleman BL, Hale G, Keogan MT, Cobbold SP, Waldmann H: Humanised monoclonal antibody therapy for rheumatoid arthritis. Lancet 1992;340:748–752.

30   Rapoport MJ, Jaramillo A, Zipris D, Lazarus AH, Serreze DV, Leiter EH, Cyopick P, Danska JS, Delovitch TL: Interleukin-4 reverses T cell proliferative unresponsiveness and prevents the onset of diabetes in nonobese diabetic mice. J Exp Med 1993;178:87–99.

31   Trembleau S, Penna G, Gregori S, Gately MK, Adorini L: Deviation of pancreas-infiltrating cells to Th2 by interleukin-12 antagonist administration inhibits autoimmune diabetes. Eur J Immunol 1997;27:2330–2339.

32   Finck BK, Linsley PS, Wofsy D: Treatment of murine lupus with CTLA4Ig. Science 1994; 265:1225–1227.

33   Elias D, Cohen IR: Peptide therapy for diabetes in NOD mice. Lancet 1994;343:704–706.

34   Maki T, Ichikawa T, Blanco R, Porter J: Long-term abrogation of autoimmune diabetes in nonobese diabetic mice by immunotherapy with anti-lymphocyte serum. Proc Natl Acad Sci USA 1992;89:3434–3438.

35   Chatenoud L, Thervet E, Primo J, Bach JF: Anti-CD3 antibody induces long-term remission of overt autoimmunity in nonobese diabetic mice. Proc Natl Acad Sci USA 1994;91:123–127.

36   Chatenoud L, Primo J, Bach JF: CD3 antibody-induced dominant self-tolerance in overtly diabetic NOD mice. J Immunol 1997;158:2947–2954.

37   Gotfredsen CF, Buschard K, Frandsen EK: Reduction of diabetes incidence of BB Wistar rats by early prophylactic insulin treatment of diabetes-prone animals. Diabetologia 1985;28:933–935.

38   Muir A, Peck A, Clare-Salzler M, Song YH, Cornelius J, Luchetta R, Krischer J, MacLaren N: Insulin immunization of nonobese diabetic mice induces a protective insulitis characterized by diminished intraislet interferon-gamma transcription. J Clin Invest 1995;95:628–634.

39   Zhang ZJ, Davidson L, Eisenbarth G, Weiner HL: Suppression of diabetes in nonobese diabetic mice by oral administration of porcine insulin. Proc Natl Acad Sci USA 1991;88:10252–10256.

40   Daniel D, Wegmann DR: Protection of nonobese diabetic mice from diabetes by intranasal or subcutaneous administration of insulin peptide B-(9-23). Proc Natl Acad Sci USA 1996;93:956–960.

41   Kaufman DL, Clare-Salzler M, Tian J, Forsthuber T, Ting GSP, Robinson P, Atkinson MA, Sercarz EE, Tobin AJ, Lehmann PV: Spontaneous loss of T-cell tolerance to glutamic acid decarboxylase in murine insulin-dependent diabetes. Nature 1993;366:69–72.

42   Elliott JF, Qin HY, Bhatti S, Smith DK, Singh RK, Dillon T, Lauzon J, Singh B: Immunization with the larger isoform of mouse glutamic acid decarboxylase (GAD67) prevents autoimmune diabetes in NOD mice. Diabetes 1994;43:1494–1499.

43   Tisch R, Yang XD, Singer SM, Liblau RS, Fugger L, McDevitt HO: Immune response to glutamic acid decarboxylase correlates with insulitis in non-obese diabetic mice. Nature 1993;366:72–75.

44   Elias D, Reshef T, Birk OS, Van Der Zee R, Walker MD, Cohen IR: Vaccination against autoimmune mouse diabetes with a T-cell epitope of the human 65-kDa heat shock protein. Proc Natl Acad Sci USA 1991;88:3088–3091.

45   Elias D, Meilin A, Ablamunits V, Birk OS, Carmi P, Konenwaisman S, Cohen IR: Hsp60 peptide therapy of NOD mouse diabetes induces a Th2 cytokine burst and downregulates autoimmunity to various beta-cell antigens. Diabetes 1997;46:758–764.

46   Tian, J, Lehmann PV, Kaufman DL: Determinant spreading of T helper cell 2 (Th2) responses to pancreatic islet autoantigens. J Exp Med 1997;186:2039–2043.

47   Isobe M, Yagita H, Okumura K, Ihara A: Specific acceptance of cardiac allograft after treatment with antibodies to ICAM-1 and LFA-1. Science 1992;255:1125–1127.

48   Pearson TC, Madsen JC, Larsen CP, Morris PJ, Wood KJ: Induction of transplantation tolerance in adults using donor antigen and anti-CD4 monoclonal antibody. Transplantation 1992;54:475–483.

49   Wofsy D, Seaman WE: Reversal of advanced murine lupus in NZB/NZW F1 mice by treatment with monoclonal antibody to L3T4. J Immunol 1987;138:3247–3253.

50   Wood KJ: Transplantation tolerance with monoclonal antibodies. Semin Immunol 1990;2:389–399.

51 Shizuru JA, Taylor-Edwards C, Banks BA, Gregory AK, Fathman CG: Immunotherapy of the nonobese diabetic mouse: Treatment with an antibody to T-helper lymphocytes. Science 1988; 240:659–662.
52 Plain KM, Chen J, Merten S, He XY, Hall BM: Induction of specific tolerance to allografts in rats by therapy with non-mitogenic, non-depleting anti-CD3 monoclonal antibody: Association with TH2 cytokines not anergy. Transplantation 1999;67:605–613.
53 Thomas JM, Neville DM, Contreras JL, Eckhoff DE, Meng G, Lobashevsky AL, Wang PX, Huang ZQ, Verbanac KM, Haisch CE, Thomas FT: Preclinical studies of allograft tolerance in rhesus monkeys: A novel anti-CD3-immunotoxin given peritransplant with donor bone marrow induces operational tolerance to kidney allografts. Transplantation 1997;64:124–135.
54 Leo O, Foo M, Sachs DH, Samelson LE, Bluestone JA: Identification of a monoclonal antibody specific for a murine T3 polypeptide. Proc Natl Acad Sci USA 1987;84:1374–1378.
55 Boitard C, Yasunami R, Dardenne M, Bach JF: T cell-mediated inhibition of the transfer of autoimmune diabetes in NOD mice. J Exp Med 1989;169:1669–1680.
56 Hutchings PR, Cooke A: The transfer of autoimmune diabetes in NOD mice can be inhibited or accelerated by distinct cell populations present in normal splenocytes taken from young males. J Autoimmun 1990;3:175–185.
57 Yasunami R, Debray-Sachs M, Bach JF: Ontogeny of regulatory and effector T-cells in autoimmune NOD mice; in Shafrir E (ed): Frontiers in Diabetes Research. Lessons from Animal Diabetes. III. London, Smith-Gordon, 1990, pp 88–93.
58 Lepault F, Gagnerault MC: Characterization of peripheral regulatory CD4(+) T cells that prevent diabetes onset in nonobese diabetic mice. J Immunol 2000;164:240–247.
59 Herbelin A, Gombert JM, Lepault F, Bach JF, Chatenoud L: Mature mainstream TCR alpha beta+CD4+ thymocytes expressing L-selectin mediate 'active tolerance' in the nonobese diabetic mouse. J Immunol 1998;161:2620–2628.
60 Hirsch R, Gress RE, Pluznik DH, Eckhaus M, Bluestone JA: Effects of in vivo administration of anti-CD3 monoclonal antibody on T cell function in mice. II. In vivo activation of T cells. J Immunol 1989;142:737–743.
61 Ferran C, Sheehan K, Dy M, Schreiber R, Merite S, Landais P, Noel LH, Grau G, Bluestone J, Bach JF, Chatenoud L: Cytokine-related syndrome following injection of anti-CD3 monoclonal antibody: Further evidence for transient in vivo T cell activation. Eur J Immunol 1990;20: 509–515.
62 Alegre M, Vandenabeele P, Flamand V, Moser M, Leo O, Abramowicz D, Urbain J, Fiers W, Goldman M: Hypothermia and hypoglycemia induced by anti-CD3 monoclonal antibody in mice: Role of tumor necrosis factor. Eur J Immunol 1990;20:707–710.
63 Durez P, Abramowicz D, Gerard C, Van Mechelen M, Amraoui Z, Dubois C, Leo O, Velu T, Goldman M: In vivo induction of interleukin-10 by anti-CD3 monoclonal antibody or bacterial lipopolysaccharide: Differential modulation by cyclosporin A. J Exp Med 1993;177: 551–555.
64 Yoshimoto T, Paul WE: CD4pos, NK1.1pos T cells promptly produce interleukin-4 in response to in vivo challenge with anti-CD3. J Exp Med 1994;179:1285–1295.
65 Abramowicz D, Schandene L, Goldman M, Crusiaux A, Vereerstraeten P, de Pauw L, Wybran J, Kinnaert P, Dupont E, Toussaint C: Release of tumor necrosis factor, interleukin-2, and gamma-interferon in serum after injection of OKT3 monoclonal antibody in kidney transplant recipients. Transplantation 1989;47:606–608.
66 Eason JD, Cosimi AB: Biologic immunosuppressive agents; in Ginns LC, Cosimi AB, Morris PJ (eds): Transplantation. Malden, Blackwell Science, 1999, pp 196–224.
67 Blanas E, Carbone FR, Allison J, Miller JF, Heath WR: Induction of autoimmune diabetes by oral administration of autoantigen. Science 1996;274:1707–1709.
68 Harrison LC, Honeyman MC, Steele C, Wright M, Gellert SA, Colman PG: Intranasal insulin trial (INIT) in preclinical type 1 diabetes. 10th International Congress of Mucosal Immunology, Amsterdam, 1999. Immunol Lett 1999;69:72.
69 Keller RJ, Eisenbarth GS, Jackson RA: Insulin prophylaxis in individuals at high risk of type I diabetes. Lancet 1993;341:927–928.

70 Kobayashi T, Nakanishi K, Murase T, Kosaka K: Small doses of subcutaneous insulin as a strategy for preventing slowly progressive beta-cell failure in islet cell antibody-positive patients with clinical features of NIDDM. Diabetes 1996;45:622–626.

71 Van Lier RA, Boot JH, de Groot ER, Aarden LA: Induction of T-cell proliferation with anti-CD3 switch-variant monoclonal antibodies: Effects of heavy chain isotype in monocyte-dependent systems. Eur J Immunol 1987;17:1599–1604.

72 Hirsch R, Bluestone JA, de Nenno L, Gress RE: Anti-CD3 F(ab')₂ fragments are immunosuppressive in vivo without evoking either the strong humoral response or morbidity associated with whole mAb. Transplantation 1990;49:1117–1123.

73 Alegre ML, Peterson LJ, Xu D, Sattar HA, Jeyarajah DR, Kowalkowski K, Thistlethwaite JR, Zivin RA, Jolliffe L, Bluestone JA: A non-activating 'humanized' anti-CD3 monoclonal antibody retains immunosuppressive properties in vivo. Transplantation 1994;57:1537–1543.

74 Woodle ES, Xu D, Zivin RA, Auger J, Charette J, O'Laughlin R, Peace D, Jollife LK, Haverty T, Bluestone JA, Thistlethwaite JR Jr: Phase I trial of a humanized, Fc receptor nonbinding OKT3 antibody, huOKT3gamma1(Ala-Ala) in the treatment of acute renal allograft rejection. Transplantation 1999;68:608–616.

75 Bolt S, Routledge E, Lloyd I, Chatenoud L, Pope H, Gorman SD, Clark M, Waldmann H: The generation of a humanized, non-mitogenic CD3 monoclonal antibody which retains in vitro immunosuppressive properties. Eur J Immunol 1993;23:403–411.

76 Routledge EG, Falconer ME, Pope H, Lloyd IS, Waldmann H: The effect of aglycosylation on the immunogenicity of a humanized therapeutic CD3 monoclonal antibody. Transplantation 1995;60: 847–853.

77 Friend PJ, Hale G, Chatenoud L, Rebello P, Bradley J, Thiru S, Phillips JM, Waldmann H: Phase I study of an engineered aglycosylated humanized CD3 antibody in renal transplant rejection. Transplantation 1999;68:1632–1637.

78 Mori Y, Suko M, Okudaira H, Matsuba I, Tsuruoka A, Sasaki A, Yokoyama H, Tanase T, Shida T, Nishimura M, Terada E, Ikeda Y: Preventive effects of cyclosporin on diabetes in NOD mice. Diabetologia 1986;29:244–247.

79 Laupacis A, Stiller CR, Gardell C, Keown P, Dupre J, Wallace AC, Thibert P: Cyclosporin prevents diabetes in BB Wistar rats. Lancet 1983;i:10–12.

80 Calafiore R, Basta G, Falorni A, Pietropaolo M, Picchio ML, Calcinaro F, Brunetti P: Preventive effects of azathioprine on the onset of diabetes mellitus in NOD mice. J Endocrinol Invest 1993; 16:869–873.

81 Cook JJ, Hudson I, Harrison LC, Dean B, Colman PG, Werther GA, Warne GL, Court JM: Double-blind controlled trial of azathioprine in children with newly diagnosed type I diabetes. Diabetes 1989;38:779–783.

82 Miyagawa J, Yamamoto K, Hanafusa T, Itoh N, Nakagawa C, Otsuka A, Katsura H, Yamagata K, Miyazaki A, Kono N, Tarui S: Preventive effect of a new immunosuppressant FK-506 on insulitis and diabetes in non-obese diabetic mice. Diabetologia 1990;33:503–505.

83 Murase N, Lieberman I, Nalesnik MA, Mintz DH, Todo S, Drash AL, Starzl TE: Effect of FK-506 on spontaneous diabetes in BB rats. Diabetes 1990;39:1584–1586.

84 Baeder WL, Sredy J, Sehgal SN, Chang JY, Adams LM: Rapamycin prevents the onset of insulin-dependent diabetes mellitus in NOD mice. Clin Exp Immunol 1992;89:174–178.

85 Nicoletti F, Meroni PL, Di Marco R, Grasso S, Barcellini W, Borghi MO, Lunetta M, Mughini L, Menta R, Schorlemmer HU, et al: The effects of deoxyspergualin on the development of diabetes in diabetes-prone BB rats. Scand J Immunol 1992;36:415–420.

86 Wang Y, Pontesilli O, Gill RG, La Rosa FG, Lafferty KJ: The role of CD4+ and CD8+ T cells in the destruction of islet grafts by spontaneously diabetic mice. Proc Natl Acad Sci USA 1991;88:527–531.

87 Hutchings PR, Simpson E, O'Reilly LA, Lund T, Waldmann H, Cooke A: The involvement of Ly2+ T cells in beta cell destruction. J Autoimmun 1990;3(suppl 1):101–109.

88 Arreaza GA, Cameron MJ, Jaramillo A, Gill BM, Hardy D, Laupland KB, Rapoport MJ, Zucker P, Chakrabarti S, Chensue SW, Qin HY, Singh B, Delovitch TL: Neonatal activation of CD28 signaling overcomes T cell anergy and prevents autoimmune diabetes by an IL-4-dependent mechanism. J Clin Invest 1997;100:2243–2253.

89 Debray-Sachs M, Carnaud C, Boitard C, Cohen H, Gresser I, Bedossa P, Bach JF: Prevention of diabetes in NOD mice treated with antibody to murine IFN-gamma. J Autoimmun 1991;4:237–248.

90 Nicoletti F, Meroni PL, Landolfo S, Gariglio M, Guzzardi S, Barcellini W, Lunetta M, Mughini L, Zanussi C: Prevention of diabetes in BB/Wor rats treated with monoclonal antibodies to interferon-gamma. Lancet 1990;336:319.

91 Boitard C, Bendelac A, Richard MF, Carnaud C, Bach JF: Prevention of diabetes in nonobese diabetic mice by anti-I-A monoclonal antibodies: Transfer of protection by splenic T cells. Proc Natl Acad Sci USA 1988;85:9719–9723.

92 Boitard C, Michie S, Serrurier P, Butcher GW, Larkins AP, McDevitt HO: In vivo prevention of thyroid and pancreatic autoimmunity in the BB rat by antibody to class II major histocompatibility complex gene products. Proc Natl Acad Sci USA 1985;82:6627–6631.

93 Lenschow DJ, Ho SC, Sattar H, Rhee L, Gray G, Nabavi N, Herold KC, Bluestone JA: Differential effects of anti-B7-1 and anti-B7-2 monoclonal antibody treatment on the development of diabetes in the nonobese diabetic mouse. J Exp Med 1995;181:1145–1155.

94 Like AA, Rossini AA, Guberski DL, Appel MC, Williams RM: Spontaneous diabetes mellitus: Reversal and prevention in the BB/W rat with antiserum to rat lymphocytes. Science 1979;206:1421–1423.

95 Atkinson MA, MacLaren NK, Luchetta R: Insulitis and diabetes in NOD mice reduced by prophylactic insulin therapy. Diabetes 1990;39:933–937.

96 Fuchtenbusch M, Rabl W, Grassl B, Bachmann W, Standl E, Ziegler AG: Delay of type I diabetes in high-risk, first-degree relatives by parenteral antigen administration: The Schwabing Insulin Prophylaxis Pilot Trial. Diabetologia 1998;41:536–541.

97 Maron R, Palanivel V, Weiner HL, Harn DA: Oral administration of schistosome egg antigens and insulin B-chain generates and enhances Th2-type responses in NOD mice. Clin Immunol Immunopathol 1998;87:85–92.

98 Song HY, Abad MM, Mahoney CP, McEvoy RC: Human insulin B chain but not A chain decreases the rate of diabetes in BB rats. Diabetes Res Clin Pract 1999;46:109–114.

99 Tisch R, Liblau RS, Yang XD, Liblau P, McDevitt HO: Induction of GAD65-specific regulatory T-cells inhibits ongoing autoimmune diabetes in nonobese diabetic mice. Diabetes 1998;47:894–899.

100 Tian JD, Atkinson MA, Claresalzler M, Herschenfeld A, Forsthuber T, Lehmann PV, Kaufman DL: Nasal administration of glutamate decarboxylase (GAD65) peptides induces Th2 responses and prevents murine insulin-dependent diabetes. J Exp Med 1996;183:1561–1567.

101 Elias D, Cohen IR: Treatment of autoimmune diabetes and insulitis in NOD mice with heat-shock protein 60 peptide p277. Diabetes 1995;44:1132–1138.

102 Pennline KJ, Roque-Gaffney E, Monahan M: Recombinant human IL-10 prevents the onset of diabetes in the nonobese diabetic mouse. Clin Immunol Immunopathol 1994;71:169–175.

103 Moritani M, Yoshimoto K, Ii S, Kondo M, Iwahana H, Yamaoka T, Sano T, Nakano N, Kikutani H, Itakura M: Prevention of adoptively transferred diabetes in nonobese diabetic mice with IL-10-transduced islet-specific Th1 lymphocytes – A gene therapy model for autoimmune diabetes. J Clin Invest 1996;98:1851–1859.

104 Yamada K, Nonaka K, Hanafusa T, Miyazaki A, Toyoshima H, Tarui S: Preventive and therapeutic effects of large-dose nicotinamide injections on diabetes associated with insulitis. An observation in nonobese diabetic mice. Diabetes 1982;31:749–753.

105 Vague P, Picq R, Bernal M, Lassmann-Vague V, Vialettes B: Effect of nicotinamide treatment on the residual insulin secretion in type 1 (insulin-dependent) diabetic patients. Diabetologia 1989;32:316–321.

106 Sadelain MW, Qin HY, Lauzon J, Singh B: Prevention of type I diabetes in NOD mice by adjuvant immunotherapy. Diabetes 1990;39:583–589.

107 Shehadeh N, Calcinaro F, Bradley BJ, Bruchlim I, Vardi P, Lafferty KJ: Effect of adjuvant therapy on development of diabetes in mouse and man. Lancet 1994;343:706–707.

108 Sadelain MW, Qin HY, Sumoski W, Parfrey N, Singh B, Rabinovitch A: Prevention of diabetes in the BB rat by early immunotherapy using Freund's adjuvant. J Autoimmun 1990;3:671–680.

109 Bonn D: Q fever vaccine on trial for type I diabetes. Mol Med Today 1999;5:143.

110 Toyota T, Satoh J, Oya K, Shintani S, Okano T: Streptococcal preparation (OK-432) inhibits development of type I diabetes in NOD mice. Diabetes 1986;35:496–499.
111 Satoh J, Shintani S, Oya K, Tanaka S, Nobunaga T, Toyota T, Goto Y: Treatment with streptococcal preparation (OK-432) suppresses anti-islet autoimmunity and prevents diabetes in BB rats. Diabetes 1988;37:1188–1194.
112 Balasa B, Krahl T, Patstone G, et al: CD40 ligand-CD40 interactions are necessary for the initiation of insulitis and diabetes in nonobese diabetic mice. J Immunol 1997;159:4620–4627.

Dr. Lucienne Chatenoud, INSERM U25 – Hôpital Necker,
161, rue de Sèvres, F–75743 Paris Cedex 15 (France)
Tel. +33 144 495 371, Fax +33 143 062 388, E-Mail chatenoud@necker.fr

IA-2β, *see* Phogrin
ICA69
 antigen characterization 266
 autoantibodies in diabetes 265, 266
 humoral and cellular immune response
  255
 localization 255
ICA512, *see* IA-2
Idd loci, *see* Nonobese diabetic mouse
IDDM1, diabetes susceptibility locus 15, 16
IDDM2, diabetes susceptibility locus
 16–18
Immunotherapy
 clinical trials
  β-cell antigens 342, 343
  CD3 antibodies 343
 cytokine therapy 338
 diabetes phase in relation to immunointervention 335, 336
 end-points for evaluation 334, 335
 goals 333, 334
 immunosuppressants 336
 interleukin-4 role in tolerance 322, 323
 monoclonal anti-T cell antibodies 336,
  337
 nonobese diabetic mouse studies
  autoantigen-induced tolerance 340
  CD3 antibody-induced tolerance
   340–342
  intranasal and aerosolized antigen
   delivery 323
  overview of studies 338–340
 oral insulin 321, 322
 regulatory T cell induction with islet cell
  autoantigens 321, 322
 RIP-LCMV mouse testing
  DNA vaccines 110
  efficacy 107
  oral antigens 108–110
 therapeutic window for intervention 334,
  344
Immunotoxin, tolerance induction in
 transplantation 293, 294
Imogen 38
 humoral and cellular immune response
  255
 localization 255

InsHA mouse
 islet antigen tolerance 131, 132
 peripheral T cell tolerance 128–130
Insulin
 antibody radioimmunoassay 256, 273
 antigen characterization 257
 autoantibodies in diabetes 257, 259–261
 complications of treatment 334
 gene
  mouse vs human 39
  polymorphisms 17
  regulation of expression 17, 18, 20
  variable number tandem repeat in
   diabetes 16–18, 70, 259
 humoral and cellular immune response
  255
 intranasal and aerosolized antigen
  delivery in nonobese diabetic mouse
  323
 localization 255
 processing 257
 rat insulin promoter transgenic mice, *see*
  RIP-LCMV mouse; RIP-OVA mouse
 regulatory T cell induction with oral
  tolerance 321, 322
 three-dimensional structure and antibody
  binding 259, 260
Intercellular adhesion molecule-1, blocking
 for tolerance induction in transplantation
 300
Interferon-α, diabetes therapy 208
Interferon-γ
 β-cell destruction mediation 150, 151,
  158–160
 candidate gene in diabetes 18, 19
 diabetes role in nonobese diabetic mouse
  56, 202
 pancreatic levels with viral infection 78,
  79
 RIP-LCMV mouse diabetes role 105
Interleukin-1
 β-cell destruction mediation 150, 156,
  157
 candidate genes in diabetes 18, 19
 diabetes type 1 role in rodent models
  198, 199
 diabetes type 2 role 270–272